PROCEEDINGS OF THE 9TH INTERNATIONAL CONFERENCE ON TEXTILE COMPOSITES

TEXCOMP9

Recent Advances *in* Textile Composites

October 13–15, 2008
University of Delaware John M. Clayton Hall
Newark, Delaware, USA

Edited by

Suresh G. Advani
Associate Director of Center for Composite Materials and
George W. Laird Professor of Mechanical Engineering
University of Delaware, Newark, DE 19716, USA
advani@udel.edu

John W. Gillespie, Jr.
Director, Center for Composite Materials
Donald C. Phillips Professor of Civil and Environmental Engineering
Department of Materials Science and Engineering
Department of Civil and Environmental Engineering
University of Delaware, Newark, DE 19716, USA
gillespi@udel.edu

DEStech Publications, Inc.

Recent Advances in Textile Composites

DEStech Publications, Inc.
439 North Duke Street
Lancaster, Pennsylvania 17602 U.S.A.

Printed in the United States of America
10 9 8 7 6 5 4 3 2 1

Main entry under title:
Recent Advances in Textile Composites (Proceedings of the 9th International Conference on Textile Composites—TEXCOMP9)

A DEStech Publications book
Bibliography: p.
Includes index p. 533

ISBN: 978-1-932078-81-7

Table of Contents

TOPIC III—INDUSTRIAL APPLICATIONS AND CASE STUDIES

TOPIC IV—FINITE ELEMENT MODELING OF TEXTILE COMPOSITES

TOPIC VI—3D TEXTILE COMPOSITES: MECHANICAL BEHAVIOR

TOPIC VII—TEMPERATURE AND 3D EFFECTS

TOPIC VIII—3D TEXTILE COMPOSITES: MANUFACTURING

TOPIC IX—GEOMETRY AND MECHANICAL BEHAVIOR OF DRY TEXTILES AND PREFORMS

Preface

The papers contained in this volume were presented at the 9th International Conference on Textile Composites (TEXCOMP9) which was held at the University of Delaware October 13 through 15, 2008. The goal of TEXCOMP is to promote knowledge in the field of textile composites throughout the world. The conference brought together scientists and engineers active in a variety of disciplines and provided a dedicated forum for discussions and reports on recent advances in textiles and their composites. It was the ninth conference in this series, the first one was in 1992 in Leuven, Belgium, chaired by Professor Ignaas Verpoest. The conference has rotated between Europe, Asia and North America in the intervening years. TEXCOMP has an international audience, attracting world leading researchers in composite materials.

The proceedings have been grouped into eleven topic areas and cover research areas that include mechanical design and modeling, advanced manufacturing processes, net-shape 3D textile preforms, nano-fibers and composites, elastic and failure behavior, textile modeling and simulation and industrial applications and case studies.

We thank Mrs. Corinne Hamed of the Center for Composite Materials at the University of Delaware for her invaluable help in compiling this volume and organizing this conference.

We hope you will find these papers that advance and expand the science and engineering of the textile composites industry interesting and instructive.

Suresh G. Advani and John W. Gillespie, Jr.
Chairmen
TEXCOMP9

TOPIC I

TEXTILE IMPREGNATION TECHNOLOGY AND APPLICATIONS

Influence of Textile Architectures on the Compaction and Saturated Permeability Spatial Variations

S. COMAS-CARDONA, S. BICKERTON, M. DELÉGLISE,
W. A. WALBRAN, C. BINÉTRUY and P. KRAWCZAK

ABSTRACT

Composite manufacturing techniques are subject to a range of disturbances, due to material and processing variability. Fiber reinforcement compaction response and permeability are very influential, as is variability in these quantities. A compaction study of two reinforcements is presented, spatial stress variations quantified using the Tekscan distributed pressure measurement system. Basic statistics of spatial variation over repeated experiments are considered. The noted spatial variability has been used to specify a model permeability field for a single layer of twill weave. Random fluctuations have been imposed on this field, and 1000 permeability fields have been generated for a 4-layer preform. The upper and lower permeability bounds have been used to predict injection pressure histories for unidirectional filling, results correlating well with observed experimental variability. This study shows that if reinforcements are well known in terms of statistical variations, a simulation can be more predictive in terms of injection responses.

INTRODUCTION

Liquid composite molding (LCM) processes are strongly dependent on fluctuations in various parameters (e.g., temperature, material batches, preforming consistency), which can induce poor process reproducibility. Fiber reinforcement properties such as compaction response and permeability govern mould filling, and variability (global and spatial) in these quantities can be very influential [1]. Average compaction response is complex [2,3], and little work has been done on the spatial distribution of compaction stress. As an initial investigation into this issue, a compaction study on two different glass fiber reinforcements is presented.

Sebastien Comas-Cardona, Mylène Deléglise, Christophe Binétruy, Patricia Krawczak, Department of Polymers and Composites Technology & Mechanical Engineering, Ecole des Mines de Douai, 59508 Douai Cedex, France.
Simon Bickerton, William A. Walbran, Centre for Advanced Composite Materials, The University of Auckland, Private Bag 92019, Auckland 1020, New Zealand.

In parallel a permeability variability study has been realized on one of the same reinforcements (a twill weave fabric), using a constant injection pressure permeability bench. From initial observations of the spatial compaction response, a spatially varying permeability field for a single layer of fabric has been proposed. Random fluctuations have been imposed on the parameters that define this field, and the permeability field for a 4-layer sample has been found through a numerical stacking process. Variability of nesting has been accounted for by allowing random variation in layer placement, this process being repeated 1000 times to establish bounds on the average permeability response. The long-term goal of this work is to more clearly establish the link between spatial variability in compaction response and permeability, and better understand their influence on processing.

COMPACTION STUDIES

A series of compaction experiments have been performed using parallel plates mounted in an Instron universal testing machine. The Tekscan distributed pressure measurement system has been used to measure spatial variation of compaction stress exerted on the platen, during a period of constant speed compaction (1.0 mm/min) to a target fiber volume fraction (V_f). This study utilized square sensors (56 mm by 56 mm) having a fine grid of 44 by 44 sensels, and a pressure range of 0 to 2125 kPa. 100 mm square samples of the reinforcement were cut, the Tekscan sensor being placed at the centre of the sample, fixed to the lower mold platen. Two E-glass fiber reinforcements have been considered, a 450 g/m^2 chopped strand mat (CSM), and a 1500 g/m^2 twill weave fabric (TWF). Experiments have been performed using samples composed of 1, 3, 5 and 10 layers. Ten repeats were performed for 1 and 5 layers, and three repeats for 3 and 10 layers. The target V_f values were calculated from the nominal areal weight.

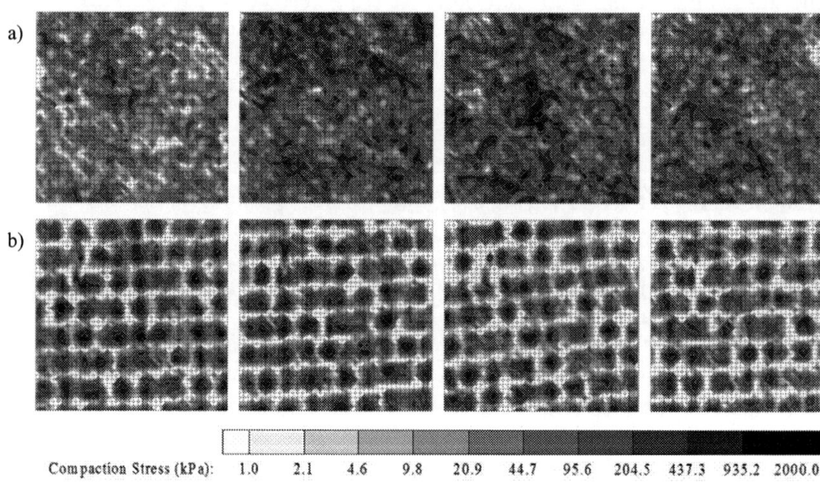

Figure 1. Sample compaction stress distributions from 5 layer samples, a) CSM, and b) TWF.

Repeatability between Individual Layup Events

The compaction stress distributions considered are at the instant the target V_f was reached (0.45 for CSM, 0.575 for TWF). To demonstrate typical variability between layups, four distributions recorded for 5 layers of CSM and TWF are shown in Figure 1. Note that compaction stress is plotted using an exponential scale, as peak stresses visually dominate if a linear scale is used. The influence of layup variations between experiments is clear. The random nature of CSM is evident, peak compaction stresses following no geometric pattern. Due to large variability in sample mass, significant variation in the spatially averaged stress were generated (mean=159.2 kPa, standard deviation=35.3 kPa). Stress frequency plots have been constructed for each experiment, and data from the 1 and 5 layer experiments are presented in Figure 2. Figure 2b highlights the variability between the 5 layer CSM tests, primarily for local compaction stress levels above 300 kPa.

While the periodicity of the TWF has been captured, the variability in magnitude and position of the peak compaction stresses is significant. This is due to variability in the structure of a single layer, and variability introduced by subsequent placement of additional layers. For the TWF, maximal nesting was introduced into each layup, the warp and weft tows of each additional layer being positioned between the previous. Despite these relatively controlled conditions, significant variability was noted between experiments. The spatial averages were of reduced variability compared to CSM, with a mean of 73.9 kPa across the 10 tests, with a standard deviation of 11.7 kPa. This is confirmed by Figure 2d, which shows the much closer agreement between the frequency plots of the TWF experiments.

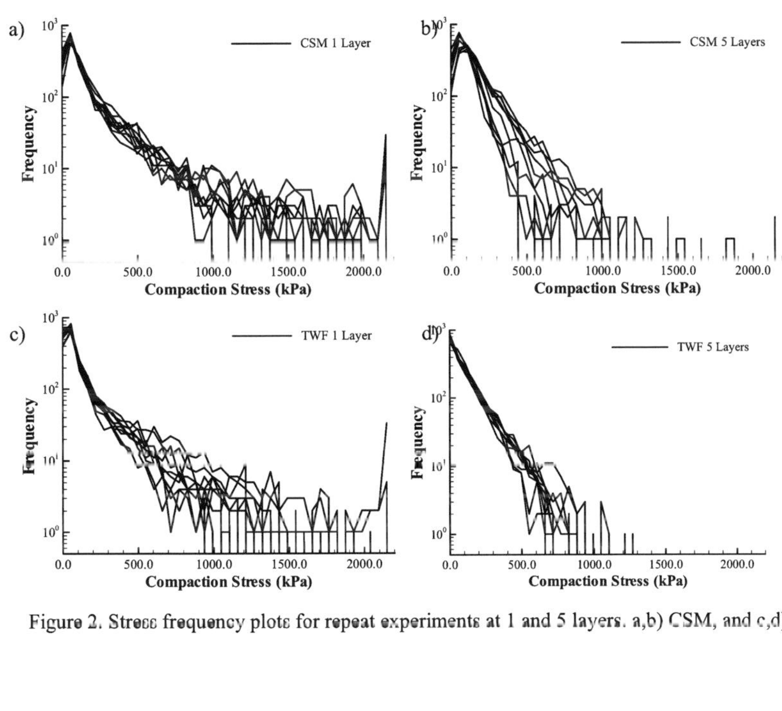

Figure 2. Stress frequency plots for repeat experiments at 1 and 5 layers. a,b) CSM, and c,d) TWF.

The Effect of a Varying Number of Layers

Figure 3 presents typical stress distributions (using an exponential scale) recorded as the number of layers was increased. Single layer distributions for both materials exhibit strong spatial variation in both the magnitude and location of peak stresses. While the TWF has an ordered structure, strong spatial variability exists, which is tempered to some extent as additional layers are added. Average stress frequency plots have been constructed for each set of experiments, by averaging across the experiments at each compaction stress level. This frequency data is presented in Figure 4, demonstrating how the distribution of stress magnitudes changes with additional fabric layers.

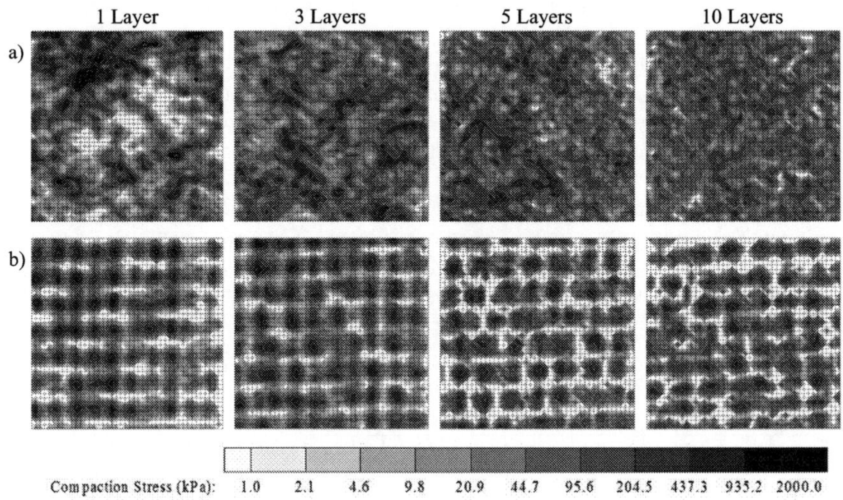

Figure 3. Compaction stress distributions for samples of increasing layers, a) CSM, and b) TWF.

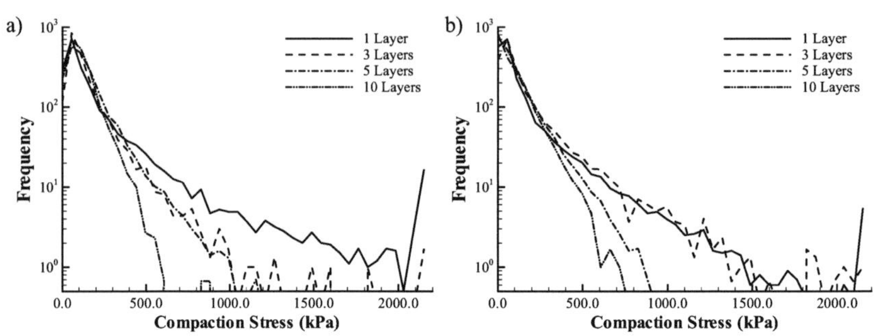

Figure 4. Stress frequency plots, averages taken between each set of repeat experiments at each compaction stress interval. a) CSM, and b) TWF.

Examining Figures 3a and 4a, a single layer of CSM exhibits localized regions of high stress, due to higher density fiber clusters, and large areas in which stress was below the measurable threshold. As additional layers are added, the fiber clusters in each layer are distributed across the sampling area, and the associated peak stresses reduced by lower fiber density in adjacent layers. The reduction in frequency of higher stresses is confirmed by Figure 4a, and the increased smoothing of the stress distribution with increasing layers demonstrated by Figure 3a.

Considering Figure 4b the TWF exhibits similar behavior, additional layers serving to reduce the influence of high stress regions generated by a single layer. For stress levels below 400 kPa the frequency plots are very similar, while above this level significant reductions in higher stresses are evident with increasing layers. Figure 2 shows that a significant reduction in variability between layups occurs when moving from 1 to 5 layers of TWF. The opposite trend is noted for CSM, particularly for stress levels above 200 kPa. For the CSM, this behavior is strongly influenced by the significant variability in sample mass.

INFLUENCE OF PERMEABILITY VARIATION ON MANUFACTURING

The first part of this study has shown that fiber reinforcement architectures exhibit strong differences in terms of local variations and average levels of stress response when submitted to unidirectional transverse compression. The question that arises is: what would be the effect of these variations when infusing (during resin transfer molding for instance) a preform composed of such fiber reinforcements? When injecting a resin through a fibrous material, resin pressure p and interstitial flow velocity v are related through Darcy's law;

$$v = \frac{-K}{\phi\mu} \nabla p \tag{1}$$

where K is the reinforcement permeability, ϕ the porosity and μ the resin viscosity. Equation 1 shows that permeability variations directly impact on the resin pressure p when constant flow velocity (flow rate) is used as an injection strategy.

The study focuses on the unidirectional (x-direction) injection of a preform composed of four plies of the TWF defined earlier. Due to the nature of fibrous reinforcements, the compression response and therefore the permeability can vary:

- within a single ply,
- and from one ply to another.

First, concerning the variability within a single ply, the following permeability distribution in the (x, y) plane can be proposed. This is reasonable, due to the reinforcement's periodic architecture.

$$\frac{K(x,y)}{K_o} = \frac{\alpha}{2} \cdot \left[\sin\left(\frac{-2\pi}{\lambda_x}(x - x_{off}) \right) + \sin\left(\frac{-2\pi}{\lambda_y}(y - y_{off}) \right) \right], \tag{2}$$

where K_o is the average permeability of the ply, α its amplitude of variation, λ_x and λ_y the periodicity of the fabric in the x and y directions respectively. Also, x_{off} and y_{off} represent the possible offsets that occur between layers during the lay-up.

TABLE I. DISTRIBUTION, BOUNDS, MEANS AND STANDARD DEVIATIONS OF THE PARAMETERS CHARACTERIZING THE TWF.

Parameter	Distribution	Bounds or Mean / Standard Deviation	Values
K_o	Gaussian	Mean	2.9×10^{-10} m^2
		Standard Deviation	4.9×10^{-11} m^2
λ_x	Gaussian	Mean	6.7×10^{-3} m
		Standard Deviation	1.5×10^{-3} m
λ_y	Gaussian	Mean	7.1×10^{-3} m
		Standard Deviation	1.5×10^{-3} m
x_{off}	Uniform	Min	0 m
		Max	6.7×10^{-3} m
y_{off}	Uniform	Min	0 m
		Max	7.1×10^{-3} m

The variability from one ply to another can be obtained by generating random (uniform or Gaussian distribution) values of the parameters of Equation 2 [4]. Table I describes distributions that have been chosen, their adequate bounds or mean, and the standard deviation. K_o has been measured from several fluid injection experiments. The fabric wavelengths, $\lambda_{x,y}$, were obtained from the compression tests run with the Tekscan sensor, measuring the length between maxima in both directions of the fabric. Permeability and wavelengths are considered normally distributed. Offsets between layers are due to the randomness of how an operator cuts and lays up the fabric. The amount of offset varies within the fabric wavelengths because of its periodicity. Therefore, the best mathematical description is obtained with a uniform distribution of offsets in both directions.

A sampling of each parameter is generated for each fabric ply. An example of the resulting permeability field is shown in Figure 5. The preform permeability is then calculated while averaging across the width direction, and through the four plies in the preform as follows;

$$K_{preform}(x) = \frac{1}{4}\sum_{i=1}^{4} K^i(x,y) \tag{3}$$

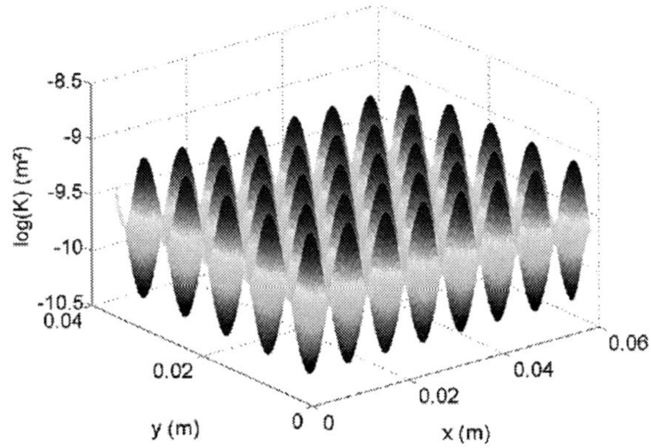

Figure 5. Generation of the permeability within a ply.

TABLE II. INPUT PARAMETERS USED FOR BOTH NUMERICAL AND EXPERIMENTAL INJECTIONS.

Input parameters	Values
Length L	0.295 m
Width l	0.18 m
Thickness h	4.5×10^{-3} m
Flow rate Q	1.67×10^{-6} m^3/s
Viscosity μ	0.1 Pa.s
Fiber volume fraction V_f	0.51

A finite difference method solver has been written to solve Equation 1 in the case of a unidirectional (x-direction) injection with constant flow rate and variable local permeability. Table II shows the input parameters that have been applied here. The output of the simulation is the evolution of resin inlet pressure with respect to the flow front position.

The filling simulation was repeated 1000 times while performing permeability sample generation. The inlet pressure evolutions, for all cases, fall in between the bounds depicted by the broken lines in Figure 6. For comparison, five different experimental injection pressure responses are also presented in Figure 6. The experimental injections exhibit variability that is not negligible and can be captured. The maximum injection pressure response is predicted to vary between 0.16 and 0.27 MPa and the experiments show responses in the range 0.15 to 0.22 MPa.

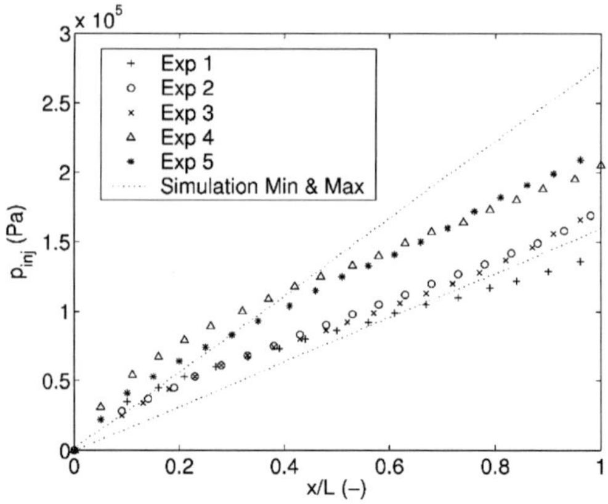

Figure 6. Experimental pressure profiles compared to the predicted pressure bounds.

CONCLUSIONS

Strong spatial variability in compaction response has been demonstrated for two glass fiber reinforcements. Large peak stresses are generated by a single fabric layer, these peak values being reduced with the addition of layers. Compaction data has helped to specify a model permeability field for a single layer of twill weave, which has been used to realize 1000 fields for a 4 layer preform. This study has shown that when a reinforcing material is well known in terms of statistical variations, a simulation can be more predictive in terms of injection responses. Realistic bounds on fluid pressure and reinforcement compaction response could be used, for instance, to help designers optimize the stiffness and strength of LCM moulds to withstand maximum applied stresses that could occur during resin injection. Future work will focus on relating compaction response to local fiber volume fraction and then to permeability, as an alternative to using a mathematical permeability law. Therefore, the permeability mapping could be much more realistic, and better depict the experimentally observed variations.

REFERENCES

1. K. Hoes, D. Dinescu, H. Sol, R. S. Parnas, and S. Lomov. 2004. "Study of Nesting Induced Scatter of Permeability Values in Layered Reinforcement Fabrics", *Composites Part A*, 35(12):1407–1418.
2. A. A. Somashekar, S. Bickerton, and D. Bhattacharyya. 2007. "Exploring the Non-Elastic Compression Deformation of Dry Glass Fibre Reinforcements", *Composites Science and Technology*, 67(2):183–200.
3. S. Comas-Cardona, P. Le Grognec, C. Binetruy, and P. Krawczak. 2007. "Unidirectional compression of fibre reinforcements. Part 1: A non-linear elastic-plastic behaviour", *Composites Science and Technology*, 67(2):507–514.
4. G. E. P. Box and, M. E. Muller. 1958. "A Note on the Generation of Random Normal Deviates", *Annals Math. Stat.*, 29:610–611.

Experimental Determination of Permeability of Woven Fiber Glass during the Vacuum Induced Preform Relaxation (VIPR) Process

J. B. ALMS, J. L. GLANCEY and S. G. ADVANI

ABSTRACT

An investigation to determine the permeability in the principal directions of woven fiber glass when using the Vacuum Induced Preform Relaxation (VIPR) process is conducted. The VIPR process is a port based resin delivery approach for vacuum infusion processes coupled with an external rigid chamber called the VIPR chamber. The VIPR chamber consists of a rigid plate, a stand-alone regulated vacuum source, and a closed cell rubber foam seal which is capable of forming an airtight chamber above the non-porous film used in vacuum infusion processes. The VIPR chamber is typically positioned over an injection port in the mold surface. In this configuration, a relatively small amount of vacuum pressure can significantly relax the fabric underneath the chamber thus altering the permeability characteristics. Subsequently, when the vacuum pressure is released and the VIPR chamber is removed, the permeability decreases, returning to its original state. The ability to modify the permeability of the fabric in real time during infusion will allow the VIPR process to potentially be used as a flow control mechanism in vacuum infusion processes such as Vacuum Assisted Resin Transfer Molding (VARTM). In this study, the permeabilities in the principal directions of the fabric are determined through the use of a three dimensional analytical model of the flow from a mold port. Results from radial flow experiments are used with the model to quantify the effects of vacuum level within the VIPR chamber on fabric permeability.

Justin B. Alms, James L. Glancey, and Suresh G. Advani[1], University of Delaware, Center for Composite Materials, Newark DE, 19711, U.S.A.

[1]George W. Laird Professor and Corresponding Author, E-mail: advani@udel.edu

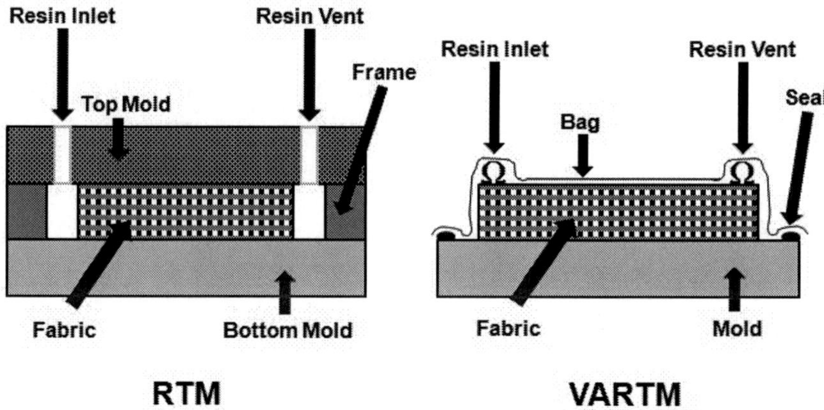

Figure 1. A cross sectional schematic of the RTM and the VARTM process. Left: An RTM mold consisting of rigid mold top and bottom. Right: A VARTM mold composed of only a single sided tooling surface at the bottom with bag serving as the top surface to ensure mold closure.

LIQUID COMPOSITE MOLDING

Liquid Composite Molding (LCM) processes are being adopted by many industries which require lighter and stronger structural parts at low cost. The characteristic feature of an LCM process is the primarily the use of a closed mold in which a dry fibrous preform material, such as glass or carbon fiber mat, is placed inside the mold and a low viscosity thermosetting polymeric resin is injected under constant pressure or flow rate. The injection is discontinued when the preform is saturated and the resin is allowed to cure forming a completely solid composite. There are many variations in the class of LCM filling processes but there are two distinctly different LCM processes which represent the extremes of the molding technique. They are: Resin Transfer Molding (RTM) and the Vacuum Assisted Resin Transfer Molding (VARTM) process. A schematic of the RTM and VARTM process is presented in Figure 1 highlighting the similarities and the differences.

Many variants of the RTM and VARTM processes have been developed to improve them and provide the most suitable injection solution depending on new industrial applications and needs. Among the variants is the Compression Resin Transfer Molding (CRTM) and the most common variant of the VARTM process, the Seemann's Composite Resin Infusion Molding Process (SCRIMP)[1-4]. Since it is difficult to manipulate the flow direction in the event of an unpredicted filling pattern in the existing variants of the VARTM process, we are introducing the Vacuum Induced Preform Relaxation (VIPR) process as a form of online flow control for any of the Vacuum Infusion Processes (VIP).

VIPR PROCESS

The VIPR process introduces a vacuum chamber system into the VARTM process and is designed to accomplish three main goals. The first goal is to decrease the mold filling times relative to the VARTM process, secondly to increase flexibility in flow control relative to SCRIMP, and lastly to utilize tool side injection ports. The combination of the VIPR process and resin delivery ports was inspired by our recent work in the area of resin delivery called Flow Flooding Chamber (FFC)[5] where an external rigid vacuum chamber is used to seal against the flexible bag surface of a vacuum infusion mold. In the FFC process a higher vacuum gauge pressure is applied to the chamber than is applied to the mold which stretches the bagging film. The purpose of the FFC method of resin delivery is to stretch the bag away from the fabric layers to create a highly permeable gap. The FFC method of resin delivery provides a means to quickly introduce resin into a vacuum infusion mold; however, the resin flow within the empty gap is difficult to control.

The VIPR process is similar in that it utilizes a similar external rigid chamber, but the goal is to increase the preform permeability and affect the flow behavior *without* stretching the bag and creating a flow flooding gap which normally suppresses flow control capability in the FFC process. In the VIPR process, the chamber is placed over a selected region of the mold and a vacuum pressure within the range 25.4-127 mmHg (1-5 in Hg) is applied to the chamber; a full vacuum, ~1 atmosphere, is applied to the mold. These conditions allow one to locally increase the permeability while maintaining the maximum vacuum pressure within the mold.

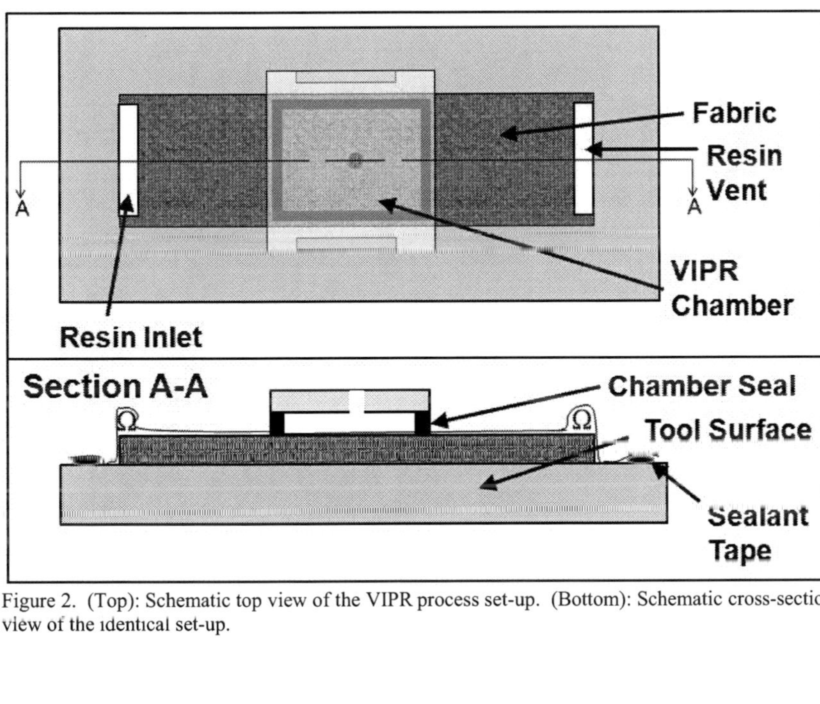

Figure 2. (Top): Schematic top view of the VIPR process set-up. (Bottom): Schematic cross-section view of the identical set-up.

A schematic representation with cross sectional view of the VIPR process set up is shown in Figure 2.

Infusion of resin using the VIPR chamber under these conditions is particularly attractive for flow control because the increase of permeability effect is local and temporary. With this type of system, the filling pattern can be manipulated during the infusion to ensure a complete filling of the mold regardless of geometry and preform variations by applying the chamber to only the desired region of the mold, and further have the ability to turn this effect off when it is not needed. The use of the VIPR chamber is especially useful in combination with resin injection ports. When the chamber is applied to the mold directly above an open port, the through thickness permeability of the fabric is drastically increased allowing for the possibility of changing the filling pattern within a very short time.

MODELING

An essential step in controlling the VIPR process is the ability to model and simulate the filling process so one can plan for the control methodology. Flow prediction also provides the opportunity to select proper processing parameters such as VIPR pressure, VIPR chamber placement and resin gelation time. In this work, we use an existing finite element and control volume analysis package called Liquid Injection Molding Simulation (LIMS)[6] to study the VIPR process. The numerical results produced by LIMS with the appropriate simulation execution conditions are then compared with experiments. The most important aspect noticed during

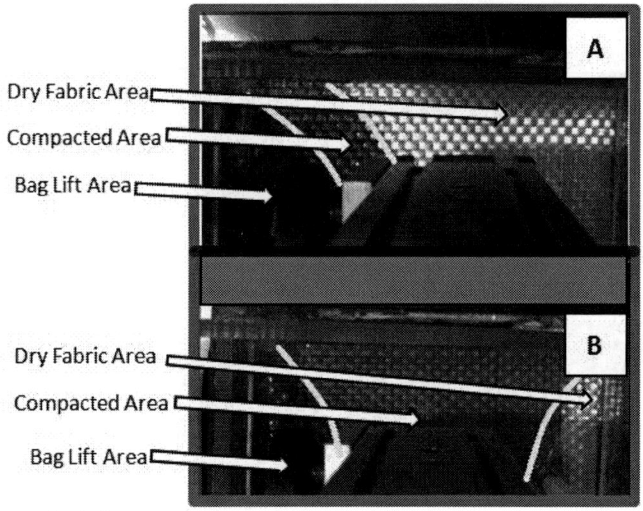

Figure 3. Photograph A shows a small compacted area between the two flow fronts where the Vacuum Pressure within the VIPR chamber is 100 mmHg (4.0 inHg). Photograph B shows a much larger compacted fabric area where the Vacuum Pressure within the chamber is 80 mmHg (3.25 inHg).

extensive experimentation is the existence of a dual flow front type of behavior, shown in Figure 3. Through the use of a specially formulated mesh and proper simulation execution, this process can be accurately modeled.

Mesh Construction

Finite Element / Control Volume method is used to advance the resin in fibrous preforms in LIMS. The most relevant modeling consideration is that the height of the preform significantly changes when the VIPR chamber is under vacuum. To model the flow phenomenon depicted in Figure 3, a special FEA mesh as shown in Figure 4 is created.

The fabric domain is constructed with two dimensional elements and these elements are labeled "Fabric". The initial permeabilities of these elements are set to be the permeabilities of the fabric experimentally measured without the presence of the chamber. At each time step of the simulation, the permeability in both the in-plane and through thickness directions is adjusted. The adjusted permeability of each element is modeled as a linear function of pressure. The "Inlet" elements are highly permeable and used to deliver resin to the entire left face of the cross-sectional mesh.

The most relevant special feature is the pair of elements that are connected to each node along the top surface of the "Fabric" domain. Experimentally, we see that when the pressure within the mold reaches a certain limit, the bag will separate from the preform as shown in Figure 3. This effect is modeled by monitoring the pressure at the nodes along the top surface of the "Fabric", if one of those nodes becomes greater than the injection pressure minus the VIPR pressure, the permeability of the "Switch" element connected to that node is changed from being impermeable to highly permeable. This action allows the resin to flow into the "Gap" elements, thus representing the action of the bag lifting. Once enough resin has entered the mold to fill the entire "Fabric" domain, the injection is discontinued.

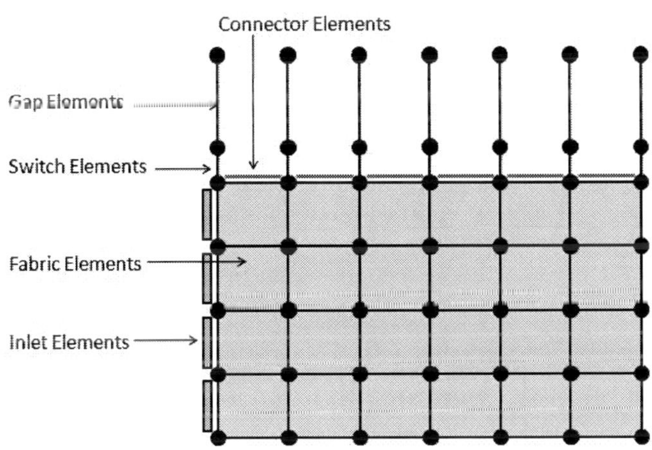

Figure 4. Type of elements used in the finite element mesh to model the mold cross-section of the VIPR process using the LIMS software.

Then the resin which entered the "Gap" elements is allowed to flow into the "Fabric" domain. At this point the "Connector" elements at the base of each of the filled "Gap" elements are also set to be permeable. These "Connector" elements allow the resin to move within the area where the bag has lifted, however, since there is nearly uniform pressure within this domain there is very little resin movement. The simulation is finally stopped when the "Fabric" domain is completed filled.

PREFORM PERMEABILITY CHARACTERIZATION

Some of the features that are incorporated into the VIPR model are independently measurable, specifically the height of the chamber represented by the length of the "Gap" elements and the vacuum applied to the chamber. The initial in-plane permeability, K_{xxi}, and the initial through thickness permeability, K_{yyi}, of the fabric were independently measured using the one dimensional measurement technique, the values of which are given in Equation 1.

It was assumed that the permeability would change linearly as a function of pressure as one introduces the vacuum in the chamber to relax the preform. A constitutive equation of the form

$$K_{xx} - AP + K_{xxi} \quad (1)$$

$$K_{yy} = BP + K_{yyi} \quad (2)$$

is proposed which couples the pressure field with permeability. The constants *A* and *B* in Eq. (1) and (2) will depend on the fabric architecture and compressibility and need to be determined before one can introduce this form in the numerical simulation of the VIPR process.

The numerical model developed to describe the VIPR process is compared with experiments to find the constants *A* and *B*. This is accomplished by creating a model of one dimensional flow in the simulation and selecting values of *A* and *B* that match the numerical flow front with the experimental results of the VIPR process. The value of *B* has little or no effect on the time required to inject the desired amount of resin. Hence, this time is used to find the constant *A* by matching the simulation time with the experimental time. The constant *B* is found by minimizing the error between the experimental and numerical flow front positions shown in Figure 5. The values of *A* and *B* for this particular fabric were found using this methodology and are listed in

Table II.

Table I. INITIAL PERMEABILITY VALUES MEASURED IN A 1-D FLOW.

Permeability	Value (m^2)
K_{XXi}	3.30E-11
K_{YYi}	1.50E-12

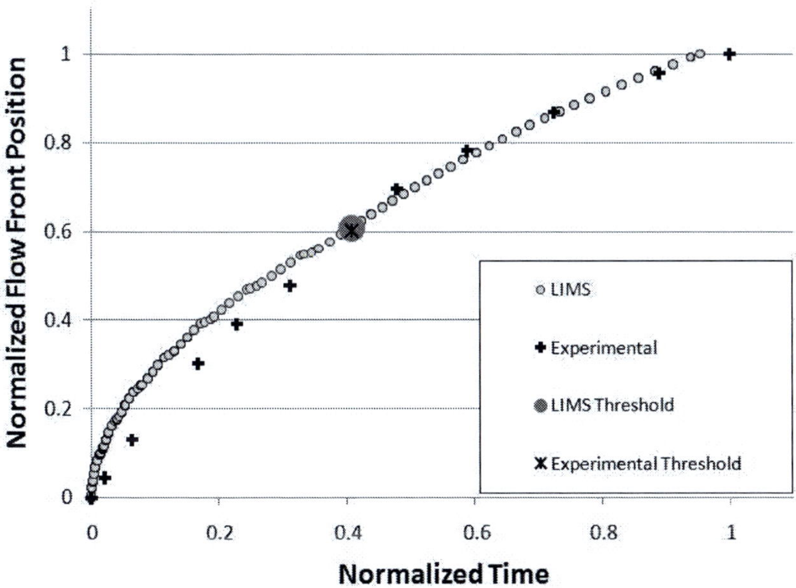

Figure 5. Comparison of the Experimental and Numerical Normalized Flow Front Position versus Normalized Time.

TABLE II. PARAMETER VALUES FOR PERMEABILITY CHARACTERIZATION.

Parameter	Value (m^2/Pa)
A	1.18E-15
B	5.00E-16

SUMMARY AND CONCLUSIONS

The VIPR process is a new flow enhancement mechanism that shows great potential for online flow control in LCM processes. The process has been extensively tested and the dual flow front behavior was identified. A novel modeling technique was used with the LIMS software to model the VIPR process. However this process requires one to dynamically change the value of the through thickness and in plane permeabilities due to the relaxation of the pressure on the preform. A linear change in permeability was assumed with pressure and introduced into the model. The constants needed to describe the linear change were determined by matching the flow front location in the simulation with the measured values from a one-dimensional infusion experiment. The values of the constant suggest that the in plane permeability and the through thickness permeability for the fabric used, 24 ounce per square yard St. Gobain woven E-Glass, increased ~3.5 times in-plane when the vacuum pressure on the chamber was 100 mmHg (4 in Hg) thus providing guidelines for the development of flow control methodology.

ACKNOWLEDGEMENTS

The authors would like to thank Pavel Simacek for his technical assistance with the LIMS software and the Office of Naval Research for funding this research under the AMIPC project Grant no: N00014-06-1-1000.

REFERENCES

1. Marsh, G., *Putting SCRIMP in context.* Reinforced plastics, 1997. 41(1): p. 22–26.
2. Seemann, W.H., *Plastic transfer molding techniques for the production of fiber reinforced plastic structures*, US Patent No. 4, 902, 215; 1990.
3. Seemann, W.H., *Plastic transfer molding apparatus for the production of fiber reinforced plastic structures*, US Patent 5,052,906; 1991.
4. Seemann, W.H., *Unitary vacuum bag for forming fiber reinforced composite articles*, US Patent 5,316,462; 1991.
5. Alms, J. and S.G. Advani, *Simulation and experimental validation of flow flooding chamber method of resin delivery in liquid composite molding.* Composites Part a-Applied Science and Manufacturing, 2007. 38(10): p. 2131–2141.
6. Simacek, P. and S.G. Advani, *Desirable features in a Mold Filling Simulation for Liquid Molding Process.* Polymer Composites, 2004. 25(4): p. 355–367.

Development of 3D Woven, Resin Transfer Molded Fan Blades

B. DAMBRINE, J.-N. MAHIEU, J. GOERING and K. OUELLETTE

ABSTRACT

Snecma (SN) and Albany Engineered Composites (AEC) have been working to develop composite fan blades for jet engines that utilize 3D woven preforms to provide the fiber reinforcement. These preforms are processed into composite components by resin transfer molding. Advantages of composite materials for this application include increased thrust to weight ratio by lowering the mass of the fan. Specific advantages of the 3D woven preform/RTM approach include improved damage tolerance and reduced touch labor.

The fiber preform is woven to near net shape and formed into a complex aero shape that includes twist and sweep. The formed preform is then molded to net shape on the aerodynamic surfaces, and requires few machining operations.

Preferred procedures for weaving, forming, and molding these components have been developed. An approach for designing and evaluating the necessary fiber preform has also been developed. This paper describes the activities that have been conducted in these areas over the past eight years.

INTRODUCTION

The use of composite materials in jet engine applications has steadily increased over the past decade. The specific strength and stiffness of these materials makes them good candidates for use in rotating structures like the fan. The use of wide chord composite fan blades (WCCFB) results in reduced rotating mass and a reduced number of blades. These features, in conjunction with a 3D aerodynamic design, lead to improved fuel burn and reduced noise.

Bruno Dambrine and Jean-Noël Mahieu, Snecma Safran Group, 77550 Moissy Cramayel, France

Jonathan Goering and Ken Ouellette, Albany Engineered Composites, Inc., 112 Airport Drive, Rochester, NH 03867-1718, U.S.A

The use of composite material for the fan blade further reduces rotating mass, as in the GE90 engine. Snecma and Albany Engineered Composites, Inc. have worked together to develop composite fan blades that utilize a 3D woven fiber preform that is resin transfer molded (RTM) into the final blade geometry.

3D weaving is a textile process that places fiber primarily in the radial and chord-wise directions of the fan blade, and includes through-thickness components that lock the fibers together into a single piece. 3D woven/RTM fan blades provide all the benefits of composite blades and offer additional benefits relative to the laminated solution.

First, the 3D woven architecture has an interlocking pattern that has been designed to eliminate delamination, which is the primary damage mechanism in laminated composites. This more damage tolerant composite structure (Figure 1) is the consequence of the micro mechanical design of the weave pattern.

Second, the 3D woven preform can be fabricated to be near net shape, eliminating the ply cutting, and collating that must be done to layup a conventional laminate, and reducing the amount of final machining that must be done. This reduction in touch labor can lead to cost savings for the final component

In addition, the formability of composite materials facilitates the use of advanced 3D aero shapes. This leads to modern, efficient fan configurations that have fewer, wider blades than conventional designs.

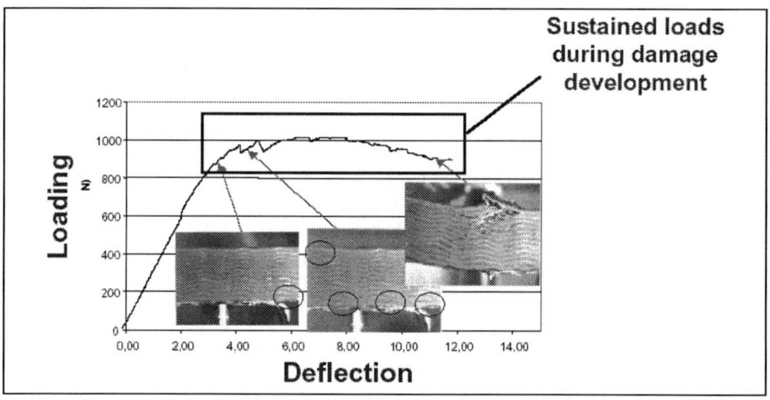

Figure 1. 4 point bending test result.

PREFORM DESIGN

The highly curved shape of the fan blade is transformed into a “flat” configuration consisting of columns of warp (radial) and weft (chord-wise) fibers. The thickness, total fiber volume, and warp/weft ratio distributions are defined so that the proper distributions will be achieved into the geometry of the final part.

Weaving patterns, such as the Snecma ply-to-ply interlock pattern shown in Figure 2, have been developed and patented to achieve : high damage tolerance along with good formability.

Figure 2. Typical Snecma ply-to-ply interlock pattern.

The part thickness variations are obtained by changing the number of layers that are woven.

The process for developing the flattened configuration is an iterative procedure. Changing any one variable, such as the weft column spacing, changes the total fiber volume fraction and the warp/weft distribution. The design process therefore becomes an optimization problem, where the objective function is the average difference between the target and feasible properties.

Factors included in the iterative design process are shown schematically in Figure 3. A series of design and analysis methods have been developed by SN and AEC that automate this process. These tools allow us to evaluate many different configurations analytically, prior to the actual fabrication of prototype components. The structure definition design of the fan blade has been developed to account for the weaving capabilities.

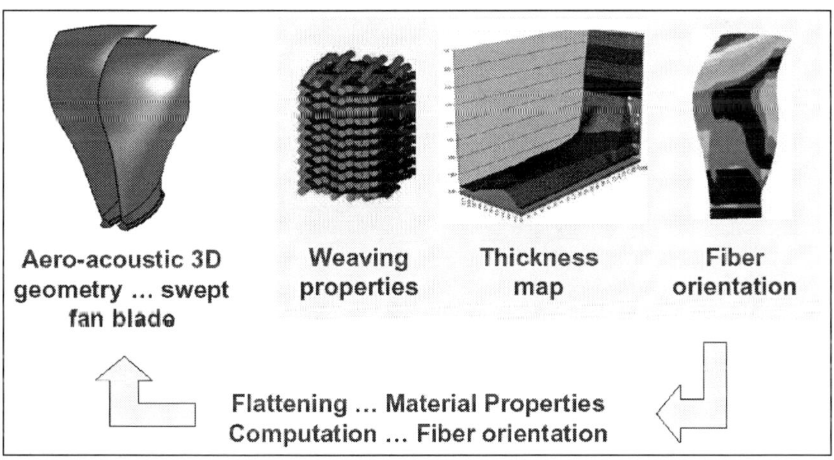

Figure 3. Design iteration loop.

PREFORM FABRICATION

Fan blade preforms are woven on purpose-built Jacquard looms that are designed and fabricated by AEC (Figure 4). A fundamental characteristic of a Jacquard loom is that each warp fiber can be controlled independently. This means that each warp fiber can take a unique path through the preform. This capability facilitates the manufacturing of variable thickness fan blade preforms.

The root of the blade is the thickest part of the preform, and is usually woven slightly oversize and machined into the final shape. The shank provides a transition from the root to the thin airfoil;part of the blade which has tapers in both the chord-wise and radial directions.

Depending on the geometry of the blade preform being woven, the loom is setup to have more or less warp fibers in each warp column.

Fibers that "weave out" no longer interlock with the main body of the preform, and are trimmed away from the preform after it is removed from the loom. The portions of the warp that do not weave into the main body are referred to as "floats", and the process of removing them is referred to as "float trimming." Design rules have been generated to distribute the "floats" so that they do not accumulate along a few radial positions, thereby reducing the structural integrity of the blade.

By its nature, the weaving process produces a preform that has a constant width, albeit with variable thickness. The perimeter of the preform is then cut from this rectangular shape to match the part geometry.

Figure 4. Custom Jacquard looms at AEC.

Figure 5. Fan blade preform after float and contour trimming.

FORMING AND MOLDING

Prior to molding, the 3D woven preform is formed into the proper shape. The 3D woven preforms are designed to have a certain amount of flexibility (drape), facilitating the advanced 3D aero shapes used in current fan blade designs. These designs require the preform to be sheared into an "S" shape in the plane of the preform, and then twisted about the longitudinal axis by as much as 60° (measured from the root to the tip).

Forming is done in a tooling that is also used to secure the preform as it is dried, prior to resin transfer molding. Tracers can be used to align the preform on the various trimming and forming fixtures, and can be used as witness of the final shape of the part.

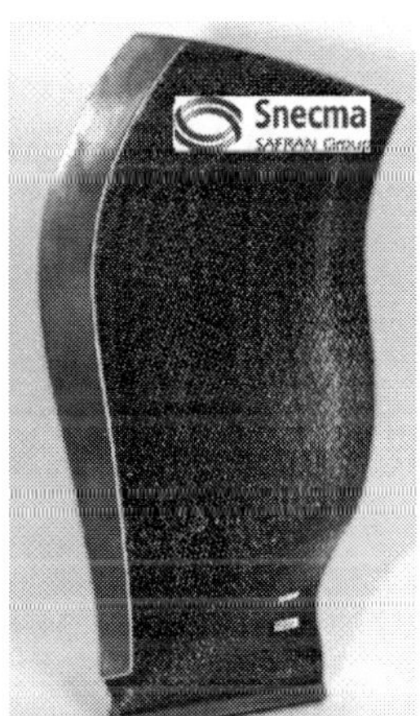

Figure 6. 3D woven/RTM fan blade.

Molding is accomplished using a matched RTM tooling. These tools are designed to produce composite parts with aero surfaces that are net shape and require no additional machining. Initial prototypes and low rate production components have been molded using tools that are heated in an oven, but self heated tools can also be used in a press. A photograph of one of these prototypes is shown in Figure 6.

PERFORMANCE

A comprehensive battery of tests has been successfully conducted on coupons, sub-elements, and full scale components to validate the performance of 3D woven/RTM fan blades. These tests include static strength, impact, low cycle fatigue, high cycle fatigue, bird ingestion (Figure 7). Absence of ply delamination and damage confined mainly to the surface with very few fiber failure have been confirmed throughout.

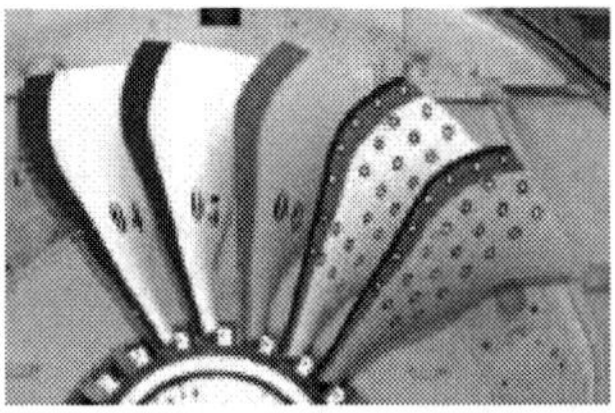

Figure 7. Bird strike test setup for woven composite fan blades.

CONCLUSIONS

The results of this program have demonstrated the viability of 3D weaving and resin transfer molding as manufacturing methods for composite highly loaded parts such as fan blades in jet engines

Processing Technique to Impregnate Glass Fibre Mats for Textile Reinforced Cementitious Composites

R. OLIVIER, J. WASTIELS, B. BELKASSEM, T. TYSMANS, H. CUYPERS and S. BOSSUYT

ABSTRACT

This research will study the quality of an innovative manufacturing process designed for the mechanical impregnation of Chopped Strand Mats by comparing laminates manufactured in such a way to laminates impregnated by Hand Lay Up means. The basic principles of this mechanical impregnator as well as the manufacturing of the cement based laminates will be explained. For comparison, uni-axial and bi-axial tensile tests and three-point bending tests are performed on specimens produced by both production techniques. All tests are monitored with extensometer, LVDT or Digital Image Correlation measuring techniques.

INTRODUCTION

Although they offer attractive properties, cementitious materials, as well as ceramics and glasses, are inherently brittle and present a modest and unreliable tensile strength. As such, they are unsuited for structural applications. Cements have a crucial advantage above ceramics and glasses: they are "cold setting", which means they can easily be strengthened by various reinforcements. A conventional mean of reinforcement is strengthening this brittle material with steel rebars leading inevitable to mostly massive building elements that often eliminate sleek and light architectural design, a concept re-introduced in the past decades with both architectural and structural benefits. Examples of this concept, like lightweight steel structures or tensile architecture, are often neglected due to their weak insulating properties and limited fire resistance. A merge of both technologies were a cement based matrix gets reinforced with diverse textile fibrous materials could in some cases be a solution.

Remy Olivier, Jan Wastiels, Bachir Belkassem, Tine Tysmans, Heidi Cuypers, Vrije Universiteit Brussel, Faculty of Engineering Sciences, Department of Mechanics of Materials and Constructions (MeMC), Pleinlaan 2, 1050 Brussels, Belgium

Future applications of glass fibre textile reinforced cementitious composites (GTRC) all require a high fibre volume fraction in order to resist to considerable high tensile forces and providing the ceramic composite a substantial strain hardening in tension. Common manufacturing processes for cement composites like spray-up and pultrusion seem unable to satisfy this requirement, and thus keep the use of GTRC for building applications limited to non or limited load bearing elements like wall panels, exterior siding, etc.. Hand lay-up, with its known disadvantages of low volume production, seems at the moment the only valid processing technique to obtain a high fibre content.

In order for GTRC to provide a substantial strain hardening and to be produced in an industrial attractive way the Vrije Universiteit Brussel developed a self compacting impregnator (SCI) to pre-process the impregnation of a fibre textile with a cement matrix. The resulting wet prepreg contains a fibre volume fraction which can reach 25%, 3 to 10 times higher than the conventionally used spray-up technique for cement composites.

SELF COMPACTING IMPREGNATOR

If the pultrusion and the compression moulding manufacturing process for cement composites are analysed, it can be seen that they both share similar problems. Due to a high viscosity of the non-Newtonian cement matrix, fibre bundle impregnation can often not be obtained in a satisfactory way. Furthermore, the fibre structure will activate a filtering process between liquid and solid components of the matrix, leading to non equilibrated distribution of the reactivity and mechanical properties of the matrix. These two phenomena cause a weak fibre bound with early fibre debonding and pull-out as a failing mechanism eliminating high fibre stresses with low composite failure stresses as a consequence.

The SCI eliminates these issues to impregnate the textile reinforcements by pressing the matrix in a controlled and continuous way through the fibre structure.

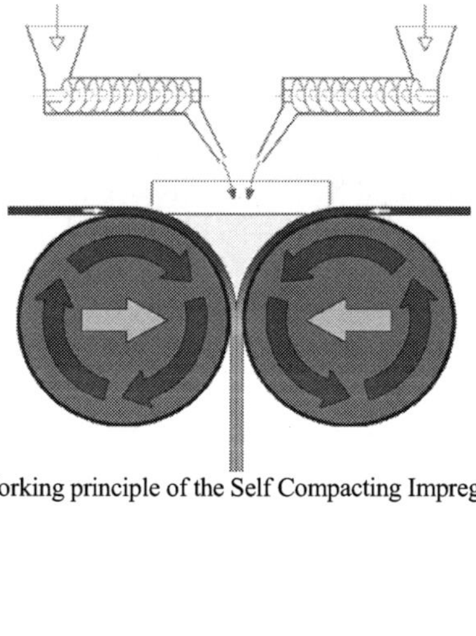

Figure 1. Working principle of the Self Compacting Impregnator (SCI).

The working principle is shown in Figure 1. Two or more textile structures are impregnated by a local compression (cf. hand lay-up production technique). The pressure is obtained by two contra rotating cylinders which control the supply of the textiles and thus the speed of production.

After impregnation the semi-finished product, or "wet prepreg", can then be pultruded or compressed without encountering the problems mentioned above. [1]

MATERIALS FABRICATION

Used Materials

The laminates used in the experiments are glass fibre reinforced cement composites. The matrix component is IPC (Inorganic Phosphate Cement), a cement developed at Vrije Universiteit Brussel and commercialised under its brand name "VUBONITE®". It is composed of an inorganic phosphate liquid and a calciumsilicate solid. The matrix has a density of 2050 kg/m³ in wet phase and a Young modulus of approximately 18 GPa. One of the advantages of using IPC rather than any other cement is that, due to its neutral pH after hardening, it does not attack standard E-glass fibres. As reinforcement a Chopped Strand Mat with a surface density of 300 g/m² is used in all laminates (VETROTEX M5).

Fabrication Of Uni-Axial Test Specimens

For uni-axial tests (tension and bending) 2 laminates have been produced: one using the conventional hand lay-up production technique and one using the self compacting impregnator. For evaluation means, the matrix-fibre ratio has been kept constant for all laminates: a fibre ratio of 20% (±1) in volume was pre-determined.

For hand lay-up production, the approx. 1200 by 400 mm fibre mats were impregnated with the matrix layer by layer, using a standard de-airing hand roller. The self compacting impregnator produced two 1200 mm by 400 mm laminates of 8 individual fibre mats by compacting 4 sets of pre-impregnated mats (Figures 2,a-b-c). The top-side of the laminates were carefully covered with a thin plastic foil to exclude air entrapments and to avoid water evaporation during hardening. In wet state, all specimens were carefully cut using a standard stanley knife (Figure 3).

After production, the laminates were covered by hardboard plates, on which weights (±100 kg) were applied to obtain a minimum of pressure. All specimens are produced at ambient conditions. The curing conditions were equivalent for all laminates: after 1 day of curing in ambient conditions, the weights were removed and the laminates were postcured at 60°C in a Heraeus oven for 24 hours while evaporation was avoided.

Figure 2,a. SCI

Figure 2,b. SCI

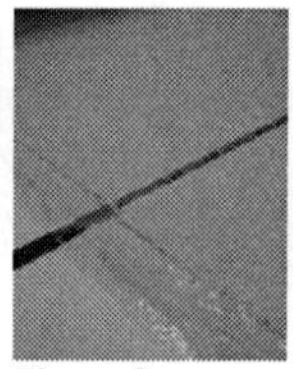
Figure 2,c. laminate

Figure 3. cutting in wet state

Fabrication of Bi-Axial Test Specimens

A geometry optimization of the cruciform specimen was studied numerically. The aim of this optimization was to obtain a geometry that allows high biaxial stresses in the middle of the specimens before one of the arms of the specimen would fail. The optimezed shape was 3 dimensional and thus required a lay-up of 3 plain crusiforms. The geometry is fully described by Tysmans et al. in "THE INFLUENCE OF BIAXIAL STRESS STATES ON THE STIFFNESS OF GLASS TEXTILE REINFORCED CEMENTITIOUS COMPOSITES", presented at TEXCOMP9; Delaware (Figure 5).

In total 4 specimens have been produced: two using the conventional hand lay-up production technique and two using the self compacting impregnator. The matrix-fibre ratio has again been kept constant for all laminates: a fibre ratio of 20% (±1) in volume was pre-determined. For both the hand lay-up as the SCI specimens, 2 laminates of 1200 mm by 400 mm were produced. From each laminate the three layers of the designed test specimen were cut, cured and post cured, all in a similar way as previously mentioned (Figure 4). In a specific lay-up the produced sub-samples were glued together with an epoxy glue (Araldite AW 136 H + hardener HY 994) to obtain the final shape (Figure 5).

Figure 4. Cruciform lay-up.

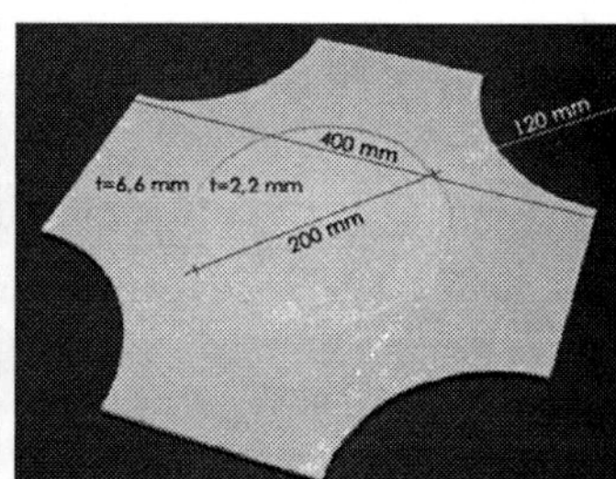

Figure 5. Cruciform specimen.

TEST RESULTS AND DISCUSSION

Uni-Axial Tensile Tests

The tensile tests have been performed on an INSTRON 5885 universal test machine with a constant displacement rate of 2 mm/min. A 50 mm double averaging clip-on extensometer determined the strain level whereas a load cell was used to monitor the stress level. The dimensions of samples were approximately 500 by 75 mm with a thickness of about 5 mm. The samples are pneumatically clamped at both ends.

In table 1 and figures 6 and 7 the results are given. The pre-cracking stiffness is dominated by the stiffness of the matrix and it has been computed in a range from 1 MPa up until 3 MPa, as in this range the behaviour was seen linear for all tested specimens. After cracking the behaviour of the composite is essentially dependent on the characteristics of the fibres. To calculate the slope of the stress-strain curve in post-cracking stage, a range from 20 up until 30 MPa was defined.

The maximum composite stress is the highest stress value that could be reached. The maximum fibre stress is the stress present in the fibre on failure.

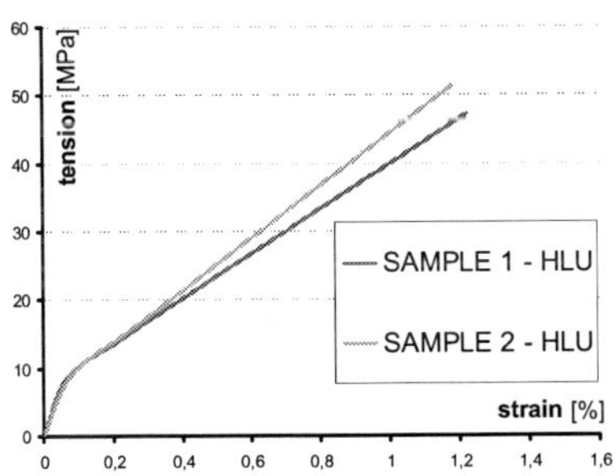

Figure 6. Stress-strain curves for HLU manufacturing process.

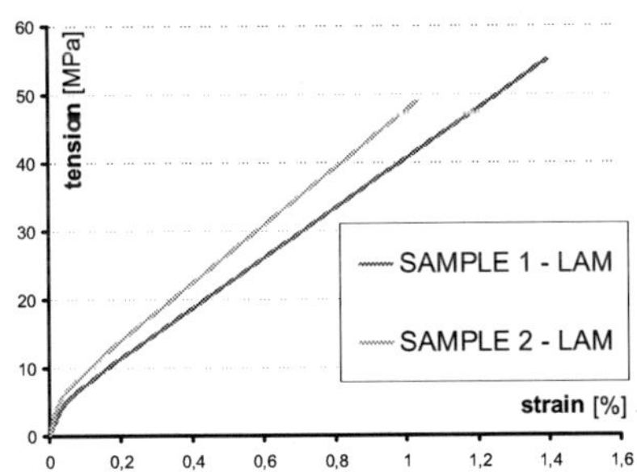

Figure 7. Stress-strain curves for SCI manufacturing process.

TABLE 1: Tensile test results.

	dimensions [mm]		stiffness's [GPa]		failure stress [MPa]	
	width	thickness	pre-cracking	post-cracking	composite	fibre
Sample 1 HLU	75,27	4,94	15,88	3,29	47	744
Sample 2 HLU	74,93	4,98	13,29	3,85	51	816
Sample 1 LAM	75,20	4,75	17,59	4,20	49	742
Sample 2 LAM	76,10	4,70	12,43	3,65	55	826

It can be seen that the maximum composite stress and the post-cracking stiffness are not greatly influenced by the production technique.

As one can see both laminates differ themselves in cracking behaviour. To quantify the experimental difference in behaviour during this multiple cracking phenomena all test results have been curve fitted with a theoretical model. Extending a proposal of Curtin et al [2], Cuypers et al ([3,4]) proposed a statistical treatment of matrix crack evolution in textile reinforced cements, assuming a Weibull distribution function to describe the stochastic distribution of the matrix strength. The Stochastic Cracking model (SC) is based on a two parameter Weibull distribution to describe the matrix strength of textile reinforced cements. The actual crack spacing x at a composite stress σ_c is related to the final (fully saturated) crack spacing X by the following formula:

$$x = X\left[1 - \exp\left(-\left(\frac{\sigma_c}{\sigma_{Rc}}\right)^m\right)\right]^{-1} \tag{1}$$

with:

σ_{Rc} = reference cracking composite stress: first Weibull parameter, which is a measure of ‘average’ multiple cracking stress in the composite;

m = Weibull modulus: second Weibull parameter, indicating the width of the strength distribution.

Another calculated parameter is $\tau X/r$, which describes the force transmitted (shear stress = τ) between two cracks (crack distance = X) per unit fibre (fibre radius = r). It is a measure for the intensity of load transfer between fibres and matrix.

TABLE 2: Tensile test results.

	σ_{Rc} [MPa]	**m**	**$\tau X/r$**
Sample 1 HLU	11,2	5	305
Sample 2 HLU	11,7	4,9	220
Sample 1 LAM	7,7	4,7	190
Sample 2 LAM	6,2	3,6	160

The parameter mostly influenced by the production process seems to be σ_{Rc}. The specimens produced by SCI clearly share a lower reference cracking stress. This could be caused by a weaker fibre matrix bound compared to the one achieved by hand lay-up. The difference is also indicated in the $\tau X/r$ parameter: the intensity of load transfer is clearly higher for specimens produced by hand lay-up means.

Uni-Axial Flexural Test

A 3-point flexural test with a support span of 150 mm has been performed on laminates produced by both manufacturing techniques to evaluate the inter-laminar behaviour of the specimens for a qualitative comparison. The bending tests have been performed on an INSTRON 1195 universal test machine with a constant

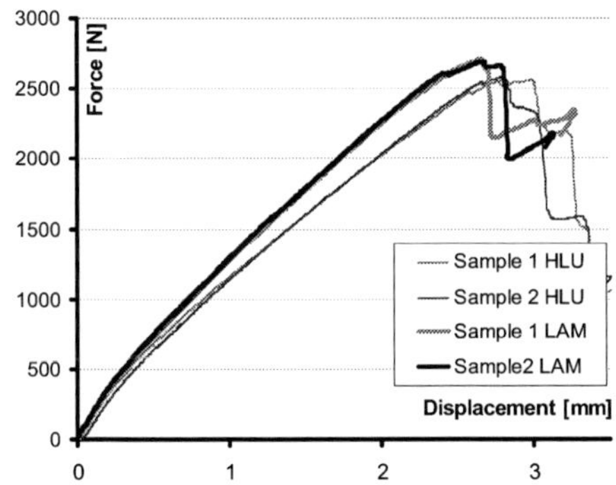

Figure 8. Force-displacement curve for uni-axial bending test.

displacement rate of 1 mm/min. An LVDT determined the midpoint displacement strain level whereas a load cell was used to monitor the stress level. All test specimens had a cross section of 50 mm in width and 15 mm in height. The force-displacement curve of all tested specimens is plotted in figure 8.

The failure mode for all tested specimens has been observed to be equal: delamination occurred in all laminates. Regardless a small difference in stiffness which also has been indicated in uni-axial tensile tests, one can see in figure 8 this delaminations occurs for all specimens around the same force.

Bi-Axial Tensile Test

The biaxial experiments are performed on cruciform shaped samples in a purposely built setup developed at the Vrije Universiteit Brussel (Figure 9). In order to avoid second order effects the plane biaxial test device has four independent servo-hydraulic actuators. The use of hydraulic actuators represents a very versatile technique for the application of the loads. When only one actuator per loading direction is used, the centre of the specimen will move. This causes a side bending of the specimen, which results in undesirable non-symmetric strains. Systems -like the one used by the authors- with four actuators [5] with a close loop servo control using the measured loads as feedback system, allow the centre of the specimen standing still. The device has a maximum capacity of 100 kN in both directions, but it is restricted to tensile loads.

The specimen is loaded with a force ratio of 1/1 at a speed of 5 kN/min. Theoretically, strains in the centre of the cruciform thus should be equal. Digital Image Correlation (DIC) was used to obtain displacement fields. The experimental distribution and direction of the principal strains, observed with DIC under a load of 20 kN in both biaxial perpendicular directions, is shown in Figure 10. The advantage of using the DIC technique is that these strains can be visualized over the whole surface, contrarily to measures with strain gages. It also shows that the experimental strain distribution is not fully symmetric, as a result of the inherent imperfections of the GTR-IPC specimen.

Figure 9. Setup bi-axial experiment developed at VUB.

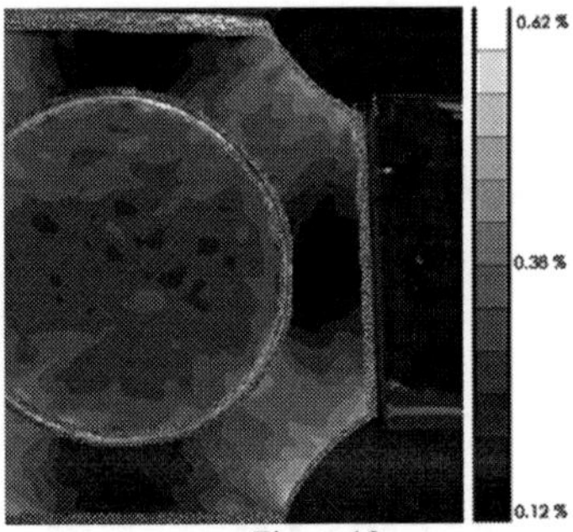

Figure 10.
Principle strains at 20 kN.

From the strain data obtained every 0,5 kN with digital image correlation, a force-strain curve is extracted for the principal strains in the center point of the cruciform specimen (Figure 11).

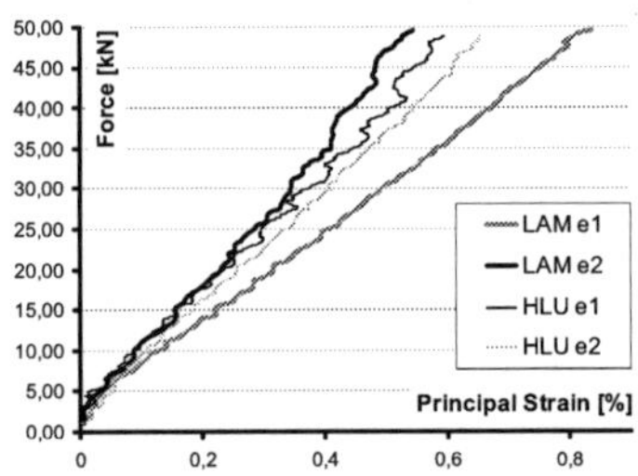

Figure 11. Force-strain curve for the principal strains in the center point of the cruciform specimen.

It can be seen that the principle strains for the hand lay-up (thin curves on figure 11) match very well one an other. The principle strains for the specimen produced by SCI (thick curves on figure 11) show more deviation from each other.

One tends to be somewhat stiffer than the HLU specimen (black curve), one somewhat less stiff (grey curve. A plausible explanation could be that the SCI uses a unidirectional sense of production while with hand lay-up the pressure is applied randomly. Further testing is necessary to draw solid conclusions.

CONCLUSIONS

The mechanical behaviour of laminates produced by the Self Compacting Impregnator has been compared to the one of laminates produced by the standard hand lay-up. As it is observed in a limited amount of tests it can be said that the quality reaches on many fields the quality obtained by hand lay-up. Maximum stress in both composite and fibre and pre- and post-cracking stiffness's are comparable. Without any after treatment like compression moulding or pultrusion, it must be said that the bounding achieved by this prototype is less than the one obtained using hand lay-up. Nevertheless, as a first step in a production process it delivers satisfactory results; in an industrial attractive way a clean and well balanced impregnation of the fibres has been achieved.

REFERENCES

1. EP 07115558.4.
2. Curtin, W.A., Ahn, B.K., Takeda, N. (1998), 'Modeling Brittle and Tough Stress-strain Behaviour in Unidirectional Ceramic composites', Acta Mater. 10:3409–3420.
3. Cuypers, H. (2002), 'Analysis and Design of Sandwich Panels with Brittle Matrix Composite Faces for Building Applications', PhD Thesis, Vrije Universiteit Brussel.
4. Cuypers, H., Wastiels, J., (2006), 'Stochastic Matrix-cracking Model for Textile Reinforced Cementitious Composites under Tensile Loading', Materials and Structures 39:777–786.
5. Smits A., Van Hemelrijck D., Philippidis T.P., Cardon A. "Design of a cruciform specimen for biaxial testing of fibre reinforced composite laminates". Composites Science and Technology 66(2006), pp. 946–975.

Pseudo Reinforcement Effect of Multiwalled Carbon Nanotubes in Epoxy/Glass Fibre Hybrid Composites

V. C. S. CHANDRASEKARAN, M. H. SANTARE
and S. G. ADVANI

ABSTRACT

Interlaminar shear strength (ILSS) of glass fiber reinforced multi-walled carbon nanotube (MWCNT) modified epoxy composites is compared with un-modified epoxy/glass fiber composites. MWCNT was dispersed in epoxy to prepare a suspension by combining the techniques of high speed mechanical stirring and ultrasonic agitation. The composites were manufactured by the injected double vacuum assisted resin transfer molding (IDVARTM) method. Compression shear tests were conducted on the manufactured samples to determine ILSS. Various suspension preparation methods and their effect on ILSS were studied. Theoretically, adding carbon nanotubes to the epoxy matrix can increase the mechanical properties. Though at first carbon nanotube seemed to enhance the ILSS in comparison with un-modified epoxy when the un-modified epoxy was taken through the same processing steps used for dispersion, ILSS improvements for un-modified epoxy were at a par with carbon nanotube modified epoxy. Therefore, the increase in ILSS can be attributed more to the process which modifies the epoxy matrix than the reinforcement effect of carbon nanotubes as reported in literature. These pseudo reinforcement effects of carbon nanotube were investigated and are explained in this paper.

INTRODUCTION

Since their discovery by Iljima in 1991, carbon nanotubes have been the focus of considerable research. Many researches have reported mechanical properties that exceed those of previously existing materials. Nanotubes have exceptionally high elastic modulus (> 1 TPa), high strength (2–28.5 GPa), and remarkable resilience,

Chandrasekaran, V.C.S., Santare, M.H., and Advani, S.G.
Department of Mechanical Engineering and Center for Composite Materials, University of Delaware. Newark, DE 19716

sustaining large, nonlinear deformation before fracture. Their high specific stiffness and strength, combined with their exceptionally large aspect ratio, has stimulated the development of nanotube-reinforced composites.

Laminated fiber reinforced polymer-matrix composite materials have been used successfully in many industries such as aerospace, automobile, marine, military etc. The limiting factor in their application is often their interlaminar shear strength (ILSS). In traditional fiber reinforced polymer-matrix composites, the ILSS is largely governed by the matrix properties. Improving ILSS has long been an important goal in the fiber reinforced composites field, and to this end, different approaches have been tried. Some of the more effective attempts have been weaving fibers in the thickness direction or using Z-pins to connect the laminae. However, these techniques are labor intensive and require additional manufacturing processes that can greatly increase the cost of the resulting components[1].

Many researchers have explored other ways to improve the ILSS by using nanoparticles and carbon nanotubes (CNT) [2–10]. Zhou et al. [3] found that adding 2 wt % carbon nanofibers (CNF) to glass fiber reinforced epoxy composites increased the ILSS by 22.3%. In 2005, Gojny et al. [9] gained a 20% improvement in ILSS by adding 0.3 wt % double-walled carbon nanotube (DWNT) into fiber reinforced epoxy composites. Wichmann et al. [4] got similar enhancement of ILSS, as well as electrical conductivity by using the DWNT. Yokozeki et al. [5] found that carbon fiber reinforced composites can benefit from dispersion of cup-stacked carbon nanotubes (CSCNTs) between fiber mats and that these can delay the onset of matrix cracking. Siddiqui et al. [6] found that adding organoclay nanoparticles increases both crack growth resistance and fracture toughness in carbon fiber reinforced polymer composites. Zhu et al. [13] proposed an innovative way to create CNT/fiber/polymer composites by coating SWNTs on the surface of glass fibers before filling the mold with vinyl ester and got a 35% enhancement in ILSS. Veedu et al. [10] extended this method by growing MWNTs on the SiC woven fabrics to make 3D hybrid composites and achieved remarkable improvement in interlaminar fracture toughness (424%), hardness, delamination resistance, and thermal and electrical conductivities.

Though all these results are encouraging, realization of the superlative mechanical properties in fiber reinforced composites has not yet been achieved outside the research laboratory. Major difficulties include, obtaining a homogeneous dispersion of CNTs in the matrix, processing complexity due to the increase of viscosity with CNT addition to the matrix and achieving a good interfacial adhesion between fibers and modified matrix.

Nano-scaled particles such as CNT exhibit an enormous surface to volume ratio, several orders of magnitude larger than that of conventional fillers. This surface area acts as an interface for stress transfer, but is also responsible for the strong tendency of the CNTs to form agglomerates. Achieving efficient exploitation of their properties in polymers is therefore related to their homogeneous dispersion in the matrix, a break-up of the agglomerates and a good wetting with the polymer. Current methods of improvement include sonication, high shear mixing, surfactant addition, chemical modification through nanotube functionalization, wrapping the nanotubes with polymer chains or combinations of these methods [11].

In this study, we tested the effects of ultrasonic agitation method used and its duration on the ILSS of epoxy/glass fiber/MWNT composites. A small amount of MWNT is dispersed in epoxy resin and the modified epoxy is injected into glass preforms by vacuum assisted resin transfer molding (VARTM). Unmodified epoxy was also made to undergo same processing history as the epoxy resin with MWNT and tested for ILSS. The ILSS of these hybrid composites and control samples were measured using compression shear test (CST). Interestingly, we found a pseudo reinforcement effect of carbon nanotubes, which is discussed in this paper.

EXPERIMENT

Materials

To keep the material costs relatively low, multi walled carbon nanotubes (MWNT) were chosen over single-walled carbon nanotubes. The MWNT were made by CVD (>95%) and purchased from the ILJIN nanotech company, Korea. The diameters of the nanotubes were of the order of 10-20 nm and the length was between 0.5 and 15 μm. The woven glass fiber mats were supplied by Saint-Gobain vetrotex company and had a fiber density of 800 g/m^2. The resin used was epoxy SC-15 from Applied Polymeric Inc.

Manufacturing of Glass Fiber/Epoxy/MWNT Hybrid Nanocomposites

The processing of glass fiber/epoxy/MWNT hybrid nanocomposites consists of two stages:

1. Preparation of MWNT suspensions
2. Injection and Double Vacuum Assisted Resin Transfer Molding (IDVARTM)

Preparation of MWNT Suspensions

In this study, we used two ultrasonic agitation techniques, namely bath sonication and tip sonication for dispersion of carbon nanotubes in epoxy. We studied sonication times of 6 hours and 20 hours for both, bath and tip agitation techniques thus creating four suspensions that had undergone different preparation methods. All preparation contained 250 mg of multi walled nanotubes and were measured and added to 38.5 g of Epoxy SC-15 resin part A in a beaker. The beaker was placed inside a Branson 1510 ultrasonic bath sonicator and sonicated for 6 hours and 20 hours respectively to prepare two of the suspensions. Similarly, the other two suspensions were prepared by sonicating the suspensions for 6 hours and 20 hours in Sonics vibra cell tip sonicator. The beaker was then kept in bell jar under vacuum for degassing. A schematic of the process is shown in Fig.1.

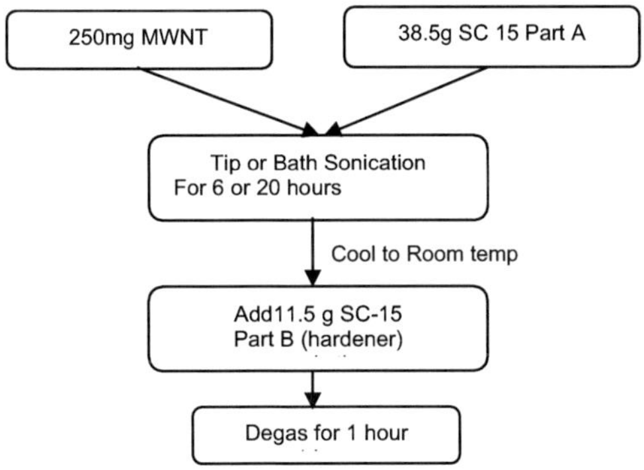

Figure 1. Flowchart of the MWNT suspension preparation method.

INJECTION AND DOUBLE VACUUM ASSISTED RESIN TRANSFER MOLDING (IDVARTM)

Eight layers of 5 cm X 5 cm woven glass fiber mat were sealed with a vacuum bag and the air was evacuated by applying the first vacuum, which is connected to the outlet of the mold; the fiber mats compacted under the external atmospheric pressure. The outlet was then clamped to discontinue the first vacuum and the MWNT/epoxy suspension was injected into the closed mold; excess resin was injected to over saturate the space between the bag and the preform. The inlet was clamped and the entire closed mold was placed into a second vacuum chamber for 30 min. This process releases the external atmospheric pressure induced by the first vacuum and allows the eight compressed fiber mats to recover some of their original thickness. The recovery expands the gaps between the glass mats making it easier for the MWNT/epoxy suspension to infuse into the larger channels between the glass fiber mats by intrinsic capillary pressure. After the MWNT/epoxy suspension fills the gaps between the glass fiber mats, it starts to infuse across into each glass fiber mat by capillary pressure. The second vacuum is then released to reapply atmosphere pressure to the mold for 30 min to further squeeze the MWNT/epoxy suspension into the fiber mat. Although some MWNTs are still being filtered at this stage, this turns out to be an advantage because the higher concentration MWNT/epoxy suspension will remain between the fiber mats where it is most needed to improve the ILSS. After the MWNT suspension fills the empty regions within and between the fiber mats, the first vacuum is reapplied to draw off the excess MWNT/epoxy suspension and recompress the mold to obtain the desired higher fiber volume fraction. Table 1 lists the processing steps.

1	Cut 8 layers of the fiber mat of dimensions 5cm X 5cm
2	Apply release agent to mold plate and stack the 8 layers on it
3	Apply sealant close to the boundary of the stacked preform and place omega tube at both ends

4	Cover the assembly with plastic and ensure it is leak proof.
5	Clamp the omega tubes to prevent leakage and ensure vacuum holds.
6	Mix MWNT suspensions with 11.5g hardener in a beaker and degas.
7	Inject the resin into the mold via the omega tube.
8	Cut off the vacuum supply and place the set-up inside a vacuum chamber for 30 minutes to induce resin impregnation.
9	Remove the setup from the vacuum chamber and reconnect it to the vacuum supply.
10	Maintain the composite under vacuum until it cures (we left it for a day). Post cure it at 80C for 4 hours

Table 1. Processing steps to manufacture the composite.

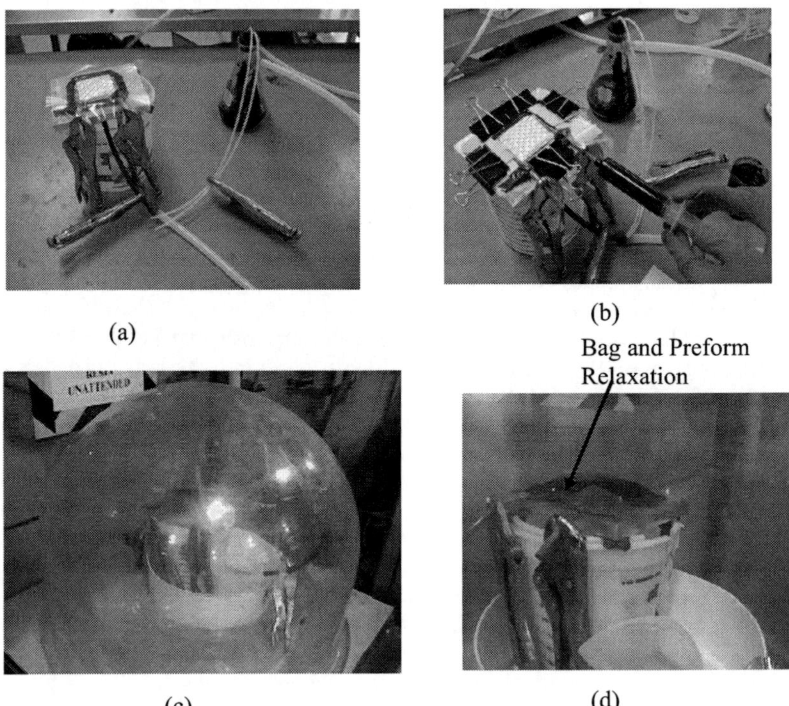

Figure 2. (a) VARTM Setup (b) Injection (c) Second vacuum under a vacuum chamber (d) Relaxation of the assembly as the pressure releases on the preform due to the second vacuum.

INTERLAMINAR SHEAR STRENGTH CHARACTERIZATION

To measure the interlaminar shear strength ILSS, a compression shear test method was used. The samples were cut from the manufactured composites. Approximate size of the samples was 0.4" in length, 0.2" in width and 0.15" in thickness. The shear test fixture is shown in Fig 3. The left clamp contacts the left half of the top surface, which is perpendicular to the laminate planes. Symmetrically, the right clamp contacts the right half of the bottom surface of the specimen. Two vertical guides support the left and right sides of the specimen to prevent it from rotating or moving horizontally. The right clamp is fixed in position, and the left clamp moves down to generate direct interlaminar shear along the central plane of the specimen until failure. The load on the specimen at failure is then divided by the area of the failure plane to determine the interlaminar shear strength (ILSS) of the specimen. The results of these ILSS tests for each of the four different specimen preparations, along with the control samples, prepared using the identical process, but without the addition of nanotubes, are shown in Fig. 4. The first results also shows the results for the composite samples prepared using the resin but without taking that resin through the preparation steps used for the nanotube suspensions.

Figure 3: Compression shear test fixture with sample loaded.

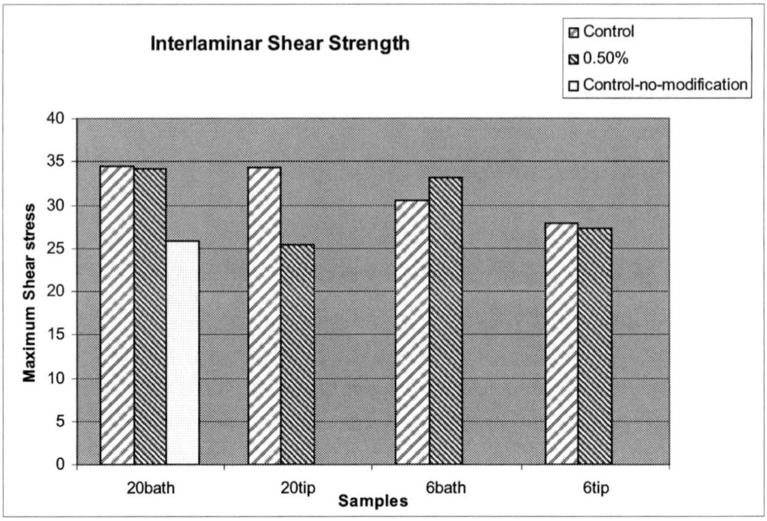

18 Samples were tested for each category and standard deviation is around 9%.

Figure 4. ILSS comparison with the control samples without CNT which were subjected to the 0.5% CNT suspension preparation method before injection. The first bar chart also includes the results from unmodified epoxy composite samples not subjected to the preparation methods of CNT suspension.

RESULTS AND DISCUSSION

The first result to note is that the unmodified epoxy composite samples demonstrate much lower ILSS than the epoxy containing nanotubes and the epoxy that was subjected to the same preparation steps as the nanotube suspension. Hence, if one had not subjected the epoxy to the sonication, one would have drawn the conclusion that the nanotubes had enhanced the ILSS substantially.

Among the different preparation methods, we see that for 0.5% CNT epoxy composites, the 6 hour bath sonication shows slight increase in ILSS due to the nanotube additions while 20 hour bath and 6 hour tip show no significant enhancement. On the other hand, the 20 hour tip gives a reduction in ILSS due to the addition of the nanotubes. These results point to the importance of the role of the preparation method of the MWNT suspension.

We hypothesize that in many instances in the literature, researchers neglect to subject the control samples of resin with no nanotubes through the same identical process history as the resin suspension containing the nanotubes. Therefore, the improvements in ILSS containing the nanotubes are as likely to be the result of the additional processing steps that the resin undergoes as seen in Figure 4. Hence, we call it the pseudo reinforcement effect.

The possible explanation of why the processing history of resin preparation before infusion may influence the ILSS is that when we sonicate the epoxy resin, bubbles form in the resin system due to small scale cavitation. With additional ultrasonic energy, these bubbles are alternately compressed and enlarged. At some point, they collapse giving rise to a localized burst of energy. This energy release and high speed agitation heats the resin resulting in a significant decrease in the viscosity. The rise in local temperature and acceleration of molecular movement in the epoxy resin decreases the viscosity and surface tension of the epoxy resin system, resulting in better resin infiltration into the preform and hence improved ILSS properties [12].

CONCLUSIONS

The improvement in ILSS for control samples which were taken through the identical processing history were similar to those of samples containing 0.5% CNT. Therefore, the improvement in ILSS is not due to the reinforcing effect of CNTs but rather due to the decrease in viscosity and increased molecular movement of epoxy resin molecules caused by the sonication giving rise to an apparent reinforcing effect. Since the improved ILSS is due to the processing steps and not the addition of CNT we refer to this phenomenon as a pseudo-reinforcement effect of CNT.

REFERENCES

1. Fan Z, Advani SG, Santare MH. Interlaminar shear strength of glass fiber reinforced epoxy composites enhanced with multi walled carbon nanotubes Composites: Part A (39) 2008 540–554.
2. Hsiao KT, Alms J, Advani SG. Use of epoxy/multiwalled carbon nanotubes as adhesives to join graphite fibre reinforced polymer composites. Nanotechnology 2003;14(7):791–3.
3. Zhou Y, Pervin F, Rangari VK, Jeelani S. Fabrication and evaluation of carbon nanofiber filled carbon/epoxy composite. Mater Sci Eng A 2006;426:221–8.
4. Wichmann MHG, Sumfleth J, Gojny FH, Quaresimin M, Fiedler B, Schulte K, et al. Eng Fract Mech 2006;73:2346–59.
5. Yokozeki T, Iwahori Y, Ishiwata S. Matrix cracking behaviors in carbon fiber/epoxy laminates filled with cup-stacked carbon nanotubes (CSCNTs). Compos Pt A: Appl Sci Manuf 2006;38:917–24.
6. Siddiqui NA, Woo RSC, Kim JK, Leung CCK, Munir A. Mode I interlaminar fracture behavior and mechanical properties of CFRPs with nanoclay-filled epoxy matrix. Compos Pt A: Appl Sci Manuf 2006;38:449–60.
7. Zhu J, Imam A, Crane R, Lozano K, Khabashesku V, Barrera EV. Processing a glass fiber reinforced vinyl ester composite with nanotube enhancement of interlaminar shear strength. Compos Sci Technol 2006;67:1509–17.
8. Shahid N, Villate RG, Barron AR. Chemically functionalized alumina nanoparticle effect on carbon fiber/epoxy composites. Compos Sci Technol 2005;65:2250–8.
9. Gojny FH, Wichmann MHG, Fiedler B, Bauhofer W, Schulte K. Influence of nano-modification on the mechanical and electrical properties of conventional fibre-reinforced composites. Compos Pt A: Appl Sci Manuf 2005;36:1525–35.
10. Veedu VP, Cao A, Li X, Ma K, Soldano C, Kar S, et al. Multifunctional composites using reinforced laminae with carbonnanotube forests. Nat Mater 2006;5(6):457–62.
11. Thostenson ET, Chou TW, Compos. Sci. Technol. 61 (2002) 1899–1904
12. Liu L, Huang YD, Zhang ZQ, Jiang b, Nie J Ultrasonic modification of Aramid-Epoxy Interface. Journal of Applied Polymer Science, Vol. 81, 2764–2768 (2001).
13. Zhu J, Khabashesku V, Imam A, Crane R, Lozano K, Barrera E. Processing and properties of polymer composites reinforced by functionalized SWNTs. In: Pricm 5: the fifth pacific rim international conference on advanced materials and processing, Pts 1–5, vols. 475–479; 2005. p. 1059–62.

TOPIC II

ADVANCED PROCESSES MANUFACTURING

Meso-macro Approach for Textile Composite Forming Simulation

P. BOISSE, N. HAMILA, P. BADEL and E. VIDAL-SALLE

ABSTRACT

Composite textile reinforcement forming simulations permit to determine the processing conditions for a quality part and to find the position of the fibers after forming. Because the textile composite reinforcements are multiscale materials, continuous (macro) approaches and discrete (meso) approaches have been developed. The finite elements presented in this paper for textile fabric forming are composed of woven unit cells. The mechanical behaviour of these unit cells is analyzed by 3D computations at the mesoscale regarding biaxial tensions, in plane shear and bending. These elements are efficient because they are close to the physics of the woven cell while avoiding the very large number of unknowns in the discrete approach.

INTRODUCTION

The increase in the usage of textile composites, especially in aeronautic applications, due to their high mechanical properties leads to develop adequate design and analysis methods. Among the different codes developed in this framework, draping simulation plays a large part. First, the simulation can give the conditions that will make the forming of the textile reinforcement possible (for instance loads on the tools, initial orientation, type of material etc) and also describe possible defects after forming (wrinkles, porosities, yarn fractures etc). Secondly, and unique to composite forming analysis, is the need to know at any point the directions and density of the fibres after forming. These directions and densities are very important for analyzing how the composite part will behave in use, with regards to stiffness, damage, fatigue etc. They also strongly influence the permeability of the textile reinforcement and must be taken into account in case of LCM Processes [1].

P. Boisse, N. Hamila, P. Badel, E. Vidal-Salle
Laboratoire de Mécanique des Contacts et des Solides, INSA de Lyon, France

The mechanical behaviour of textile is a multi-scale problem. The macroscopic behaviour is very much dependent on the interactions of fibres at the meso-scale (scale of the woven unit cell) and at the micro-scale (level of the fibres constituting yarns). Despite a great amount of work in the field, there is no widely accepted model that accurately describes all the main aspects of fabric mechanical behaviour. A first family of models is obtained by homogenizing the mechanical behaviour of the underlying meso-structure and considering the fabric as an anisotropic continuum [2,3]. If these models can easily be integrated in F.E. classical shell or membrane elements, the definition of the continuous constitutive model of the textile (at large strains) involves many difficulties

On the other hand, the multi-scale nature of fibre reinforcements allows discrete approaches where each fibre or yarn is modelled [4]. Because of the large number of fibres in a fibrous reinforcement, this kind of approach is generally limited to a small domain study (woven cell, knitted loop…). If the simulation of a complex part is made with such approaches, the mesoscopic modelling has to be too coarse (usually rods and springs, …) for an accurate result.

In the semi-discrete approach [5,6], the components at meso scale are considered as part of finite elements that fix the kinematics in these discrete components in function of nodal displacements of the elements. That leads to some continuity within the elements and especially two initially superimposed warp and weft fibers stay superimposed during the draping. This semi-discret approach is close to the physics by considering mesoscopic entities and accounting only for their significant mechanical properties. For instance, in the present paper, only fiber tensions and in plane shear rigidities will be taken into account. The identification of the necessary mechanical characteristics from classical tests (tensile and bias tests) is simple and straightforward.

MECHANICAL BEHAVIOUR OF TEXTILE REINFORCEMENTS

In the present approach the textile composite reinforcements under consideration are composed of a discrete number of woven unit cells such as shown on figure 1.

The diameter of the fibres constituting the textile is extremely small in comparison to their length in case of continuous fibres. Consequently the main loading that can be applied to these fibres is a tensile load in the fibre direction. The yarns are made of thousands of fibres, these fibres can slide relatively and the tensile stiffness is still much larger than the other rigidities. (The textile reinforcement under consideration is dry i.e. with no matrix). The loads on this unit cell are assumed to result in two tensions T^{11} and T^{22} in the warp and weft directions $\underline{f}_1$ and $\underline{f}_2$ and a torque $C_s\,\underline{f}_3$ which is due to the internal actions from warp yarn on weft yarns in reaction to in plane shear ($\underline{f}_3$ is perpendicular to $\underline{f}_1$ and $\underline{f}_2$) .

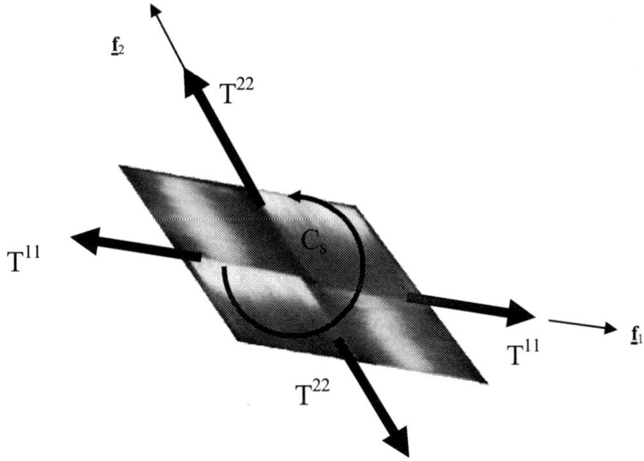

Figure 1. Unit woven cell submitted to tension and in plane shear.

Because the yarns are composed of many fibers with a very small diameter, the shear stiffness is small and the torque C_s as well. Nevertheless, it is important to consider this in-plane shear behaviour because it plays a significant role in some cases and especially in wrinkle development which is an important aspect in draping simulations. From these loads on a unit woven cell, the interior virtual work in a textile reinforcement in a virtual displacement field $\underline{\eta}$, can be written:

$$W_{int}(\underline{\eta}) = \sum_{p=1}^{ncell} {}^{p}\varepsilon_{11}\left(\underline{\eta}\right){}^{p}T^{11}\,{}^{p}L_1 + {}^{p}\varepsilon_{22}\left(\underline{\eta}\right){}^{p}T^{22}\,{}^{p}L_2 + \sum_{p=1}^{ncell} {}^{p}\gamma\left(\underline{\eta}\right){}^{p}C_s \tag{1}$$

where ncell is the number of unit woven cell of the domain under consideration, $\varepsilon_{11}\left(\underline{\eta}\right)$ et $\varepsilon_{22}\left(\underline{\eta}\right)$ are the linear axial strain in the warp and weft directions $\underline{f}_1$ and $\underline{f}_2$ in the virtual field $\underline{\eta}$. $\gamma\left(\underline{\eta}\right)$ is the angle change between warp weft direction in the virtual field $\underline{\eta}$. L_1 and L_2 are the length of the unit cell in warp and weft direction. ${}^{p}A$ means that the quantity A is considered for the cell number p.

It has been shown by experiments that the tensions are reasonably independent of the shear angle. As the tensile behaviour is biaxial it will be in the form $T^{11}(\varepsilon_{11}, \varepsilon_{22})$, $T^{22}(\varepsilon_{11}, \varepsilon_{22})$. It will also be assumed that the shear torque does not depend on the axial strain and will be in the form $C_s(\gamma)$. This is probably less true [7,8] but there is not enough available data and models concerning the influence of tensions on shear stiffness to take it into account.

Mesoscale Analyses

The mechanical properties $T^{11}(\varepsilon_{11}, \varepsilon_{22})$, $T^{22}(\varepsilon_{11}, \varepsilon_{22})$ and $C_s(\gamma)$ can be determined by 3D F.E. analyses of the unit cell submitted to biaxial tension on one hand and in plane shear on the other hand. The advantage of those mesoscale analyses is firstly that it is easy to vary geometrical parameters or boundary conditions. Secondly they give 3D

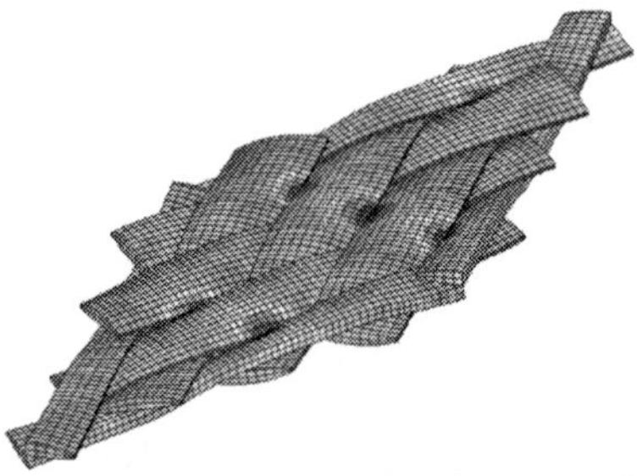

Figure 2. Deformed geometries of the 2x2 twill unit cell for a shear angle of 40°.

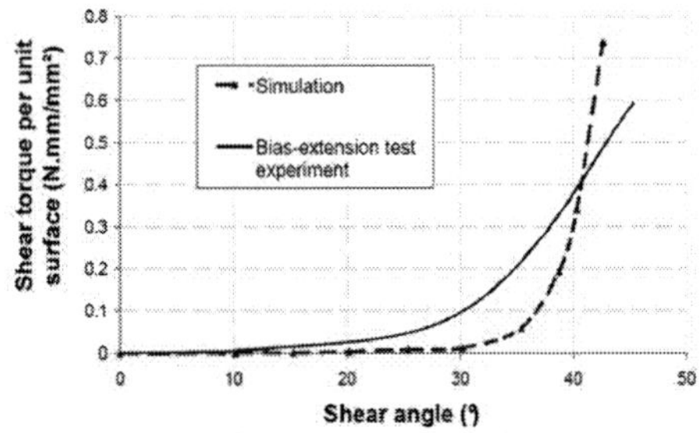

Figure 3. Experimental and numerical shear curve.

field results, such as the local transverse strains, that are very difficult to obtain from experiments. Finally they can be performed on textile reinforcement before its manufacturing which is very important at the design stage.

The analysis is not standard, mainly because the yarn is made of several thousand fibres and its mechanical behaviour is very specific. It is especially necessary to use an approach that strictly follows the fibre direction. For that, a rate constitutive equation based on an objective stress rate using the fibre rotation has been developed [3]. Another important aspect concerns the periodicity that has to be taken into account in boundary conditions. The detail of these analyses can be found for in plane shear in [9]. In the case of a 2x2 twill, an in plane shear analysis shown figure 2 and 3 gives the in plane shear torque $C_s(\gamma)$.

THREE NODE MESO-MACRO FINITE ELEMENT

The three node triangle shown figure 4 is composed of ncelle woven cells. The material vectors $\underline{k}_1$, $\underline{k}_2$ are defined from the coordinates r^1, r^2 in the warp and weft directions:

$$\underline{k}_1 = \frac{\partial \underline{x}}{\partial r^1} \qquad \underline{k}_2 = \frac{\partial \underline{x}}{\partial r^2} \qquad \text{and consequently} \quad \underline{k}_1 = \underline{AM}_2 \quad \underline{k}_2 = \underline{BM}_3 \tag{2}$$

The virtual deformations in warp and weft directions are:

$$\varepsilon_{\alpha\alpha}\left(\underline{\eta}\right) = \left(\left(\underline{\underline{\nabla}}^s(\underline{\eta})\right)\cdot \underline{f}_\alpha\right)\cdot \underline{f}_\alpha = \frac{\partial \underline{\eta}}{\partial r^\alpha}\cdot \frac{\underline{k}_\alpha}{\left\|\underline{k}_\alpha\right\|^2} \tag{3}$$

The elementary tensile virtual work can be written :

$$W_{int}^{te}(\underline{\eta}) = \boldsymbol{\eta}_n^{eT}\mathbf{F}_{int}^{te} = \sum_{p=1}^{ncelle} {}^{p}\left(\frac{\partial\underline{\eta}}{\partial r^1}\cdot\frac{\underline{k}_1}{\|\underline{k}_1\|^2}\right){}^{p}T^{11}\,{}^{p}L_1 + {}^{p}\left(\frac{\partial\underline{\eta}}{\partial r^2}\cdot\frac{\underline{k}_2}{\|\underline{k}_2\|^2}\right){}^{p}T^{22}\,{}^{p}L_2 \quad (4)$$

Consequently, the tensile nodal interior loads are:

$$\left(F_{int}^{te}\right)_{ij} = ncelle\left(B_{1ij}T^{11}(\varepsilon_{11},\varepsilon_{22})\frac{L_1}{\|\underline{k}_1\|^2} + B_{2ij}\,T^{22}(\varepsilon_{11},\varepsilon_{22})\frac{L_2}{\|\underline{k}_2\|^2}\right) \quad (5)$$

In the same way, , the shear nodal interior loads are:

$$\left(F_{int}^{se}\right)_{ij} = ncelle\,B_{\gamma ij}C_s(\gamma)\,. \quad (6)$$

The detail of the calculations can be found in [10] in the case of the presented three node element.

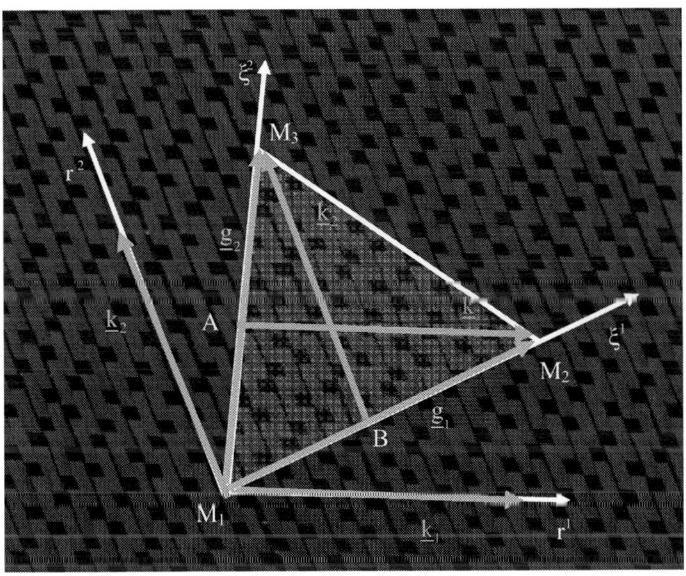

Figure 4. Three node finite element made of woven cells.

NUMERICAL SIMULATIONS

Picture Frame

In this test (Fig. 5), a hinged frame with four bars with equal length is assembled in a tensile testing machine. A tensile force is applied across diagonally opposing corners of the picture frame rig causing the picture frame to move from an initially square configuration into a rhomboid. Consequently, le specimen within the picture frame is theoretically subjected to a pure shear strain field. This test has been intensively studied and is a main issue in textile reinforcement mechanical behaviour [11]. Figure 5 shows the deformed mesh. It can be seen that the nodal loads on the frame are well perpendicular to fibres (there is no tension in the fibres in case of pure shear). Many continuous model proposed for textile behaviour at large strain fail to this requirement. The computed load on the grip of the tensile machine (F on figure 5) is compared to the theoretical load corresponding to $C_s(\gamma)$:

$$F(\gamma)=C(\gamma)\frac{\text{ncell}}{a}\frac{2}{\sqrt{2}}\left(\cos\frac{\gamma}{2}-\sin\frac{\gamma}{2}\right)^{-1} \tag{7}$$

The theoretical and computed curves are exactly the same (figure 5 right).

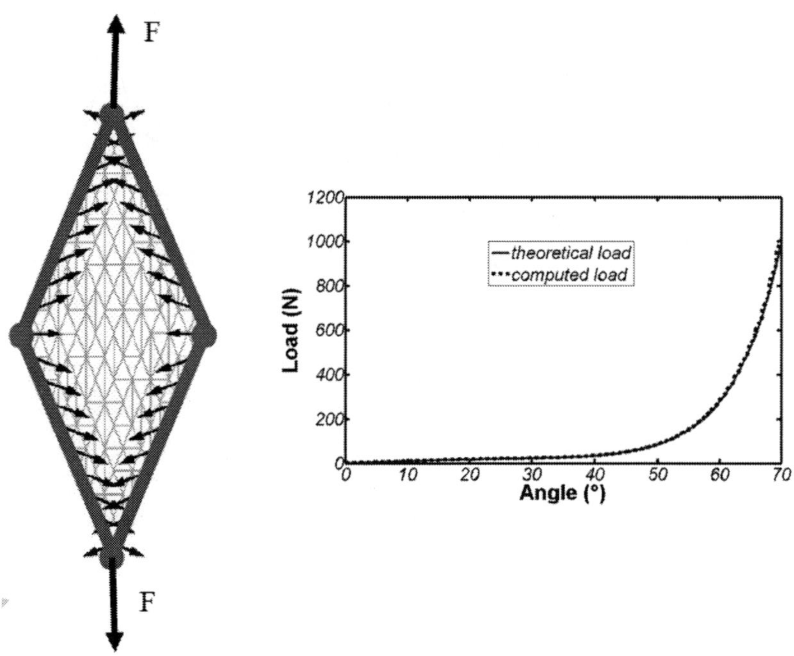

Figure 5. Simulation of the picture frame.

Hemispherical Forming Of An Unbalanced Composite Fabric

The simulation of a very unbalanced fabric is presented Fig. 6 (The simulation is performed on a quarter of the part for symmetry reasons). The warp yarns are much stiffer than the weft yarns (ratio =250). The blank holder is a 6 kg ring submitted to its own weight. The experimental forming has been performed in Nottingham University [12]. The final shape obtained after forming is very asymmetrical. There is a large axial strain in the weft direction (horizontal) and large displacements with very small axial strain in the warp direction (vertical). This final shape is well obtained by the simulation. The ratio of the lengths after deformation l_{weft}/l_{warp} is equal at the top of the hemisphere to 1.8 in experiments and in simulation as well. There are many wrinkles, especially along the vertical axis. They are fairly well obtained by the simulation.

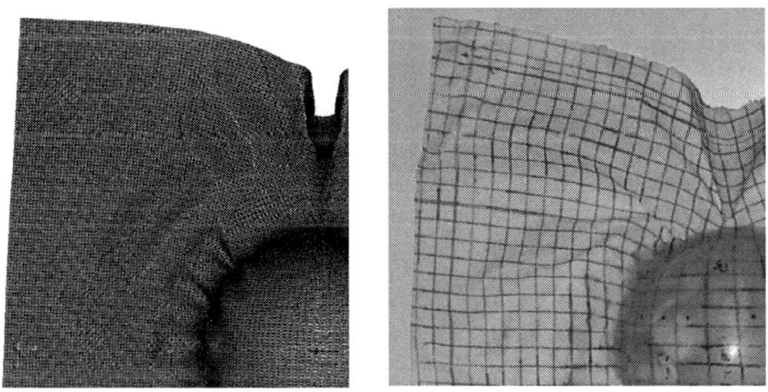

Figure 6. Hemispherical forming of an unbalanced composite fabric. Simulation (left) and experiment (right).

CONCLUSIONS

The finite elements proposed in this paper are made of woven unit cells in biaxial tensions and in plane shear. These elements are efficient because they are simple and close of the mechanical behaviour of woven materials. The biaxial tensile and in plane shear properties can be obtained by 3D simulations at the mesoscale level. The membrane assumptions (no bending stiffness) used in the present approach leads to wrinkles when the shear angles are large. Nevertheless the shape and the number of those wrinkles depend on the bending stiffness of the fabric. The next step in the improvement of the element is to take into account the bending rigidity [13].

ACKNOWLEGEMENTS

The work reported here has been carried out in the scope of the projects ITOOL (European Commission) and MACODEV (Rhône- Alpes region).

REFERENCES

1. Loix, P. Badel, L. Orgéas, C. Geindreau, P. Boisse, Woven fabric permeability : from textile deformation to fluid flow mesoscale simulations, Compos. Sci. and Tech. 68 (2008) 1624–1630.
2. Akkerman R and Lamers E A D. Constitute modelling for composite forming. In: Composite forming technologies, Woodhead Publishing, 2007; Chapter 2: 22–45.
3. P. Badel; S. Gauthier, E. Vidal-Salle; P. Boisse; Rate constitutive equations for computational analyses of textile composite reinforcement mechanical behaviour during forming, Composites Part A, On line, doi : 10.1016/j.compositesa.2008.04.015.
4. G. Creech, A. K. Pickett, Meso-modelling of Non-Crimp Fabric composites for coupled drape and failure analysis, J Mater Sci (2006) 41:6725–6736.
5. Boisse P, Zouari B, Gasser A. A mesoscopic approach for the simulation of woven fibre composite forming. Compos. Sci. Technol. 2005; 65 (3–4): 429-436.
6. Boisse P., Zouari B., Daniel J. L., Importance of In-Plane Shear Rigidity in Finite Element Analyses of Woven Fabric Composite Preforming. Composites A. 2006; 37(12): 2201-2212.
7. Lomov S.V., Verpoest I.: Model of shear of woven fabric and parametric description of shear resistance of glass woven reinforcements. Composites Science and Technology 66, 919–933 (2006).
8. Launay J., Hivet G., Duong A. V., Boisse P., Experimental analysis of the influence of tensions on in plane shear, Composites Science and Technology, 68 (2008) 506–515.
9. P. Badel; E. Vidal-Salle; E. Maire; P. Boisse; Simulation and tomography analyzis of textile composite reinforcement deformation at the mesoscopic, Composites Science and Technology, On line, doi: 10.1016/j.compscitech.2008.04.038.
10. N. Hamila, P. Boisse, Simulations of textile composite reinforcement draping using a new semi-discrete three node finite element, Composites Part B, 39 (2008) 999–1010.
11. J. Cao, R. Akkerman, P. Boisse, J. Chen, H.S. Cheng, E.F. de Graaf, J.L. Gorczyca , P. Harrison, G. Hivet , J. Launay,, W. Lee, L. Liu, S.V. Lomov, A. Long, E. de Luycker, F. Morestin, J. Padvoiskis ,X.Q. Peng, J. Sherwood, Tz. Stoilova, X.M. Tao, I. Verpoest, A. Willems, J. Wiggers, T.X. Yu, B. Zhu, Characterization of mechanical behavior of woven fabrics: Experimental methods and benchmark results, Composites: Part A 39 (2008) 1037–1053.
12. J.L. Daniel, D. Soulat, F. Dumont, B. Zouari, P. Boisse, A.C. Long, Forming simulation of very unbalanced woven composite reinforcements, Int. J.l of Forming Processes, 6, N° 3-4., 2003, pp 465-480.
13. N. Hamila, Simulation de la mise en forme des renforts composites mono et multi plis, PhD thesis, INSA Lyon, 2007.

Stiffened Spherical Panels—Utilization of Modern Braiding Processes

M. GLOWANIA, J. STÜVE, F. F. KRUSE and T. GRIES

ABSTRACT

In many applications composite parts curved shell constructions do form the surface. Those parts need curved stiffener profiles to avoid buckling effects. Especially the aerospace industry has a big need for stiffener profiles as a light weight design made of Carbon Fibre Reinforced Plastic (CFRP) and the corresponding economic production methods. Newly investigations at Institut fuer Textiltechnik, RWTH Aachen University (ITA), Aachen developed new production methods for the economic manufacture of curved complex stiffener preforms realized as sandwich design. The setup of hat-shaped cross sections necessitates the development of special production devices. Additionally two different braiding technologies—the 3D-rotary-braiding and the overbraiding technology—have been part of the research to manufacture cross sections. There is a great demand of those cross sections to realize crossed stiffener constructions. The production of the preforms using both braiding technologies will be explained in the paper. Comparing both braiding technologies referring to the preform quality, economic viability and mechanical performance of the final stiffener the advantages as well as the limitations of each braiding technology in respect of this special application are shown. As result of the investigations the most economic way to manufacture preforms for curved crossed stiffener structure realized as sandwich design is identified.

INTRODUCTION

The realization of complex three-dimensional textile structures in near-net-shape form a repeatable and cost-effective production of reinforces structures for parts made out of carbon fiber reinforced plastic (CFRP) is of great importance. Current research projects develop one-step and multi-step-processes to create textile preforms [1–3].

Michael Glowania, Jan Stüve, Frank Felix Kruse, Prof. Thomas Gries, Institut fuer Textiltechnik, RWTH Aachen University, Eilfschornsteinstrasse 18, 52062 Aachen, Germany

In an integrated project founded by the German Research Foundation, the Institut fuer Textiltechnik, RWTH Aachen University (ITA), the RWTH Aachen University Department of Aerospace and Lightweight Structures, and the Institute of Plastics Processing at RWTH Aachen University (all Aachen, Germany) the necessary knowledge for a repeatable manufacturing of large, complex curved stringer stiffened panels made of carbon fiber reinforced plastics (CFRP) was investigated. The contribution of ITA is the investigation of different preforming methods for the production of these composite parts. For example, textile technologies like multi-axial warp knitting, braiding and stitching have been analyzed. The research objective of this work was to identify the suitable textile production technologies for the realization of the different preform shapes for the subpreforms and preforms of complex curved stringer stiffened panels.

Such structures are used today in many applications e.g. boat hulls where a stiff lightweight shell with a smooth surface is needed. The structures that are focus of this work can be used as body panels in automotive or railway applications (Figure 1). For the production of the subpreforms, the following technologies have been identified. The plane of the panel is carried out by multi-axial warp knitting, the stiffeners are realized using different braiding technologies. The joining of these subpreforms is done by stitching [4].

In the following, the two different braiding technologies to create the stiffener preforms will be described. First, a rough estimation of the costs of each technology is done. Second, the results of the bending stiffness of the stiffener profiles are discussed. Moreover two different ways of the realization of stiffener crossings with a continuous yarn course over the crossing point are established.

Figure 1. Curved stringer stiffened panel.

Figure 2. Cross-section of braided stiffener.

REQUIREMENTS ON BRAIDED STIFFENERS

The stiffeners are manufactured as sandwich parts with a hat-shaped cross-sectional form around a foam core. For the connection of the stiffeners with the multi-axial warp knitted fabric, flange areas are needed. These are included into the stiffener preform (Figure 2). Part of the research program was the investigation of the 3D-braiding technology as well as the overbraiding technology for the realization of this had-shaped preform.

3D-BRAIDED STIFFENERS

The structure of the 3D-braided stiffener is a combination of an overbraided foam core and strand braids that form the flanges. The integration of the flanges into the braided structure is accomplished with the 3D-rotary braiding technology that has been invented by ITA and realized in close cooperation with August Herzog Maschinenfabrik GMBH & Co. KG (Herzog), Oldenburg, Germany. For the realization of the flange areas, braiding needles are needed to avoid prebraids.
For the production of a curved stiffener with the 3D-rotary braiding machine a special guiding device is necessary in addition to the linear take up of the machine. This device is essential for a repeatable take-up movement, in order to realize an orientation of the cross section of the braid at the braiding point parallel to the machine plate at any time. Figure 3 shows the principle of the device and the realization on the machine.

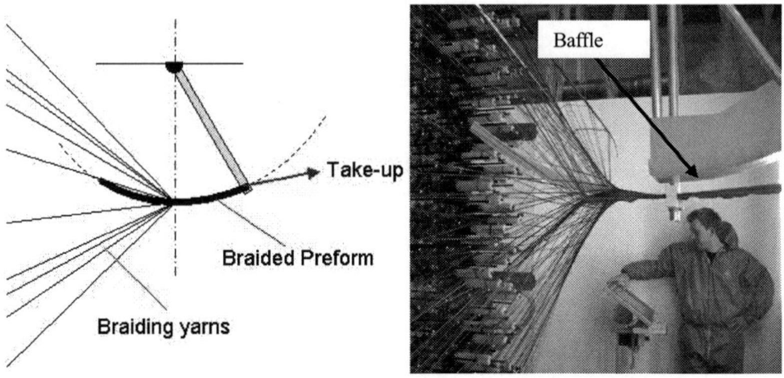

Figure 3. Principle of take-up device (left) and realization of it (right).

OVERBRAIDED STIFFENERS

An additional forming step is indispensable for the realization of the hat-shaped cross-sectional stiffeners using the overbraiding process. The tubular braid has to be draped into a had-shape. The braid has to be fixed to it due to the fact a special yarn orientation has to be guaranteed in relation to the foam core. This requirement leads to the need of an auxiliary mandrel. The design and realization is shown in Figure 4 in combination with a radial braiding machine built by the company Herzog.

This mandrel realizes a tubular braid with sufficient perimeter to form the hat-profile after braiding and gives the stiffener preform its curved shape. The foam core is placed inside the mandrel and a combined removal of the foam core and the braid fixed to each other can be accomplished. For this purpose, the perimeter of the auxiliary mandrel can be decreased. Thus, the fixed braid can be removed.

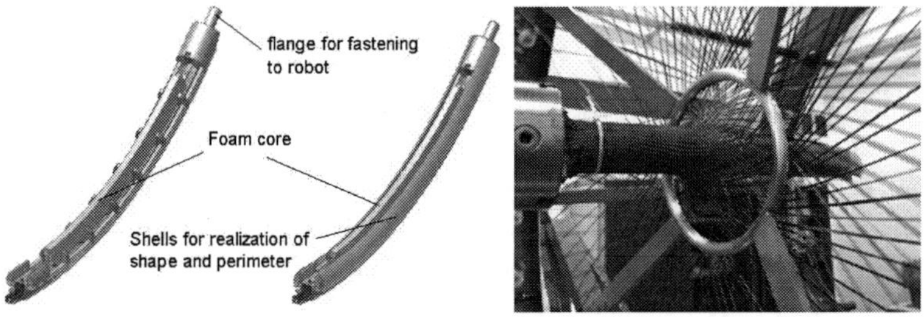

Figure 4. Design of auxiliary mandrel for overbraiding (left) and realization of it (right).

COMPARISON OF BRAIDING TECHNOLOGIES

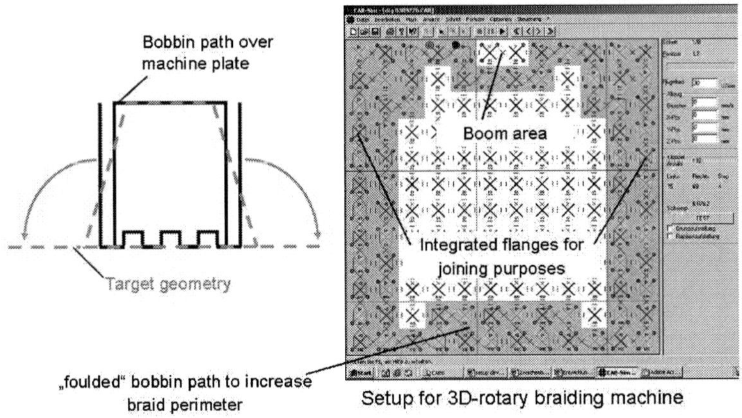

Figure 5. Setup for the 3D-rotary braiding machine.

For the production of the stiffener preform described above both braiding technologies can be used. The 3D-rotary braiding technology permits the realization of a complex preform shape in a single production step whereas the production using the overbraiding technology is a multi-step process. Using the 3D-rotary braiding the yarn orientation can be achieved very exactly. This orientation remains unchanged during the further process steps. The draping and forming of the overbraided preform into its final shape involves the risk to cause defects in the final outline. The production of the overbraided preform is much faster than the production using the 3D-rotary braiding technology that is currently installed at ITA. The production time to create the final stiffener preform using overbraiding technology is more than half of that using the current 3D-braiding technology (15 minutes using overbraiding technology). In addition it causes less yarn damage than the 3D-braiding process by reason of its special machine design. The 3D-rotary braiding machine at ITA is designed for research purposes. A realistic reduction of the production time down to 50 % of the current value can be realizable if a 3D-braiding machine with a much simpler design would be extracted from the current complex design.

The setup for the 3D-rotary braiding machine for the manufacturing of hat-shaped stiffeners is quite complex (Figure 5). The bobbin path is packed somehow into the operation area of the machine.

The manufacturer Herzog could easily built a machine that generates the needed shape directly with a much simpler algorithm of bobbin movement. The outcomes of such a machine would be a much higher production speed, and a manufacturing of the complex stiffener geometry in one single production step. This idea becomes more important if the mechanical properties of the final stiffeners are taken into account. First 4-point-bending tests carried out at stiffeners manufactured using both braiding technologies. The results show, that the 3D-braided design is significantly stiffer than the overbraided design (Figure 6)

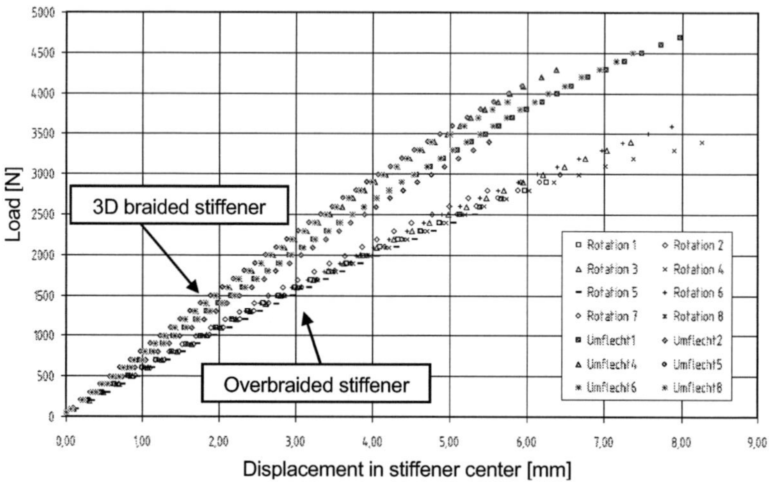

Figure 6. Comparison of stiffness of braided stiffeners.

The results are not fully understood yet, especially because the same amount of yarns, the same yarn count and the same yarn type have been used for both stiffener versions. Obviously, the slight difference in construction (the overbraided version consists of a double layer in the flange areas) leads to this significant difference. Further investigations have to be carried out to be able to describe the effects in detail.

STIFFENER JUNCTIONS USING BRAIDING TECHNOLOGIES

For the realisation of large size stringer stiffened panels, a crossed stiffener architecture is needed. Therefore a method has to be found to manufacture stiffener junctions (Figure 7). It is important to realize a continuous yarn course over the crossing point of the junction to achieve a high stiffening effect. Especially the stretched and continuous yarn course of the 0°-yarns over the crossing point is important. Both braiding technologies, 3D-rotary braiding and overbraiding, provide the production of structures that meet these requirements.

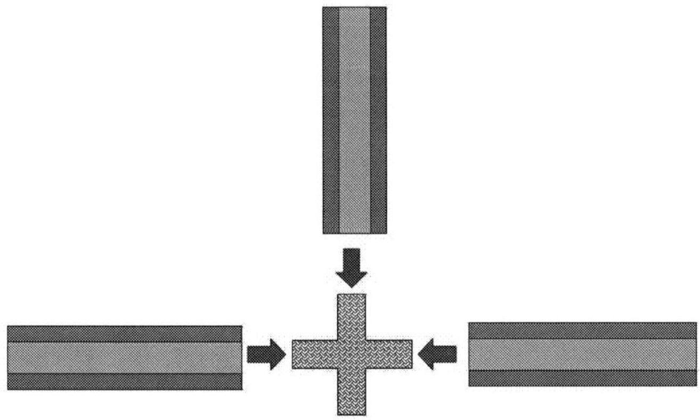

Figure 7. Principle of stiffener junctions.

As discussed above the stiffeners are realized in sandwich design around a foam core. The junctions should be realized in the same sandwich design as well. Hence, the stiffeners are attachable to the junction by removal of a certain length of the initial foam core and replacement by one arm of the junction.

This stiffener junction can be realized by using the overbraiding technology. The cross section consists out of foam core. This core is overbraided 4 times, for- and backward in each direction (Figure 8).

During the manufacturing process the foam core has to be removed from the support, twisted and attached to the support again. 0°-yarns can be integrated on top of the beams. Thus, they are running continuously over the crossing point.

Another way to create junctions is to use the 3D-rotary braiding technology. This technology enables the production of two tubular braids that can be joined to produce a compact area and which can be divided again to form the two separate braided tubes (Figure 9). Compared to that, foam cores can be introduced and therefore overbraided during the production of the tubular braids. 0°-yarns can be included in the area where needed (here at the bottom). A continuous yarn course of the 0°-yarns over the crossing point can be achieved by exchanging the respective yarn carriers on the machine plate. The entire process runs automatically except the introduction of the foam cores. The final prototype is shown in Figure 9.

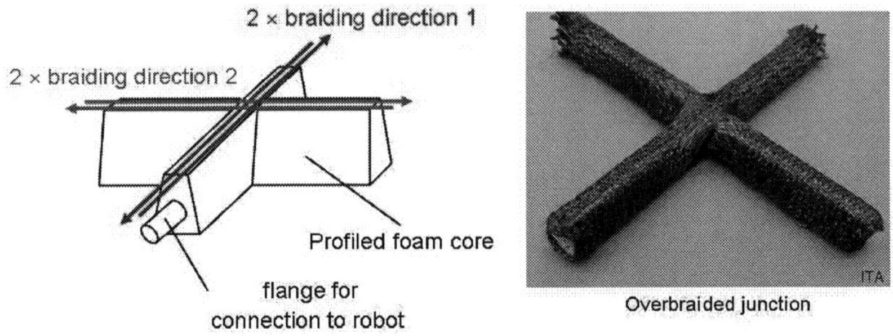

Figure 8. Production principle of overbraided junction (left) and the prototype (right).

Figure 9. production of 3D-braided junction (left), final 3D-braided prototype (right).

Thus it appears that the centre of the 3D-braided junction appears uneven. For this special geometry the 3D-rotary braiding technology is not the production technology of choice. As soon as a sandwich design of the stiffener junction is chosen, the overbraiding technology is the more suitable technology to meet the requirements, even if the production time regarding this product sample is nearly identical. If a junction between e.g. a stringer in sandwich design and a rib in compact C-cross-section is needed, the 3D-rotary braiding technology could be very beneficial. This is due to the fact that this technology can easily combine both design principles in one production process.

CONCLUSION AND OUTLOOK

Braiding technologies are very capable methods for an automated production of load-conform stiffener preforms. The decision for a most suitable braiding technology depends on the geometry, the design, and the requirements during application of the product. Further, investigations have to be carried out to define the mechanical capability of braided structures and to find out more about the capabilities and limitations of the braiding technologies.

REFERENCES

1. Drechsler K. and Gessler A. 2005. "Textile Fertigungsverfahren für optimierte Faserverbundstrukturen in Proc," presented at the 2nd WING Conference Aachen, November 09–11, 2005.
2. Hermann A.S., Book M., Dressler G. and Piepenbrock J. 2004. *Verbundprojekt PRO-CFK*. Deutscher Luft- und Raumfahrtkongress Jahrbuch 2004 Band I + II.
3. Ogale A. and Mitschang P. 2004. "Tailoring of textile preforms for fibre-reinforced polymer composites", *Journal of Industrial Textiles Volume*, 34(2): 77–96.
4. Hufenbach, W. 2007. Textile Verbundbauweisen und Fertigungstechnologien für Leichtbaustrukturen des Maschinen- und Fahrzeugbaus. SDV—Die Medien AG.

Processing and Mechanical Properties of Pultruded Moldings with Braiding Technique

Y. HIDEKUMA, A. OHTANI, A. NAKAI and H. HAMADA

ABSTRACT

In the pultrusion process, usually the unidirectional fiber reinforced composites are fabricated. One of the problems of the unidirectional fiber reinforced composites is the anisotropic of mechanical properties. In this study, the unidirectional fibers were covered with the braided yarns to reduce the anisotropic. Moreover, the new molding method which had continuity was established by combining braiding machine and pultrusion process. The impregnation states and the flexural properties of the pultrusion moldings in longitudinal and lateral directions were investigated. The fractured surfaces of the specimens after flexural tests were observed, in order to clarify the difference of the fracture mechanism between unidirectional moldings and braided one. As the result of the experiments, braided moldings with good impregnation state could be obtained in case that fiber volume fraction was enough high. It was clarified that specimens with braiding technique had high isotropic, because the braided moldings showed the much higher flexural properties in lateral directions than the only unidirectional moldings. Besides, it was found that the braiding layer restrained the development of the cracks.

INTRODUCTION

The pultrusion process has become increasingly broader use for the production of composite [1]. The pultrusion is an automated process for manufacturing reinforced materials into continuous, constant cross section profiles. The basic operation consists of pulling a continuous reinforcing fiber through a facility to impregnate the reinforcement with resin, and then shaping and curing the profile in various dies. In addition to producing complex sections, the pultrusion process has an operating flexibility enabling optimum efficiency of reinforcement to be realized at high speed.

Yuya HIDEKUMA, Kyoto Institute of Technology, Gosyokaido-cho, Matsugasaki, Sakyo-ku, Kyoto, 606-8585, JAPAN
Akio OHTANI, Asami NAKAI, Hiroyuki HAMADA, Kyoto Institute of Technology

That is the advantages of pultrusion process are good moldability and high mechanical properties depend on continuous processing. These and other advantages make pultrusion such an efficient process that it has been said that, if possible, a composite should be designed so that it can be pultruted [2].

The unidirectional fibers are used as general reinforcement materials for pultrusion molding, because of their greatly mechanical properties of longitudinal direction. Normally, only longitudinal properties were considered, because the mechanical properties of other directions were very weak in compared with longitudinal direction. In other words, it is anisotropic material. Therefore, in order to extend further applications, the pultrusion moldings should design the stiffness at all other direction according to the requirements. To improve the mechanical properties, usually the mats were used as reinforcement form.

In this study, braided techniques were used instead of the mats. The schematic drawing of braided tube was shown in Fig. 1. In this figure, the braided angle means the bundle orientation of the braidings, and the middle end fiber means the unidirectional fibers which along the longitudinal direction in between braided fibers. The braiding is the textile technology which promises improvements in properties of composite material and being to meet requirements of automated fabrication [3]. Since brandings have fiber bundles which continued in longitudinal and circumferential direction, it can be expected that mechanical properties would be improved. In addition, the braidings can be designed their mechanical properties by changing the braided angle and applying the various kind of fiber to by one of the braiding fiber and the middle end fiber. Therefore, the purpose of this study is to improve anisotropic of the unidirectional pultrusion molds with braiding technique.

The schematic drawing of pultrusion processing line with braiding technique is shown in Fig. 2. The braiding machine was inserted between glass rovings and resin bath. That braiding machine can braid the cylindrical tube. Besides, continuous of molding line which is one of the advantages of pultrusion process was not lost in new molding line with braided technique. As it was shown in Fig. 3, unidirectional fibers were covered with braiding yarns. And then it was pulled into the die after providing the resin. The impregnation states and the flexural properties of the pultrusion moldings were investigated. Since the purpose of this study is to improve anisotropic, anisotropic parameter was introduced.

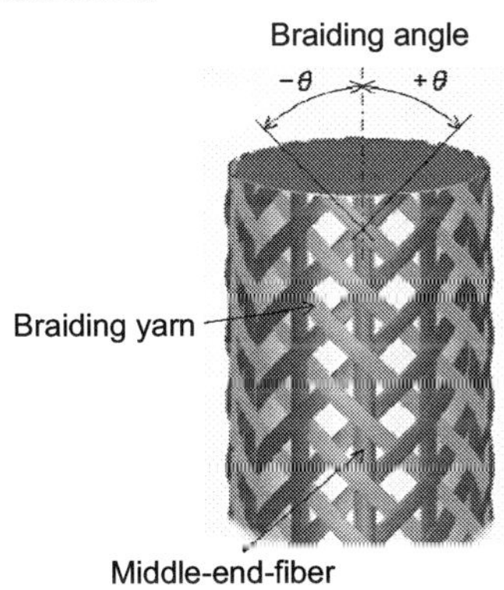

Figure 1. The schematic drawing of braided tube.

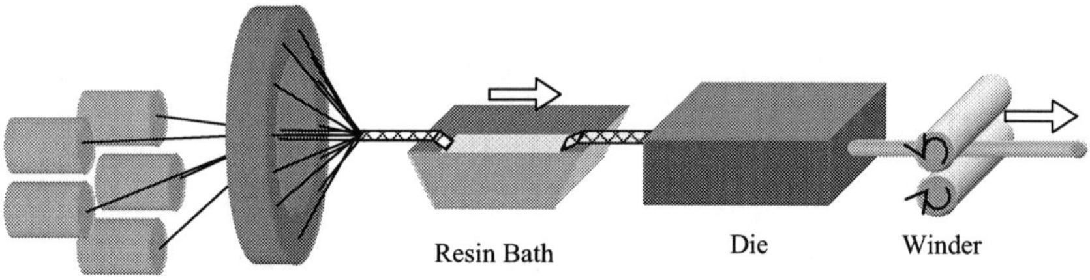

Figure 2. New pultrusion molding line with braiding technique.

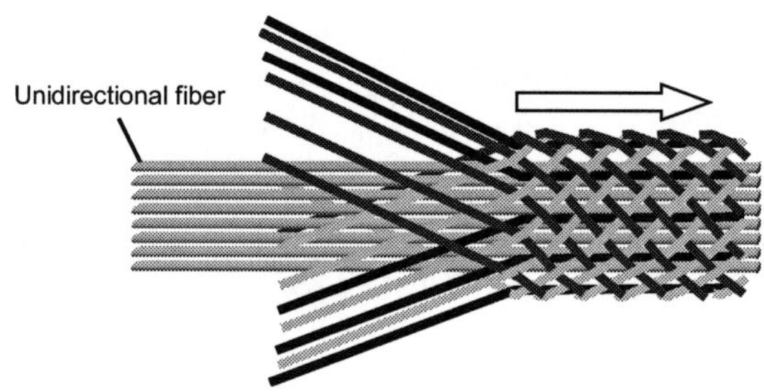

Figure 3. The schematic drawing of braided molding.

EXPEREMENTAL PROCEDURE

Material And Molding Condition

As matrix, an unsaturated polyester (UP) resin was used. And as reinforcement fiber, the glass fiber (SAINT GOBAIN VETROTEX Ro99 4400 P122) was used. Same type of fibers was used both for unidirectional and braided fiber.

The die had a rectangle section which size was 32×6 mm, length of die was 800 mm. The temperature of die was controlled every three parts, from entrance they were 80, 140, 160 degree Celsius. The pultrusion velocity was 200 mm/min.

Effect Of Middle-End-Yarn On Impregnation State

Three types of moldings were manufactured. One was the braided molding without middle-end-yarn, next one was the braided molding with middle-end-yarn and last one was the only unidirectional molding. The braided moldings were compared with the unidirectional molding. The braided angles of two braided moldings were 30 degree. All moldings had 50% fiber volume fraction (Vf). In order to investigate the effect of middle-end-yarn on impregnation state, the surface and the cross section of the each specimen was observed.

Effect Of Supply Method Of Resin On Impregnation State

Two types of supply method of resin were examined for getting the good impregnation state. One was a single resin bath like Fig. 2. The other was double resin bath. In this method, two resin baths were inserted before and after braiding machine. Therefore, the unidirectional fibers were soaked in the resin before they were covered with braiding fibers, and then they were soaked in the resin again after braiding. Those braided angle was 25 degree, and Vf was 48%. In order to investigate the effect of supply method of resin on the impregnation state, the surface and the cross section of the each specimen were observed.

Effect Of Fiber Volume Fraction On Impregnation State

To investigate the relationship between fiber volume fraction and impregnation state, the void contents of three specimens which had different fiber volume fraction were measured. The Vf of three specimens were 48%, 50% and 52%. In order to control the Vf, the number of the unidirectional fiber was changed, in this study these were 12, 14 and 16. The surface and the cross section of the each specimen were observed, and then the void contents were calculated by the cross section photos which were obtained by microscope.

Flexural Tests

To investigate the flexural properties, three point bending tests in longitudinal direction and lateral direction were performed by using Instron universal testing machine (Model 55R4206). The span lengths were 96mm for longitudinal direction and 24mm for lateral direction. The crosshead speed was 1mm/min.

Since the purpose of this study is to improve anisotropic, anisotropic parameter which was divided a characteristic of 0 degrees direction by 90 degrees direction's one was introduced. If this parameter is close to 1, that material is close to isotropic.

Theoretical Calculation

In order to compare between experimental value and theoretical value, the theoretical flexural modulus of the braided composite was calculated by using the laminate theory. The braided composite was presumed as laminate plate which consisted of some unidirectional fiber reinforced plates. The laminated constitution was +25/0/-25/0/+25/0/-25, this constitution was similar with actual braided composite. And, the thicknesses of each layer were obtained from cross sectional photograph of braided composite.

RESULTS AND DISCUSSIONS

Effect Of Middle-End-Yarn On Impregnation State

The aspects of these moldings were shown in Fig. 4. From this figure, it was found that the width was decreased during molding in braided molding without middle-end-yarn, because when braided moldings was subjected to tension in longitudinal direction, it shrunk in lateral direction by changing braiding angle. In the case of braided molding with middle-end-yarn, a decrease in lateral direction did not occur since middle-end-yarns prevented the shrinking. It was clarified that the middle-end-yarns were essential for braided moldings. However that molding had many voids on the surface. In the case of unidirectional molding, the voids on the surface and the change in width did not occur.

The cross section of each molding was shown in Fig. 5. From this figure, it was found that there were big unimpregnated region in the middle of braided molding without middle-end-yarn, because the resin was pushed out from unidirectional fiber bundle by shrinking of braided bundles. Such big unimpregnated regions were not observed in the braided molding with middle-end-yarn and unidirectional molding. However, there were many voids in the braided molding with middle-end-yarn, especially they were observed in the resin rich regions which were in between unidirectional fibers which were shown in Fig. 5 as red area and braided yarns.

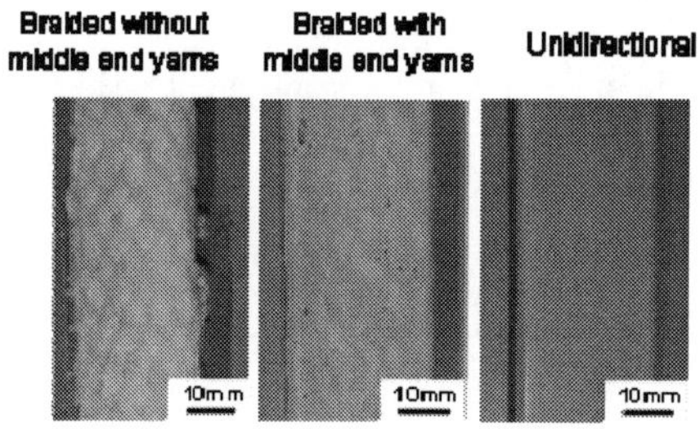

Figure 4. The aspects of moldings.

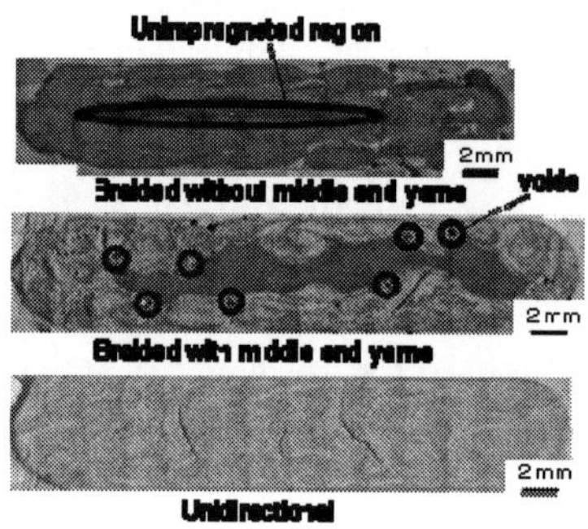

Figure 5. The cross section of each molding.

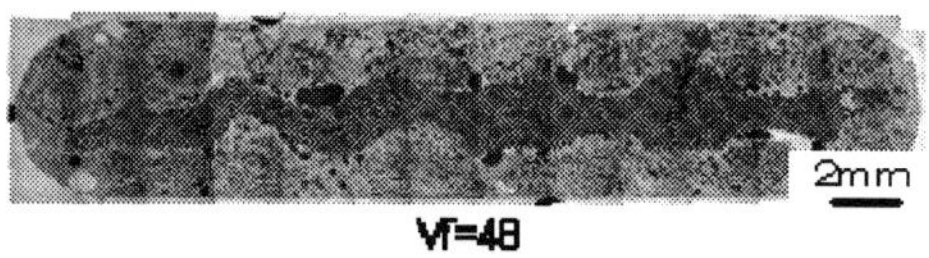

Figure 6. The cross section of molding with double resin bath.

Effect Of Supply Method On Impregnation State

In Fig. 6, the cross section photo of braided specimen with double resin bath was shown. From this figure, many voids were still observed and the impregnation state was not improved with double resin bath. The double resin bath was not effective, but by using double resin bath, different resin may be applied to unidirectional fibers and braiding fibers. That is to say, the possibility of materials design was increased.

Effect Of Fiber Volume Fraction On Impregnation State

The aspects of moldings which had different Vf were shown in Fig. 7. From this figure, it was found that the number of voids on the surface decreased as Vf increased. The cross section of moldings which had different Vf was shown in Fig. 8. From Fig. 6 and Fig. 8, voids contents were calculated by dividing sum of area of voids by area of cross section. The relationship between voids contents and Vf was shown in Fig. 9. From this figure, the voids contents decreased linearly as Vf increased. Therefore, to increase the Vf improved the impregnation state.

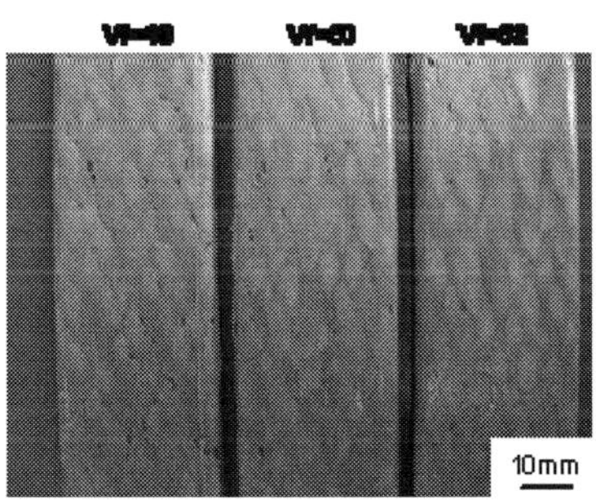

Figure 7. The aspects of moldings with high Vf.

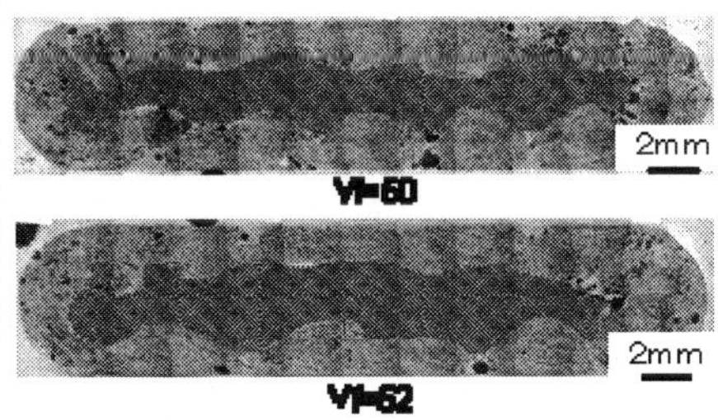

Figure 8. The cross section of moldings with high Vf.

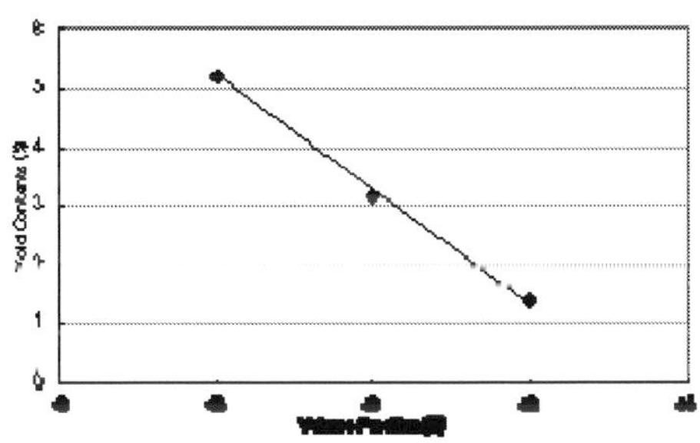

Figure 9. Relationship between voids contents and Vf.

Flexural Tests

The relationship between flexural stress and deflection in longitudinal direction was shown in Fig. 10. In this figure, the stress of unidirectional specimen decreased sharply after the maximum stress, but the stress of braided specimens decreased gradually. The bending modulus, strength and theoretical modulus were shown in Table I . Here, theoretical modulus was calculated with laminate theory, and the values in parenthesis were the achievement ratio of experimental modulus. The unidirectional specimen had higher modulus and strength than the braided specimens. The modulus and strength of braided specimen increased as Vf increased. The achievement ratio of the braided specimen was increased with increase in Vf and that with Vf of 52% was superior to the unidirectional specimen's one.

The relationship between flexural stress and deflection in lateral direction was shown in Fig. 11. In this figure, the stress of unidirectional specimen decreased sharply after the maximum stress, but the stress of braided specimens decreased gradually as same as results in longitudinal direction. The bending modulus, strength and theoretical modulus were shown in Table II. The braided specimen had higher strength than the unidirectional specimens. The modulus and strength of braided specimen increased as Vf increased. The modulus and the achievement ratio of the braided specimen which had 50% and 52% as Vf were superior to the unidirectional specimen.

The anisotropic parameters of modulus and strength which were provided by the flexural tests were shown in Table III. From this table, it was found that all braided specimens had higher isotropic than unidirectional specimen. In addition, as much as Vf rose, the anisotropic parameters of the modulus came toward 1. In other words, the braided technique improved the anisotropic of unidirectional pulustusion moldings. As it was mentioned in introduction, braidings can be changed the braided angle and be applied the various kind of fiber to by one of the braiding fiber and the middle end fiber, therefore the braiding plutrusion can control the anisotropic parameter of products.

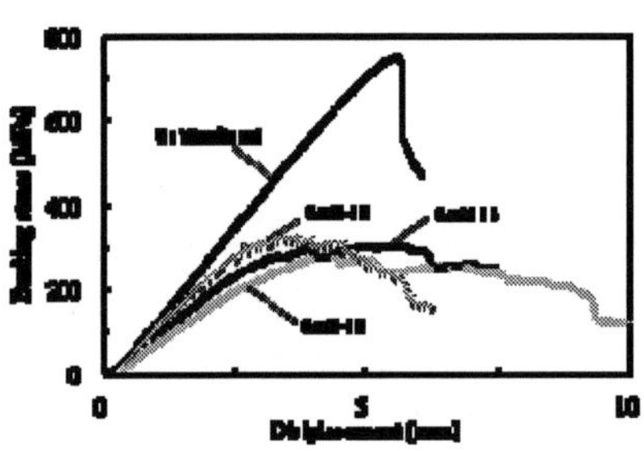

Figure 10. The relationship between flexural stress and deflection of longitudinal direction.

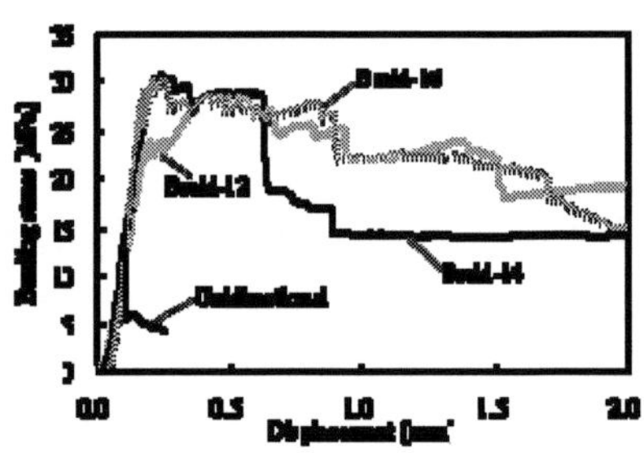

Figure 11. The relationship between flexural stress and deflection of lateral direction.

Table I The bending modulus, strength and theoretical modulus in longitudinal direction

	Bending modulus (GPa)		Bending strength (MPa)
	Exp.	Cal (%)	Exp.
Unidirectional	37.39	44.0 (85.0)	765.3
Braided-12	21.14	29.7 (71.1)	273.7
Braided-14	26.47	29.9 (88.7)	307.3
Braided-16	28.73	30.0 (95.9)	320.4

Table II The bending modulus, strength and theoretical modulus in lateral direction

	Bending modulus (GPa)		Bending strength (MPa)
	Exp.	Cal.	Exp.
Unidirectional	3.74	12.29 (30.5)	16.1
Braided-12	2.51	13.46 (20.7)	33.0
Braided-14	3.56	13.50 (29.1)	34.5
Braided-16	4.62	13.54 (37.8)	35.6

Table III The anisotropic parameters of modulus and strength

	E_0/E_{90}	σ_0/σ_{90}
Unidirectional	10.0	47.6
Braided-12	8.4	9.2
Braided-14	7.4	9.8
Braided-16	6.2	10.0

Theoretical Calculation

The theoretical values were shown in Table II and Table III. In these tables, the bracketed values were achievement ratio of experimental value against the theoretical value. From those tables, it is found that the achievement ratio of unidirectional specimens was higher than braided one. This reason was seemed that the unidirectional specimens had fewer defects such as voids and unimpregnated regions. The achievement ratio of braided specimens increased as Vf increased because of less defect.

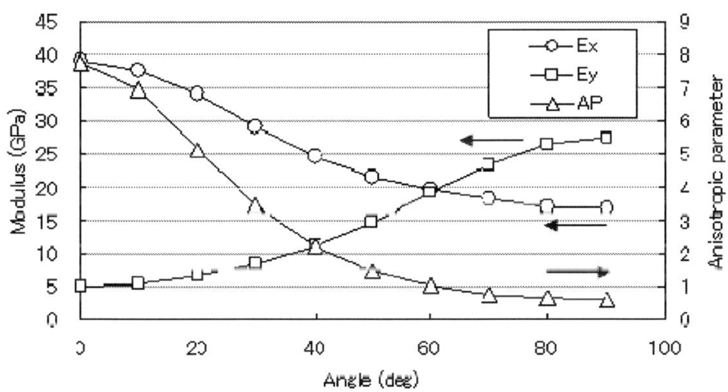

Figure 12. Relationship between theoretical modulus, anisotropic parameter and braided angle.

In order to consider the relationship between braided angle and anisotropic parameters, the theoretical flexural modulus of braided composite in longitudinal direction and lateral direction were calculated at every 10 degree, from 0 to 90 degree. The results were shown in Fig. 12. From this figure, the modulus in longitudinal direction decreased and that of lateral direction increased with increase in braided angle. And at 60 degree of braided angle, the modulus of lateral direction grew bigger than that of longitudinal direction. This result means that the anisotropic parameter at 60 degree is 1, and if the braided angle is bigger than 60 degree, the anisotropic will become smaller than 1.

CONCLUSIONS

To improve anisotropic, unidirectional fibers were covered with braided yarns in pultrusion moldings. Besides, continuous of molding line which is one of the advantages of pultrusion process was not lost in new molding line with braided technique. The main results of the molding, the cross section observations and the flexural tests in longitudinal direction and lateral direction were shown blow.

- For braided moldings, the middle-end-yarns were essential.
- To obtain good impregnation state, Vf had to high enough, in this time it was 52%.
- The braided technique improved the anisotropic of unidirectional pulustusion molding.
- From the theoretical calculation, it was clarify that the braiding plutrusion can control the anisotropic parameter of products.

REFERENCES

1. V. P. STAVROV, E. N. TSVIRKO, "PULTRUSION MECHANICS OF FIBER-REINFORCED THERMOPLASTIC COMPOSITES", Mechanics of Composite Materials 31 (1995) 402.
2. W. J. TOMLINSON, J. R. HOLLAND, "Pultrusion and properties of unidirectional glass fibre-polypropylene matrix composites", Mater. Sci. 13 (1994) 675.
3. A. FUJITA, Z. MAEKAWA, H. HAMADA and A YOKOYAMA, "Mechanical Behaviour and Fracture Mechanism in Flat Braided Composites", Journal of Reinforced Plastics and Composites 11 (1992) 600.

Robotic Preforming of Thick Near-Net Composites

T. SHARIF and P. POTLURI

ABSTRACT

Electronic Jacquard looms in conjunction with modified take-up systems can be used to produce relatively thick fabrics, with a variety of weave styles including orthogonal, angle interlocked and layer-to-layer interlocked structures. These fabrics have a number of applications such as ballistic plates and billets for machined composites. However, 3D woven composites have a number of limitations including limited formability, lack of bias fibers and significant material wastage (if the parts have complex geometry with significant taper). A robotic preforming approach is presented here to overcome the limitations of conventional 3D weaving. A large gantry robot, along with fiber deposition end-effectors, has been built to deposit a dry carbon yarn directly on a (double-curvature) mould surface. Hence thick curvilinear parts can be produced using this technique. The carbon yarn can be deposited in any preferred orientation, with increased density in certain critical regions. The component can have a complex boundary in addition to thickness taper. Additional feature of this process is the incorporation of through-thickness reinforcement.

INTRODUCTION

Traditionally aerospace composites have been produced using broad-good prepreg material. The prepreg sheet is spread on an automated cutting table, and then cut into required shapes in preferred orientations. Individual prepreg panels are then stacked and draped over a mould surface; this is a relatively slow and labor intensive process. Additionally, it is difficult to drape flat prepreg panels on double curvature surfaces. Buckingham and Newell [1], Zhang and Sarhadi [2] developed automated systems for handling broad prepreg panels. However, these systems are complex and as a result have not been in commercial use. However, automated placement of prepreg tapes

Textile Composites Group, School of Materials (North West Composites Centre), University of Manchester, Manchester M60 1QD, United Kingdom
Email: Prasad.Potluri@manchester.ac.uk

has become popular in the aerospace industry. Grant [3] presented an excellent review of automated processes used in composite aircraft structures. These machines developed by machine tool industry are classified as Fiber Placement (FP) and Automated Tape Layers (ATL). Fiber Placement (FP) machines deposit multiple narrow prepreg tapes (1/8 inch) on complex contoured surfaces. Automated Tape Layers (ATL) deposit wide prepreg tapes (3 ~6 inches) on relatively flat or gently curved surfaces. Prepreg tape has inherent tackiness that helps with the deposition on the tool surface. Prepreg materials deposited by FP or ATL machines and subsequently cured by autoclaves produce excellent quality parts. However the major drawback is the cost – FP/ATL machines and autoclaves are highly capital intensive. Prepreg material is significantly more expensive than dry fibers.

In recent years, there is a significant interest in the development of dry fiber preforms in conjunction with out-of-autoclave processes such as RTM (Resin Transfer Molding) and RFI (Resin Film Infusion). Textile technology, in particular 3D weaving has been viewed as a promising technology for manufacturing thick preforms. However 3D weaving has a number of limitations including,

- Fiber orientations are limited to orthogonal directions; off-axis orientations not possible
- Thick preforms have very limited ability to drape
- 3D weaving offers limited near-net capabilities in terms of changing the preform width and thickness
- Warp preparation, loom setting are relatively slow processes

This paper reports the development of a preforming technique that combines the advantages of 3D weaving and Fiber Placement Processes.

ROBOTIC PREFORMING

The major technical problem in Fiber Placement of dry tape/ tow is the lack of tackiness to hold the fibers on the mold/ tool surface. Rudd et al [4] deposited dry carbon tow on a flat sticky surface. The tow is compacted on the surface with the aid of rollers. One of the limitations with this technique is that subsequent draping of the preform is inhibited by the sticky substrate. Mills [5] proposed Bindered Tape Placement (BTP) concept. Bindered Tape is produced by sticking a tackified polyester thread on a flat carbon tow, and the resulting Bindered Tape is deposited with a roller arrangement attached to a robot. Ahrens [6], Mahieux [7], Lamontia et al [8] reported thermoplastic fiber placement processes. The thermoplastic tow is heated during deposition on the tool surface – this is a relatively slow process. The preforming approach presented here is distinctly different from the previous approaches. This is essentially a textile preforming technique in which individual tows are handled by a Robotic system. Individual warp layers are produced by the robot and deposited on a flat or curved surface. We utilize pins to mechanically clamp the tow ends –these pins will be removed once all the layers are held together by resin or z- fibers.

Fiber Placement Robot Design

A modular gantry robotic system has been developed in order to deposit fibers in a completely flexible manner. This gantry robot consists of 4 –axis of freedom i.e. X-axis, Y-axis, Z-axis and Rotation axis; however, additional axes can be added if required. The work envelope is 3m x 2m x 0.6 m. Additionally, a number of pneumatic valves are available for controlling cutters and other devices.

.

Figure 1. A four axis Robot for performing.

Hardware and Software Control

The Robotic control system consists of four AC servo drives and several I/O ports. The main software platform used here is CoDeSys [10] stands for Controlled Development System. This software supports all languages described by the standard IEC-61131. i.e. Instruction List (IL), Structural text (ST), Sequential function chart (SFC), Function block diagram (FBD), Ladder Diagram (LD) and Continuous Fuction (CFC). The motion of the drives can be controlled by different methods like by Function blocks, CNC editor or CAM editor, however, CNC editor provides multi-dimensional motions graphically and textually. To interpret these motions, their specified libraries must be included in the CoDeSys program. These libraries will automatically create the corresponding data structures (CNC Data, CAM Data), which can be accessed by the IEC program. However, for the communication of this IEC program with the hardware, structure of drives is first mapped in the PLC configuration and the appropriate parameters of these drives will be set. This structure then will be made accessible for the application with the aid of the Drive Interface libraries. The internal library SM Drivebasic.lib. provides IEC data structures and global variables, which will represent the drives, axisgroups and bus interfaces which have been configured in the PLC Configuration. With the aid of <BusinterfaceName> Drive.lib the data get exchanged in the structure and hardware. Figure 2 presents the general structure of the control system.

Figure 2. Soft motion components overview.

Fiber Placement Concept

Robotic preforming aims to overcome the limitations of 3D weaving by achieving the following objectives;

- To produce preforms with any arbitrary geometry lengthwise and widthwise,
- To produce a perform with a large number of layers in 0°, 90° and $\pm\theta^{\circ}$ directions,
- To produce a preform with single or double curvatures,
- To produce taper in thickness in any direction, and
- To incorporate through thickness reinforcement.

The robotic system has an end-effector for depositing carbon yarn continuously on a mould surface. The yarn is supplied from a spool attached to the robot arm (figure 3). Unlike prepreg material, dry carbon yarn does not have any tackiness to stick on the mould surface. Alternate concepts are required for securing individual tow. Cahuzac [9] proposed the use of pins for securing the yarns – similar approach has been adopted here. It can be seen from figure 4 that carbon fibers are deposited around pins in order to create a large number of warp, weft and bias layers. The component need not be flat and the fibers can be deposited on a curved tooling. In a 3D weaving process, individual warp layers are produced off-line and drawn painstakingly through the loom parts. In addition, all the ends and picks have the same length, and as a result, 3D weaving produces a rectangular preform. In the present approach, individual warp layers are produced in-situ. Each end/pick can have different length. The tows are deposited on a tool surface and hence preforms with single/double curvature can be produced. Once all the layers are deposited on the tool surface, they are stitched or tufted together with a tufting head. It may be noted here that, due to the textile nature of the preform, it can be further formed to into a final shape.

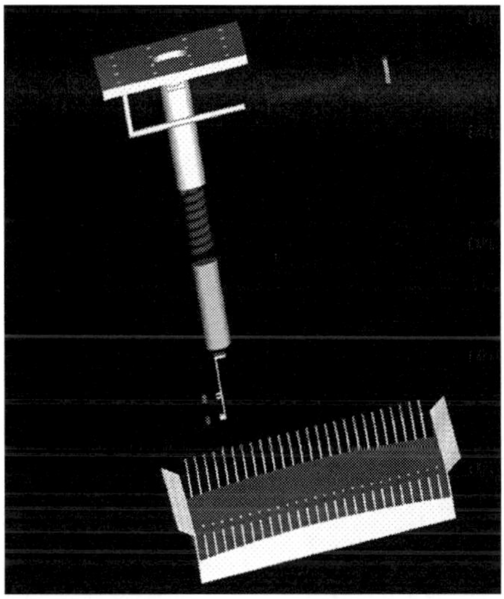

Figure 3. Tow deposition head.

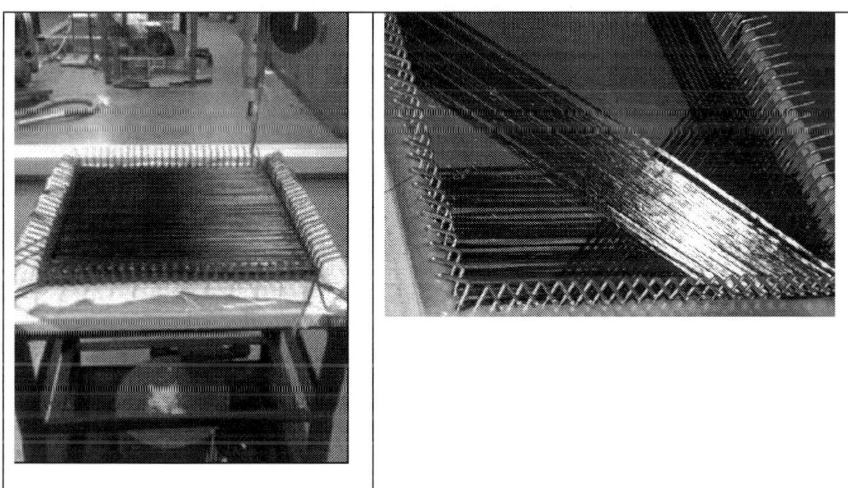

Figure 4. Deposition of carbon fiber tows around pins.

Panel Manufacturing and Resin Impregnation

The most important advantage of using fiber placement is a design freedom, this is achieved by fiber orientations, fiber type and their volume fraction, thus avoiding the over design of the product. The carbon fibers are deposited around the angular pins which hold them in place and prevent it from possible unhooking. The fibers are placed in 0 °, $\pm$ 45 ° and 90° direction to get mechanical properties in all direction (Fig 4), unlike conventional looms where in-plane yarns can only be aligned at angles 0 and 90°, hence, the produced 3D fabric consist of highly anisotropic properties and low shear and torsion properties[10].

After tow placement, the preform is impregnated using vacuum infusion process (Figure 5 a). The tufting head is still under development. As a result, the preform has to be infused in-situ. Once the tufting process is carried out, the preform can be easily handled and placed in an RTM mold. The release film was placed on the mold and pins prior to robotic tow placement to protect the laminate from sticking to their surfaces. After the laminate is cured the pins were pulled off from mould surface and carbon composite is extracted out of it. However, pins may be taken out from the mould when the laminate is semi-cured stage. Fig 5b shows a flat carbon composite panel produced using this method.

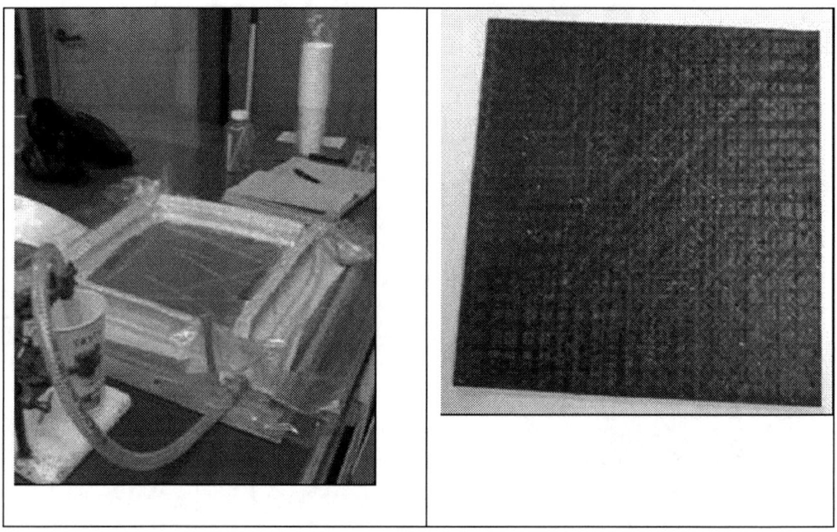

Figure 5 a) Vacuum bagging resin Infusion b) Flat carbon composite panel.

Figure 6 shows a circular panel with tapered edge. The pins are arranged in a circular fashion and individual layers are deposited around these pins. There are additional pins in the tapered region to hold warp layers that terminate on the tapered section.

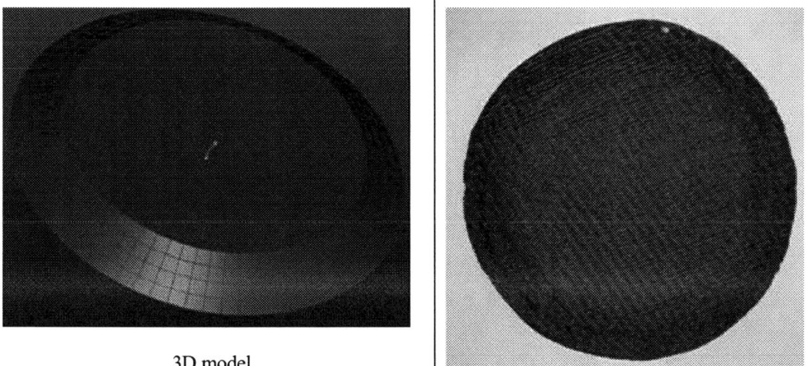

3D model

component

Figure 6. Circular panel with tapered edge.

DISSCUSSION

This paper presents a Robotic approach to textile preforming of 3D near-net shapes. Number of limitations of 3D weaving process has been overcome with this technique. Once the tufting head is in place, a 3D near-net preform with optimum fiber orientations, with in-situ curvatures and containing through thickness reinforcement can be produced.

REFERENCES

1. Buckhingahm R.O. and Newell GC, "Automating the manufacture of composite broadgoods," *Composites Part A*, 1996, 27(3):191–200.
2. Zhang Z, and Sarhadi M., "An integrated CAD/CAM system for automated composites manufacture," *Journal of Material Processing Technology*, 1996, 61(1-2):104–109.
3. Grant C., "Automated processes for composite aircraft structure," *Industrial Robot*, 2006, 33(2): 117–121.
4. Rudd C.D., Turner M.R.,Long A.C. and Middleton V., "Tow placement studies for liquid composite moulding," Composites A, 1999, 30, 1105–1121.
5. Mills A., "Automation of carbon fibre preform manufacture for affordable aerospace applications," *Composites A*, 2001, 32:955–962.
6. Ahrens M., "Robotic based Thermoplastic fibre placement process," Proc. of 1998 IEEE International Conference on Robotics & Automation, Leuven, Belgium, May 1998.
7. Mahieux C.A., "Cost effective manufacturing process of thermoplastic matrix composites for the traditional industry," *Composite Structures*, 2001, 52:517–521.
8. Lamontia M.A., Gruber M.B., Waibel B.J. and Cope R.D, "Conformable compaction system used in automated fiber placement of large composite aerospace structures." Proc. of the 23rd SAMPE Europe Conference, Paris, April 2002.
9. Cahuzac G, US patent No. 5759321, June 1996.
10. Festo (GB) Ltd, Northampton, UK.

Effects of Weaving on S-2 Glass™ Tensile Strength Distribution

A. ABU OBAID[1], S. M. ANDERSEN[1], J. W. GILLESPIE, JR[1,2], B. DICKINSON[3], A. WATSON[3], G. CHAPMAN[3] and R. A. COFFELT[3]

ABSTRACT

Tensile tests were conducted on dry and impregnated S-2 Glass ™ fiber tows extracted from control spools (types of 250, 750 and 1250) and from different directions (weft, warp and Z) of 3D woven S-2 Glass fabrics having an areal density of 100 oz/yd^2. Strength distribution, failure probabilities and retention in strength were quantified. Results showed that with respect to control samples, the weaving process in general shifted the distribution and failure probabilities of the dry tows to lower strength by different levels depending on the type and location of the of tows (weft, warp and Z). Weft and warp yarns exhibited the highest strength retention (77-100%). The dry Z-tows exhibited the lowest retention in strength (29-35%). Z-tows impregnated with vinyl ester showed a significant increase in strength retention up to 75%, confirming that the impact of weaving process on the performance of the fibers is of less concern when used in composites where interlaminar properties remain far superior to 2-D woven fabric composites.

Keywords: Impregnated tows, Tensile strength of S-2 Glass, 3 D Woven fabric

INTRODUCTION

Three dimensional (3D) woven fabrics often consist of in-plane warp and weft yarns interlaced in the through-thickness direction by Z yarns. Fabrics can be used dry or as preforms for composites. They have several advantages over 2D fabrics; 3D fabric can be manufactured in complex shapes having different types of fibrous materials (e.g. aramid, glass, carbon) and the Z fibers can improve impact damage tolerance by suppressing interlayer delamination in composites. Furthermore, they can eliminate usage of multiple plies of 2D reducing the associated labor costs,

[1]Center for Composite Materials, University of Delaware, Newark, DE 1971
[2]Departments of Civil and Environmental Engineering and Materials Science and Engineering
[3]3TEX, Inc, 109 MacKenan Drive, Cary, NC 27511

especially for large scale structures [1-3]. Valerie et al. [4] reported a ten-fold increase in interlaminar toughness and 25% improvement in in-plane fracture toughness measured for 3D woven fabric based composite, with respect to 2D laminates.

However, the effects of weaving on the translation of tow strength from spools into fabrics are not well documented. Although tows are sized to reduce the surface abrasion, some tow abrasion still occurs. Tows are subjected to tension and bending during weaving that may contribute to fiber breakage [1, 5]. The level of degradation will vary for different weaving machines. Very little information is available on strength translation in 2-D laminates from commercial suppliers. In a recent university study, Mouritz et. al. [1] used a Jacquard loom to weave 3D orthogonal E-glass fabric and reported that yarn/yarn and yarn/loom abrasion resulted in a loss of sizing, broken fibers and a 30-50% reduction in yarn strength.
In the present study, the translation of strength from S-2 Glass tows into 3D fabrics is quantified using 3TEX's commercial equipment to weave 3D orthogonal fabric (3WEAVE® S-2 fabric). Tow strength distribution and failure probability for weft, warp and Z yarns extracted from 3TEX woven fabric are compared to the as-received tows. In addition, tows impregnated with vinyl ester were evaluated to gain insight into the effect of weaving on the translation of strength into unidirectional composites.

MATERIALS

Materials tested were S-2 Glass tows of types of 250, 750 and 1250 yield with linear densities of 18,040 den, 6,104 den and 3,614 den, respectively. These tows were cut from the spools as control samples and also extracted from 3D woven fabrics of areal density of 100 oz/yd^2 (type: 3WEAVE® S-2 Glass with vinyl ester compatible sizing – type 463). The 3D woven fabric is a single symmetric layer consisting of 5 discrete non-crimp sub-layers that is woven together with a Z tow. The sub-layer sequence and nomenclature used in the results section are weft-top/warp top/weft middle/warp-bottom/weft-bottom and through thickness Z tows. The S-2 tow types of 250, 750 and 1250 yield were considered as control samples for warp, weft, and Z tows. One 3TEX loom was used to fabricate the 3D-fabrics where tow routing was altered to investigate effect on strength retention. The two 3-D fabrics are designated as original and modified fabrics in the discussions below.

EXPERIMENTAL

Specimen Preparation

For both dry and impregnated samples, the tows were carefully cut from the control spools. The sample length was 1.651 m and 0.381 m long for dry and impregnated tows, respectively. Warp and weft tows were easily extracted from the 3D non-crimp woven fabric without inducing additional fiber damage. Great care was taken during extracting of Z-tows from the woven fabrics, applying the following procedure: strips (25.4" width) were cut along the warp direction, 2 warp tows on both sides of the strip were removed and the Z-tows were easily collected by gentle shaking the strip by hand. From each strip, 2 tows were collected. For linear density measurements, tows (98 mm long) of control spools and of 3D woven fabrics fabric were cut and individually weighed within an accuracy of 0.001 mg.

Impregnation Procedure

Z-tows were impregnated with vinyl ester resin. The vinyl ester system prepared was a mixture of Derakane 8084 (8-10% strain to failure), Trigonox 239A (1.5% by weight) and 6%-Cobalt Naphthalene solution (0.3% by weight). Impregnated tows were allowed to cure at RT overnight followed by postcuring at 79C for 5hrs. Specimens were then shortened to 25.4 cm, individually weighed and end-tabbed using cardboard and RT adhesive.

Tensile Testing Measurements

DRY TOWS

Tensile tests were conducted on dry tows applying ASTM standard D 2256-02. An Instron test frame with a 1000 lb load-cell was used for all warp and weft tows. Z-tows were testing using a 100 lb load cell. Testing was carried out for all specimens with a 25.4 cm gauge length and a crosshead speed of 304.8 mm/minute. Specimens were mounted in a fiber tow fixture (tow is wrapped around the cylindrical fixture to build tow friction under tension). The advantage to using the tow fixture is that stress concentration at the ends is reduced and a uniform stress distribution along the fiber can be obtained. This approach will generate accurate strength data. However, it is very difficult to obtain accurate strain data directly from cross-head displacement due to the high compliance of the test fixturing. At least 30 replicates were tested for each tow type.

IMPREGNATED TOWS

Tensile testing measurements were performed on the Z-tows impregnated with vinyl ester according ASTM standard D 4018-99. For each test, the impregnated tow was mounted in the clamps of the Instron machine with 1000 lb cell, where 25.4 cm gauge length was applied. All tests were carried out at 250 mm/minute. At least 30 samples were tested.

DATA REDUCTION

From load-displacement data obtained for each dry tow tested, the tensile strength σ was calculated as:

$$\sigma = P/LD \qquad (1)$$

Where P is failure load and LD is the linear density of the tow. Equation 1 was also applied to calculate the strength of the impregnated tows assuming the load carried

by the matrix is negligible according to the ASTM standard. Based on uniform strain assumption in the fiber direction and typical properties for the constituents and tow fiber volume fraction, the load carried by the matrix was estimated to be less than 2.5% allowing one to use simplified data reduction (i.e. Equation 1).
For all test results of dry and impregnated tows, Minitab software was used to identify the statistical distribution (Weibull or 3-parameter Weibull) that best describes the experimental data and calculate the failure probability as a function of strength (the shape, threshold and scale fitting parameters are given in the plots). From failure probability plots measured for control and 3D woven fabric tows, the retention in strength was calculated at 10% (R-10%) and 90% (R-90%) failure probability. Also, the retention in strength (Rm) was calculated based on the mean strength value. The retention in strength was defined as the ratio of the strength for the samples of the 3D fabric normalized by the strength of their corresponding control samples.

RESULT AND DISCUSSION

DRY TOWS

Dry tows extracted from each layer (warp-top, warp-bottom, weft-top, weft middle and weft-bottom) of both original and modified fabrics were tested in tension. The results (strength distribution and failure probabilities) of these tows were compared to their control samples in Figures 1-3. Based on the strength values of control samples, the retention in strength (Rm, R-10% and R-90%) are listed in Table I for all types of dry tows. The average values are used for discussion purposes. Overall the retention of strength for in-plane non-crimp sub-layers is in the range 77-100% range (0-23% strength reduction) which is far superior to the results reported in Reference [1]. The Z-tows undergo local bending at the fabric surface show strength retention in the range of 29-35%. The 3D weaving process used for the modified fabric is superior providing higher overall translation of strength for the various sub-layers.

TABLE I. RETENTION IN STRENGTH MEASURED FOR DRY WARP, WEFT AND Z-TOWS OF ORIGINAL AND MODIFIED 3D WOVEN FABRIC.

	Original Fabric (%)			Modified Fabric (%)		
Type	Rm	R-10%	R-90%	Rm	R-10%	R-90%
Warp-Top	84	81	86	78	79	77
Warp-Bottom	83	81	83	77	76	77
Weft-Top	80	86	77	96	100	89
Weft-Middle	94	~100	89	94	100	86
Weft-Bottom	100	100	94	100	100	94
Z-Tows	29	26	31	35	34	35

Figures 1-3 show the data for the warp tows for the top and bottom layers. In Figure 1, the Weibull distribution shows that the top and bottom strength distributions are comparable for the original fabric (83-84% retention). In Figure 2 the warp tows of the modified fabric are compared to the control (77-78% retention). This figure illustrates the shift to lower strength and a narrowing of the strength distribution. Figure 3 shows the probability of failure for the warp layers from the original and modified fabric versus the control data. This plot clearly shows the shift to lower strengths for a given probability. These plots were used to construct the entries in Table 1.

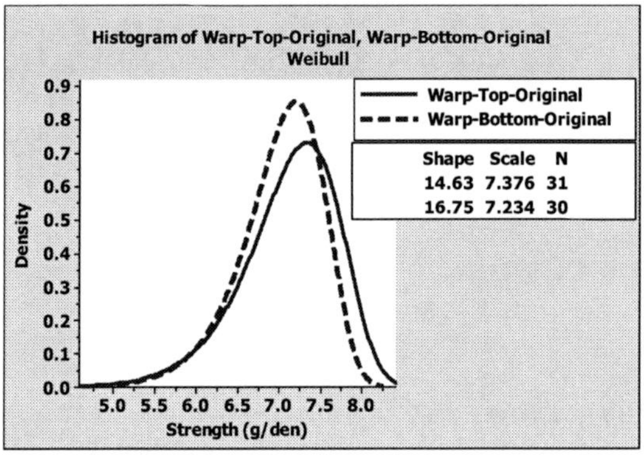

Figure 1. Strength Distributions of Warp Tows for Original fabric.

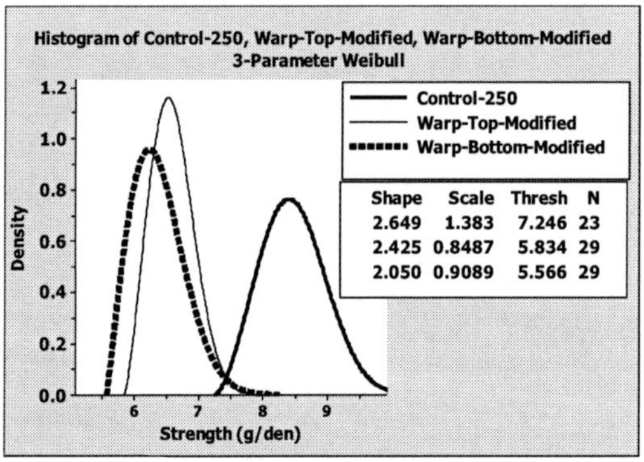

Figure 2. Strength Distributions of Control and Warp Tows of Modified fabric

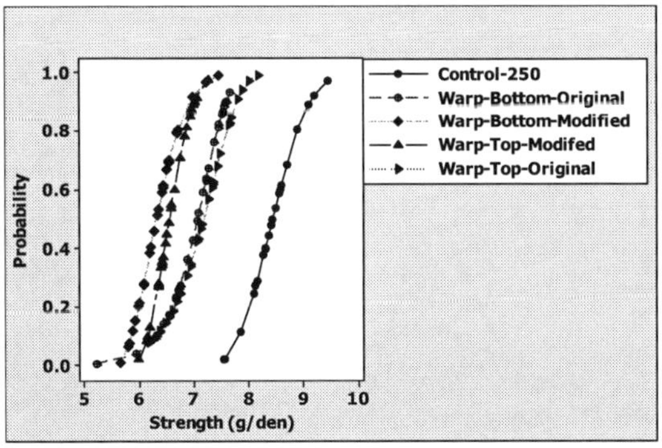

Figure 3. Failure Probabilities of Control and Warp Tows of Original and Modified Fabric.

Figures 4-6 provide the strength distribution results for each of the 3 weft sub-layers in the original and modified fabrics. With respect to control samples, the strength retention is in the range of 80-100% for the original fabric (see Figure4). Interestingly, the strength retention is highest on the bottom weft sub-layer. Figure 4 shows the highest shift for the top weft sub-layer. However, all weft sub-layers undergo a narrowing of the strength distribution (even for the case of 100% retention). From Figure 5, 3 weft sub-layers in the modified fabric exhibit insignificant differences in the peak locations of their distributions (94–100% strength retention). However, a narrowing of the distributions is seen similar to the original fabric results. Figure 6 provides the probability of failure for all weft sub-layers for both the original and modified fabrics. In general (with the exception of the top weft sub-layer from the original fabric that has a clear shift to lower strength), the effect of weaving is to narrow the distribution and rotate the probability of failure curve counter clockwise. In this case a small increase in applied stress will more rapidly increase the probability of failure at that stress level. These trends are quantified by the Weibull shape, scale and threshold fitting parameters given in the figures.

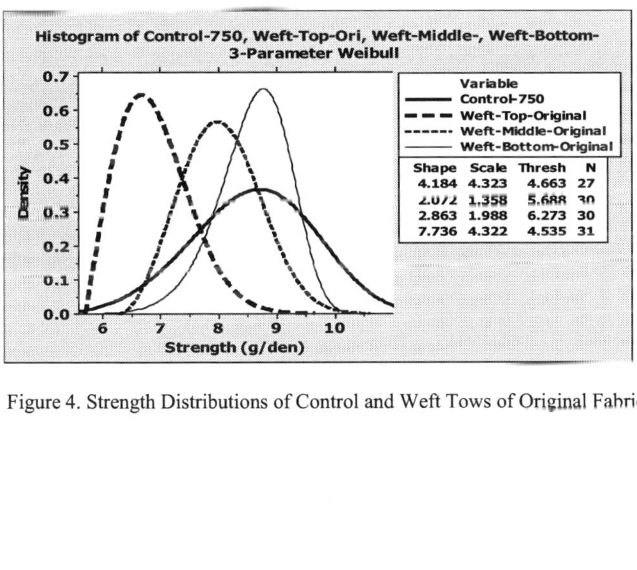

Figure 4. Strength Distributions of Control and Weft Tows of Original Fabric

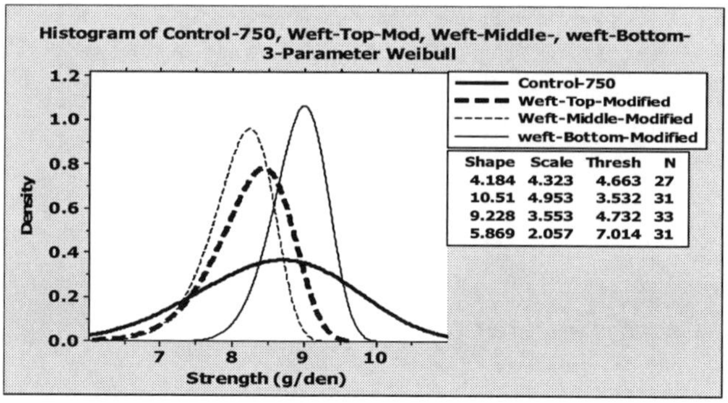

Figure 5. Strength Distributions of Control and Weft Tows of Modified Fabric.

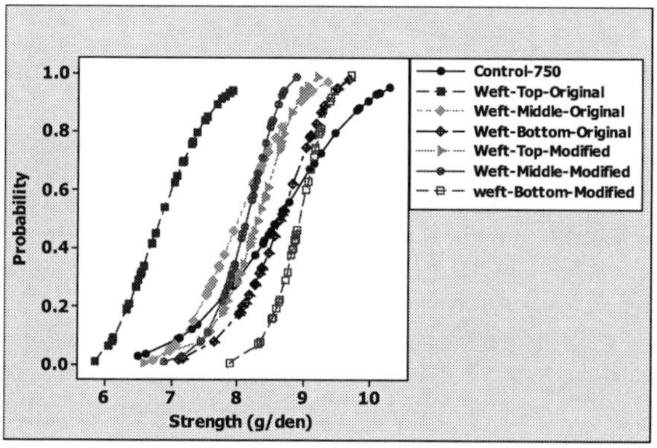

Figure 6. Failure Probabilities of Control and Weft Tows for Original and Modified Fabric.

Strength distributions of Z-control tows (type 1250), Z-tows of original and modified fabric are depicted in Figure 7a and 7b. As summarized in Table 1, the Z tows exhibit the greatest strength reduction among all other types of tows. This is more likely due to higher levels of abrasion during the weaving process and the extreme tow bending at the fabric surface. Figure 7 indicates that the commercial weaving equipment used for the modified fabric has improved strength retention slightly. In Figure 8, the probability of failure for the Z-tows clearly shows the significant shift to lower strengths. For comparison, Mouritz reported a ~50% strength degradation for Z-tows of 3D woven E-glass fabric woven in the laboratory.

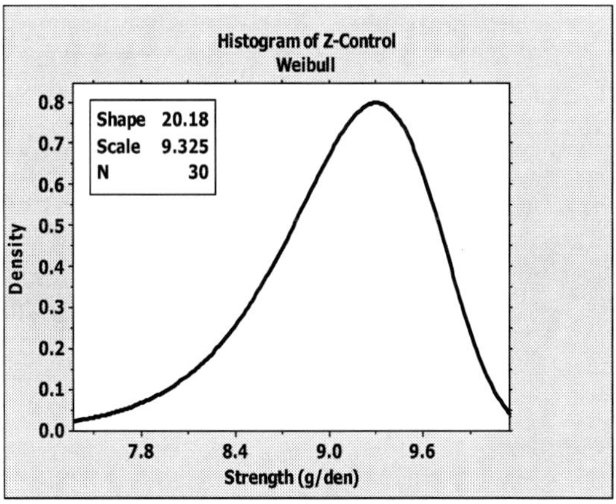

Figure 7a. Strength Distributions of Control Z Tows.

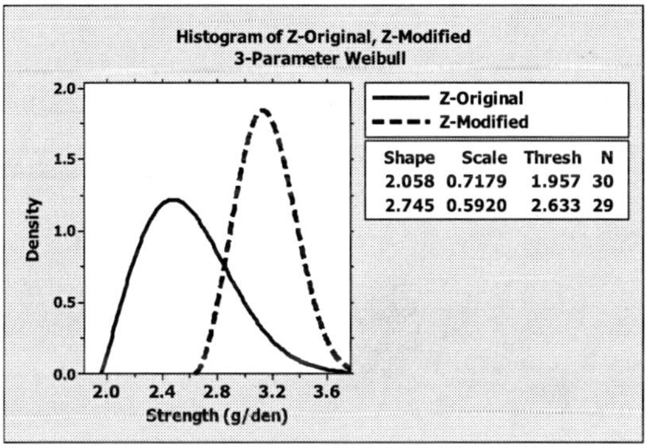

Figure 7b. Strength Distributions of Z Tows of Original and Modified Fabrics.

Impregnated Z-Tows

The results presented above describe the translation of tow strength into dry fabrics. In composites, the matrix material provides an opportunity to transfer load amongst the individual filaments that may have been damaged. Consequently, impregnated tow strengths are typically much higher than dry tow test results. This is shown clearly in Figures 8 and 9 for the Z tows where the impregnated tow data is compared to the dry tow test results. The impregnated control tow strength shifts to higher levels by approximately 75% compared to the dry control samples. The effect of weaving on impregnated tow tests is shown in Figures 8 and 9 where strength retention was found to be in the 77 to 86% range. This is significantly higher than the dry tow case (29-35% strength retention). Clearly, the presence of a high elongation matrix in composites reduces the sensitivity of weaving on unidirectional composite properties. These results indicate that nearly full translation of in-plane warp and weft tows can be expected in 3WEAVE fabrics used as composite preforms.

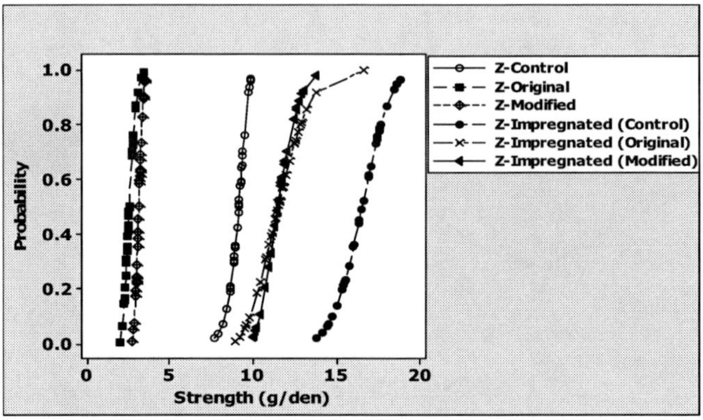

Figure 8. Failure Probabilities measured for Z-Tows Impregnated with Vinyl ester.

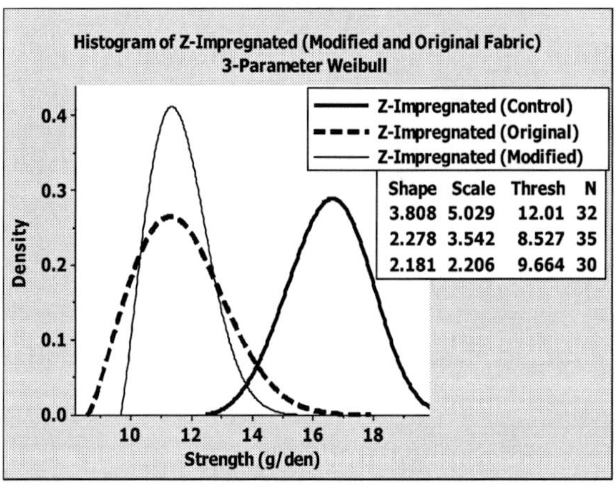

Figure 9.Strength Distributions of impregnated Z-Tows Impregnated with Vinyl ester.

CONCLUSIONS

The effect of 3D weaving on dry tow strength retention has been quantified by comparing control samples to tows extracted from 100 oz/yd^2 3D fabrics consisting of five sub-layers. Results show that the level of degradation can vary by sub-layer and warp or weft direction and that weaving can affect the entire strength distribution (typically a shift to lower strength and a narrowing of the distribution). In-plane results showed strength retentions in the range of 77-100%. The highest degradation was seen in the Z tows with strength retention in the range of 29-35%. It is important to note that this material was 100% S-2 Glass, which is not ideal for the Z-yarn because of the extreme bending in the final architecture. To improve the strength retention in the Z-yarn, alternate fiber types with higher tensile strength and higher damage tolerance to the weaving process can be used, such as aramid or polyethylene fiber. One of the greatest benefits of 3WEAVE is that it can be tailored to meet specific properties. If higher are values are needed only in the one direction, the fibers, the size of the fibers or the density can be altered to increase these values.

Future work will more closely investigate the degradation mechanisms. The effects of 3D weaving on the strength of impregnated Z-tows were also studied. Strength retention of impregnated tows was nearly 75% of the impregnated control sample, a marked increase over the dry tow results. Clearly, the presence of a high elongation matrix in composites reduces the sensitivity of weaving on unidirectional composite properties. These results indicate that nearly full translation of in-plane warp and weft tows can be expected in 3WEAVE fabrics used as composite preforms.

REFERENCES

1. S. Rudov-Clark, A. P. Mouritz, L. Lee and M. K. Bannister. 2006. "Fiber damage in the manufacture of advanced three-dimensional woven composites," *Composites Part A: Applied Science and Manufacturing*, 24 (10): 963–970.
2. http://www.3tex.com/3weave.html, Brochure on 3 Weave® Products.
3. J. Brandt, K. Drechsler, F. J. Arendts. 1996. "Mechanical performance of composites based on various three-dimensional woven-fiber preforms", *Composites Science and Technology*, 56 (3): 381–386.
4. A. G. Valerie, T. W. Chou and J. W. Gillespie Jr. 1989. "Toughness properties of a three-dimensional carbon-epoxy composite," *journal of materials Science*, 24 (11): 4168–4175.
5. L. Lee, S. Rudov-Clark, A. P. Mouritz, M. K. Bannister, I. Herszber. 2002. "Effect of weaving damage on the tensile properties of three-dimensional woven composites," *Composite Structures*, 57 (1-4): 405–413.

A Model to Describe Thermoplastic Resin Impregnation in a Unit Cell of a Fiber Tow within a Textile Preform

C. STEGGALL, P. SIMACEK, S. YARLAGADDA
and S. G. ADVANI

ABSTRACT

A novel process has been developed to manufacture continuous fiber thermoplastic matrix composites. In this process, continuous fiber fabrics are placed in a special mold and impregnated with thermoplastic powder. Then it is consolidated using autoclave or compression molding. A model is formulated to describe the consolidation phase in which the molten thermoplastic powder is forced into adjacent fiber bundles under applied pressure. A non-dimensional parameter, which groups the influence of material and process variables, is identified. The model is applied to a selected unit cell geometry of a textile preform and can predict the void content as a function of the dimensionless parameter.

INTRODUCTION

Currently, there are many methods being used to manufacture thermoplastic matrix composites. For applications that require continuous fiber reinforcement most methods usually consolidate a precursor material at elevated temperature and pressure in an autoclave or a heated hydraulic press. The precursor material form could be commingled fabrics, pre-impregnated tapes, or thin films of resin in between layers of dry fibers. Commingled bundles combine reinforcing fibers (glass, carbon, aramid) in the same bundle as a matrix material (i.e. nylon, polypropylene, polyethylene, etc.) These bundles can then be knit, woven, or stitched to form the commingled fabric. Pre-impregnated tape consists of fabrics or aligned fibers pre-impregnated with resin. There are also some less traditional methods of manufacturing thermoplastic matrix composites. These include melt impregnation and thermoplastic powder deposition. In one melt impregnation

---

Claire Steggall, Pavel Simacek, Shridhar Yarladda, and Suresh G. Advani[1], University of Delaware, Center for Composite Materials, Newark DE, 19711, U.S.A.

[1] George W. Laird Professor and Corresponding Author, E-mail: advani@udel.edu

process fiber bundles are passed over a pin submerged in a polymer melt under tension to spread the bundle and impregnate it [1]. Powder deposition processes can be generally classified into wet and dry. Vodermeyer et. al. developed a version of wet particle deposition. They introduced particles in a continuous fiber preform by running bundles over a spreader and through an aqueous bath containing thermoplastic powder prior to forming the final preform [2]. Ye et al. made fiber preforms from bundles that were constructed by combining powder and fibers inside a thermoplastic sheath [3]. In an example of dry deposition, forced air is used to deposit thermoplastic particles onto the fiber bundle. Then the polymer/fiber combination is passed briefly through a furnace to soften the matrix material and adhere it to the bundles. Finally the powder-impregnated bundle is cooled and wound onto a spool and used as precursor material [4].

The majority of the methods available today to manufacture thermoplastic matrix composite materials involve extensive labor to manufacture thick section structures. The process described in this paper can also be used with three dimensional or thick fabrics and is an attractive feature of this proposed manufacturing process.

MANUFACTURING PROCESS

The proposed manufacturing technique is a two-step process. It uses a modified vacuum assisted resin transfer molding (VARTM) method in the first step to impregnate the fabrics with the thermoplastic powder. Then the matrix material is consolidated with the fabric by using the traditional compression molding or autoclave in the second step.

During the first step, the reinforcement fabric is placed on a rigid tooling surface surrounded by a rubber gasket. The gasket has a thickness equal to that of the fabric plus an additional height to leave a gap between the top plate of the mold and the fabric. There are vacuum vents in the bottom plate and inlets for injection in the top plate. Once the fabric and gasket are in place and the mold is closed, a vacuum is applied. Once full vacuum is achieved, the inlet for injection is opened. The gap between the top of the fabric and the mold is flooded with a water/plastic powder slurry before the fabric is infiltrated with the water and powder mixture due to the pressure gradient present across the fabric. A filter layer placed beneath the fabric layers helps build up a concentration of powder in the fabric because it allows only water to exit the mold through the vacuum vents. After the total volume of the mixture has been filtered through the fabric, the powder-impregnated fabric is removed from the mold and placed in an oven at low temperature to evaporate the water. At this point the macropores in the fabric between the bundles are filled with the powder while the micropores within the fiber bundles are completely dry as the size of the powder particles and the diameter of most reinforcing fibers are nearly equal.

Figure 1 is a schematic of the glass-fabric powder-impregnation process and

Figure 2 is a photograph of the mold used to impregnate fiber preforms for a 0.09 m^2 (1 ft^2) panel.

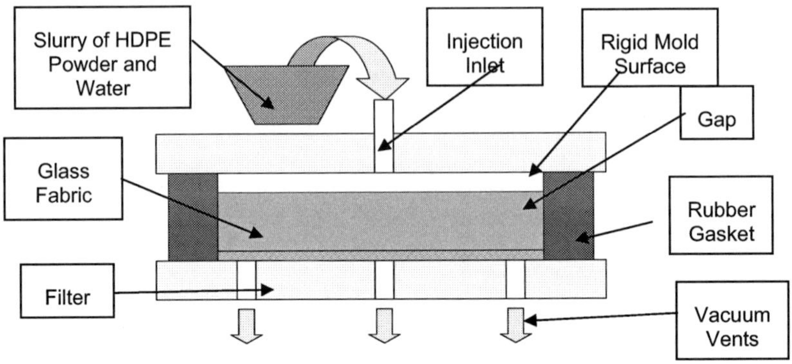

Figure 1. Diagram of Powder Impregnation Mold.

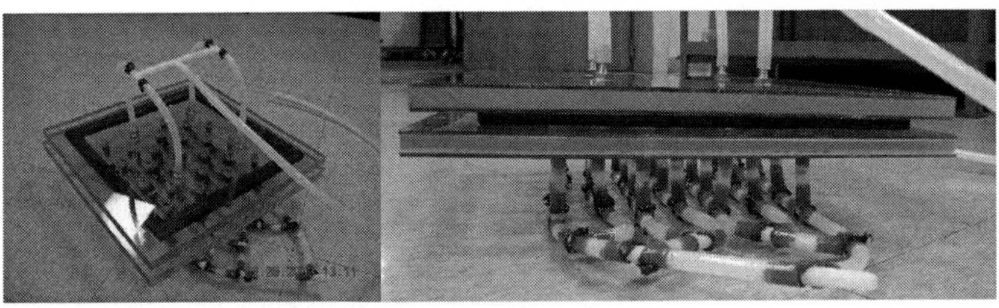

Figure 2. Photographs of Powder Impregnation Mold.

During the second step of the process, the powder-impregnated fabric is placed in a rigid aluminum plate and frame mold for compression molding.

Figure 3 shows an example autoclave recipe that is divided into four phases and can be used for consolidation. In Phase 1, heat is applied to melt the powder while the mold is held at atmospheric pressure. Phase 2 starts once the powder is in the melt state and the pressure on the assembly is increased to facilitate the impregnation of the fiber bundles with the molten resin. During Phase 3 the temperature and pressure held constant to allow additional dwell time for saturation of fiber tows and consolidation of the composite. Finally, in Phase 4, the pressure is held constant while the composite is cooled, solidifying the resin in place. The rigid mold is necessary for this step to ensure that the molten thermoplastic will not flow out of the sides of the fabric layers instead be forced to flow into the fiber bundles. This work will model the Phase 2 and Phase 3 of the process to predict the void content within a fiber bundle during this consolidation process.

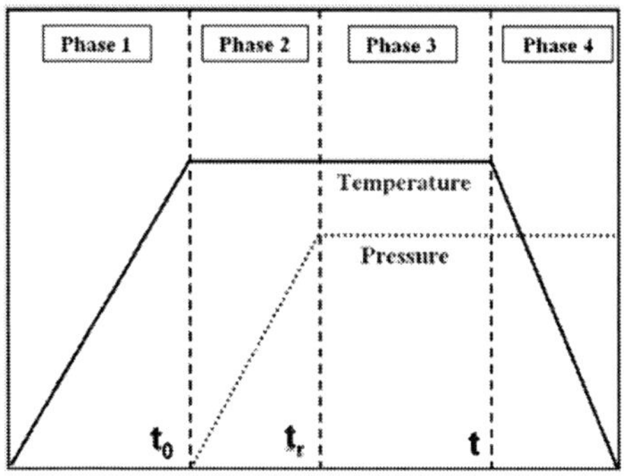

Figure 3. Time, temperature and pressure methodology used in the Autoclave to consolidate the powder impregnated thermoplastic.

MATERIAL SELECTION

It is easy to build thickness with only a few layers of material with this two-step wet particle deposition process because it can successfully impregnate even thick fabrics. In order for this process to work effectively, the matrix material used should be available in a powder form. Many thermoplastics such as polypropylene (PP), polyethylene (PE), and polyether ether ketone (PEEK) are already available in the powder form. The surface of the powder must be treated to increase wettability and lower surface energy. The treatment facilitates the dispersion of the powder in a carrier fluid, such as water. The resulting mixture of water and powder is a low viscosity slurry which can then be injected into the fabric preform. The matrix material used in the experiments is manufactured by Inhance®/Fluoro-Seal Ltd. It is a high density polyethylene (HDPE) powder with an average particle diameter of 20 microns.

Both 2-D and 3-D fabrics can be used as reinforcements however most of the work done to date with this process has been with a 2-D fabric. The fabric used is a 3637 g/m^2 (108 oz/yd^2) biaxial glass fiber fabric manufactured by Vectorply™ Corporation. It's non-woven and the fiber bundles are held in place by a network of polyester stitching with empty channels in between and parallel to the fiber bundles. In traditional LCM processes with thermosetting resins the channels would function as a pathway for resin to racetrack [5] and impregnate the preform. In this process, these channels serve as a collection area for the matrix powder that is later melted and forced into the neighboring fiber bundles in the second step of compression molding or autoclave processing. Any fabric that is used in this manufacturing process must have sufficient free space around the fiber bundle network to accumulate the powder so that once the fabric is compressed both the compressed macropores and micropores within in the fiber bundles will be filled with the matrix to the desired degree.

ANALYSIS

The analysis focuses on the impregnation of the fiber tows during Phase 2 and 3 from the molten matrix surrounding the tow. As simplified in

Figure 4, one can consider the flow to be one dimensional in which the resin flows from the macropore into the fiber tows. In reality, the macropore shrinks under the compression pressure as the resin impregnates the fiber tows. In our model, we assume that as the resin leaves the macropore, the volume of macropore reduces accordingly and hence there will be no voids remaining in the macropores. It is also assumed that the fiber bed doesn't deform as pressure is applied and that permeability of the fiber tow is constant.

H_{g0} represents one half of the macropore containing the resin while d represents one half of the fiber tow.

Equation (1) describes the flow of resin into the fiber tow using Darcy's Law. Here L is the location of the resin front into the fiber tow, P_a is the applied pressure, φ is the tow porosity, μ is the viscosity and K is the fiber tow permeability. The expression has been formulated in terms of the unit cell geometry (H_{g0}, d) which can be measured from the textile preform architecture.

$$\frac{dL}{dt} = \frac{K}{\mu\phi}\frac{P_a}{L}\left(\frac{H_{g0} + d - L\phi}{H_{go}}\right) \quad (1)$$

During Phase 2 of bundle impregnation in the autoclave processing, P_a is changing with respect to time as seen from figure 3 and is explicitly described by the expression $B*t$. B is the constant ramp rate that describes the rate at which the pressure is increasing with time.

Separating the variables in Equation (1) and integrating using the initial condition of L equal to 0 at time equal to 0 provides an implicit expression to describe the change in L with time during the ramp period:

At the end of the ramp period, t_{ramp} one can find the L_{ramp}, the position of the flow front inside the tow (which provides the fraction of fiber tow impregnated with resin). During Phase 3 as seen from

Figure 3 where the applied pressure, P_a is held constant over a dwell period to promote bundle impregnation, one can integrate Equation (1) directly which results in:

$$-\frac{H_{g0}L}{\phi} + \frac{H_{g0}^2}{\phi^2}\ln\left(\frac{H_{g0} + d}{H_{g0} + d - L\phi}\right) + \frac{H_{g0}d}{\phi^2}\ln\left(\frac{H_{g0} + d}{H_{g0} + d - L\phi}\right) = \frac{BK}{2\mu\phi}t^2 \quad (2)$$

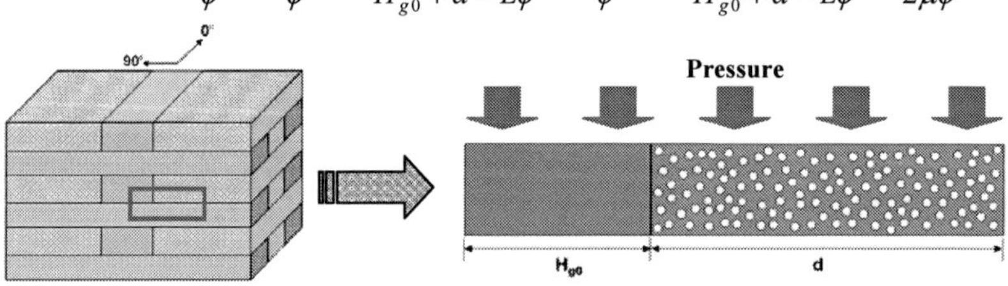

Figure 4. Unit cell of idealized preform architecture considered in resin flow model.

$$-\frac{H_{g0}L}{\phi}+\frac{H_{g0}^2}{\phi^2}\ln\left(\frac{H_{g0}+d}{H_{g0}+d-L\phi}\right)+\frac{H_{g0}d}{\phi^2}\ln\left(\frac{H_{g0}+d}{H_{g0}+d-L\phi}\right)=\frac{P_aK}{\mu\phi}t \quad (3)$$

The position of the flow front at any time in this phase can be found by adding L_{ramp} found using Equation (2) with the value for L (at the desired time) found with Equation (3).

RESULTS AND DISCUSSION

The theoretical void fraction resulting from the model can be calculated using the following equation:

$$v_v=\frac{V_v}{V_r+V_f+V_v}=\frac{(d-L)\phi}{H_{g0}+d} \quad (4)$$

The minimum possible void fraction can be calculated based on the known dimensions of the tow and the tow gap. As stated before, it is assumed that the only resin available to fill the tow is the resin in the tow gap. Using Equation (4) and the value for the maximum possible position of the flow front L_{max} (which is calculated by dividing H_{g0} by the porosity of the tow) we can find a theoretical minimum possible void fraction. For the experimental parameter values presented in TABLE the minimum possible void fraction equal to 0.0723 under any pressure as there is no more resin available in the macropore to impregnate the tows.

TABLE shows the results from the model which uses the material and process parameters listed in TABLE and its comparison with experimental data. Having validated the model, a parametric study was carried out to investigate the role of the geometry of the unit cell of the textile architecture on the final void fraction.

TABLE I. MATERIAL PROPERTIES AND UNIT CELL GEOMETRY.

Material Property [units]	Value
Tow Permeability: K [m^2] at 0.345 MPa & 0.690 MPa	2 * 10^{-13}
Tow Permeability: K [m^2] at 1.379 Mpa	1 * 10^{-13}
Molten HDPE Viscosity: μ [Pa·s]	2750
Bundle Porosity: φ	.38
H_{g0} [m]	0.000735
d [m]	0.002561

TABLE II. MODEL AND EXPERIMENTAL RESULTS

Pressure [MPa]	Model	Experimental	Model	Experimental
	30 minutes		60 minutes	
0.345	0.1611	0.1677 ± 0.0487	0.1139	0.1364 ± 0.0128
0.690	0.1194	0.0564 ± 0.0591	0.0723*	0.036 ± 0.0092
1.379	0.0723*	0.0439 ± 0.0125	-	-

Effect of Unit Cell Geometry on Properties

It can be seen from Equation (3) that the model dependents on the geometry of the unit cell of the fabric and on the macroscopic gap between the fiber tows that can be filled with powder before consolidation.

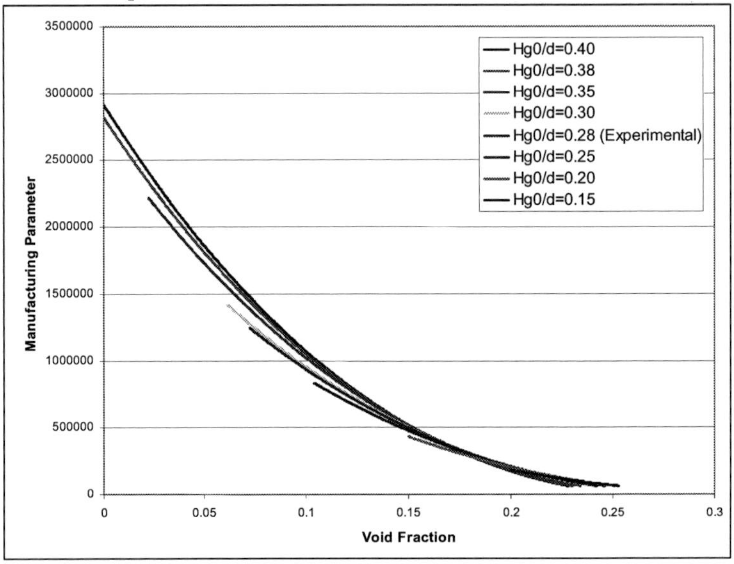

Figure 5 is a plot of several different H_{g0}/d ratios representing different tow size versus gap size within a unit cell of the fabric architecture of a 2-D Vectorply. A higher H_{g0}/d ratio potentially can provide more resin for impregnating the tow than a fabric with a lower H_{g0}/d ratio. The ratio of H_{g0}/d equal or greater than 0.38 represents the situation where, in theory, full fiber bundle impregnation can be achieved using Equation (4). The vertical axis is a non dimensional manufacturing parameter, $\frac{P_a(t-t_r)}{\mu}$. The horizontal axis is the void fraction within a fiber tow at the end of Phase 3. Each point on the H_{g0}/d ratio curves provides a value for the manufacturing parameter to achieve the corresponding void fraction for the available resin gap to tow size ratio assuming the gap is entirely filled with the resin in step 1 of the process. The value of H_{g0}/d ratio for the 2-D fabric for the experiments conducted was around 0.28 as determined from optical micrograph measurements. The case of H_{g0}/d equal to 0.40 is representative of when the resin gap can store more matrix material than is needed to fill a tow (with porosity of less than the H_{g0}/d ratio). Comparison of H_{g0}/d=0.38 to H_{g0}/d=0.40 curve shows that the higher ratio has a larger MP to achieve an identical void fraction. This is because of the assumption in the model that all of the applied pressure is transferred as the resin pressure and the effective resin pressure in Darcy's Law is given by P_r/v_r. For

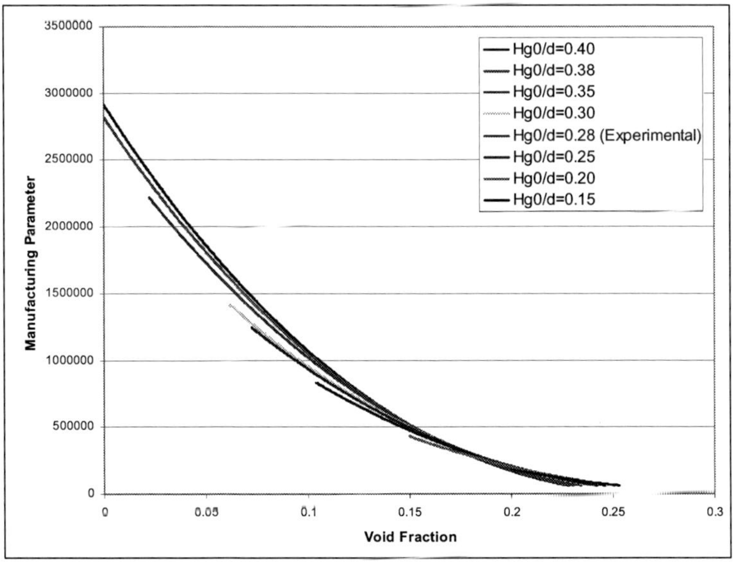

Figure 5. Plot Demonstrating Effect of Unit Cell Geometry on Void Fraction.

higher H_{g0}/d, the volume of resin is higher and hence the effective applied pressure is reduced thus requiring a larger applied pressure to achieve the same void fraction.

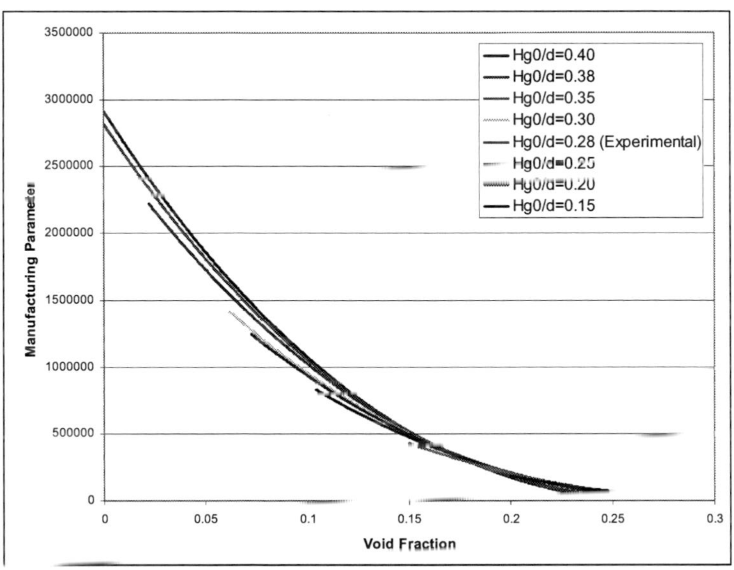

Figure 5 helps us further understand the important roll of unit cell geometry on the properties of the final composite. It has been shown in the early stages of our mechanical testing that in applications where energy absorption properties are critical that one must choose a balance between mechanical and energy absorption properties [6]. Through the testing completed by Steggall et. al. it was seen that the low matrix volume fraction plates manufactured using the powder-impregnation process described in this paper performed better in the energy absorption testing then its traditional E-glass/HDPE pre-preg counterpart. Though, when it came to mechanical testing the traditional pre-preg outperformed the powder-impregnated samples in both tension and compression tests. Since the lower matrix volume fraction samples performed better than a fully filled sample in the energy absorption testing it is important to understand how this manufacturing process can be tailored by the manufacturing parameters and the unit cell geometry to obtain the desired level of void fraction for applications in which energy absorption is important. Future work on this project will include a model to describe particle deposition during the powder-impregnation phase of the manufacturing process.

ACKNOWLEDGEMENTS

The authors would like to acknowledge the Department of the Navy, Office of Naval Research grant number N00014-06-1-1000 for funding this work. They would also like to thank Touy Thiravong, John Thiravong, Dr John Tierney, and Jason Etherington for their generous help with this project. Thanks also goes to Vectorply and Inhance/Fluoro-Seal Ltd. for supplying material.

DISCLAIMER

Any opinions, findings, and conclusions or recommendations expressed in this material are those of the author(s) and do not necessarily reflect the views of the Office of Naval Research.

REFERENCES

1. Bijsterbosch, H. and R.J. Gaymans. 1993. "Impregnation of Glass Rovings with a Polyamide Melt. Part 1: Impregnation Bath", *Composites Manufacturing*. 4(2):85–92.
2. Vodermeyer, A.M., and, J.C. Kaerger, and G. Hinrichson. 1993. "Manufacture of High Performance Fiber-Reinforced Thermoplastics by Aqueous Powder Impregnation", *Composites Manufacturing*. 4(3):123–132.
3. Ye, L., and K. Friedrich. 1995. "Processing of Thermoplastic Composites from Powder/Sheath-Fiber Bundles", *Journal of Materials Processing Technology*. 48:317–324.
4. van Hattum, F.W.J., J.P. Nunes, and C.A. Bernardo. 2005. "A Theoretical and Experimental Study of New Towpreg-Based Long Fibre Thermoplastic Composites" *Composites Part A*. 36:25–32.
5. Bickerton, S. and S.G. Advani. 1997. "Experimental Investigation and Flow Visualization of the Resin Transfer Mold-Filling Process in a Non-Planar Geometry",*Composites Science and Technology*, 1997. 57(1):23–33.
6. Steggall, C., S.G. Advani, and S. Yarlagadda. 2008. "Mechanical Properties and Energy Absorption of Powder-Impregnated Continuous Fiber Thermoplastic Matrix Composites", presented at the SAMPE Conference, May 19–23, 2008.

TOPIC III

INDUSTRIAL APPLICATIONS AND CASE STUDIES

5 Axis Weaving—Technology for the Next Generations of Aircraft—Mechanical Performance of Multiaxis Weave Structures

J. P. QUINN[1], A. T. McILHAGGER[1], R. McILHAGGER[1], S. WILSON[2], D. SIMPSON[2] and W. WENGER[1]

ABSTRACT

Weaving of preforms for the manufacture of structural aerospace composites has been at the forefront of engineering textile technology since the development of composites for structural applications on commercial aircraft. The advantages available through the use of not only preforms but engineered textiles, promises reduced cycle times for part manufacture and increased specific performance in terms of impact and damage tolerance. 5 axis weaving (SpaRC) weaving holds the most promise at present as a viable technique for the manufacture of engineered reforms with tailored properties including off axis reinforcement. This paper outline some of the result form a development project completed by University of Ulster and Bombardier Aerospace (Shorts Bros Plc). Results are presented from various mechanical tests performed on preforms woven on a bespoke automatic 5 axis loom. In addition to this discussion is presented on the future applications available to 5 axis preforms.

INTRODUCTION

Developments in the use of composite components on the emerging aircraft and other transportation vehicles have necessitated the examination of composite processing methods that offer flexibility and speed. The combination of dry near net-shape textile preforms along with liquid moulding processes offer significant production advantages [1]. Alternative manufacturing processes exist for preform manufacture including advanced textile processes such as weaving, stitching, braiding and weft and warp knitting and, in addition, the newer non-crimp fabrics.

[1]Engineering Composites Research Centre, University of Ulster, Newtownabbey, Co Antrim, BT37 0QB
[2]Bombardier Aerospace UK, Airport Road, Belfast, BT3 9DZ

The production of near net-shape preforms, e.g. T pieces and I beams, provides the opportunity to engineer specific properties into the reinforcement by control of tow placement [2]. Selection of an appropriate weave architecture at the design stage means different fibre contents can be laid in the different principal strain directions built up in a loaded component [3]. This approach allows the production of more efficient structures.

Mohamed et al [4] have discussed the advantages of 3D (orthogonal) weaving. The authors have concluded that thick preforms can be produced thus dramatically reducing labour time since one or several plies can replace dozens of 2D plies required to manufacture a component of the correct thickness. 3D weaving leads to a reduction in the cutting pattern complexity, eliminating the need for polymeric binders or tackifiers for preform stabilisation and reduces waste of a high value-added material such as prepreg.

3D woven reinforcements have also been shown to have superior mechanical performance in comparison to plied or 2D woven composites [5]. Significant improvements in impact resistance, compression-after-impact performance, reduced damage area and hence greater damage tolerance have been demonstrated.

While composites manufactured from 3D woven reinforcements go some way towards the improvement in mechanical properties and reduced manufacturing costs, these composites are compromised to some extent in their in-plane shear characteristics due to the lack of off-axis ($\pm 45^{\circ}$) tow alignment. To overcome this limitation, a weaving machine has been developed to manufacture multi-axis, multi-layer reinforcements for composite applications [6]. These Spatially Reinforced Composites (SpaRC) are designed to have tows oriented in 5 axes, viz. 0°, 90°, $\pm 45^{\circ}$ (off axis) and through-the thickness (or z) directions.

Initial assessment of the mechanical properties of these multi-axis, multi-layer reinforced composites where consolidating binder tows with a 10 x 10 mm spacing were used has been reported [7]. The properties of these composites have been compared with similar 3D reinforced composites. This work has shown that, by increasing the amount of through-the-thickness tow binder (up to 0.2% by weight) in 3D (orthogonal) structures, the damage area after impact was reduced. In addition it has been shown that low-crimp, multi-axis, multi-layer woven structures offer significant potential for improved damage tolerance [7]. The work presented here extends the initial study, utilising a higher binder density.

COMPOSTIES MANUFATCURE

For this paper 5 specific architectures are presented; these were woven and cut for flat plaque sample manufacture. These textile structures represent a variation around a theme of quasi-isotropic preform design. Variation is incorporated into the designs in a number of ways including—layer sequence and % fibre in each direction.

A complex weaving (SpaRC) machine was used to produce a range of advanced reinforcements for composites [7]. Samples of low-crimp, multi-axis, multi-layer woven fabric, 315 mm with binder tow spacing was reduced from of 5 mm in both the warp (longitudinal) and weft (transverse) directions, giving binder tow densities

Table 1. Reinforcement details for the mutliaxis performs.

Sample No.	Weave Code	Warp Tow %	± 45 Tow %	Weft Tow %	Binder Density Stitch/cm^2	Binder Tow Spacing mm	Architecture
1	BL401	24.3	51.5	24.2	4	5x5	90(2)/0/45/45/90(4)/0/90(4)/45/45/0/90(2)
2	BL402	24.3	51.5	24.2	4	5x5	90(2)/0/90(4)/45/45/0/45/45/90(4)/0/90(2)
3	BL403	24.3	51.5	24.2	4	5x5	90(2)/45/45/0/90(4)/0/90(4)/0/45/45/90(2)
4	BL405	34.4	48.4	17.2	4	5x5	90(2)/45/45/0/0/90(4)/0/0/45/45/90(2)
5	BL407	34.4	48.4	17.2	4	5x5	90(2)/0/45/45/0/90(4)/0/45/45//0/90(2)

The architecture can be interpreted from Table 1 in conjunction with Figure 1 for sample number 5 (BL407).

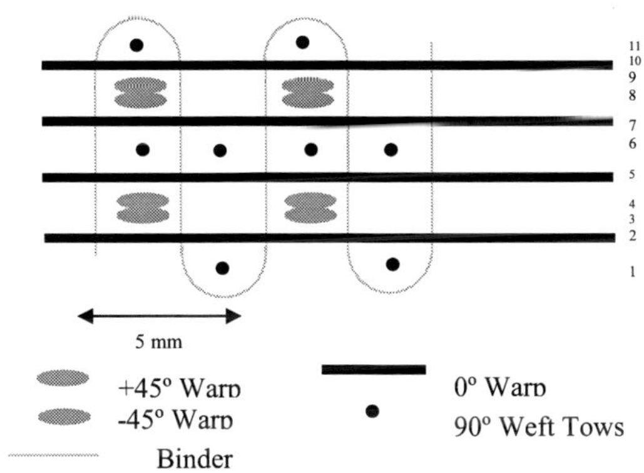

4 binder tows (stitches) per square centimetre. The machine has been completely automated being controlled by a computer-aided design and manufacturing system in conjunction with an electronic Jacquard. These samples were impregnated with an aerospace certified epoxy resin system (RTM6) using a resin transfer process. Details of the reinforcements used in this investigation are summarized in Table 1.

Figure 3. BL 407 Fabric.

All preforms from Table 1 were impregnated with a single component epoxy (RTM6) resin, using a 2040x600 mm fixed-cavity tool. A cavity thickness of 3.8mm was used so that a fibre volume fraction of approximately 55% was achieved. Multiple preforms were impregnated simultaneously.

MECHANICAL TESTING

Following preparation to the relevant standard each of the sample were tests for flexural [8], interlaminar shear [9] and tensile properties [10]. In the case of the flexural and interlaminar tests CRAG standards were followed. An open hile tension test was used to assess tensile properties.

RESULTS

The completed test results are present in Figures 2, 3 and 4.

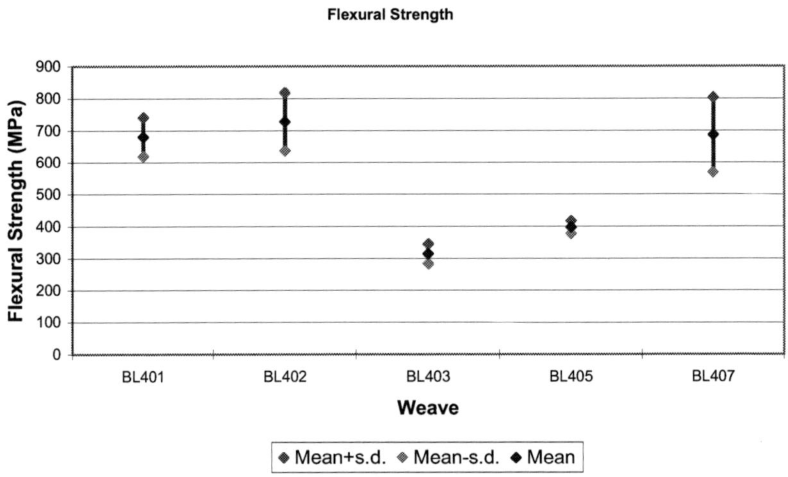

Figure 2. Flexural Strength.

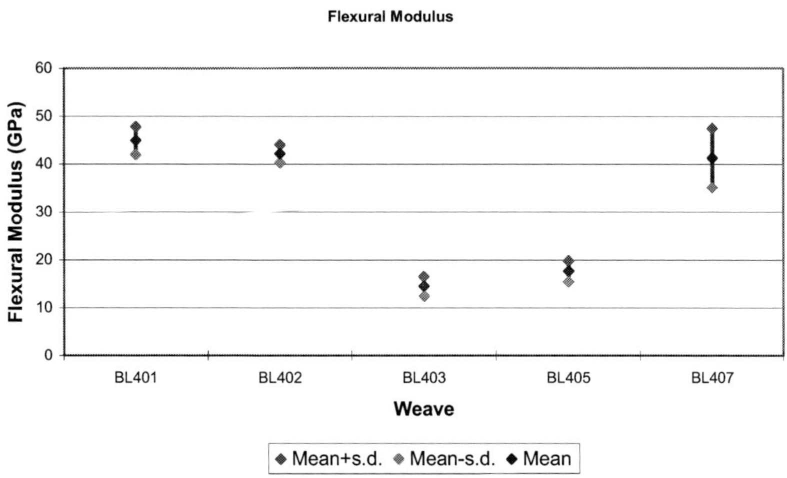

Figure 3. Flexural Modulus.

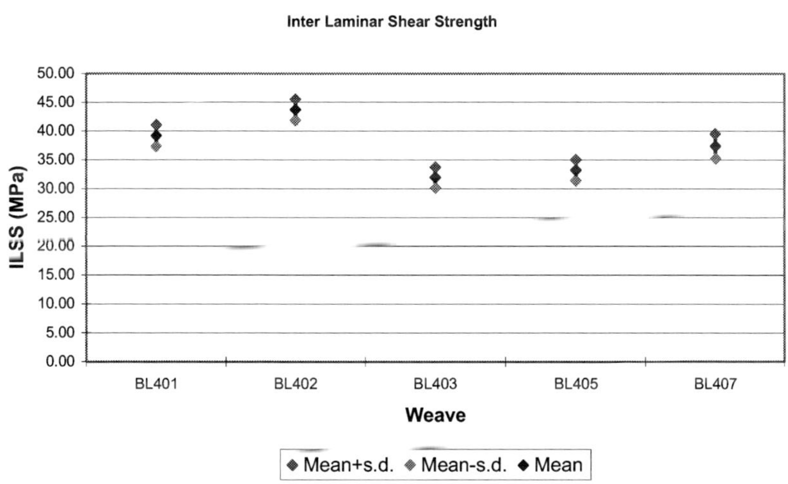

Figure 4. ILSS.

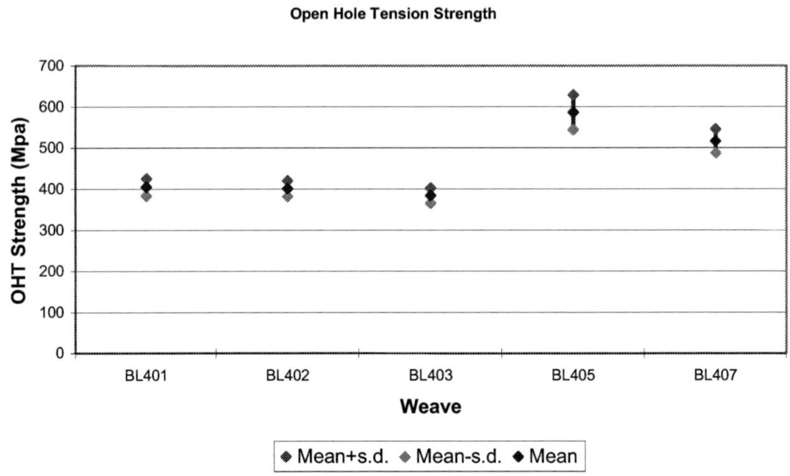

Figure 5. Flexural Modulus.

REFERENCES

1. 1 Miravete A. Introduction: Why are 3-D textile technologies applied to composite materials. 3-D reinforcements in composite materials In Miravete A, editor, Cambridge: Woodhead Publishing Ltd., 1999, p. 1–2.
2. Limmer L, Weissenbach G, Brown D, McIlhagger R, Wallace E. 3-D woven composites for lower leg prosthesis. Composites A 1996; 27A(4):271–278.
3. Quinn J, McIlhagger R, McIlhagger AT. A modified system for design and analysis of 3D woven preforms. Composites A 2003; 34A(6):503–510.
4. Mohamed MH, Bogdanovitch AH, Habil, Dickinson LC, Singletary JN and Bradley Lienhart R. A New Generation of 3D woven Fabric Preforms and Composites, SAMPE Journal 2001; 37(3):8–17.
5. Brandt J, Dreschler K, Mohamed M and Gu P. Manufacture of and Performance of Carbon/Epoxy 3D Woven Composites, In: Proceedings of 37th Int SAMPE Symp., Anaheim, CA, 1992. p. 864–877.
6. Wilson S, Wenger W, Simpson D, Addis S, In: Proceedings of the 43rd International SAMPE Symposium, Anaheim ,1998. p. 1330.
7. Ganeshmal M, Wang L, Brown D, McIlhagger R, Wilson S, Simpson D and Wenger W. Performance Analysis of 'SpaRC', 5 axis, 3D woven Carbon Composites. In: Proceedings of 36th Int SAMPE Technical Conference—Materials & Processing: Sailing into the Future, San Diego, CA, 2004.
8. 8 CRAG Standard Test Method 200 "Method and test for the Flexural Strength and Modulus of Multi-directional Reinforced Plastics", Royal Aerospace Establishment, Edited by PT Curtis, 1988.
9. 9 CRAG Standard Test Method 100 "Method and test for Interlaminar Shear Strength of Multi-directional Reinforced Plastics", Royal Aerospace Establishment, Edited by PT Curtis, 1988.
10. Airbus Industrie Test Method (AITM) AITM 1.0007 Open-Hole Tension Test.

Analysis of NCF-Preforms for Helicopters Composites Parts

F. DUMONT, M. GÖTTINGER and C. WEIMER

ABSTRACT

With a growing interest the LCM processes are investigated in aeronautical industry. The Non Crimp Fabrics (NCF) have demonstrated their ability to be used in the manufacturing of such semi-products called preforms for structural parts. But their industrialisation brings new problems in comparison with the classical Hand Lay-Up of pre-impregnated materials. We are here analysing the kinds of specific deformation modes of NCF which occur on experimental preforms for structural composites parts for helicopters. On a representative 'Z' profile, we analyse capabilities of a simulation tool of draping. In order to assess the reliability of the results we proceeded to an experimental measurement study with a Preform Measurement System in use by Eurocopter Group. The comparison between the experimental orientation of fibres and the one simulated gave interesting results about the important parameter to be taken into account for the forming of the NCF.

DEFINITION OF DEFORMATION MODES ON NCF PREFORMS

The preforming is the preparation and the forming of a temporarily product. This semi-finished product is dry, without resin. As a consequence, it's fragile, and need special handling, assembly and forming methods along with new quality insurance methods. We are presenting here the observed defects, during the manufacturing of NCF-Preforms for helicopter parts. These observation and definition are quite independent from the mechanical characteristics of the material. The definition of these characteristics is not treated here.

Material Used For The Preforms

The reinforcement used in the study is a bi-diagonal carbon fibres non-crimp fabric (NCF), defined in the Table I.

Francois Dumont, Marco Goettinger, Christian Weimer, EUROCOPTER Deutschland GmbH, Donauwoerth, Germany.

TABLE I. DEFINITION OF THE NCFs USED IN THE STUDY.

NCF A	*NCF B*	
Warp and Stitching direction → 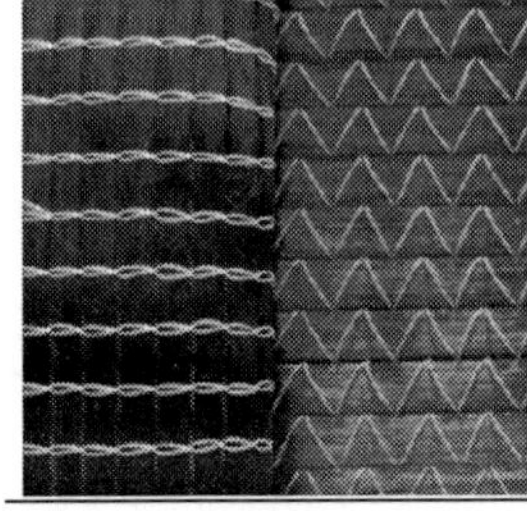	Stitching direction → 	Type : NCF Loop Chain; HTS Fibers 12K, without Torsion. Linear Weight 400/450 tex Density (tow/mm) : $n_{warp/weft°}$=2,84 Stitching Pattern & Yarns: Loop Chain per 5mm Polyester (PES 48dtex) 4 g/m^2. Aeral densities(g/m^2) : $W_{warp°}$= $W_{weft°}$= 125 W_{NCF} = 254 Supplier : *Saertex GmbH*

Two types of NCF-plies are supplied from the supplier. Each ply consists of two layers of unidirectional aligned fibres. The two layers are sewed together according to a tricot loop allowing the handling of the material. For the NCF A the stitching yarns are parallel to the chain fibres. For the NCF B they are diagonal with the fibres. This parameter has an influence of the behaviour of the material by forming.

The material has the specifity to be very thin, in order to fit the fineness in design required for the parts of helicopters (choice of the most optimal thickness). In comparison to the pre-preg technology new types of deformation modes appear during the manufacturing of a dry preform. Those deformation modes may drive to defect in the structure, and have thus to be analysed. The consequence in term of possible design can thus be drawn. The known defects inherent to the material (fish-eyes, foozles) are not treated here.

Observed Deformation Modes During Preforming

The following deformation modes have been identified during manufacturing of different kinds of NCF-Preforms:

1. Fibre Buckling (FB): planar sliding of fibres creating gaps between yarns or even planar waviness of a layer relative to another. The counter layer is in the meantime submitted to a planar compaction (perpendicularly to the fibers)
2. Structure Decomposition (SD): longitudinal sliding of fibres & destruction of internal structure of fibre and NCF (inclusive stitched yarns)
3. Compaction and Stretch (C&S): complementary deformation modes implying They are presented and partially explained in the Table 2.
4. Intra-Ply Slip (IPS): A special interaction of the NCF with the shape can conduct to different stress states in the fibres of a ply.

The TABLE II sums up and illustrated those modes of deformation. Those deformations may lead off course to defect in the preform, and have to be taken into account.

TABLE II. TYPES OF DEFORMATION MODES OBSERVED EXPERIMENTALLY.

	Deformation	Causes	First Layer	Counter-Layer
1	1 Fiber Buckling (FB) [3]	Ondulation of fibers & Compression of counter fibers due to localized in-plane-shear. Possibly compareable with wrinkles of woven fabrics.	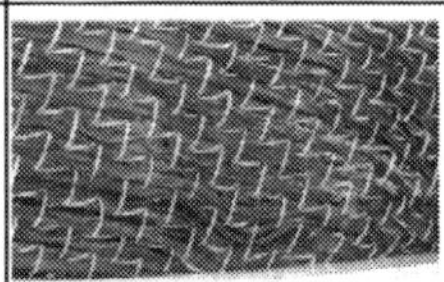	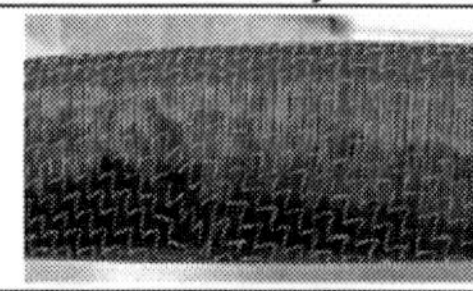
2	2 Structure Decomposition (SD)	Strong Gradient in the path length of fibres. Implies the destruction of the fibre and ist neighbours. Implies change of orientation on the other layer	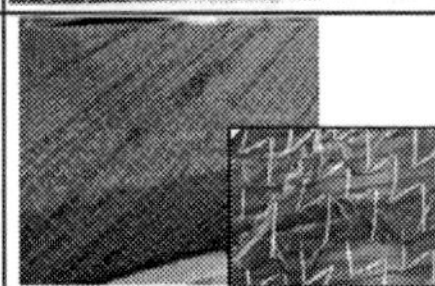	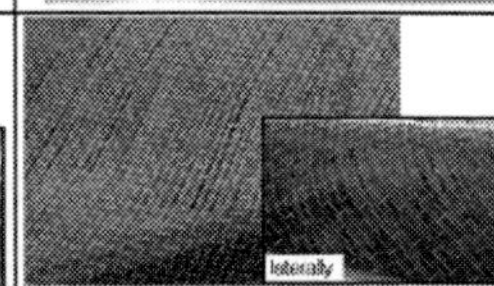
3	3 Compaction & Stretch (C&S)	Compaction of one fibre direction and stretch of the counter-fibre direction. Caused by a small radius of a curved part.		
4	4 Intra-Ply Slip (IPS)	The UD-layers of a ply slip relativ to each other. Caused by the difference of fibre pathlengths in each direction on a curved profile.	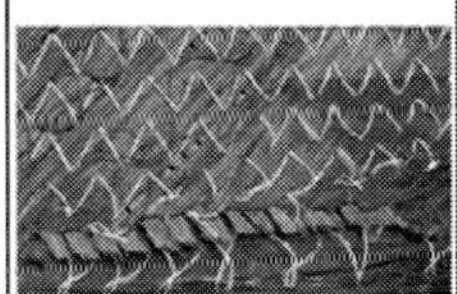	

PREFORM MEASUREMENT SYSTEM & PREFORM ANALYSE SOFTWARE

The present experimental investigation was conducted with the measurement system available at the ECD Facility in Ottobrun (Munich) dedicated to quality management during the preforming [1]. It's composed of hardware (PMS) and software (PAS).

This system carries out planar surface inspection of dry preforms made out of fibres contact-less and non-destructively. It measures the fibre orientation, fuzzballs, gaps and stitching yarn defects at the horizontal surfaces of preforms. In serial production this system will allow for an early detection of divergent fibre orientation relative to the target value, and ensures an on-line quality check. For this study the system allows the comparison of the experimental fibre orientation with the one found by the simulation.

Digital image data is obtained from an optical camera sensor installed on a robotised portal (see Figure 1) which moves parallel to the measured surface. This automatic positioning ensures the sharpness of the photos. The camera is equipped with a special illumination system in order to avoid any reflection.

Figure 1. The Preform Measurement System in use.

The fibre orientation is calculated through the analysis of digital images. A statistical approach initially proposed by Tamura et al. [2] is adapted here. It is based on the texture direction histograms built for parcels of the images after a digital treatment. This treatment intends to avoid influence of stitching yarns, gaps and fuzzle in the calculation of the main fibre orientation. This method is implemented in an own software called Preform Analysis Software (PAS).

The 2D version of the set up (used here) is presently under test phase before its soon introduction in the serial production quality control. A 3D-version is still under development.

DRAPING TESTS & SIMULATIONS

Studied Shape

The draping tests & simulations have been conducted on a complex non developable shape composed of 3 main planar surfaces. The section is a 'Z', with a variable height. This shape is representative of the most complex geometrical features present on aeronautical structural parts (see Figure 2)

We use a curvilinear axis starting from the corner edge O, along the separation edge between two of the surfaces of the shape (see the black **X** axis between upper and vertical surface). A convention will be held to express the orientation of the fibres relative to the preform according to this axis: 0° means along **X**, 90° perpendicular clockwise.

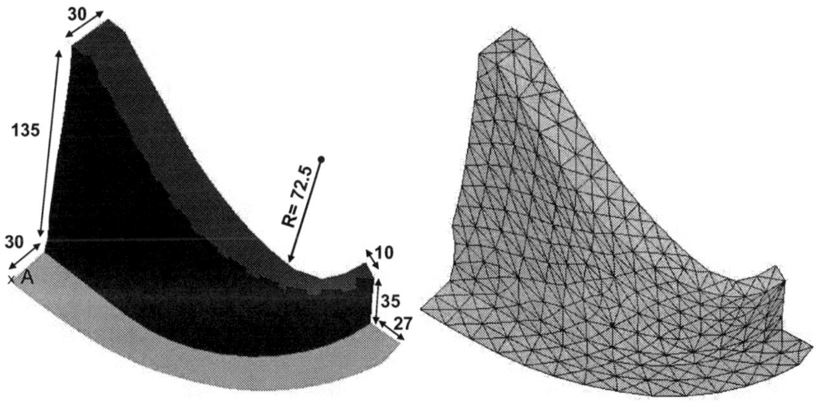

Figure 2. The forming shape out of 3 surfaces (with the dimensions and the curvilinear axis) and the mesh used for the Draping simulation.

Four types of preforms have been manufactured, matching the four different possible orientations of the fibres relative the axis X. for each four orientations of the fibres relative to axis X (see Figure 2): +45° & -45° (135°) with NCF A; 0° & 90° with NCF B. They correspond to the real lay-up designed for the part. On top of that, we've conducted one simulation of woven fabric using the classical mapping of woven. As the consequence, the stitched yarns (SY) adopted alternatively the orientation 0°, 45° or 90° relative to the axis X. The influence of this parameter proved to be of utmost importance.

The materials have been formed on a metallic tooling with the help of a coated GRFP parts. This type of draping is common in the industry where the press forming process (with matrix, die and stretchers) is not largely spread. The first point of the tooling to be in contact with the flange is the point A (see Figure 2).

The tooling has been then positioned with its upper surface perpendicularly to the camera axis.

Draping Simulations

The draping simulation uses a classical Fishnet-Algorithm, based on the kinematical approach of Van der Weeën [3] and implemented in much commercial software. Based on the hypothesis of non-extensible fibres and free-in-rotation nodes this algorithm proved to be efficient for the simulation of the draping of woven fabrics. The main drawback is the impossibility to take into account the mechanical behaviour of the material, and the influence of its environment (sliding and force on the blank-holders, stresses in the fibres etc…). But it can be considered that the NCF material has more similarity with unidirectional as with woven fabrics. That's why we intended to use a variant of this method based on a geodesic approach which allows simulating of the mapping of aligned fibres materials [4]. In this approach the distance between two parallel fibres according to perpendicular geodesic has a minimal value (equivalent to width of fibre). This distance may become greater and doing so creates a

'Stretch effect' within the material, corresponding to the Deformation n°3 'C&S' (see TABLE II). This numerical mapping for unidirectional fibres has been originally developed from Gutowski et al. [5] and implemented in (among others) ESI-Software QUIK-FORM that we used in this study. These calculations are purely geometric in nature and are mainly governed by the non-extensibility of the fibres [6].

Computational mappings were conducted accordingly to the orientations of the manufactured preforms.

The mesh of the shape consists of 1489 Elements and 1593 nodes. For the computation a "starting point" has been defined in A (see Figure 2). The mapping of the layer begins at this point and grows circularly until the whole surface is covered or the solver can't get a result. Further, the drape direction has to be set.

COMPARISON EXPERIMENT & SIMULATION

Two types of result are compared: firstly the fibre orientations, and secondly the gaps between parallel fibres (known as stretch ratio).

Fibres Orientation On The Upper Surface

The orientations of the fibres of NCF A & B at the upper surface are plotted according to the distance to the corner edge along with the **X**-axis.

The size of each picture taken by the PMS is 100 x 100 mm^2, and the size of study parcels of the PAS processing was set to 10x10 mm. alike the element size of the grid. The orientations obtained for each study parcel were plotted in an excel matrix which had the same format as the matrix for the assembled experimental results.

For the NCF B the orientations depicted with PMS are in good accordance with the ones obtained by the simulation, within the precision of the PMS (see Figure 3).

No influence of the orientation of the stitching yarns has been depicted.

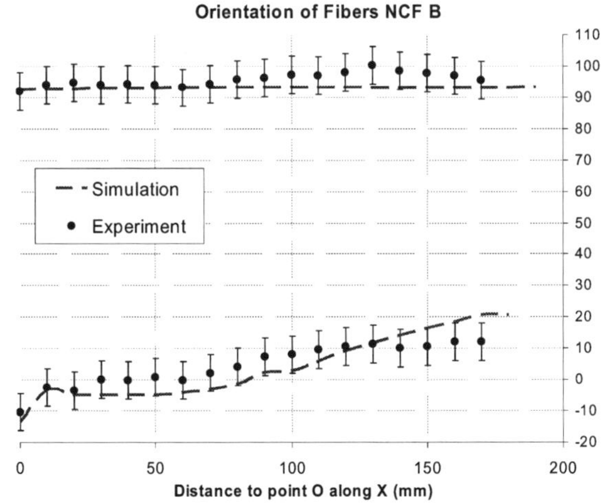

Figure 3. Orientation of the Fibres for the NCF B.

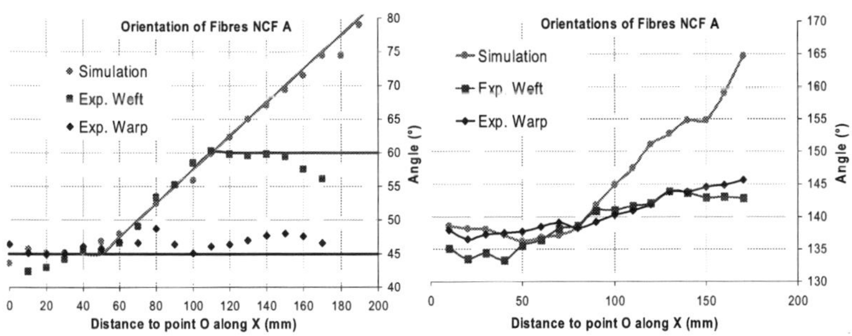

a b

Figure 4. Fibre orientation for the NCF A.

For the NCF A, the results need to be interpreted. The measured and simulated orientations are quite divergent and depend on the type of fibres (warp and weft) and on the initial orientation (+45° or 135°). Four experimental curves are thus presented. They are representative from the global forming results on the whole shape and are dependant of the orientation in the other surfaces of the part.

The comparison shows an independent behaviour between warp and weft. Oriented initially to 45° (see Figure 4-a), the angles present a 3-domain curve. Until 50mm along the X-axis, the 3 results are equivalent. From 50mm to 110mm, the weft orientation follow the change given by the computation (15°). From 110mm till the edge the weft fibres don't show any change in their orientation. Responsible are the perpendicular stitching yarns (in the warp direction) which blocked the 'stretch' (see following paragraph) of fibres. The rotation at the upper surface is constricted. Meanwhile the orientation of the warp fibres doesn't change. Typical deformations are to be seen at the vertical surface of the part (blue surface on Figure 2). Fibre Buckling together with Compaction and Stretch are noticed and impeach the warp fibres to rotate during the draping.

For the fibres initially oriented at 135°, we observe an increasing difference between simulation and experiment. The planar flexural stiffness of the warp fibre and the interaction with the counter-layer (through the stitched yarns) may explain this divergence. This comparison showed convergence between simulation and the experiment for the NCF B and a divergence for the NCF A when the deviation reaches a high value (above 15°).

Gap Analysis With The Parameter 'Stretch'

The QUIK-FORM solver considers the distance between parallel fibres as a degree of freedom. This 'stretch' parameter (derivative) can be depicted at the surface of the part. It corresponds to experimental observation. As seen before, this parameter has a great influence on the disorientation of fibres on the whole part (not just locally). To verify this stretch behaviour of the material, pictures were taken from the lower surface (green on Figure 2). The standard distances between parallel fibres were compared with the distance after the drape process and the results on the preforms showed a very good agreement in the localisation (in the TABLE III).

TABLE III. STRETCH AND SHEAR PARAMETERS.

Stretch of UD	Shear of UD	Shear of woven fabric
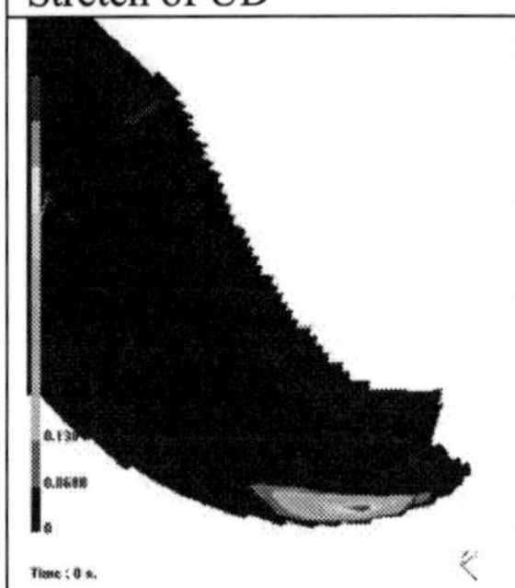	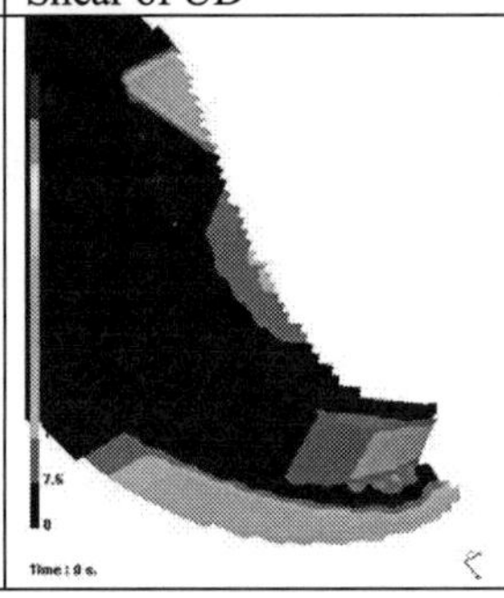	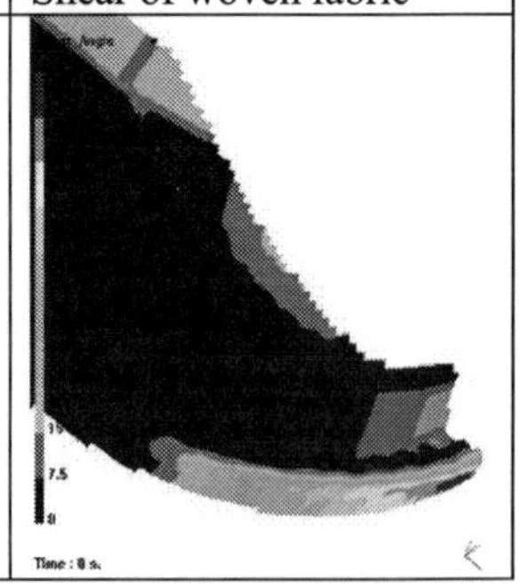

On the Table III man sees the great interaction between 'stretch' and 'shear' fields for the aligned fibres (UD). The apparition of stretch in one layer of the NCF will the let the value of shear decrease relative to the one on woven fabrics.

CONCLUSION

We identified through experimental observation the deformation modes enhanced in the preforming with NCF. We launched a comparison between experiment and simulation on a 'Z' shape which is a geometrical feature representative of structural parts in use by Eurocopter. The result of this study showed that unidirectional fibres, unlike woven fabrics, do not follow the fish-net mapping. The simulation gave partially good results. We've noticed particularly the occurrence of the special modes of deformation inherent to NCFs. The introduction of the 'stretch' parameter in the mapping is in agreement with the observed change in interfiber spacing. Nevertheless it proved to be not sufficient to represent completely the draping of NCF on strongly non-developable shapes. The flexural stiffness of the fibres as well as the action of the stitching yarns have to be taken into account.

REFERENCES

1. Miene, A. Göttinger, M. 2007. "Digital Image Analysis for Quality Assurance in the Preforming and Draping Process" presented at the CFK-Valley Stade Convention 2007, June 13–14, 2007.
2. Tamura, H., Mori, S., and Yamawaki, T. 1978. "Textural features corresponding to visual perception", IEEE Transactions on Systems, Man and Cybernetics, 8(6):1264–1274.
3. Van der Weeen, F. 1991. "Algorithms for draping fabrics on doubly-curved surfaces", *Int. J. for Numerical Method in Engineering,* 31, 1414–1426.
4. Tucker, C.L. 1997. "Forming of Advanced Composites", in *Advanced Composites Manufacturing,* T. G. Gutowski, eds. John Wiley & Sons, Inc., pp. 297–372.
5. Gutowski, T. G. et al. 1991. "Differential geometry and the forming of aligned fibre composites", *Compos. Mfg.*, 2, 147–152.
6. Boisse P. 2006. "Finite element analysis of composite forming", in: *Composites forming technologies*—Chapter 3, ed. A.C. Long, Woodhead Publishing Limited, pp. 46–79.

The authors would like to thank Ms L. Lacroix and M. J. Hader for their active participation.

Material Characterization of Non-Crimp Braided CFRP for Aeronautic Applications

P. MIDDENDORF, B. VAN DEN BROUCKE, A. GEẞLER,
F. MAIDL and C. METZNER

ABSTRACT

It is well known that the circular braiding technique provides an innovative and flexible method for the production of certain types of fiber reinforced polymeric parts. A major drawback of this production method is that it produces materials with possible largely crimped yarns, hence causing a 'non-ideal' orientation of the fibers inside the structure. The mechanical performance of a braided part would be significantly lower than an equivalent non-crimped fabric (NCF) draped or tape manufactured variant. However, recent developments performed at EADS Innovation Works have resulted in a non-crimped braided CFRP or so-called unidirectional braid (UD-braid) which solves the yarn crimp problem.

The presented paper is organized in four different sections which describe different details of the non-crimped braids. An introduction describes the new manufacturing method itself and provides a comparison with other manufacturing techniques. Experimental investigations on the determination of geometrical, mechanical and damage characteristics are presented in the second part. Afterwards an overview of the meso-mechanical finite element model of a material representative volume element is presented. This model is used in non-linear numerical simulations to investigate the mechanical as well as the damage behavior. In a last section, the paper shows the applicability of the method to aeronautic structures by means of a set of possible validation parts.

UNIDIRECTIONAL BRAIDING TECHNIQUE

A circular braiding machine generates an interlacing pattern between yarns by moving bobbin carriers in two opposite directions. For conventional braids all carriers are populated with bobbins containing the same material and both carrier directions contain the same amount of bobbins. An additional inlay yarn can be optionally added in braiding direction to obtain a tri-axial braid. The interlacing pattern of these materials results in a large crimp and rotation of the roving. Especially when carbon is used this leads to a significant reduction of the composites mechanical performance compared to unidirectional material.

EADS Deutschland GmbH, Innovation Works, D-81663 Munich, Germany

A new manufacturing method invented in 1992 [1] solves this problem. The asymmetric braiding technique was further refined by Geßler et al. [2] during the past years. In this technique the carriers moving in the first direction are populated with bobbins containing the carbon reinforcement roving. The carriers moving in the opposite direction are partially populated with bobbins containing auxiliary yarns (figure 1, left). The function of the auxiliary yarns is to support the reinforcement yarns and keep them in place on the mandrel as well as flattening of the roving during manufacturing of the fabric. Once the part is infiltrated, the auxiliary yarn does not have any further function and hence the used amount is kept as low as possible. In the resulting non-balanced textile structure (figure 1, right) the carbon roving has only very limited crimp and torsion due to much higher bending stiffness and larger dimensions compared to the auxiliary yarn.

Recent developments have improved the process by using a special polymeric auxiliary yarn which has a similar molecular structure as the epoxy resin. During curing and infiltration of the reinforcement textile, the auxiliary yarn material partially dissolves and finally cross-links with the epoxy resin resulting in a non-crimped braid or unidirectional braid. Another advantage of the use of auxiliary yarns is the possibility to incorporate a resin modifier which could enhance the properties of the final composite.

Experimental investigations performed at EADS Innovation Works have shown that, compared to regular braid reinforced composites, the non-crimped braid variant shows an increase of 45% in tensile and 25% in compressive strength, respectively [3, 4].

The non-crimped braid for which the results are presented in this paper are manufactured using a Tenax-E HTS roving and a Grilon co-polymeric auxiliary yarn.

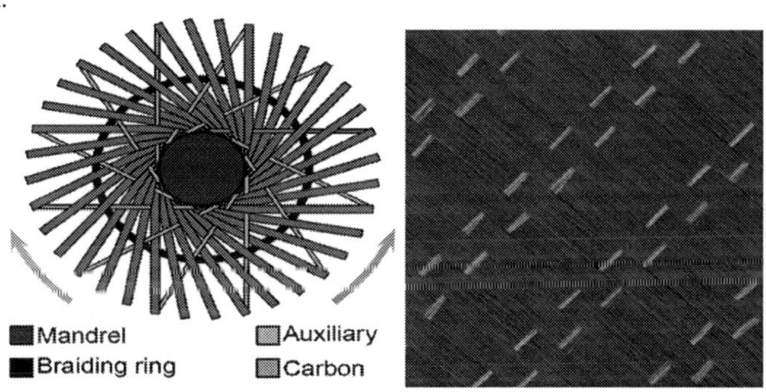

Figure 1. Left: Unidirectional braiding technique; Right: resulting non-crimped braid.

MATERIAL CHARACTERIZATION

The non-crimped braid is characterized geometrically, mechanically and its damage behavior is investigated by Eisenhauer [5].

A set of basic textile measurements have been performed for the generation of a geometric model of a textile structure. The basic properties (i.e. shape, width, thickness, spacing, and orientation of the yarns) are determined using high-resolution scans of the textile or microscopic cross-section analysis. The results of the geometrical characterization are used to generate geometrical and finite element models using the WiseTex software tools [6].

As described in the next section, the Ladevèze damage model [7] is used to simulate the damage behavior of unidirectional reinforced composite materials. This damage model requires a set of parameters which can be determined by a five dedicated mechanical tests:

1. A cyclic tensile test on a $[0]_8$ stacking sequence determines the elastic and damage behavior of the composite in fiber directions. This cyclic test is performed at low strain rate and a low cycle number with increasing amplitude. Typically five to ten cycles are performed before failure of the sample.
2. A simple compressive test on the same $[0]_8$ stacking sequence determines the compressive properties.
3. A cyclic tensile test on a $[\pm 45]_{2s}$ stacking sequence is used to determine the shear damage behavior of the material.
4. A cyclic tensile test on $[+45]_8$ stacking sequence determines the transversal behavior of the composite.
5. Finally a cyclic tensile test on a $[\pm 67.5]_{2s}$ stacking sequence characterizes a possible coupling between the transversal and shear damage behavior.

From these series of experiments both elastic as well as Ladevèze damage parameters are derived and compared to values found in literature. The results from the experiments are used as input for the material data in the FE model and as validation for the meso-mechanical simulation.

MESO-MECHANICAL MODELLING AND SIMULATION

Model Preparation

Although a geometrical model of the non-crimped braid is generated using the WiseTex software tool, it was found that the shape of the fibrous material after infiltration significantly deviates from the geometrical model. Therefore it was decided to manually model a Representative Volume Element (RVE) based on the experimental available data.

Due to the fact that the auxiliary yarn dissolves in the resin and the same yarn can not be located in any of the microscopic images generated from the composite, it is assumed that the auxiliary yarn does not have a significant influence on the mechanical performance of the material and is therefore not modeled. The carbon roving is modeled and meshed with linear iso-parametric hexagonal elements whereas pure resin zones located in-between the carbon rovings are meshed with linear iso-parametric tetrahedron elements.

At macroscopic level, the average fiber volume fraction is $V_f = 60\%$, however the local fibre volume fraction within the roving is at about 65% due to resin pockets. Changes of the mechanical properties caused by a variation in fiber volume fraction are based on the micro-mechanical mixing formulas described by Chamis [8]. The fiber orientation for the carbon material is kept constant in the direction of the roving.

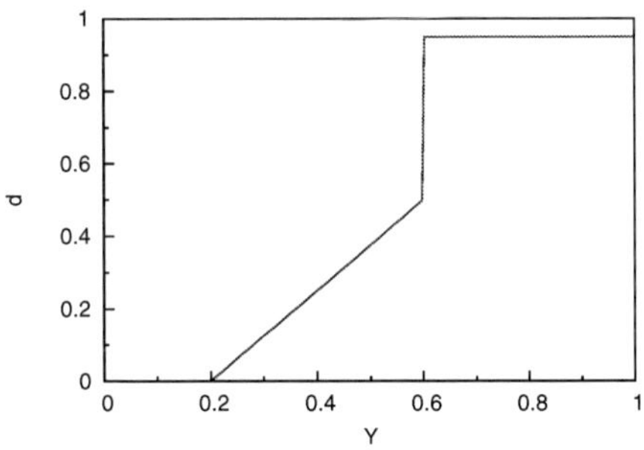

Figure 2. Evolution of the damage parameter.

Incorporation of Damage

To be able to predict the damage behavior inside the detailed finite element model of the material RVE, the Ladevèze damage model is implemented in a user-subroutine for solver MSC.Marc. This damage model takes into account longitudinal, transverse and shear damage by introducing three damage variables d_l, d_t and d_s, respectively. The damage variables deteriorate the Young's and shear moduli according to

$$E = E^0(1-d), \tag{1}$$

where E^0 is the undamaged value of the modulus. In the model, a damage variable d is written in function of a thermodynamic equivalent force Y as shown in figure 2. This function predicts a typically linear increase of the damage parameter d between the initial threshold Y_0 and an ultimate threshold Y_{LIM} after which total failure is assumed (i.e. d = d_{max}). The thermodynamic equivalent force Y is a function of the strain energy W and is written as

$$Y_l(t) = \sup_{\tau \le t} \sqrt{Z_l(\tau)}\,,\; Y_t(t) = \sup_{\tau \le t} \sqrt{Z_t(\tau)} \text{ and } Y_s(t) = \sup_{\tau \le t} \sqrt{Z_s(\tau) + bZ_t(\tau)} \tag{2}$$

where b is the transverse-shear coupling factor and

$$Z_l = \frac{\partial W}{\partial d_l},\; Z_t = \frac{\partial W}{\partial d_t} \text{ and } Z_s = \frac{\partial W}{\partial d_s}. \tag{3}$$

For the resin material a similar approach is used except that an isotropic damage model is used with only one damage parameter d.

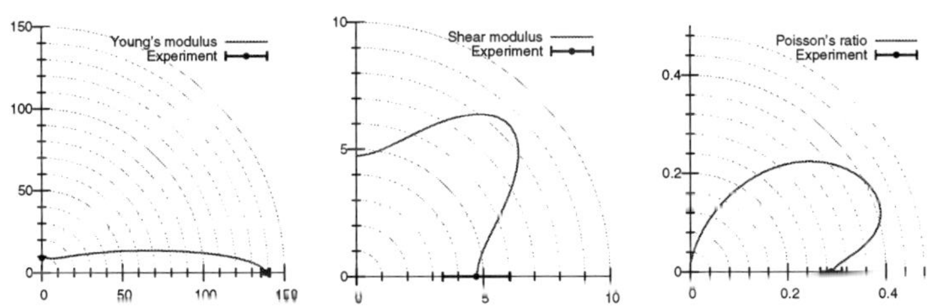

Figure 3. Comparison between calculated and measured engineering constants.

RESULTS

Homogenised Elastic Constants

The average homogenized engineering constants of the material are calculated by applying six well chosen uni-axial strain boundary conditions to the RVE. Under periodic boundary conditions the stress and strain fields are averaged according to

$$\langle \sigma_i \rangle = \frac{1}{V} \int_V \sigma_i \mathrm{d}V \text{ and } \langle \varepsilon_i \rangle = \frac{1}{V} \int_V \varepsilon_i \mathrm{d}V \tag{4}$$

and used to calculate the homogenized elastic stiffness matrix from which the engineering constants can be determined.

A comparison between the simulated and experimentally determined properties is given in figure 3. The polar graphs represent the Youngs's modulus, shear modulus and in-plain Poisson's ratio (from left to right, respectively) plotted against an in-plane rotation of the reference coordinate frame. The calculated result shows very good agreement with the experimental data points.

It should be noted that the presented method is able to calculate the full stiffness matrix of any material modeled and loaded using the methods described above. This opens a way to calculate also material properties that are very difficult if not impossible to determine experimentally.

Damage Behavior

The RVE is loaded in tension and compression in different directions under periodic boundary conditions and using a non-linear iterative solution scheme. The average stress and strain vectors are determined for each increment of the simulation with the formulas given in equation 4. The comparison between the simulated and average experimental curve is given as an example in figure 4.

The implemented model provides a prediction of the strength within the observed variation of the experimental results. The same is found true for the prediction of the ultimate strain. It should however be noted that the non-crimped braid is a rather simple material and a quasi linear stress-strain response is observed. Therefore a comparison between stress-strain curves of a tensile test on a more complex tri-axial braid is presented in figure 5. For this material the strength is predicted reasonable well and an overestimation of the ultimate strain is observed. Initially a small overestimation of the stress for the same strain level is predicted. Then two 'jumps' in the simulated curve are observed at approximately 0.6 and 0.85% of strain. These events correspond to transverse and shear failure of rovings which are oriented perpendicular and at an angle of 45 degrees to the load direction, respectively. From computer tomography experiments where micro-crack orientations are observed in the same direction, this prediction is validated.

It is assumed that the difference between the two curves is caused by the remaining stiffness of the fully damaged elements, which is in this case 5% of the original stiffness. This remaining stiffness is caused by the maximum allowable value of the damage variable (d_{max}).

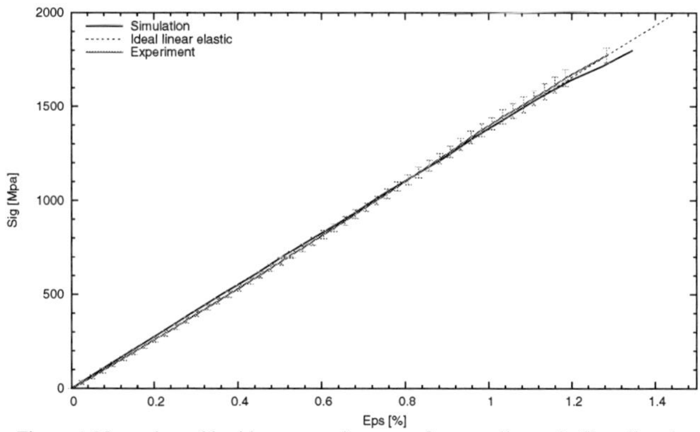

Figure 4. Non-crimped braid stress-strain curves for a tensile test in fiber direction.

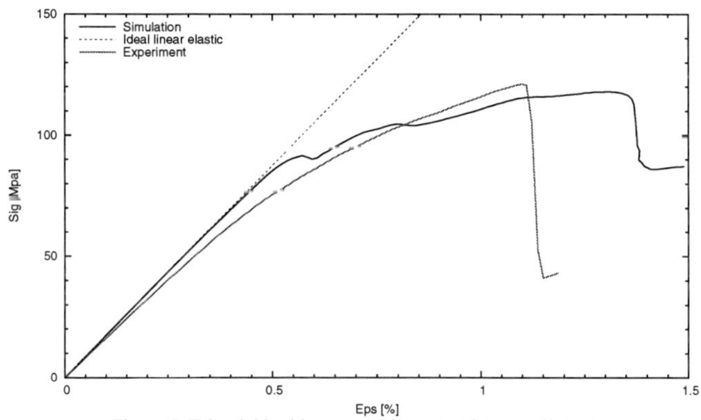

Figure 5. Tri-axial braid stress-strain curves for a tensile test.

APPLICATIONS IN AERONAUTICS

The non-crimped braid is a material with many possible application regions. Due to its quasi-uni-directional reinforcement it possesses very good mechanical properties and hence could be used for high-end applications. Moreover, the fact that beam structures, straight or curved, can be manufactured with optimized fiber orientation at each point of the beam makes the non-crimped braiding technique a very good candidate for the production of aeronautic structures. As an example the reinforcement frames and stringers of a fuselage structure can be manufactured with non-crimped braids with high efficiency. Also structures with varying cross section could benefit from this manufacturing technique. For instance multi-spar structures which have typically a decreasing cross section from inner to outer wing positions could be braided for smaller aircrafts

CONCLUSION

Within this paper, the uni-directional braiding technology is presented. The resulting material proves high potential for aeronautical applications by combining good mechanical properties with a flexible manufacturing technique.

A multi-level modeling approach is adapted to numerically evaluate the elastic as well as the damage behavior of the material. Based on homogenization algorithms the average properties of a material on a larger scale can be calculated based on a detailed analysis on a smaller scale. This technique has proven to be efficient and provide accurate results for different types of heterogeneous materials.

The non-crimped braid is analyzed in detail by meso-mechanical modeling of a representative volume element of the material. The calculation of all engineering constants is in this way enabled by use of a homogenization technique.

Using the Ladevèze damage model a method was adapted to predict damaging effects inside the material. This technique is applied to the non-crimped braided material and shown to give good results. Applied to more complex textile reinforced polymer composites the damage model has proven to provide acceptable results yet some differences are observed.

REFERENCES

1. G. Earle, A. Kruesi, J. Stockton and D. Kruesi. 1992 , "Asymmetric braiding of improved fiber reinforced products", Patent No: WO9215740A1.
2. A. Geßler and F. Maidl, 2005, "Verfahren zum Herstellen von Faserverbund-Halbzeugen mittels Rundflechttechnik", Patent No: DE102004017311A1.
3. A. Geßler, Dr. J. Brandt, F. Maidl, Ch. Breu, Dr. J. Horn, H. Schneider, "Neue Entwicklungen bei der Fertigung von Kohlenstofffaserpreforms mit der Rundflechttechnik", DGLR Tagung, 2004.
4. A. Geßler, F. Maidl and M. Schouten , 2007 , "Advancements In Braiding Technology For Textile Preforming" , Sampe Europe International Conference & Exhibition (SEICO).
5. Ch. Eisenhauer, 2006, "Characterisation of UD-Braids", Master's thesis, Institut für Textiltechnik, RWTH Aachen University.
6. I. Verpoest and S. V. Lomov, 2005, "Virtual textile composites software WiseTex: Integration with micro-mechanical, permeability and structural analysis", *Composites Science and Technology*, 65(15-16), pp. 2563–2574.
7. P. Ladevèze, O. Allix, J. Deü and D. Lévêque , 2000, "A Mesomodel for localisation and damage computation in laminates", *Computer methods in applied mechanics and engineering* , *Elsevier Ltd.*, 183(1-2), pp. 105–122.
8. C. Chamis, 1989 , "Mechanics of composite materials: Past, present and future.", *Journal of Composites Technology and Research* , 11(1), pp. 3–14.

Innovative Woven Spacer Preforms

A. R. TORUN, G. HOFFMANN and C. CHERIF

ABSTRACT

Within the scope of Collaborative Research Centre SFB 639 "Textile-reinforced composite components for function-integrating multi-material design in complex lightweight applications", woven spacer fabrics from commingled yarns are developed, pressed and the characteristics of the formed composite structures are tested. The developed spacer fabric structures have U- shaped and V- shaped connections with woven crosslinks which yield good performance under complex stresses especially under bending strength in comparison with conventional spacer fabrics. Various modifications on weaving machines are executed to enable commingled yarn processing into 3D textile preforms. These new developments extend the performance and the application areas of cost-effective textile reinforced thermoplastic composites. Integration of functional components and further near-net-shape structures are the main emphasis of future work.

INTRODUCTION

The woven, warp-knitted, weft-knitted, braided and stitch-bonded textile preforms, which can be produced from glass, carbon and aramid yarns, are on a large scale used for the production of composite materials. Spacer fabrics can be developed with weaving, warp knitting and weft knitting techniques. These conventional spacer fabrics have two outer layers which are connected through pile yarns. If spacer fabrics

Institute of Textile and Clothing Technology Technische Universitaet Dresden, 01062 Dresden Germany

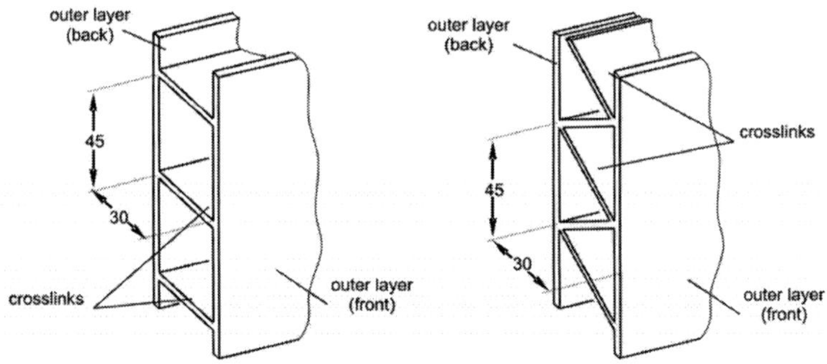

Figure 1. U-shaped (left) and V-shaped (right) spacer fabric structures.

are to be used as preforms for the thermoplastic composite materials with high flexural rigidity, then it is advantageous to connect the two outer layers with woven crosslinks instead of pile yarns. The woven crosslinks, which create the space between two outer layers, can also be easily pressed in order to produce thermoplastic composite components. However, the pressing of the spacer part is mostly not possible for the conventional spacer fabrics, which are produced through pile yarn connections. The woven crosslinks, in comparison with pile yarn connections, yield significantly higher strength against tensile, compressive and especially bending forces. Therefore, the development of spacer fabrics with woven crosslinks is necessary [1].

Woven spacer fabrics from commingled yarns are developed for the lightweight applications of thermoplastic matrix composites. Possible connection variations for the woven crosslinks are analyzed. After considering the reproducibility and press ability of the various structures, U- and V-shaped crosslinks (Fig. 1) are selected for further developments. The used commingled yarns consist of one high-strength and one thermoplastic matrix components. In particular, GF/PP commingled yarns are produced with 0.5 fiber volume fraction through air-texturizing process [2]. An over-delivery of the matrix component of hybrid yarn in comparison with the high-strength component is necessary, which to some extent prevents damage and strength reduction of textile structures during production. Simultaneous investigations are going on about the online spinning of GF/PP hybrid yarns.

DEVELOPMENT OF SPACER FABRICS

Figure 2 illustrates the production technique of the developed U-shaped spacer fabrics, which is realized in three steps. In the first step (Fig. 2A) upper and lower surfaces are produced. In the second step (Fig. 2B) half of the warp yarns make floatings while the remaining warp yarns produce the woven crosslinks. In order to create the sandwich structure, the floating warp yarns are pulled back in the third step (Fig. 2C) [4].

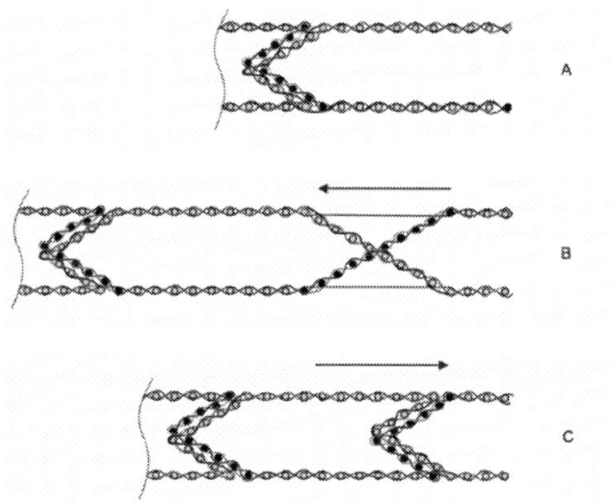

Figure 2. Production steps of the U-shaped spacer fabrics; A: weaving of the outer layers, B: weaving of crosslinks and floatings, C: pulling back.

Essentially, a terry weaving mechanism is developed to enable the transformation of flat woven structure into sandwich structure. The preliminary trials of the spacer fabric development are accomplished on the narrow weaving machine. Main elements of the modified narrow weaving machine are illustrated in figure 3 [5].

In addition to the developed terry weaving mechanism, a new fabric take-up system is developed. Aim of the system is to ensure a uniform take-up of the fabric layers and to avoid buckling which causes damage on the reinforcement material. Straight-line arranged roller pairs characterize this system, one of which can be engaged to the other roller with a defined contact force. In front of the take-up system a gripper inserts supporting bars into the hollow space of the woven fabrics. A second

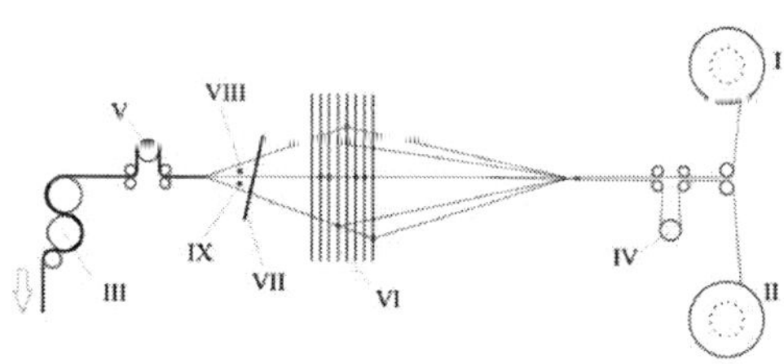

I: warp beam for crosslinks
II: warp beam for floatings
III: fabric take-up
IV: pulling-back mechanism
V: terry weaving (fabric storing) mechanism
VI: shafts
VII: reed
VIII: upper weft insertion
IX: lower weft insertion

Figure 3. Main elements of the modified narrow weaving machine.

Figure 4. Woven spacer fabric (left) and its composite (right).

gripper extricates the bars from the fabrics at the back side of the system. A controlling system is developed for the synchronization of the mechanisms of the modified narrow weaving machine and good results are achieved in flexibility concerning the dimensions of the fabrics as well as the defined movements of the mechanisms and precise actuating [3].

Based on the experiences of spacer fabric production on narrow weaving machine, the production of spacer fabrics on a double rapier velvet weaving machine is carried out [4]. The low elasticity of the produced commingled hybrid yarns requires a precise adjustment of the weaving geometry and reduces the feasible machine performance.

CONCLUSION

Within the scope of Collaborative Research Centre SFB 639 "Textile-reinforced composite components for function-integrating multi-material design in complex lightweight applications", woven spacer fabrics are developed, pressed and the characteristics of the formed composite structures are tested. Spacer fabric structures with woven crosslinks yield good performance under complex stresses especially under bending strength in comparison with conventional spacer fabrics. These new developments extend the performance and the application areas of textile reinforced thermoplastic composites. Integration of functional components and further near-net-shape structures are the main emphasis of future work.

This paper presents the results of the research program SFB 639 Subproject A3 of German Research Foundation. The authors wish to thank for the financial support.

REFERENCES

1. www.tu-dresden.de/sfb639/.
2. Choi B.-D., Offermann P., Diestel O., Mäder E. and Hübner T. "Advancement of commingled hybrid yarns thermoplastic fiber reinforced composites". Technische Textilien/Technical Textiles, Vol. 45, pp. 16–18, 2002.
3. Mühl A., Großmann K., Offermann P. and Löser M. "A new take-up motion, cutting and storing system for weaving of spacer fabrics". 8th Dresden Textile Conference, Dresden, pp. 59-61, 2006.
4. Torun A.R., Hoffmann G., Ünal A., Klug P., Cherif Ch. and Badawi S. "Technological solutions to the development of spacer fabrics including fabric structures as the spacer". 8th Dresden Textile Conference, Dresden, pp. 56–59, 2006.

Composite Rope

T. SUN, D. STENVERS, H. WRIGHT and R. C. SAMSON

ABSTRACT

Ropes made from composite materials have many advantages over wire ropes because of their lighter weight, longevity, lower operating cost, and reduced personal injury. In this paper, we present the effects of curing temperature and time, twist level and tension, and fiber loading on the tensile strength of three different composite rope constructions and their sub-building block towpreg. The resin systems used were thermoplastic in combination with cross-linkable thermosetting resin; and the fiber was Technora yarn.

The test results indicated that, for the same size, the composite ropes have comparable tensile strength as wire ropes, with less than a quarter of the weight.

INTRODUCTION

Synthetic and wire ropes have long been used in various industrial, marine, and offshore applications. Both synthetic and wire rope products today have their own limitations because of the materials used. Synthetic ropes can be made of high-performance fibers such as high modulus polyethylene fibers, aramid fibers, and liquid crystal polyester fibers. These ropes usually have great tensile strength and flexibility, but lower cut and abrasion resistances compared to wire ropes. On the other hand, while wire ropes have good dimensional stability and abrasion resistance, their applications can be limited for their heavy weight and high rigidity.

A composite rope, as strong as wire rope but more flexible and lighter, is very desirable as a wire rope replacement option in many applications. One way to produce a composite rope is to start with resin-impregnated fiber bundles called towpreg.

Samson Rope Technologies Inc., 2090 Thornton Street, Ferndale, Washington 98248, U.S.A.

Towpreg can be made by pultrusion, extrusion, commingled yarns, or other resin-impregnation processes.

References 1–3 list different patents on producing such composite ropes. The polymer resins used in these systems were thermosetting resins which, once cured, became relatively rigid and brittle and therefore could not be further processed. People have developed thermoplastic resins to pultrude more flexible rods [4,5]. However, these pultruded rods were still too rigid to make flexible ropes.

The purpose of our research was to (1) explore the possibility of producing composite ropes from flexible thermoplastic resin impregnated yarn, (2) define processing parameters, and (3) understand the physical and mechanical properties of the composite ropes.

EXPERIMENT

Raw Materials

Technora® fiber T200 supplied by Teijin Aramid, Ltd was used in this study. The resin used for the towpreg is a mixture of thermoplastic and cross-linkable thermosetting resins. Based on a solution of micronized thermoplastic, the thermoplastic resin can be cross-linked by heat. Several ends of Technora® yarns were put together and impregnated in a coating bath followed by it passing through a shaping die, to form a flexible resin-saturated fiber bundle.

Towpreg Variables

To determine the effects of cure time and temperature on the tensile strengths of the towpreg, we varied the curing temperature from 175°F to 375°F with ranging time intervals from no cure (zero curing time) to 30 minutes. We also prepared towpregs with different twists in order to study the optimal level of twist to yield the highest tensile strength, under the optimal curing conditions. Other design and process variables studied were fiber contents and curing with or without tension.

Composite Design and Process

The process for preparing composite ropes is illustrated in Fig. 1. The towpreg was twisted under tension then passed through a heat-forming die. The purpose of the forming die is to smooth the surface of the towpreg for consistent secondary processes. As shown in Fig. 2, the twisted towpreg has a smoother surface after the die than the one shown before going through the heat-forming die.

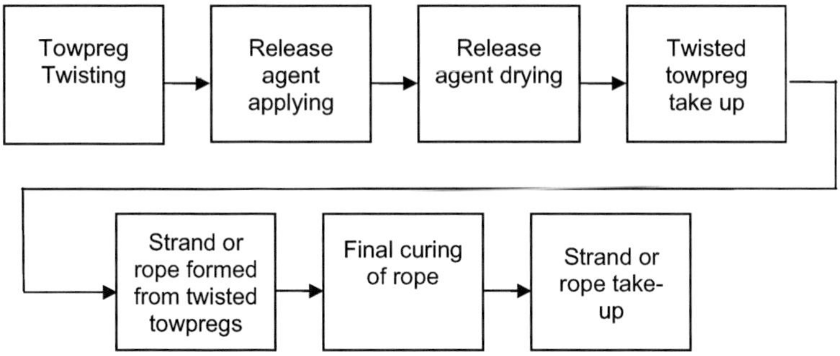

Figure 1. Composite rope fabricating flow chart.

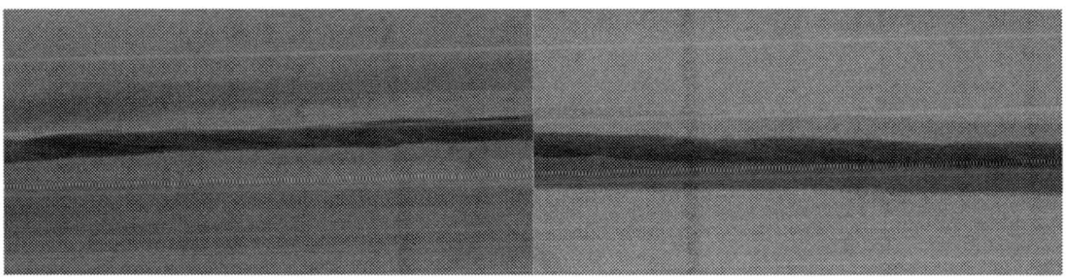

(a) (b)

Figure 2. Towpreg surface (a) before heat-forming die and (b) after.

Once the heat-die process was completed, a release agent was applied to the towpreg, followed by a drying process. This created a film that prevented the strands from bonding to each other, ensuring the flexibility of the formed composite structure. The twisted towpreg was then wound to a bobbin by a take-up unit. Depending on the constructional design, the proper number of bobbins were put onto a carousel, which rotated at a constant speed to form a strand, and proper number of strands were laid together to form a rope. The rope was cured and finally wound to a bobbin by a take-up unit.

As an initial study, three constructions were produced and tested, namely 1×7, 1×19, and 7×7. For the 1×7 and 1×19 constructions, either 6 or 18 twisted towpregs were laid around a center core to form a rope. For a 7×7 composite rope, six 1×7 sub-ropes were laid around a center 1×7 core to form a rope. The diameters of the three rope samples were 3, 5, and 10mm, respectively. An example of a 7×7 composite rope based on this process is shown in Fig. 3.

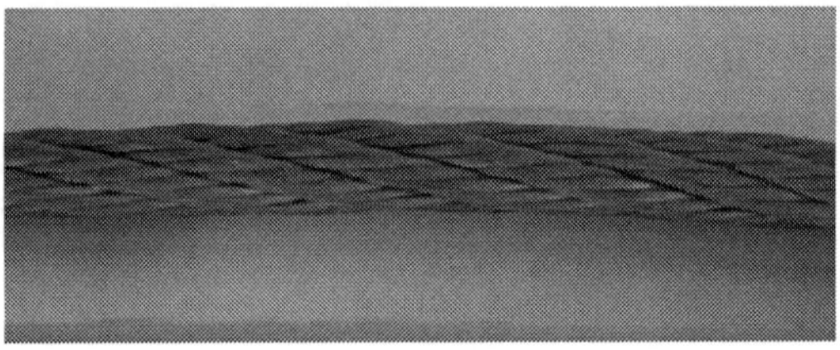

Figure 3. Example of a composite rope – 7×7 construction.

Tensile Testing

With the exception of 7×7 construction, tensile strength measurements were preformed on Admet eXpert universal test machine. These samples were tested by wrapping the samples around a capstan to secure the ends with a crosshead speed of 10 in/min.

The 7×7 composite ropes were tested with a spliced eye in each end attaching to 2" diameter pins on a Rhiele horizontal test machine. To set the splice, the samples were precycled 10 times to 50% of their estimated breaking strength then pulled to destruction, following standard spliced synthetic rope testing procedures [6]. The test was run at a crosshead speed of 13 in/min at room temperature.

RESULTS AND DISCUSSION

Effects of Curing Temperature and Time on Towpreg Strength

To find the optimal processing condition, towpregs with different curing temperatures and times were prepared in the lab oven as described in the Experiment section, followed by determination of tensile strength at room temperature. The test results are shown in Fig. 4, where the optimal curing temperature is around 300°F. The towpreg achieved the maximum break strength when cured for 5 to 15 minutes. At lower temperatures the resin was not fully cured, so the towpreg had lower strength.

A prolonged cure time or higher curing temperature degrades the resin, consequently weakening the towpreg. A thermogravimetry study was conducted and revealed that the resin began to degrade at 350°F, reaching a maximum degradation rate at 411°F. This observation is consistent with the strength curve in Fig. 4.

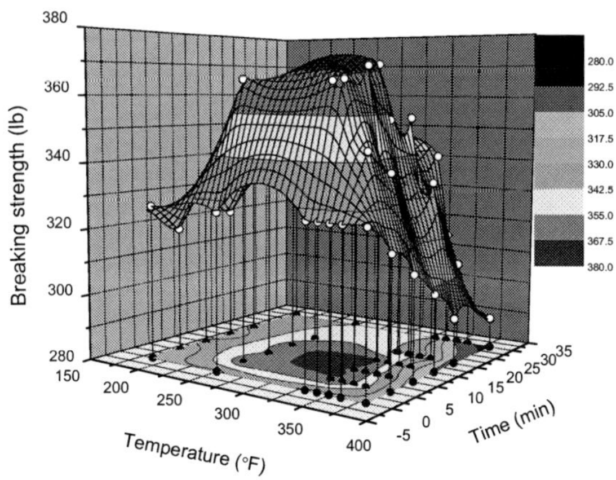

Figure 4. Effects of temperature and time on breaking strength of the towpreg samples.

With the information extracted from the curing time/temperature model, an in-line curing device was designed to study the effect of towpreg-processing conditions on towpreg strength. Figure 5 shows the effect of the curing temperature on towpreg strength. The towpreg was twisted at one turn per inch (TPI) and was cured for 5 minutes under 8 lbs of tension. The twisted towpreg had the maximum breaking strength at the tube around 300°F. The dependence of the twisted towpreg strength on the curing temperature follows the same trend as the untwisted towpreg.

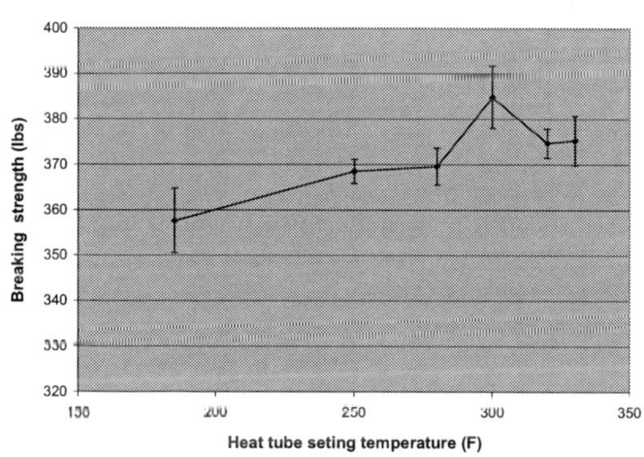

Figure 5. Twisted towpreg strength and its standard deviation under different curing temperatures.

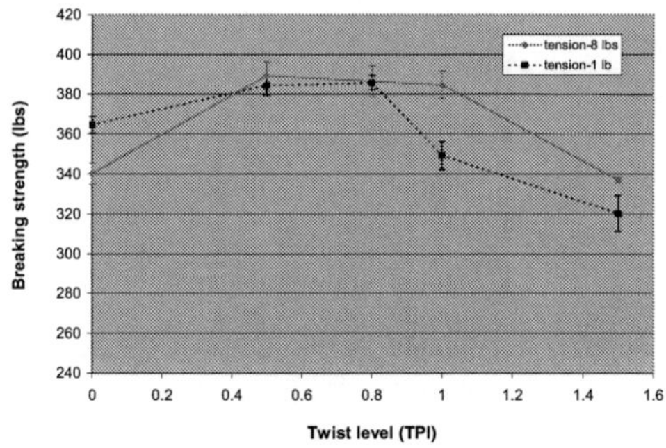

Figure 6. Tensile strength and its standard deviation of the towpreg with twist levels. Curing condition = 300°F for 5 minutes.

Effect of Towpreg Twist Level on Its Strength

It is well known that maximum yarn strength can be achieved at a certain optimum twist level. For Technora yarn, the maximum strength was achieved at about 1.9 TPI.[7]. Theoretically, the maximum breaking strength of a yarn would be expected if all the filaments aligned parallel to the yarn axis have the same strength and breaking elongation. However, due to the variations of individual filament strengths and uneven load sharing, the weak filaments have an overall lower strength than expected for an untwisted yarn. For a twisted yarn under tension, the transverse force generates frictional forces among the filaments and prevent the filaments from slipping, thus forming a whole monolithic functional yarn, where the weak filaments get support from neighboring strong filaments. A maximum tensile strength of the yarn can be reached when these factors reach an ideal combination.

Figure 6 shows the towpregs' breaking strengths at different twist levels. Compared with yarns consisting of only filaments, the towpreg achieves its maximum strength at a lower twist level.

Figure 6 also illustrates the effects of tension on towpreg strength during curing. In general, we saw that tension in curing helped the strength. However, this is not the case for untwisted towpregs. This is possibly caused by some separations between the filament and resin under higher tension, consequently reducing the level of load transfer by the resin. For towpregs with a higher twist level, we speculate that higher tension generated higher internal forces caused by twist, preventing the separation between the filaments and resin to result in more frictional forces, which helped the load transfer among the yarns.

Effects of Fiber Content on the Strength of Towpreg and Its Composite Rope

Figure 7 shows the strengths of various samples of different fiber contents. As expected, higher fiber content leads to higher strength. While it is expected that a linear relationship exists between the composite rope's tensile strength and its fiber content, it has a limit. At very high-fiber loading, there will be "dry spots" in the structure where there is not enough resin to fill the gaps between fibers, leading to poor load transfer, consequently reducing the overall strength of the structure[8–10].

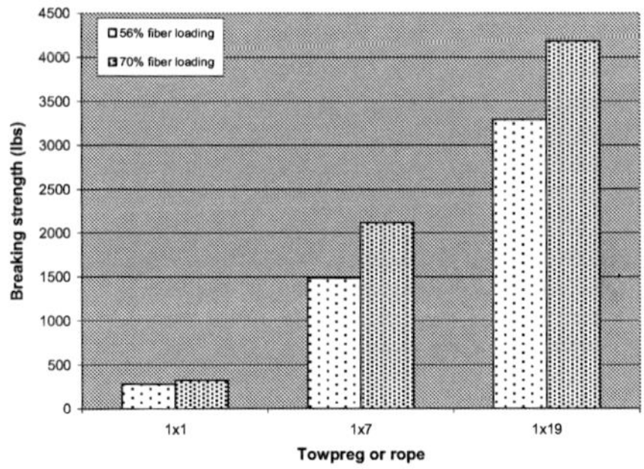

Figure 7. Strengths of two different fiber-content towpregs and their composite rope.

General Comments about Other Factors

In addition to the variables discussed so far, the strength of the towpreg and its composite ropes also depend on many other factors such as fiber type, resin chemistry, fiber/resin interface, constructions, and fiber arrangement. For example, a strong interfacial shear strength between the fiber and the resin can effectively transfer the load between fibers; hence, yielding a higher tensile strength. There are also many techniques such as fiber sizing, plasma treatment of fiber surface, chemical grafting, etc., to engineer the adhesion between the fiber and resin, but these are beyond the scope of this study.

Comparison of Weight and Strength Between Wire and Composite Ropes

Table 1 compares the tensile strength and linear weight of experimental samples with typical wire ropes [11–12]. Composite ropes have similar strength compared with wire rope, but are much lighter in weight. For all constructions, the weight of composite rope is 17–22% of the wire rope.

TABLE I. STRENGTH AND WEIGHT OF WIRE AND COMPOSITE.

Rope properties		Constructions of composite/wire ropes			
		towpreg/wire	1×7	1×19	7×7
		diameter, mm (inch)			
		1.1 (3/64)	3 (1/8)	5 (3/16)	10 (3/8)
Breaking strength, lb	Technora composite rope	340	2113	4813	10905
	Wire rope	375	2100	4410–4700	12000–14400
Weight per 100 ft, lb	Technora composite rope	0.082	0.610	1.719	4.533
	Wire rope	0.55	3.5	7.8–8.3	24.3

CONCLUSIONS

We successfully produced prototype composite rope samples with towpregs. These ropes are as strong as wire cable but more flexible and lighter, showing great potential to replace wire.

The initial study of the various design and process variables is completed, providing good methodology for product and production optimization in the next phase to develop the technology.

ACKNOWLEDGMENTS

The authors would like to thank Amy Solomon of Teijin Aramid, Ltd for her help in providing Technora fiber, and Anthony DeLima of ForteRTS for preparing the towpregs.

REFERENCES

1. Honda, K., 1987. U.S. Patent 4,677,818.
2. Matisda, S., 1991. U.S. Patent 5,060,466.
3. Ikeda, T., 1992. U.S. Patent 5,130,193.
4. Edwards, C., 2001. "Thermoplastic Pultrusion Promises New Synergies," *Reinforced Plastics*, 45(4): 34–39.
5. Lohmiller, J., 2004. "Cyclics Thermoplastics—New Possibilities in Composites," in *Proceedings IVW-Kolloquium*, Kaiserslautern, Germany, 5–6 October, pp. 76–80.
6. CI1500, "Testing Methods for Fiber Rope," Cordage Institute International Standard, May 2006.
7. Teijin Techno Products Limited Web site. http://www.teijin-technoproducts.co.jp/HP20070903/eng/tech_bussei_sonota.htm.
8. Hill, C.A.S. and A. Khalil, 2000. "Effect of Fibre Treatments on Mechanica Properties of Coir or Oil Palm Fiber Reinforced Polyester Composites," *J. Appl. Poly. Sci.*, 78(9): 1685–1697.
9. Bansal, N., July, 1996, "Effects of fiber content on mechanical properties of CVD SiC fiber-reinforced strontium Aluminosilicate Glass-Ceramic composites," NASA Technical Memorandum 107273.
10. Taha, I. and G. Ziegmann, 2006. "A Comparison of Mechanical Properties of Natural Fiber Filled Biodegradable and Polyolefin Polymers," *J. Compos. Mater.*, 40(21):1933–1946.
11. Bergen Cable Technology Web site. http://www.bergencable.com.
12. Ronstan Web site. http://www.ronstan.com/catalogue/A010-011.pdf.

TOPIC IV

FINITE ELEMENT MODELING OF TEXTILE COMPOSITES

The Source of Unrealistic Predictions of Damage Evolution in Textile Composites

L. GORBATIKH, D. IVANOV, S. LOMOV and I. VERPOEST

ABSTRACT

The modelling of damage evolution in composite materials normally requires two criteria: one is to identify failure in the undamaged material and one is to establish subsequent failure in already damaged material. However, in the materials models based on the continuous damage mechanics approach (CDM) the second criterion is often assumed to be the same as the criterion of failure initiation. The CDM approach with a single criterion is shown to predict physically unrealistic damage accumulation when applied to textile composites (such as unnaturally wide damage zones or damage evolution across yarns). The source of the unphysical damage evolution may be attributed to the fact that the principle of strain equivalence imposed by CDM is generally applicable only when damage consists of a very large number of defects that are more or less uniformly distributed throughout the considered volume. The latter is not the case in textile composites where no multiple microcracking is observed on the meso-level.

In order to shed some light on whether the replacement of a material with a crack-like defect by a fictitious continuous material with degraded properties is adequate for modelling of damage evolution, we make a qualitative comparison of failure initiation around a crack-like defect and around a fictitious inhomogeneity with particular interest in the failure direction and the failure mode. We also investigate whether the Murakami-Ohno model adopted for calculation of properties of the fictitious inhomogeneity contributes to the unphysical results.

INTRODUCTION

Damage accumulation in textile composites on the meso level is commonly accompanied by brittle fracturing in the form of intra-yarn cracking that

Department of Metallurgy and Materials Engineering, Katholieke Universiteit Leuven, Heverlee (Leuven), 3001 Belgium precedes the onset of debonding at yarn/matrix interfaces and interply delamination ([1-3]). Finite element models developed to simulate this stage of damage evolution, however, avoid direct analysis of crack nucleation and crack propagation. Instead, the modelling approach is based on two ideas: (1) failure in yarns is identified from *local stress analysis* where yarns are treated as an orthotropic medium with homogenized properties of a unidirectionally reinforced composite and (2) the problem of cracking is replaced by a fictitious problem of material property degradation following the *continuous damage mechanics* approach. According to the latter a material with defects (damaged/failed material) is replaced by a fictitious continuous material with no defects but which elastic constants are degraded. Degradation of properties is normally done according to one of the postulated procedures (for example, according to [4-6]). The method is often referred to as a property degradation method or a stiffness reduction approach.

Among commonly used criteria for material failure in fibre reinforced composites are maximum stress criterion, Hoffman criterion, Tsai-Hill criterion, Tsai-Wu criterion, Puck criterion etc [7,8]. When used in the meso-FE modelling, the chosen criterion determines elements where damage initiates and in some cases also provides information on the mode of failure from analysis of stress-to-strength ratios. Particulars of the occurred damage remain, however, unclear, as criteria do not specify whether damage occurs due to a single or multiple fractures; whether the latter are uniformly spread or localized etc.

The CDM-based methods have been quite popular for estimation of how accumulated damage changes stress-strain curves. Recent investigations on the subject include works [10-17]. Models presented in the literature are mainly for composites with orthogonal setting of yarns and that are loaded in the fibre direction. This is important to note as no significant shearing is generated in such models. In the previous works [18,19] the current authors incorporated the CDM approach to analyze damage evolution in a triaxial braided composite under tensile loading where at least one family of yarns was oriented at 45° to the loading direction (Fig. 1a). A summary of the used meso-FE modelling can be found in [20,21] that provides a road map for validation experiments. In spite of the significant effort in computational simulations, validation results indicate a number of inconsistencies between CDM modeling predictions and experimental observations. For example, tested on the triaxial braided composite, the model predicts unnaturally wide damage zones (Fig.1b) opposed to the absence of extensive microcracking in experimental setting ([16,18]). The current work is geared toward identifying the source of inconsistencies between experiment and modelling by revisiting fundamentals of the adopted damage model.

ASSUMPTIONS OF CDM AND ITS APPLICATION TO MODELLING OF DAMAGE IN TEXTILES

As noted in the literature, damage evolution predicted by the property degradation approach must be understood in a certain sense, namely, that the damaged material is equivalent to the material with degraded properties (fictitious material). These two are often used as synonyms while their equivalence depends on the problem at hand. Another related remark is that the

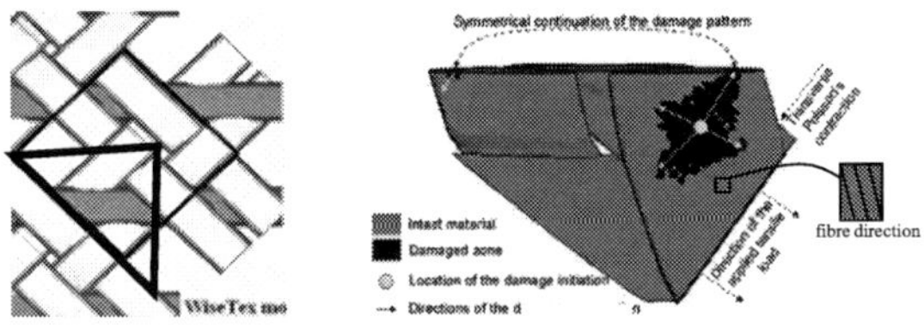

(a) (b)

Figure 1. (a) Schematics of the triaxial braided composite; (b) FE modelling artifact: damage calculated according to the Murakami-Ohno scheme does not propagate along fibres.

damage pattern obtained from CDM modelling does not necessarily reflect the way intra-yarn cracks would grow in the material [9]. Since damage in textiles on meso-level is accumulated through discrete fracturing (intra-yarn cracking) the physical meaning of the damage pattern obtained using the CDM approach becomes inconclusive.

The basic hypothesis of CDM is the principle of strain equivalence. It states that the loss of material properties due to gradual irreversible changes in the material microstructure (such as formation of microdefects) can be recorded through a certain damage parameter (scalar or tensor depending on the application) and that the damaged volume of the material under nominal stress shows the same strain response as a comparable fictitious undamaged volume under effective stress ([22]). With the use of CDM it is assumed that the degradation of the microstructure occurs volume-wide (i.e. spread through the volume) and that the damage tensor is representative of that volume wide degradation. This principle is generally applicable when damage consists of a very large number of defects that are more or less uniformly distributed throughout the considered volume. It means that material damage must be statistically homogeneous on the considered scale. With clustering of cracks or formation of a macroscopic defect, the essential assumption of CDM breaks down. The concept of CDM is originally intended for modelling of damage accumulation in the material prior to formation of detectable defects. It, however, has become quite popular for modelling of damage evolution in textile composites where the above assumptions do not necessarily hold.

A TEST PROBLEM OF A CRACK AND AN EQUIVALENT FICTITIOUS INHOMOGENEITY

Let us assume that the failure criterion is satisfied in one of the mesh elements and that failure is attributed to a single crack-like defect (dotted line in Fig. 2a). According to the modelling procedure described in the beginning of the paper this element is declared damaged and is replaced with a fictitious element of the same geometry, which elastic constants are degraded. Elastic constants are calculated according to the chosen property degradation model, for example, Murakami-Ohno. The damaged element becomes an inhomogeneity (understood in Eshelby's sense [23]). In order to evaluate whether the replacement of a material with a crack-like defect by a fictitious continuous material with degraded properties is adequate for modelling of damage evolution, a qualitative comparison must be made of the damage onset around a crack-like defect and around an inhomogeneity (Fig. 2). Particular emphasis is placed on the damage direction and the failure mode.

We calculated average stresses around the inhomogeneity (averaging is done over segments with $\Delta\theta = \pi/8$, $\Delta R = 2L$ as shown in Fig. 3a). The data is considered for the case of a carbon/epoxy unidirectional composite. Taking into account that material strengths F_1^t and F_{12}^s are normally higher than F_2^t (in the case of a carbon/epoxy unidirectional composite it is about 45 times for F_1^t and 3 times for F_{12}^s), we observe that:

(1) under σ_{22}^0, the dominant stress (stress that defines failure mode) is σ_{22}. It reaches maximum in the vicinity of $\theta = 0$ and π. If the maximal stress criterion is used these points determine direction of further damage evolution (Fig. 3b). The mode and direction of failure are consistent with the ones followed from the crack problem: under transverse tension a crack propagates in its plane;

(2) under σ_{12}^0, the dominant stress is to be determined from competition between stress-to-strength ratios σ_{22}/F_2^t or σ_{21}/F_{21}. Since for some materials these values can be quite close two scenarios of further damage evolution are possible. If σ_{21}/F_{21} dominates, it reaches maximal value in the vicinity of $\theta = \pi/2$ and $3\pi/2$. If σ_{22}/F_2^t dominates, its maximal value is observed in the vicinity of $\theta = \theta^*$ and $\theta^* + \pi$. The exact value of θ^* depends on properties of the material. These angles are to direct further damage evolution if the maximum stress criterion is used (Fig. 3c). Both scenarios contradict to the predictions followed from the crack solution, namely that a crack propagates in its plane. This would translate into $\theta = 0$ and π for an inhomogeneity problem. Thus, physics of the original problem is distorted.

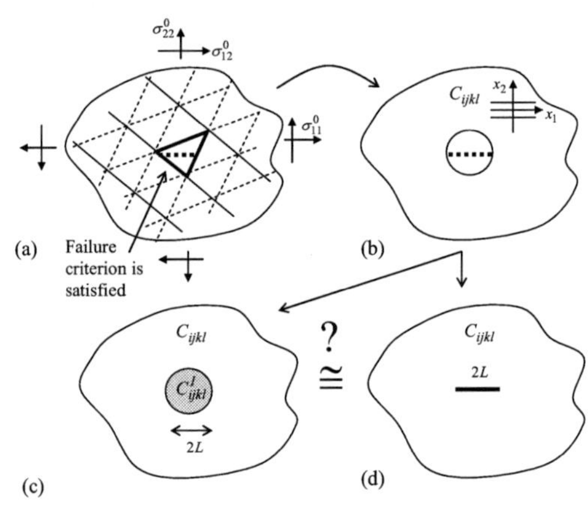

Figure 2. (a) An element where the stress criterion for material failure is satisfied (the dashed line indicates failure/crack direction); (b) a failed element modelled as a circular inclusion with damage; (c) a problem of an inhomogeneity with reduced elastic constants; (d) a problem of a crack-like defect.

This simple exercise illustrates that replacement of an element that has a crack-like defect by a fictitious inhomogeneity that has degraded elastic constants predicts a proper mode and direction of failure only when loads are normal to the crack surface. Such replacement in combination with the failure criterion formulated in stresses becomes inadequate under shear loading as it predicts unphysical direction of damage evolution as well as an incorrect mode of failure.

MURAKAMI-OHNO DAMAGE MODEL AND ELASTIC PROPERTIES

Stress fields around an inhomogeneity depend on several factors, including properties of the surrounding medium (a homogenized UD composite with orthotropic properties that are given *a priori*), the shape of the inhomogeneity (dictated by the finite element mesh) and properties of the inhomogeneity (calculated according to the chosen property degradation model). While the first two factors generally cannot be controlled, the latter one is postulated. In this section we investigate whether the model we adopted to calculate properties of the fictitious inhomogeneity may be responsible for the unphysical results described earlier.

The idea behind Murakami-Ohno's assumptions is simple: a crack cannot withstand loads in normal and shear directions therefore stiffnesses of the fictitious material in these directions are nullated. It is, indeed, true if an element has a crack-like shape ($x_2 \to 0$). If, however, its x_2-dimension differs from zero (for example, in the case of a square, a rectangle, a circle, etc) the error of such approximation may become significant. According to the general concept of CDM a damaged volume must be replaced with the fictitious volume that has the same effective elastic response. In this section we compare the change in the effective elastic response of the material due to the introduction of a crack and an inhomogeneity with properties that give the same compliance constribution as a crack.

Figure 4 illustrates the change in the material elastic properties under transverse tension and in-plane shear due to the presence of an inhomogeneity in comparison with the case of a crack (namely H^I_{2222}/H^C_{2222} and H^I_{1212}/H^C_{1212}) as a function of material properties (E_2/E_1), where components H^I_{ijkl} are introduced in [24, 25] and E_i is Young's modulus of the orthotropic matrix in

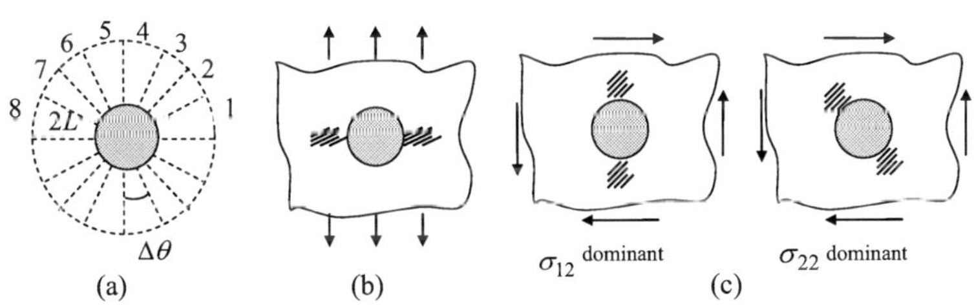

Figure 3. (a) Segments around an inhomogeneity with $\Delta\theta = \pi/8$ and $\Delta R = 2L$ for stress averaging; (b) direction of damage evolution for the case of normal loading; (c) Possible scenarios of damage evolution for the case of shear loading.

ith direction. The following observations are made:

- H_{1212} is overestimated more than 8 times for the case of an isotropy. If the property mismatch in x_1 and x_2 directions broadens, the discrepancy increases, reaching 31.4 times for the case $E_2 = 0.01E_1$.
- H_{2222} is overestimated more then twice for the case of an isotropy and it reaches 2.8 times for the case $E_2 = 0.01E_1$.
- H_{1122} and H_{1111} are not zero for the inhomogeneity as they are for the crack.

Summing up, Murakami-Ohno's damage model overestimates degradation of properties due to damage. Even though all components H_{ijkl} are affected, the impact is particularly profound on the shear constant H_{1212}. Inadequate modelling of shear properties of a fictitious inhomogeneity may explain why damage evolution is predicted incorrectly under shear. In the following section we introduce modifications to Murakami-Ohno's degradation model to see whether the results in the previous section can be improved.

CONCLUSIONS

The principle of strain equivalence imposed by CDM is generally applicable when damage consists of a very large number of defects that are more or less uniformly distributed throughout the considered volume (no clustering of cracks or formation of a macroscopic defect is allowed). This assumption does not necessarily hold for the type of damage observed in textile composites on the meso-level. Failure in volumes corresponding to finite elements appear more as a crack-like defect than volume-wide formation of multiple microcracks.

In order to shed some light on whether the replacement of a material with a crack-like defect by a fictitious continuous material with degraded properties is adequate for modelling of damage evolution, we made a qualitative comparison of failure initiation around a crack-like defect and around a fictitious inhomogeneity with particular emphasis on the failure direction and the failure mode. This simple exercise illustrates that while the model predicts proper failure mode and direction under transverse tension, it becomes inadequate under shear loading.

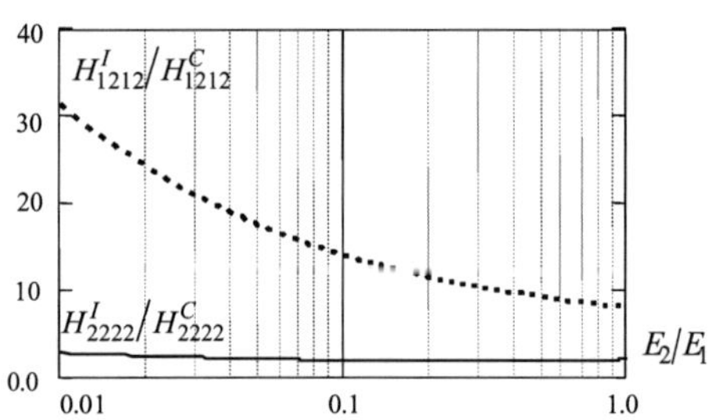

Figure 4. The change in material compliance due to the presence of an inhomogeneity normalized by compliance of a crack.

We also investigate whether the Murakami-Ohno model adopted for calculation of properties of the fictitious inhomogeneity contributes to the unphysical results. For this we compare contributions into material elastic response done by the crack and the inhomogeneity. It is illustrated that the Murakami-Ohno model significantly overestimates degradation of properties of the fictitious inhomogeneity (especially for the shear constant).

A new property degradation procedure was suggested in the place of the Murakami-Ohno model (referred as the effective elastic response model) [24]. According to the latter, elastic constants of the fictitious inhomogeneity are calculated in such a way that its contribution into material elastic response is identical to the contribution of a crack. In contrast to the Murakami-Ohno model, properties of the inhomogeneity in the EER model depend on the size of the inhomogeneity and properties of the surrounding material. From local failure analysis we show that the failure mode is now clearly established under shear loading (this was not the case when the Murakami-Ohno model was used). The failure direction, on the other hand, cannot be clearly identified. This can be an indicator that the failure criterion formulated in stresses is not sufficient to determine direction of damage evolution and that an additional criterion is required. The unphysical results seem to come not from the degradation procedure itself but from the original concept of using the CDM approach for modelling of damage that is not volume-wide formation of multiple cracks.

ACKNOWLEDGEMENTS

Support of the European Commission (ITOOL project) and the Research Council of K. U. Leuven (OT project "Concurrent multi-scale design/ engineering of textile composites" and East-European PhD grant of D.S.Ivanov) is gratefully acknowledged.

REFERENCES

1. Lomov S.V., Baudry F., Ivanov D.S., Truong T.C., Verpoest I., Vettori M., Xie H. Experimental methodology of study of damage initiation and development in textile composites, Proceedings of 27th int. SAMPE/JEC conf., PARIS, March 27-29, 2006, pp. 21–26.
2. Ivanov D., Lomov S.V.,Verpoest I., Baudry F., Xie H. Damage initiation and development in triaxial braid and fine structure of damage, Proceedings of 12th European. conf. on Composite Materials, Biarritz, France, Aug. 29–Sept. 1, 2006, CD-edition (N 160).
3. Edgren F., Mattsson D., Asp L.E., Varna J. "Formation of damage and its effects on non-crimp fabric reinforced composites loaded in tension", Composite Science and Technology 64, 2004, 675–692.
4. Blackketter D.M., Walrath D.E., Hansen A.C., Modeling damage in a plain weave fabric-reinforced composite material, Journal of Composites Technology and Research, 1993, 15 pp. 136–142.
5. S. Murakami, N. Ohno, A continuum theory of creep and creep damage. Creep in Structures, Springer-Verlag, 1981, pp. 422–443.
6. Murakami, S. Mechanical modeling of material damage. Journal of Applied Mechanics, 1988, 55: 280.
7. Failure criteria in fibre reinforced polymer composites: The World-Wide Failure Exercise, ed. M.J. Hinton. 2004: Elsevier Science Ltd.

8. Kaddour, A.S. and M.J. Hinton, The launch of WWFE-II: Benchmarking of 3D strehgth and damage theories for polymer fibre composites, in Proceedings ICCM-15. 2005: Durban. CD ed.
9. Zhao L.G., Warrier N.A., Long A.C. Finite element modelling of damage propagation in non-crimp fabric reinforced composites, Composites Science and technology 66, 2006, pp. 36–50.
10. Takeda, T., Takano S., Shindo Y., Narita F., Deformation and progressive failure behavior of woven-fabric-reinforced glass/epoxy composite laminates under tensile loading at cryogenic temperatures, Composites Science and Technology 65, 2005, pp. 1691–1702.
11. Tabiei A., Ivanov I., Materially and geometrically non-linear woven composite micro-mechanical model with failure for finite element simulations, International Journal of Non-Linear Mechanics 39, 2004, 175–188.
12. Scida D., Aboura Z., Benzeggagh M.L., Bocherens E. A micromechanics model for 3D elasticity and failure of woven-fibre composite materials, Composites Science and technology 59, 1999, pp. 505–517.
13. Nicoletto G., Riva E., Failure mechanisms in twill-weave laminates: FEM predictions vs. experiments, Composites: Part A 35, 2004, pp. 787–795.
14. Takano N., Uetsuji Y, Kashiwagi Y., Zako M, Hierarchical modelling of textile composite materials and structures by the homogenization method, Modelling Simulation Materials Science Engineering, 7,1999, pp.207–231.
15. Hoffman, O. The brittle strength of orthotropic material. Journal of Composite Materials, 1967; 1(2): 200–2006.
16. Tang X., Whitcomb J. "Progressive failure behaviours of 2D woven composites" Journal of Composite Materials, 37, No 14, 2003, 1239–1259.
17. Zako M., Uetsuji Y., Kurashiki T. "Finite element analysis of damaged woven fabric composite materials" Composite Science and Technology, 63(3-4): 2003, 507–516.
18. Ivanov D., Lomov S.V.,Verpoest I., Baudry F., Xie H. Damage initiation and development in triaxial braid and fine structure of damage, Proceedings of 12th European. conf. on Composite Materials, Biarritz, France, Aug. 29 –Sept. 1, 2006, CD-edition (N 160).
19. Lomov S.V., Ivanov D.S., Verpoest I. Meso-FE modelling of 3-axial braided composites Proceedings of 8th International Conference on Textile Composites, Nottingham, 2006, Oct. 16–18.
20. Lomov S.V., Ivanov D.S., Verpoest I., Zako M., Kurashiki T., Nakai H., Hirosawa S. "Meso-FE modelling of textile composites: Road map, data flow and algorithms" Composites Science and Technology, in Press, available online 1 December 2006.
21. Lomov S.V., Bernal E., Ivanov D.S., Kondratiev S.V., Verpoest I. "Homogenisation of a sheared unit cell of textile composites: FEA and approximate inclusion model", European Finite Elements Review, 2005, 14 (6-7): 709–729.
22. J. Lemaitre, J. Eng. Mater. Technol. 1985, 107(1), 83.
23. Eshelby J. D. The determination of the elastic field of an ellipsoidal inclusion, and related problems. Proceedings of the Royal Society, 1957, v. A241, pp. 376–396.
24. Gorbatikh L., Ivanov D., Lomov S., Verpoest I., On modeling of damage evolution in textile composites on meso-level via property degradation approach, *Composites Part A*, 38/12 pp. 2444–2453, 2007.
25. Gorbatikh L., Lomov S. and Verpoest I. Elastic compliance of a partially debonded circular inhomogeneity, International Journal of Fracture, 2005, v. 131, n. 3, pp. 211–229.

FE Modelling of Textile Reinforced Composites and Comparison with Real Testing

P. VOZKOVÁ and J. SALAČOVÁ

ABSTRACT

The woven composites are prepared by a prepreg technology. Heat-treatment and pressure form the composite to the resultant shape. A deep micro-scanning of the two-dimensional fabric reinforced composite allows an evaluation of the multi-scales of the composite structure precisely in NIS Elements. The whole structure of a composite is defined for the 2D multi-scale modelling in the levels: matrix-fibres in lengthwise and crosswise direction → matrix-fibres part with pores in both directions → three types of unit cell raised from observing the real composite microphotographs → laminas for all types of unit cells → whole composite. Studied composite was made from E-glass fibres and polysiloxane matrix.

The multi-scale geometry is used for a finite element simulation of the elastic properties in a commercial program Comsol Multiphysics™. Appropriate material properties are set obtained from the experiment or the lower scaled computed model. The reinforcing fabric is tested with tensile test, bending test and pressure test. So because meso-scale is evaluated by experiment also, comparison between them follows.

Combination of experiment and simulation gives us verified data. It helps to interpret whole orthotropy in the composite plate. The dominant factors influenced elastic properties of the composite are position of the reinforced fibres and occurrence of defects in the structure. Presence of the fibres in the matrix is fundamental for the difference of the properties of the crosswise and lengthwise bundles. From the results it is clear there are the most important lengthwise fibre bundles for longitudinal modulus and crosswise fibre bundles for the shear modulus. The results correlate well with a theory.

Pavla Vozková, Technical University of Liberec, Faculty of Textile Engineering, Department of Textile Evaluation, Studentska 2, 461 17 Liberec, Czech Republic, EU
Jana Salačová, Technical University of Liberec, Faculty of Textile Engineering, Department of Textile Materials, Studentska 2, 461 17 Liberec, Czech Republic, EU

INTRODUCTION

Composites are top materials used for special applications. In place of traditional material, e.g. steel, the composites lower its weigh much more than the traditional material. It is based on low weight of a composite itself, on specific properties of the components and on creation possibility of composite with a specific use. The combination of the materials provides specific properties of a final composite, e.g. chemical resistance, electric and thermo-insulate properties.

A complex structure of the composite has to be adapted for computing. Some levels of a structure idealisation are used from simple brick models and weave idealisation to complex multi-scales. The finite element method is than used for calculation among others (Discrete Fourier transformation and different algorithms) [1, 2]. The finite element method is a numerical tool for a simulation process of stress, strain, heat flow and more for physical model. A continuous area is discretized to the countable number of units. It is possible to apply the simulation for a long-term testing and for a detection of critically loaded area. Tested area is covered with the net of triangles or quadrangles that cover surface without overlapping. Loading in the sample body is transferred to the equivalent loading in the net structure. Each element is taken as an independent area with some boundary conditions [3].

Basics of the Geometric Models

The volume element of a structure exists as a representative of the mechanical response for the whole composite body as an assumption. The representative volume element (RVE) is based on the type of reinforcement and on the scale. Fibres and matrix are always included in the representative volume element. A stiffness and a compliance tensor describe the representative volume element with a mean stress and a mean strain in the volume element of the material [4].

One dimension reinforced composites, e.g. reinforced with parallel fibres, which should be defined by an iso-strain model. For the model in multi-component system in loading strain of all components it is the same but there is no stress. The stiffness tensor C_s for system with n components is

$$C_s = \sum_{i=1}^{n} k_i C_i \text{,} \tag{1}$$

where C_i is the stiffness tensor of i component and k_i is the volume ratio of this component. Model of iso-stress has got the same stress for all components as an assumption but there is no strain. The compliance tensor S_s for this model should be

$$S_s = \sum_{i=1}^{n} k_i S_i \cdot \tag{2}$$

The compliance tensor is an inverse for the stiffness tensor. Both models are on the highest and on the lowest limits for elastic properties of the composites. Both models are also applied straight to the laminate with longitudinal infinite fibres [5]

Non-homogenous material is replaced by homogenous for the Hooke's law in mechanics of continuum. Basically is valid that the stress σ_{ij} and the strain ε_{ij} from each component with volume ratio are sums for total stress and strain according to

$$\left[\sigma_{ij}\right]=\sum_n V_n\left[\sigma_{ij}^n\right]$$
$$\left[\varepsilon_{ij}\right]=\sum_n V_n\left[\varepsilon_{ij}^n\right] \tag{3}$$

where V_n is the volume ratio of each component. From the tensors stress σ_{ij} and the strain ε_{ij} it is possible to count array of stiffness C_{ijkl}

$$\left[\sigma_{ij}\right]=\left[C_{ijkl}\right]\left[\varepsilon_{kl}\right] \tag{4}$$

Modules of elasticity are defined in this array. Inverse to stiffness array is compliance array with coefficients of deformation. Both arrays are tensors of the fourth grade [6, 7]. Matrix material is pressumpted isotropic and homogenous with a structure close to vitreous silica. Elastic constants are not changed during rotation around axes. Pores properties are also isotropic and do not accept gas compression [4]. Reinforcement is a homogenous and isotropic glass material [4, 8].

Composite from laminas is an orthotropic material with three axes of symmetry, i.e. warp, weft and cross to fabric. Nine independent constants are necessary to find [4]. Stiffness array is than

$$\begin{bmatrix}\sigma_1\\\sigma_2\\\sigma_3\\\sigma_4\\\sigma_5\\\sigma_6\end{bmatrix}=\begin{bmatrix}c_{11}&c_{12}&c_{13}&0&0&0\\c_{12}&c_{22}&c_{23}&0&0&0\\c_{13}&c_{23}&c_{33}&0&0&0\\0&0&0&c_{44}&0&0\\0&0&0&0&c_{55}&0\\0&0&0&0&0&c_{66}\end{bmatrix}\begin{bmatrix}\varepsilon_1\\\varepsilon_2\\\varepsilon_3\\\varepsilon_4\\\varepsilon_5\\\varepsilon_6\end{bmatrix} \tag{5}$$

and the compliance array is

$$\begin{bmatrix}\varepsilon_1\\\varepsilon_2\\\varepsilon_3\\\varepsilon_4\\\varepsilon_5\\\varepsilon_6\end{bmatrix}=\begin{bmatrix}s_{11}&s_{12}&s_{13}&0&0&0\\s_{12}&s_{22}&s_{23}&0&0&0\\s_{13}&s_{23}&s_{33}&0&0&0\\0&0&0&s_{44}&0&0\\0&0&0&0&s_{55}&0\\0&0&0&0&0&s_{66}\end{bmatrix}\begin{bmatrix}\sigma_1\\\sigma_2\\\sigma_3\\\sigma_4\\\sigma_5\\\sigma_6\end{bmatrix}. \tag{6}$$

MULTISCALE MODELLING

The multi-scale modelling allows decomposition of complex heterogeneous structure to the partial levels from the simplest composition to the composition of whole sample body. They are substituted from the lowest level properties to higher and also highered up to the higher studied composite structure. The accuracy of the simulation is better because specifications and properties from each scale are combined during processing.

A consideration about the structure of composites could be done. A simulation in this chapter presents the two-dimensional model of lengthwise and crosswise fibre bundles surrounded with matrix. Defects are mentioned here, too. It is necessary to include defects in the bundles, between laminas and voids raised during pyrolysis and layering. Whole structure is a combination of geometry formations.

The lowest scale of tested structure in a microstructure with fibres and matrix, q.v. fibre bundle on Figure 1. Properties and geometry are not changed during testing for fibres and matrix too. Fibres are circular. Structure is random with volume ratio around 60%. Proportions agree with the real material [9, 10].

Figure 1. Lengthwise and crosswise section of the fibre bundle (black– fibres, white – matrix)

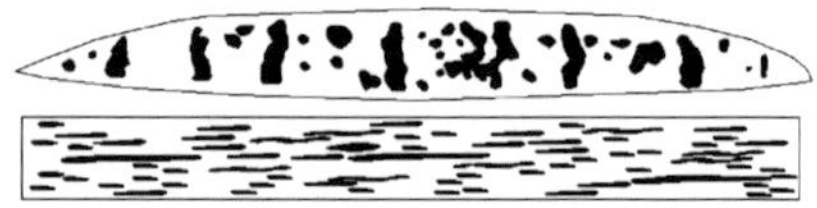

Figure 2. Lengthwise and crosswise section of the fibre bundle with pores [9] (black– pores, white – carbon part)

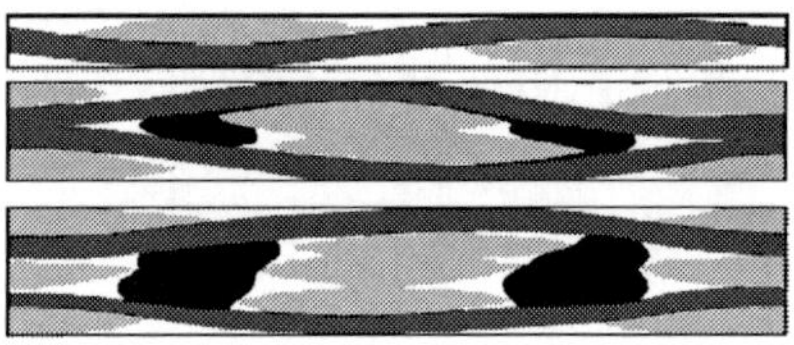

Figure 4. Structural model of Lamina [9]

Figure 3. Structural Unit cell 1,2,3 [9]

Figure 5. Whole composite structure [9]

(light grey – crosswise bundles, dark grey – lengthwise fibre bundles, black – voids, white – matrix)

Pores have risen during the production process in the fibre bundle. Figure 2. shows microstructure scale of whole bundle with pores inside the structure. Pores are filled with the air, if they are open. Closed pores contain pyrolytic gas but for simulation air was filled [9, 10].

Next scale level is a Unit cell as a meso-scale structure. Repeating of Unit cell forms whole plain weave fabric. On Figure 3. there are three Unit cells. The first is a simple Unit cell for plain weave. Two others are obtained from complete observation of microphotographs of real composite structure. These three Unit cells form lamina (Figure 4.). Paper [9] demonstrates that pores and cracks in the structure of Unit cell are necessary for precise computation. Defects rise in structure during fabrication as a lamina sliding and bridging of cross bundles [9, 10].

Laminas model whole composite structure on Figure 5. In the simulation lamina and composite structure are idealized as a bricks with identical properties in each counted from lower scale (unit cell → lamina bricks → composite bricks) [9, 10].

EXPERIEMENTAL EVALUATION

In the experiment there is a description of the results obtained from the simulation in the software Comsol Multiphysics^{TM} using FEM and PDE. The experiment is included with a partialy measured data in meso-scale level. Correlation from data from simulation and experiment and simmulation is presented.

The tested composite is six layered laminate from E-glass fibre in plain weave surrounded with vitreous silica matrix. Matrix was prepared from polysiloxane with high-temperature treatment. Matrix has got good size stability, thermo stability, chemical resistance, electrical insulation and good abrasion resistance. Bundle is made

TABLE I. VALUES OF IMPORTANT INPUT MECHANICAL PROPERTIES.

Component	**Density [kg.m^{-3}]**	**Young's modulus [GPa]**	**Poisson's ratio [-]**
E-glass fibres	2550	80	0,22
Polysiloxane matrix	1310	70	0,17
Voids - air	1,3	0,0	0,0

from parallel infinite e-glass fibres. Porosity is 20–30%. Important input properties for computation are in Table I.

The Structural Mechanical Module was used for computation in Comsol Multiphysics^{TM} 3.2. The Eigenfrequency Analysis was set as application. The structure is defined at first. The properties of components, the boundaries and the external impact are defined in Physics. The auxiliary constants and equations could be defined in Options. After Meshing and Solving the result graphs are shown in Postprocessing.

Real composite was used from three types differed by used pressure. Their properties and results from measurement are shown in Table II.

Dividing of stress and strain from each computed results gives us these values as critical elastic properties. The longitudinal modulus is different in one place value of lengthwise than crosswise bundle without defects. The difference is smaller for bundles with defects but it is still here. The unit cell is combined from various components so the difference is minimal between unit cells. Value is in the range from previous lower scales. The laminas and the whole composites modules are at same units as unit cells. The synergy effect increases modules slowly from the unit cell to the whole composite.

Table III. shows results from experimental evaluation of the real composite and simulated values from Periodic unit cells computed by software included experimental values. The thinest plate C has the highest results for all Unit cell types while plate A and plate B have got very balanced results. Compisitioned reinforcement with matrix is anyway one grade higher than the experimental warp and weft modules.

TABLE II. PROPERTIES OF REAL COMPOSITE PLATE.

Plate	**Used pressure [MPa]**	**Plate thickness [mm]**	**density [kg.m^{-3}]**	**Volume fraction of reinforcement [%]**	**Volume fraction of matrix [%]**	**Volume fraction of voids [%]**
A type	1,4	1,09	1500	47,5	22,23	30,3
B type	3	0,88	1770	58,83	20,48	20,7
C type	5,9	0,94	1660	55,07	19,85	25,1

TABLE III. RESULTS FROM MESO-SCALE.

Scale and type of the tested sample	Longitudinal modulus [GPa]	Longitudinal modulus plate A [GPa]	Longitudinal modulus plate B [GPa]	Longitudinal modulus plate C [GPa]
Experimental warp	1,64	-	-	-
Experimental weft	1,52	-	-	-
Simulated Unit cell 1	-	24,8	22,4	35,9
Simulated Unit cell 2	-	15,9	16,6	19,8
Simulated Unit cell 3	-	13,7	9,63	20,0

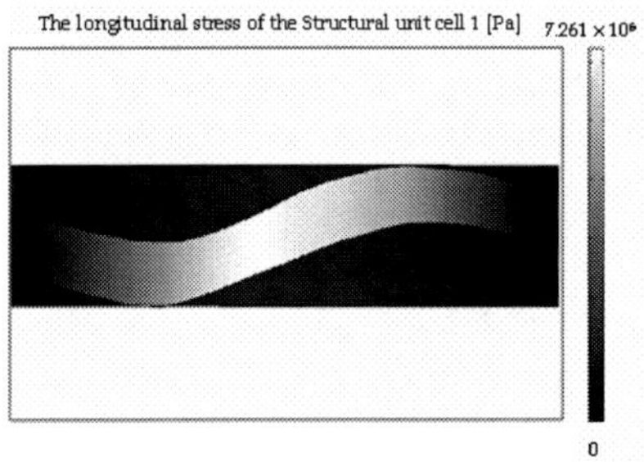

Figure 6. Longitudinal stress of PUC 1 type A.

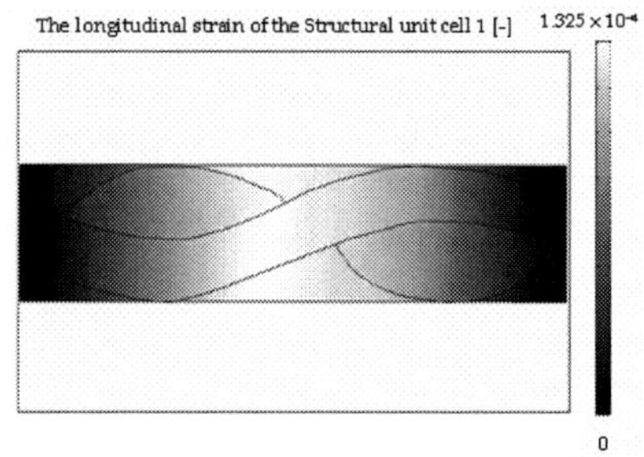

Figure 7. Longitudinal strain of PUC 1 type A.

The result response is defined for the first eigenfrequency of sample with free ends on Figure 6 and Figure 7. One node point is detected in the middle of sample. The free ends oscillate across along the lengthwise axis.The colour scale in the figure goes from minimal black through middle grey to the maximal white for value of output property in every case. On Figure 6 it is clear that the highest stress is concentrated in the lenghtwise bundle instead matrix and croswise bundles. In the crosswise bundles stress is occurred minimal but from the practise it is known some stress loaded by them so simulation is a bit idealised. Strain is holded by whole structure does not matter matrix and both fiber bundles. Results in plane are similar in all cases.

Table IV. shows longitudinal modules from simulation with plain compution and with partialy inserted values on meso-scale from experiment. All values are in tens of GPa. As a Unit cell results the highest modulus is occurred for plate C. Plate A has got the lowest longitudinal modules in both cases but the difference between plate B is not realy high.

TABLE IV. RESULTS FROM MACRO-SCALE.

Scale and type of the tested sample	Longitudinal modulus plate A [GPa]	Longitudinal modulus plate B [GPa]	Longitudinal modulus plate C [GPa]
Experimental inserted to simulation on meso-scale	29,4	31,5	56,8
Simulation on all scales	26,7	29,2	44,3

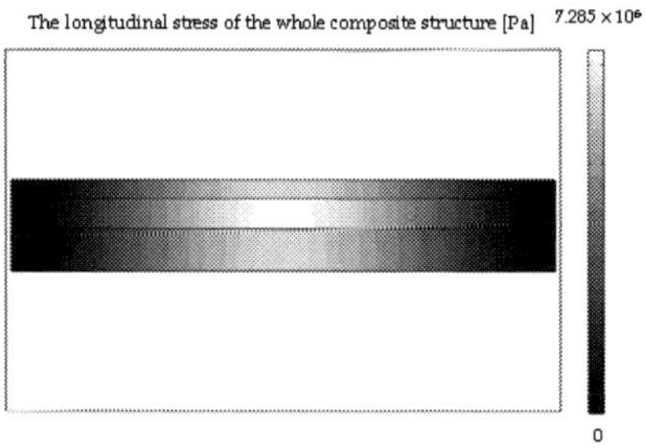

Figure 8. Longitudinal stress of the simulated composite type A.

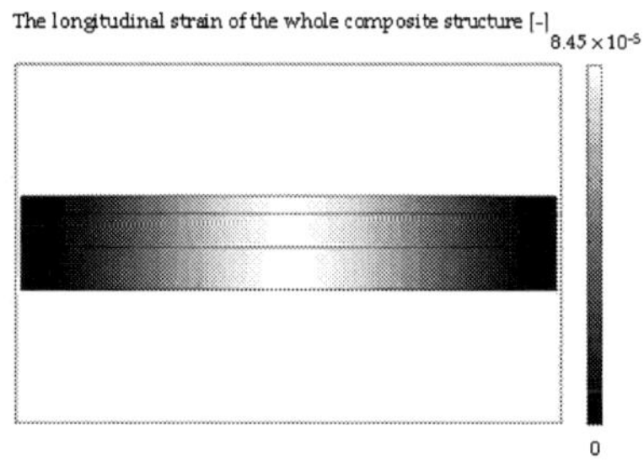

Figure 9. Longitudinal strain of the simulated composite type A.

Figure 8 and Figure 9 on the next page show example results how stress and strain are spread in plane of the sample body. Lamina from structural periodic unit cell 2 carry the highest loading in the composite. From practise it is known that it is not because it is in the middle but because of the structure of the unit cell. Lamina from unit cell 3 carry the lowest stress but difference from lamina 1 is small and statisticaly insignificant. Strain is the same for all laminas on the same possition from the ends. Results coralate well between Unit cell and whole composite body. Strain has got similar progression along. And stress in mostly occurred in the part which is from one end to other.

CONCSLUSION

The non-destructive testing is widely used for evaluation of the composites. The brittle composite is helpfuly to be tested with the non-destructive methods because critical loading is not occurred with plastic deformation. Simulations with the finite element method are fast and precise as possible. The commercial software Comsol Multiphysics™ was used for calculation with proper determination of the material properties.

The most influencing factor for the elastic properties is an amount of the lengthwise bundles for the longitudinal modulus. The position of the crosswise bundles with the matrix properties influences the shear modulus. The transversal modulus is mostly affected by the presence of the defects. Defects work upon all others modulus anyway.

The simulation was adapted well for the real composite. Combination of the multi-scale and the finite element simulation provide highly relevant data of the elastic modules which is confirmed by the used experiment. Data from experiment inserted on the meso-scale and from plain simulation correlate well.

The simulation with the multi-scale modelling is necessary to evaluate precise input values for the lowest scale. Here it is not problem with the fibre testing. Many methods are used and suitable. The matrix properties are more complicated to obtain. Testing of the matrix alone is a problem still and it should be solved soon.

Previous testing is possible to use for any material with the similar structure and the same prepreg technology with some modifications. If geometry models on the Structural unit cell scale are adapted, different technology should be evaluated with the same procedure too.

REFERENCES

1. D'Amato, E. (2001) Finite Element Modeling of Textile Composites. *Composite Structures*, Elsevier, No. 54, page numbers (467–475).
2. Reddy, J.N. (2006). *An Introduction to the Finite Element Metod.* Third Edition, Mc Graw-Hill Higher Education, ISBN 0-07-246685-5, New York, USA.
3. Hokr, M. (2006). *Application of computer modelling.* (In Czech), Study material, Technical University of Liberec, Liberec, CR.
4. Berthelot, J.-M. (1999). *Composite Materials. Mechanical Behavior and Structural Analysis.* Springer-Verlag, ISBN 0-387-98426-7, New York, USA.
5. Bogdanovich, A.E.; Pastore, C.M. (1996). *Mechanics of textile laminated composites.* Chapman and Hall, ISBN 0 412 61150 3, London, UK.
6. Miravete, A. (2000). *3-D textile reinforcements in composite materials.* Book of contributors. Woodhead Publishing Ltd Cambridge England and CRC Press, ISBN 1-85573-376-5, Cambridge, UK.
7. Vinson, J.R.; Chou T.W. (1975). *Composite materials and their use in the structure*. Applied science publishers Ltd., London, UK.
8. Stříž, B. (2002). *Textile Mechanics II. Application of continuum mechanics.* (In Czech), Technical University of Liberec, Liberec, CR.
9. Tomková, B. (2006). *Modelling of Thermophysical Properties of Woven Composites.* Dissertation thesis, Technical University of Liberec, Liberec, CR.
10. Šejnoha M. (2007). *Homogenisation of heterogeneous materials—similarity between textile reinforcement and historical masonry constructions*. Lecture for Centre of quality and reliability of manufacturing, Technical University of Liberec, Liberec, CR.

Contribution to Multiscale Modeling of Textile Reinforced Composites

B. TOMKOVÁ

ABSTRACT

The paper presents the modeling approach enabling transformation of real structure of textile reinforced composites including important structural imperfections (e.g. slip of textile plies, textile reinforcement reduction, pores and cracks, etc.) into geometry model applicable to numerical simulations of composite material response to various loads and calculations of effective properties of studied specimen. The source of structural information are digital images of composite strucure.

INTRODUCTION

Analysis of textile reinforced composites (textile composites) using structural micrographs has produced a lot of information on their structure and morphology. Digital images as one of the best sources of raw structural data enable direct evaluation of the reinforcement, matrix and voids morphology. Since the beginnig some of the following questions have been solved: what geometric parameters of the structure should be measured and how could be acquired data processed to improve the phenomenological and theoretical analysis of processing, structure and properties of composite material.

One of the first contributions to the topic was presented by *Jortner* already in 1986 [4]. Author studied and quantified the structural imperfections in multi-directional carbon-carbons, and described the composite structure from a mini-mechanical viewpoint. Here the building blocks of the composite are carbon tows, the intertow matrix, and the macroporosity (shrinkage cracks, thermal-stress cracks and bubble-like pores).

Department of Textile Materials, Technical University of Liberec, Hálkova 6, 46117 Liberec, Czech Republic
Corresponding author: blanka.tomkova@tul.cz

Progress of imaging techniques has given support to the development in quantification of textile composite structures, composite geometry has been analysed on a scale ranging from meso- to nano- [10,11]. It has resulted into realization the description of reinforcement geometry using real coordinates offers a number of chances to improve structural models, however the research has discovered many problems accompanying the approach. That is why the structural micrographs are mostly used for acquisition of basic geometry parameters of unit cells representing the periodic structure of textile reinforcement. The resultant 2D or 3D composite structures are formed due to repeating of particular unit cell. The reinforcement geometry is mostly simplified using basic mathematical formulas (e.g. "sine model"), no matter if it describes the fabric geometry or the geometry of tow cross-sections [1,9].

The lack of presented models is they in detail describe the parameters of reinforcement but mostly ignore the other parts of composite structure (textile reinforcement reduction, porosity, etc.). For example, the porosity as an important part of textile composites is discussed in variety of publications [5,12], the methods for quantification of void systems are systematically developed, even the spatial arrangement of voids has been analysed with the aid of visualization techniques [7]. Still the complexity of these imperfections makes suitable implementation of void description into structural models quite difficult unless the void geometry is maximum simplified.

The objectives of the paper have been finding the way to transform maximum structural information acquired from digital images of composite specimens into representative geometry models applicable in futher calculations of material properties.

EXPERIMENTAL WORK

Two types of textile composites have been investigated in the experiment: first, carbon-fabric-reinforced carbon composite (C/C composite), second, unidirectional basalt-fiber-reinforced silicone oxycarbide ceramic (basalt/SiOC). As for C/C composite, the main focus has been on the influence of the porous structure on thermal properties of the material. While newly developed basalt/SiOC ceramics have been analysed with the view of their electrical properties as they are designed as promising heat resistant insulating materials.

C/C Composite—Geometry Model Formulation

The specimens have been composed of eight layers of plain weave carbon fabric embeded in carbon matrix. Carbon matrix is formed of phenolic resin due to pyrolysis of the polymeric precursor within the series of carbonization/densification processes succeeded by final graphitization. Total porosity of graphitized composite rises up to 30 %, the porosity of carbon tow varies from 10 to 15 % [13].

The cross-section cuts of composite plate have been scanned in two directions, direction parallel to warp and to weft of woven reinforcement. Structural micrographs were processed with laboratory image analysis system. Further analysis of micrographs, estimation of particular structural components,

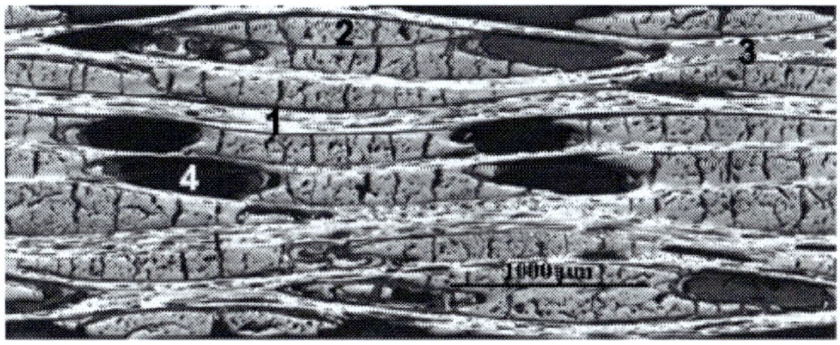

Figure 1. Basic structural components of woven reinforced C/C composite (resolution 1.7 μm/px): 1 - carbon tows, parallel cuts, 2 – carbon tows, cross-section cuts, 3 – carbon matrix, 4 – voids (optical microscope NIKON ME600 Eclipse).

and evaluation of structural parameters have been performed by using softwares LUCIA G, AdobePhotoshop, CorelDraw, Matlab R12 [13].

The reinforcement geometry has been described with respect to information presented in [6]. The shape of carbon tows in directions parallel and transverse to tow axis was derived from sets of coordinates acquired from parallel and cross-section cuts of the tows.

The analysis of void system followed [13]. First, the voids were described according to their shape and size. Considering the shape the voids were divided into four groups: extensive vacuoles (bubble pores), pores, transversal cracks, and delamination cracks. Under these observations the void systems were defined by the voids geometry and their position in the composite structure, see Figure 2.

The multiscale structural model is based on following assumptions:

- Parameters of filaments are uniquely determined and are nonchanging due to composite technology on nanoscale level.
- Matrix is forming into isotropic carbon due to composite technology, its structure and properties resemble those of glassy carbon [3].

Composite technology influences the following structural parameters: distribution of fibres in carbon tows, structure of tows in reinforcing fabrics, and ultimate amount of pores and cracks, that rise in composite structure due to phenolic resin impregnation, heat molding and pyrolysis of composite plate, respectively.

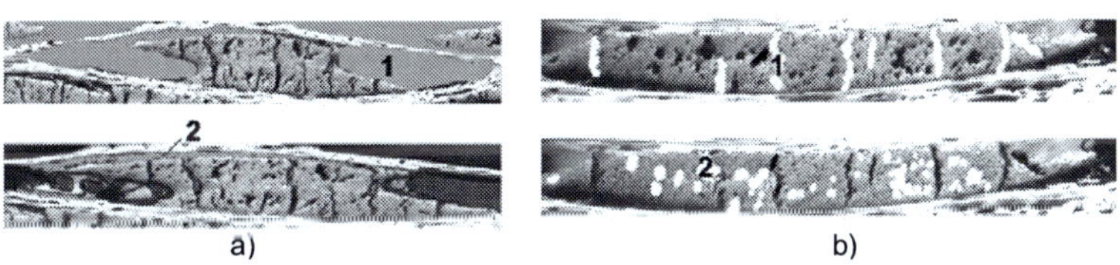

Figure 2. a) Extensive intertow voids: 1) extensive vacuoles, 2) delamination cracks, b) Intratow voids: 1) transversal cracks, 2) pores.

Basic structural components are defined successively on micro- (structure of carbon tows including intratow voids), meso- (woven reinforcement bonded with carbon matrix, including intertow voids), and macro-level (composite plate):

- Geometry of carbon tow - approximated by two-step geometry model to reflect random character of the voids distribution, see Figure 3.
- Composite unit cell - based on the assumption the carbon tows form periodic structure in woven reinforcement. Three types of periodic unit cells (PUCs) have been defined to reflect main parts of the composite structure, see Figure 4.
- Composite laminate - the laminae are formed due to translation of particular unit cell, see Figure 5. Number of PUCs per composite lamina has been defined with respect to size of composite specimen examined in experimental measurement. The composite laminate is formed as combination of selected laminae, see Figure 6.

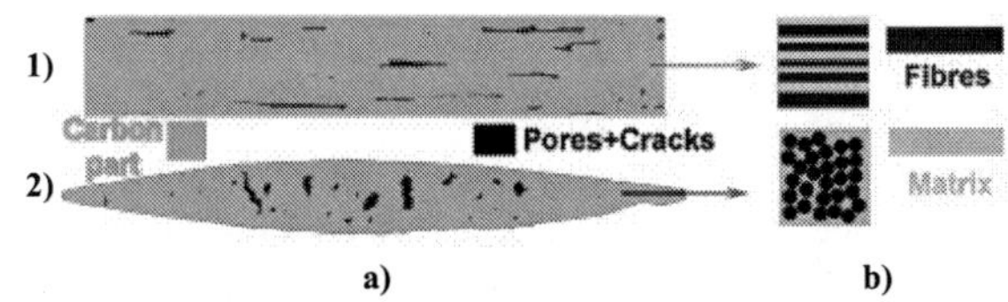

Figure 3. Two step geometry model of carbon tow structure: 1) parallel cut, 2) cross-section cut, a) carbon-void system, b) fibres-matrix system.

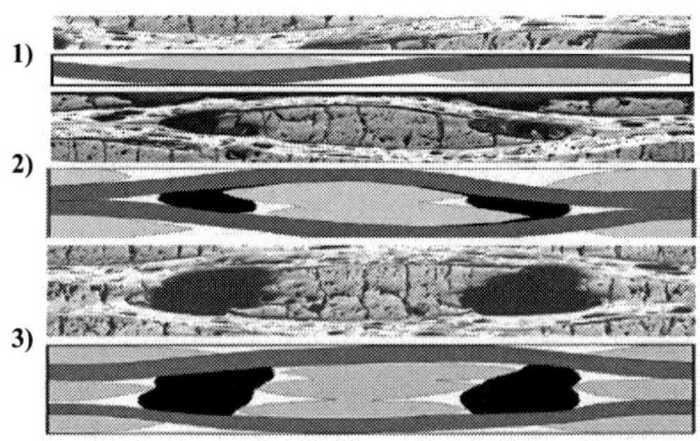

Figure 4. Periodic Unit Cells (PUCs) representing studied composite structure: 1) PUC1 composed of carbon tows and matrix only respecting the geometry of woven reinforcement, 2) PUC2 with vacuoles allied to delamination cracks representing the parts where the slip of textile plies occur, 3) PUC3 with extensive vacuoles representing the parts with the reduction of textile reinforcement due to bridging effect that appears in middle woven ply.

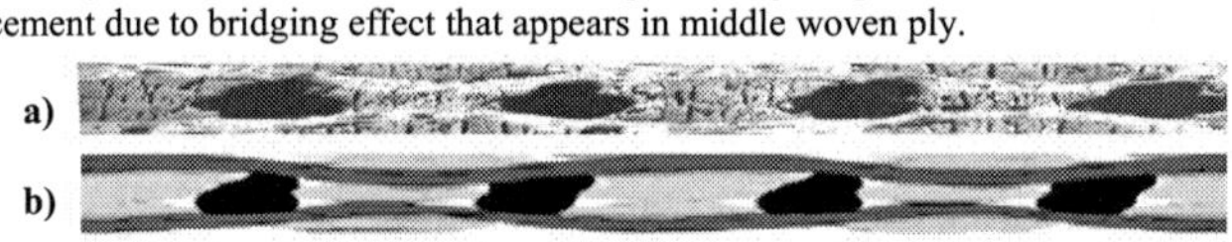

Figure 5. Structural model of composite lamina: a) micrograph of lamina, b) scheme of lamina geometry.

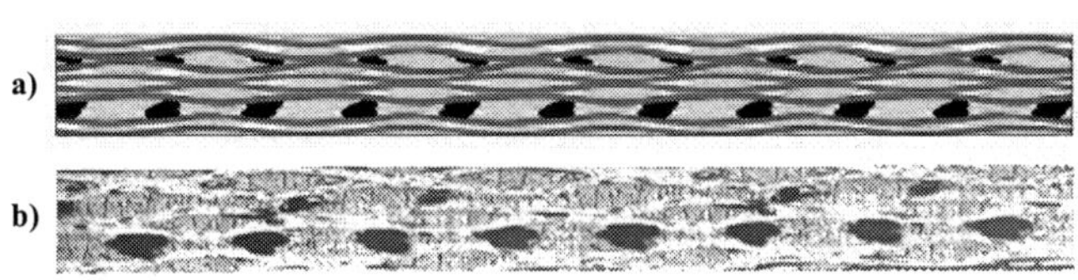

Figure 6. Scheme of generated composite plate(a) compared to the real one analyzed in structural micrograph (b).

Basalt/Sioc - Geometry Model Formulation

Specimens of basalt/SiOC composite have been reinforced with unidirectional basalt tows. Tows have been impregnated by silicone resin (polymethylsilicone resin dissolved in xylene). The specimens have been cured at the temperature of 100°C for 24 hours. The basalt-polysiloxane precursors have been pyrolysed in the air at heating rate 50°C per hour. Final temperature (heat treatment temperature) has been set on 1000°C. In the process of pyrolysis cured polysiloxane resin should be transformed into heat resistant porous silicon oxycarbide glass [2].

The mezo-structure of composite has been scanned by optical microscope and associated devices, see Figure 7. The analysis of microstructure has been carried out by use of scanning electrone microscope, see Figure 8.. The micrographs allowed complete quantification of the structure and morphology of the specimens. Multi-scale geometry model of specimen has been designed with respect to structural information obtained from composite micrographs, see Figure 9.

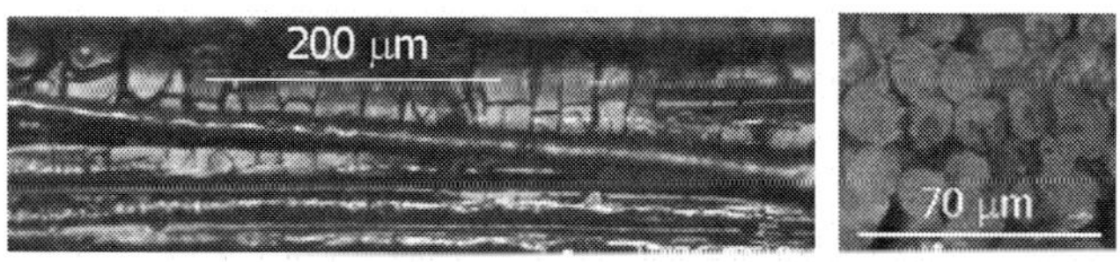

a) **b)**

Figure 7. Basalt fiber reinforced silicon oxycarbide: a) surface view, b) cross-section view (optical microscope NIKON ME 600 Eclipse).

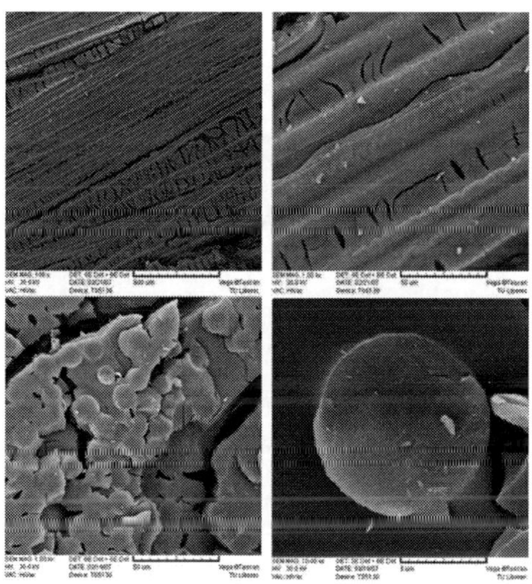

Figure 8. Surface and fracture area of basalt fiber reinforced silicon oxycarbide (scanning electron microscope VEGA-TESCAN).

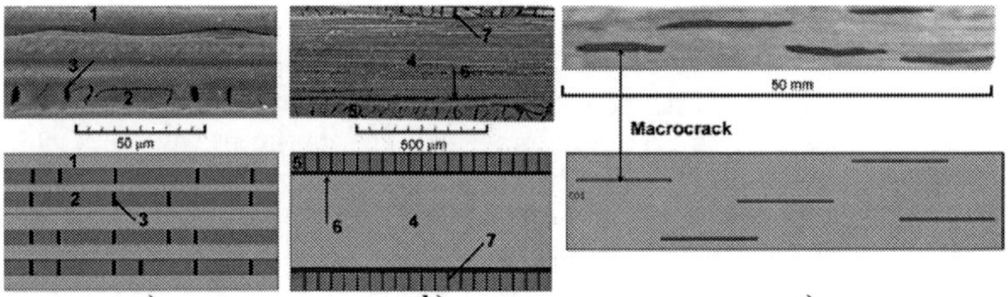

a) b) c)

Figure 9. Multiscale geometry model of studied composite specimen: a) microscale level, 1 - basalt fibre, 2 - SiOC matrix, 3 - microcracks, b) mesoscale level, 4 - part composed of microscale particles, 5 - SiOC matrix, 6 - delamination cracks, 7 - transversal cracks, c) composite plate with macrocracks (macroscale level).

Numerical Simulation Of Material Properties

Numerical simulations of heat transfer in composite laminate and successive calculations of effective thermal properties have been performed by using FEMLab software, the Heat Transfer Mode, that allows user-friendly implementation of complex geometry models and material properties. Numerical simulations of electrical conductivity have been performed by using Comsol Multiphysics software, the Electromagnetics Module, that provides a unique environment for the simulation of electromagnetic wave and field propagation in conductive media. The numerical results have been verified with the data acquired from experimental measurements. The pulse transient method has been applied for measurement of thermophysical parameters [8]. The HP 4339B High Resistance Meter has been used to estimate the electrical conductivity of the specimens [14].

RESULTS

The outputs from heat transfer numerical simulations are both: the graphs, those of temperature distribution and of heat fluxes in defined directions, and numerical, representing the values of effective thermophysical properties on defined level of multiscale structural model. The graphs of temperature fields serve as a source of information on relation between complex porous structure and thermal properties of studied material, see Figure 10. The numerical results can be also utilized for planning of experiment time schedule, as we are able to estimate heating rate of specimen or rate of temperature stabilization when heat transfer is induced into specimen, see Figure 11.

The graphs from electrical properties simulation show the influence of complex composite structure on charge transport in the specimen, see Figure 12. Subsequently the effective electrical properties are calculated. Numerical and experimental results are shown in TABLE I and II.

TABLE I. THERMOPHYSICAL PARAMETERS OF C/C COMPOSITE IN DIRECTIONS PARALLEL AND TRANSVERSE TO COMPOSITE PLATE.

Thermophysical properties	Computed	Measured
Thermal conductivity [$W.m^{-1}.K^{-1}$]	[8.3; 1.8]	[10; 1.6]
Specific heat [$J.kg^{-1}.K^{-1}$]	970	775
Thermal diffusivity $*10^{-6}$ [$m^2.s^{-1}$]	[5.9; 1.3]	[6.5; 1.3]

TABLE II. ELECTRICAL PROPERTIES OF BASALT/SIOC COMPOSITE SPECIMEN IN DIRECTIONS PARALLEL TO COMPOSITE PLATE.

Electrical properties	Computed	Measured
Electrical conductivity [$S.m^{-1}$]	$1{,}8 \,.\, 10^{-11}$	$4{,}3 \,.\, 10^{-11}$
Surface resistivity [$\Omega.m$]	$5{,}6 \,.\, 10^{10}$	$2{,}3 \,.\, 10^{10}$

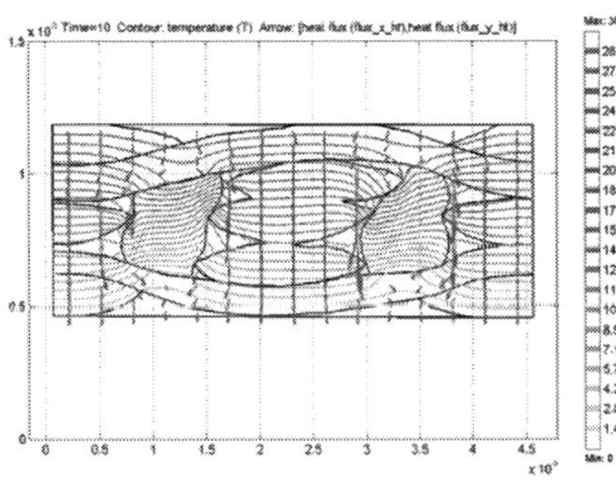

Figure 10. Temperature field and heat flux in C/C composite unit cell. Deformations of temperature field and heat flux course due to structural imperfections are clearly visible.

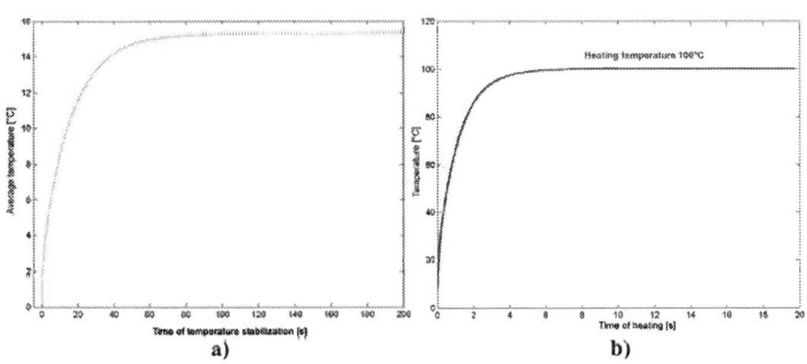

Figure 11. The C/C composite plate a) stabilization of the temperature after heat transfer was induced, b) heating of the composite plate.

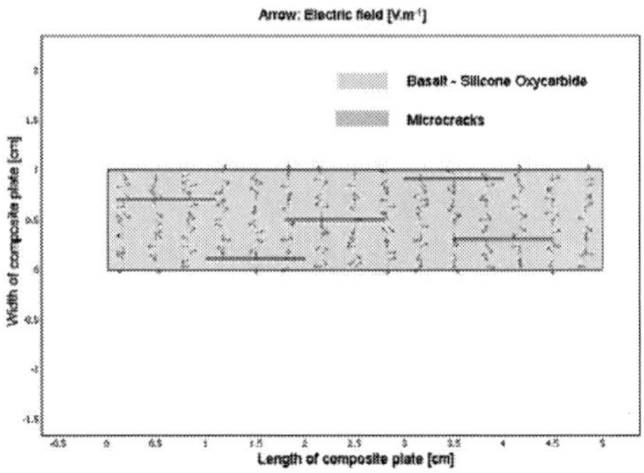

Figure 12. Behavior of electric field on the surface of the basalt/SiOC composite plate.

CONCLUSION

Complex porous structure of textile composites gives a little chance to create reliable universal geometry model. Thus the models are prepared individually with respect to real structure of the specimens. The composite structure has been scrutinized, morphology of particular components has been defined, their properties have been specified. Formulation of geometry models as well as further numerical calculations of material parameters have been performed by using of commercially available softwares.

Good agreement between the numerical results and the experimental ones indicates, the geometry models of composite structure can be applied for further simulations of specimen behaviour. Main disadvantages of presented modeling approach, that have appeared within the research, follow: time-consumption of the approach, necessity of skilled operator with knowledge in material science, neglect of the volume structure, reliable definition of input parameters (mainly in case of matrix). These problems limit convenient utilization of the modeling approach and their solution is subject for further research.

ACKNOWLEDGEMENT

Authors are grateful for the financial support by the Czech Science Foudation in the frame of the Project No. 106/07/1244. Word of thanks also belongs to Ministry of Education, Youth and Sports in the frame the Projects Textile Research Centre 1M4674788501, and The Center of Quality and Reliability 1M06047C468.

REFERENCES

1. Bigaud, D. at al. 1995. "A geometric modelling software for reinforced textile composites," *Euromech Lyon*, 334: 102–114.
2. Brus J., at al. 2001. "Structure of silicon oxycarbide glasses," *J. of Non-Crystaline Solids*, 289: 62–74.

3. Fitzer, E., Manocha, L.M. 1998. *Carbon reinforcement and Carbon-Carbon Composites*. Springer Verlag, Berlin.
4. Jortner, J. 1986. "Macroporosity and interface cracking in multi-directional carbon-carbons," *Carbon*, 24(5): 603–613.
5. Jiquao, L. at al. 2002. "Influence of porosity and total surface area on the oxidation resistance of C/C composites," *Carbon*, 40: 2483–2488.
6. Košek, M., Mikolanda, T. 2005. "Visualization of Textile Structures Using Virtual Reality," presented at Spring Conf. on Computer Graphics, Budmerice, Slovakia, April 19-21, 2005.
7. Košek, M., Seják, P. 2006. "Visualization of voids in actual C/C woven composite structure," presented at 12th European Conference on Composite Materials, Biarritz, France, August 29 - September 3, 2006.
8. Kubičár, L., Boháč, V.,Vretenár, V. 2002. "Transient methods for the measurement of thermophysical properties: The pulse transient method," *High Temperatures - High Pressures*, 34: 505–514.
9. Kuhn, J.L., Charalambides, P.G. 1999. "Modelling of planar weave fabric composite geometry," *J. of Composite Materials*, 33(3): 188–220.
10. Lomov, S.V. at al. 2001. "Textile composites: modelling strategies", *Composites: Part A,* 32: 1379–1394.
11. Piat, R., Schnack, E. 2003. "Hierarchical material modelling of carbon/carbon composites," *Carbon*, 41: 2121–2129.
12. Pollock, P.B. at al. 1993. "Crack development in a carbon-carbon laminate", *Carbon*, 31(6): 992–996.
13. Tomková, B. 2006. *Modelling of Thermophysical Properties of Woven Composites*. Doctoral thesis TU Liberec.
14. Tomková, B., Maršálková, M. 2007. "Resistivity of basalt fiber reinforced silicon oxycarbide - numerical simulation and experimental measurement," presented at 3rd International Tech. Textile Congress, Istanbul, Turkey, November 30-December 2, 2007.

Meso-scale Modeling of Unbalanced Woven Composites

P. POTLURI*, D. S. JETAVAT, A. MANAN and R. B. RAMGULAM

ABSTRACT

This paper reports the development of meso-scale FE models for unbalanced plain woven composites. Forming forces applied during preforming or consolidation can result in crimp interchange between warp and weft tows that in turn leads to unbalanced unit cell geometry. Load-deformation behavior of a plain woven carbon fabric has been predicted using an elastica-based tensile model. It can be shown that the fabric compliance is dominated by tow bending stiffness rather than tensile stiffness. Based on the observation of micrographs of laminates produced using un-stretched and stretched fabrics, meso-scale unit cells have been constructed. Experiments have been conducted to validate the meso-scale FE predictions.

INTRODUCTION

Meso-scale unit cell representation of woven composites, as found in the literature [1, 2], is almost always that of a balanced weave, with identical warp and weft tow geometry. However, warp and weft tow geometry may not be identical in most composite laminates for the following reasons:

a) Weaving parameters: warp & weft spacing and warp & weft tensions are seldom equal during the weaving process. Subsequent finishing processes tend to apply tension mainly in the warp direction.
b) Difference in tow tensions applied in warp and weft directions applied during forming results in crimp interchange.

Textile Composites Group, School of Materials (North West Composites Centre), University of Manchester, Manchester M60 1QD, United Kingdom
*Email: Prasad.Potluri@manchester.ac.uk

As a result, warp and weft crimps are seldom equal, and this results in unbalanced weave geometry. This paper reports the development of a non-ideal unit cell to represent unbalanced woven composites.

MODELLING TENSILE BEHAVIOR

Load-deformation behavior of dry woven fabrics can be predicted using a variety of analytical [3] or FE [4] analysis techniques. In this paper, an elastica-based solution has been employed to predict the load-deformation behavior of the woven fabrics, by considering the tensile, bending and transverse compaction properties of carbon tows. Detailed mechanics of elastica formulation has been presented elsewhere [5]. Geometry of the plain weave in the relaxed state can be seen in figure 1. Based on the input parameters (table 1), uniaxial tensile model was run to compute the load-extension behavior of the 12k carbon fabric. There is a good agreement between the experimental and computed curves (figure 2) in the initial region dominated by tow bending modulus. Beyond this point, tow tensile modulus computed from fiber modulus gives an over estimate. In reality, dry fibers in a tow are not perfectly aligned resulting in lower than estimated tow tensile modulus.

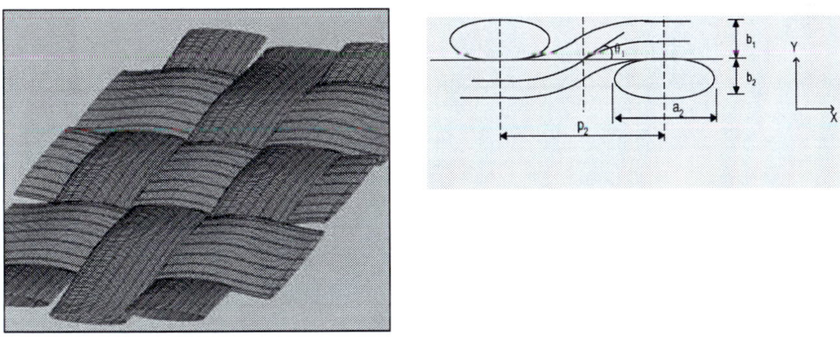

Figure 1. a) plain woven fabric b) fabric parameters

TABLE 1. FABRIC SPECIFICATIONS.

Parameter	Value
Ends/cm	3.1
Picks/cm	3.1
Tow spacing (p_1, p_2)	3.22 mm
Yarn Major Diameter (a_1,a_2)	2.75 mm
Yarn Minor Diameter (b_1, b_2)	0.39 mm
Yarn Bending Modulus	1.3 Nmm^2
Yarn Tensile Modulus	160000N/mm^2
Filament Diameter	0.0069 mm
Filament Modulus	230000 N/mm^2
Number of Filaments	12000

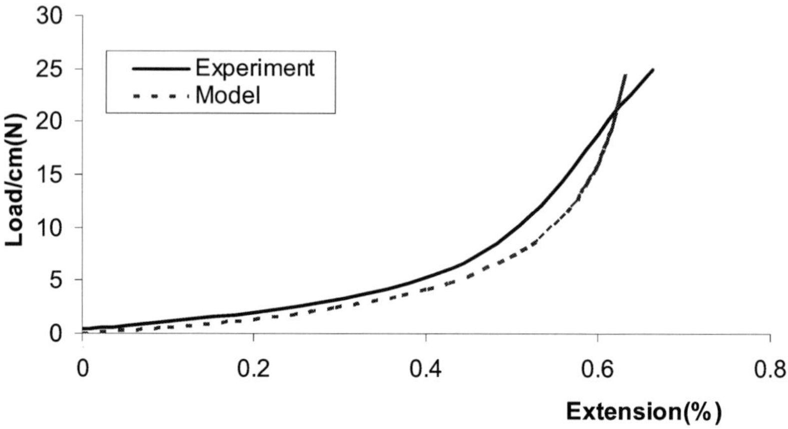

Figure 2. Comparison of model and experimental data.

PREPARATION OF UNBALANCED LAMINATES

Carbon-epoxy laminates have been prepared from un-stretched and stretched 12k carbon fabrics with the aid of vacuum infusion process. A tensile rig has been employed to apply tensile strain during the resin infusion process. Figure 3 shows the micrographs of the laminate produced with a relaxed fabric.

It can be see from figure 4 that in the laminate produced with stretched fabric, warp tows are nearly straight. Weft tows spread completely to the available gap (figure 4a) at the position between two adjacent warp tows. It is interesting to see that there is a significant overlap between weft tows at the cross-section (figure 4b) taken through a warp tow. Figure 5 shows the increase in weft crimp. Warp tows have become thicker and narrower due to the tension applied.

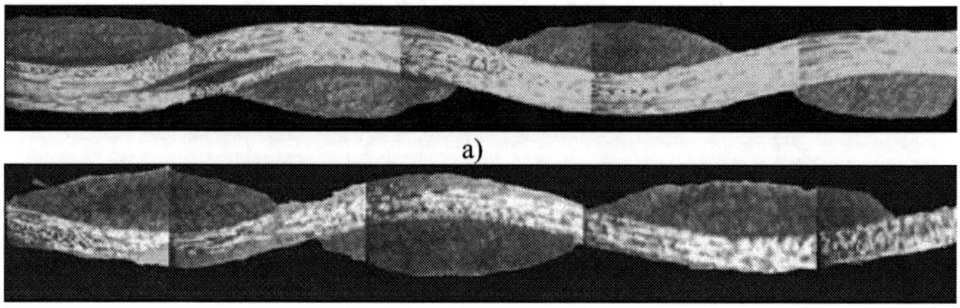

a)

b)

Figure 3. a) weft cross section b) warp cross section of unstretched laminate.

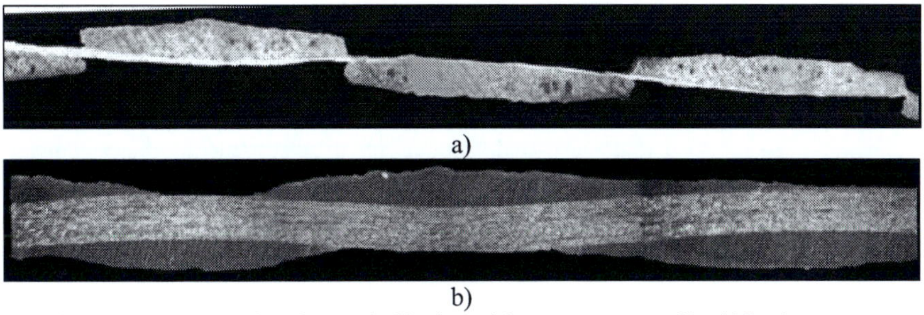

Figure 4. weft cross section of a stretched laminate a) between warp tows b) middle of warp tow.

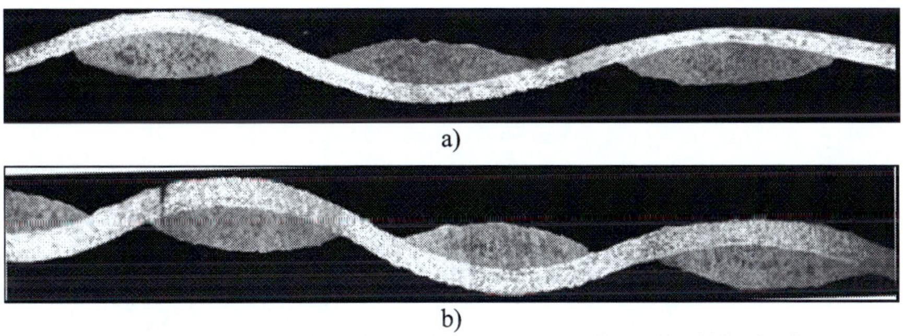

Figure 5. warp cross section of stretched laminate a) between weft tows b) middle of weft tow.

TABLE 2. TOW CROSS-SECTIONAL DETAILS.

tow	Unstretched (mm)			Stretched (mm)		
	$a_{1,2}$	$b_{1,2}$	$p_{1,2}$	$a_{1,2}$	$b_{1,2}$	$p_{1,2}$
warp	2.8	0.35	3.1	2.45	0.4	3.05
weft	2.9	0.35	3.1	4.8	0.32	3.2
weft, in between tow				3.2	0.32	3.2

FINITE ELEMENT ANALYSIS OF THE UNIT CELL

Mechanical behavior of heterogeneous materials can be simulated by homogenization methodologies by treating them as periodic structures and considering representative unit-cells with periodic boundary conditions. In this work, it is assumed that meso-scale models have heterogeneous material media. For analysis purposes, tow sub-elements are treated as transversely isotropic UD laminate while the resin pockets are taken as homogeneous.

Geometry Of The Unit Cell

To understand the mechanical behavior of fibrous reinforced interlaced network, a periodically repeating part is picked up from this network and defined as the unit cell. Within boundary of this unit cell, there are three sub-elements called warp tow, weft tow and resin pockets. Geometry for two types of unit cells is developed i.e. with and without stretching warp tow. Here in this geometry, warp and weft tows are modeled with lenticular geometry which describes its cross-section with two circular arcs. The unit cells without and with resin for stretched and un-stretched case are shown in Figure 6a and 6b and Figure 7a and 7b respectively.

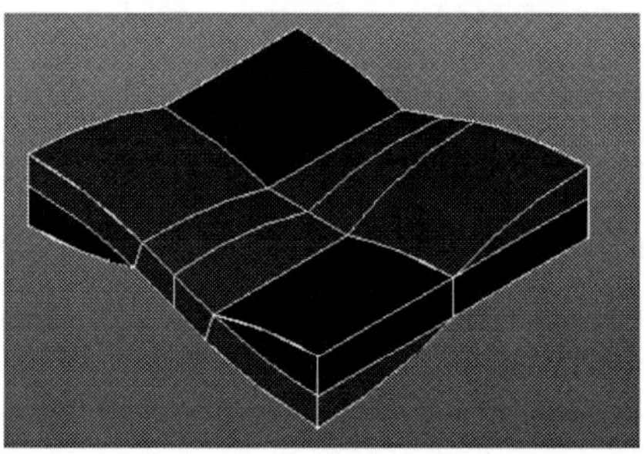

Figure 6a). Unit cell modeling-Stretched case

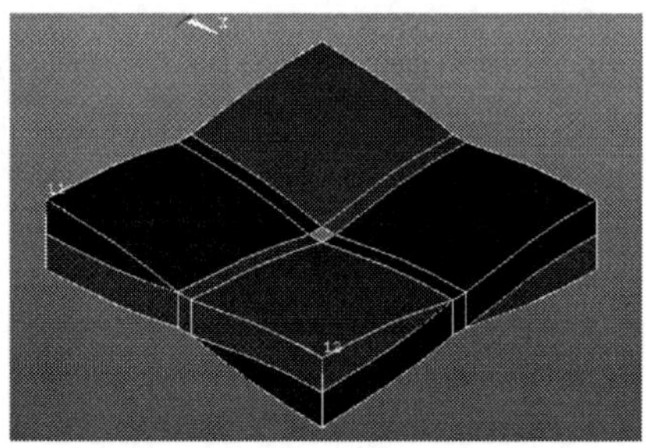

b) Unit cell modeling-Un-stretched case

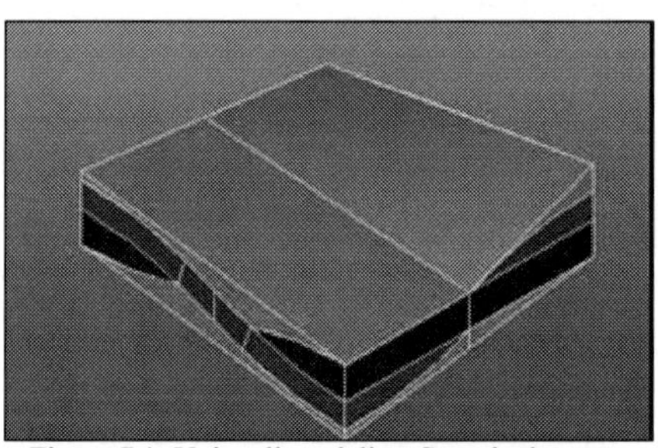

Figure 7a). Unit cell modeling-Stretched case

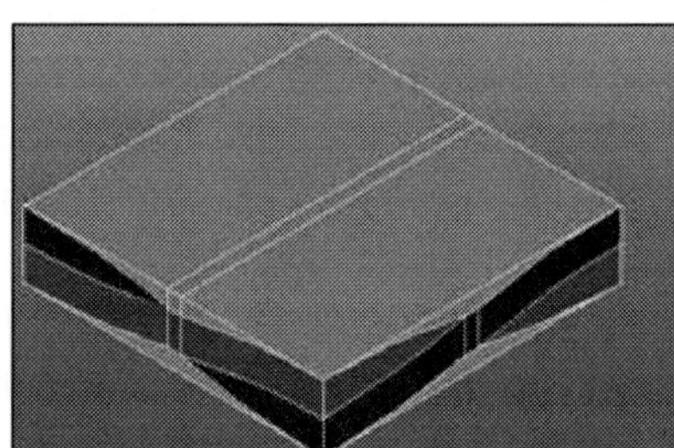

b) Unit cell modeling-Un-stretched case

Warp- Weft tows and resin material properties

Following properties of carbon fiber (in Table 3) and epoxy resin (in Table 4) are assumed

TABLE 3. CARBON FIBRE MATERIAL PROPERTIES.

Material	E_{1f}[GPa]	E_{2f}[GPa]	G_{12f} [GPa]	ν_{12f}
Carbon	230	28	50	0.23

TABLE 4. MATERIAL PROPERTIES OF EPOXY RESIN.

Material	E_m [GPa]	ν_{12m}
Epoxy	3.45	0.35

The elastic properties of the tow are determined using the equations of micro-mechanics. The tow is considered as a transversely isotropic material and hence it requires the specification of six elastic constants i.e. E_{11}, E_{22}, G_{12}, $\nu_{12,}$ ν_{23}. E_{11} and $\nu_{12,}$ can be found from well-known rule of mixture and rest can be calculated, for example, from Halpin-Tsai or Chamis model [6]. Resin is taken as isotropic material. Calculated tow properties are tabulated in Table 5.

TABLE 5. TOW PROPERTIES.

	E_{11} [GPa]	E_{22} [GPa]	G_{12} [GPa]	ν_{12}
Tow	139.38	9.119	3.982	0.278

Overall fiber volume fractions for stretched and un-stretched unit cells are given in Table 6. Fiber volume fraction decreased due to stretching as a result of increase in laminate thickness.

TABLE 6. OVERALL FIBRE VOLUME FRACTION.

	V_{fo}
Stretched case	0.33
Un-stretched case	0.38

Finite Element Analysis To Predict Effective Modulus

The finite element model of unit cell is developed in FEA software ANSYS. For resin solid 187 and for tow solid 186 structural elements with three degree of freedom at each node are used in the model. To predict the effective modulus of the unit cell in warp direction total strain energy stored in the unit cell (comprising warp tow energy part, weft tow energy part and resin energy part) is calculated with the following equation 1 under the application of external load.

$$U_{uc} = \frac{1}{2} \int_V \sigma \varepsilon dV \tag{1}$$

Where, σ is the stress field, ε is the strain and V is the volume of the unit cell. This strain energy is numerically calculated for stretched and un-stretched unit cells as described below [7].

Boundary Condition Definition On Unit Cell

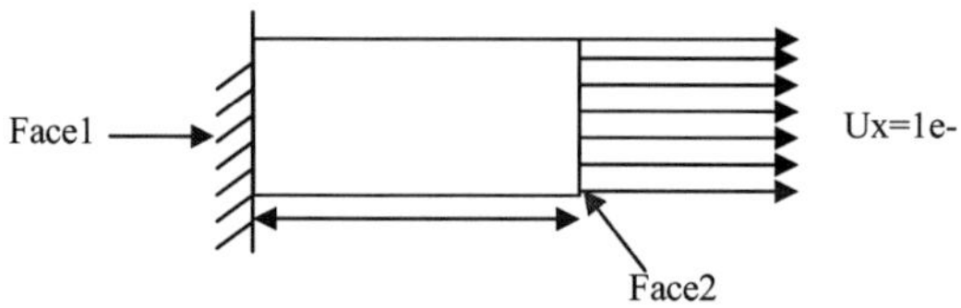

Figure 8. Schematic to calculate Young's Modulus.

A schematic illustration for applying boundary and loading conditions for numerical calculation of effective constant E_{xx} is shown in Figure 8:

We applied a uniform displacement field of U_x=1e-6 on the face 2.The initial length was L_x and applied strain was equivalent to $\varepsilon_{xx} = U_x / L_x$. We applied the constraint of U_x=0.0 on the face 1. The effective Young's modulus is calculated from equation-1 as follows:

$$E_{xx} = \frac{2(U_{uc})}{V(\varepsilon_{xx})^2} \tag{2}$$

in which U_{uc} is the strain energy produced by the applied strain ε_{xx} and V is the total volume of the unit cell. Calculated Finite Element Analysis results are tabulated in Table 7.

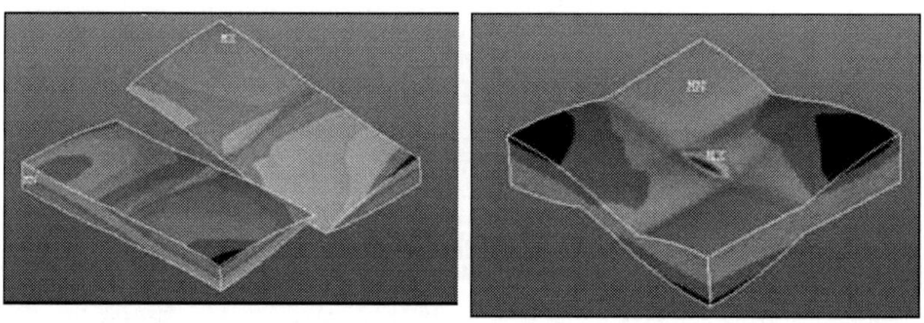

Figure 9a). Un-stretched warp tow strain plot b) Stretched case-warp and weft tow strain plot

Strain Plots for un-stretched and stretched laminates are shown in Figure 9a and 9b respectively, where displacement is applied along warp direction.

TABLE 7. FEA AND EXPERIMENTAL RESULTS.

Sample	E_{11} [GPa]	Experimental Results [GPa]
Un-stretched	24.67	27.18
Stretched	36.93	34.89

DISCUSSION

Following observations can be made from the study of unbalanced plain woven composites:

- Crimp interchange during fabric stretching was dominated by tow bending stiffness
- Due to the tensile force applied in the warp direction, inter-tow spaces increased in the warp direction and decreased (to zero) in the weft direction. This is counter intuitive to what one would normally predict. Distance between weft tows did increase as expected. However, due to absence of warp crimp, weft tows spread completely to fill any gaps. In fact, there is a degree of overlap (at cross-sections through warp centre). Conversely, warp yarns have assumed a rounder cross section due to increased weft crimp resulting in increased inter-tow spacing between warp tow.
- Geometry of the weft tow in a stretched sample (stretched in warp direction) is complex with unequal lenticular cross-section. The tow width is maximum over the warp tows and minimum in between the warp tows.
- As expected, E_{11} increased with stretching in the warp direction. This has been corroborated experimentally. Further work is required to improve the unit cell geometry and to predict the limit strengths.

REFERENCES

1. Whitcomb J.D., "Three dimensional stress analysis of plain weave composites," in *ASTM composite materials: fatigue and fracture, vol 3* (edited by O'brien TK), ASTM, 1991: 417–38.
2. Dasgupta A., Agarwal R.K. and Bhandarkar S.M., "Three dimensional modelling of woven fabric composites for effective thermo-mechanical and thermal properties," *Composite Science and Technology,* 1996, 56: 209-23.
3. Sagar T., Potluri P. and Hearle J.W.S., "Mesoscale modelling of interlaced fibre assemblies using energy method," Computational Material Science, 2003, 28(1):49–62.
4. Boisse P., Gasser A. and Hivet G., "Analysis of fabric tensile behaviour: determination of the biaxial tension strain surfaces and their use in forming simulations," Composites A, 2001, 32:1395-414.
5. Ramgulam R.B. and Potluri P., "Three dimensional elastica for modelling fibre assemblies," in *Mathematics in Industry Volume 12*, Springer Berlin Heidelberg, 2008.
6. Gibson RF. Principles of composite material mechanics. 1st ed. McGraw-Hill; 1994.
7. Zhang YC., Harding J. A numerical micromechanics analysis of the mechanical properties of a plain weave composite. Computers Structure 1990; 36(5) 839–44.

Effect of Free Edges on Stresses in Woven Composites

B. OWENS, J. WHITCOMB and J. VARGHESE

ABSTRACT

Many years ago free edge effects were studied extensively, perhaps excessively, for ordinary tape laminate composites. Various techniques were used to show that the free edge effects extended only a short distance from the free edge and that stacking sequence had a significant impact on interlaminar stresses, which could cause premature failure. In contrast, the analytical study of woven composites has focused on periodic analysis of unit cells. Since such analyses assume an infinite array of unit cells, free edge effects were not considered.

The improvement in computational resources has made it feasible to investigate the effects of tow architecture (e.g. weave type and waviness) and constituent properties on the perturbation in the stresses that arises due to a free edge. The effects of a free edge were studied for a plain weave composite and for a composite consisting of a plain weave mat and tape laminas. The significance of the free edge effect was found to be sensitive to the waviness of the plain weave mat.

INTRODUCTION

In the past, numerous investigations have been conducted into the effects of a free edge in tape laminates. These studies used a variety of techniques to gain insight into the stresses due to a free edge. Investigators have studied a wide variety of tape laminates for free edge effects [1-3].

Woven composites have been studied analytically for their complex stress states due to the weave architecture. To date, investigations of woven composites have studied periodic behavior of unit cells [4]. These studies do not account for the effects of a free edge in a laminate. Therefore, it is necessary to further investigate such effects in woven composites.

Aerospace Engineering Department, Texas A&M University, College Station, Texas 77843-3141, U.S.A.

This paper will study woven laminate and "hybrid" woven and tape laminate configurations through three-dimensional finite element analysis. The model architecture, boundary conditions, and material properties are described in the following sections. Results will compare free edge behavior in the configurations to periodic behavior of the configuration and also to comparable tape laminates. Different waviness ratios will be discussed to determine the effect of waviness ratio on free edge stresses. Finally, conclusions will be summarized.

CONFIGURATIONS

Multiple configurations were studied to gain insight into the effects of free edges in woven composites. The following sections describe the model architecture, boundary conditions, and material properties for the configurations studied.

Model Architecture

This section describes the physical dimensions, waviness ratio (WR), and tow volume fractions for each configuration. Waviness ratio is defined herein as the thickness (h) of a woven mat divided by the wavelength (λ) of a tow. The free edge models were constructed by the duplication and translation of a unit cell (see Figure 1).

PLAIN WEAVE SYMMETRIC MODEL

A four-ply plain weave symmetric free edge model was created with a tow waviness ratio of 1/3 and a tow volume fraction of 0.636. A unit cell of the free edge configuration consists of four plain weave plies arranged symmetrically about the mid-plane. The length of the configuration in the x_1 direction is one warp tow wavelength, and the width in the x_2 direction is four fill tow wavelengths. The unit cell is periodic in the x_1 direction. However, this configuration may be reduced to a 1/8 unit cell model by exploiting symmetry about $x_1 = 0$, $x_2 = 0$, and $x_3 = 0$ (see Figure 2). Both the warp and fill tows have a thickness of 0.15mm. This results in dimensions of *a*, *b*, and *c* of 0.45mm, 1.8mm, and 0.6mm respectively (see Figure 2).

The free edge effect could depend on the location at which the composite is cut at the free edge due to spatial variation of the weave architecture. Therefore, the woven architecture of the configurations was ended at the free edge in two different locations or "cuts". One cut is "Cut 1" which corresponds to a location at which there is zero undulation in the fill tow. "Cut 2" corresponds to a location at which there is maximum undulation of the fill tow. The configuration with the free edge at Cut 2 is slightly narrower, but not enough to bias the results. See Figure 3 for a description of cut locations. The model in Figure 3 is rotated to show the tow architecture at the free edge. Comparisons between woven models will compare areas of identical weave architecture.

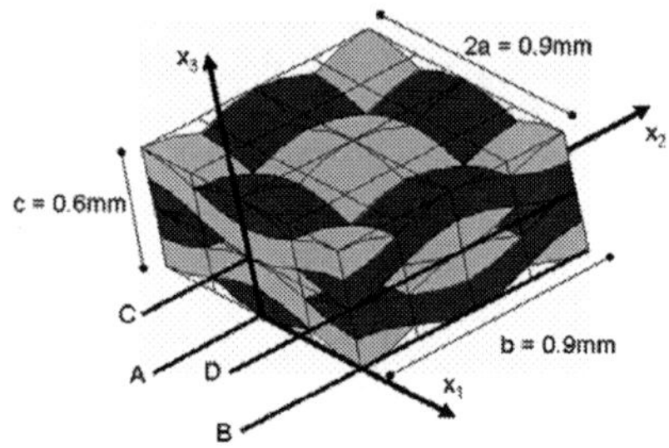

Figure 1. Top half of four-ply plain weave symmetric unit cell (matrix is transparent).

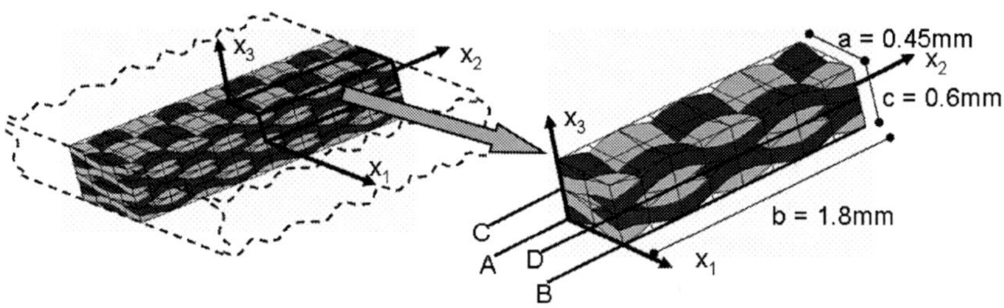

Figure 2. Plain weave free edge unit cell reduced to 1/8 unit cell model (transparent matrix).

For comparison, results were also obtained for two cases without free edges. Both cases used periodic boundary conditions to impose infinite in-plane x_1 and x_2 dimensions. One case also used periodic boundary conditions to impose infinite thickness (x_3 dimension). The periodic model had the same waviness ratio and tow volume fraction as the plain weave free edge model (see Figure 1). The model dimensions *a*, *b*, and *c* are 0.45mm, 0.9mm, and 0.6mm respectively.

Comparable tape laminate models were created for comparison to the plain weave symmetric free edge model (see Figure 4). These models are equivalent in overall dimension to the free edge laminate models. Each ply of the tape laminate has the thickness of a tow. Tape laminate models were studied due to the extensive documentation of free edge effects for tape laminates and to determine if tape laminate models can sufficiently approximate free edge effects in woven composites.

The close-up of the mesh for the plain weave free edge model in Figure 5 shows typical mesh refinement. The mesh for Cut 1 contains 64,000 twenty node quadratic elements and 267,781 nodes. All models presented in this subsection have comparable refinement.

HYBRID SYMMETRIC FREE EDGE MODEL

The combination of $\pm\theta$ and 90 degree tape plies often results in significant free edge effects because of the large differences in Poisson's ratios between the plies. A plain weave composite with some mats at zero degrees and some rotated by 45 degrees results in an architecture that is not periodic. Also, even if one approximated the architecture to make it periodic, the meshing is quite cumbersome. For this effort it was decided to examine a hybrid laminate that

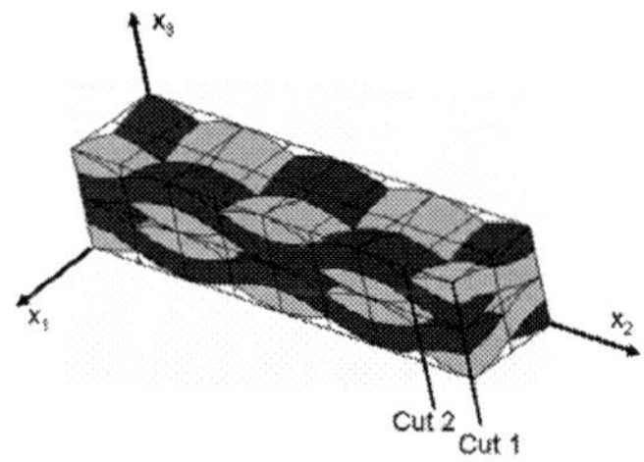

Figure 3. Free edge cut locations.

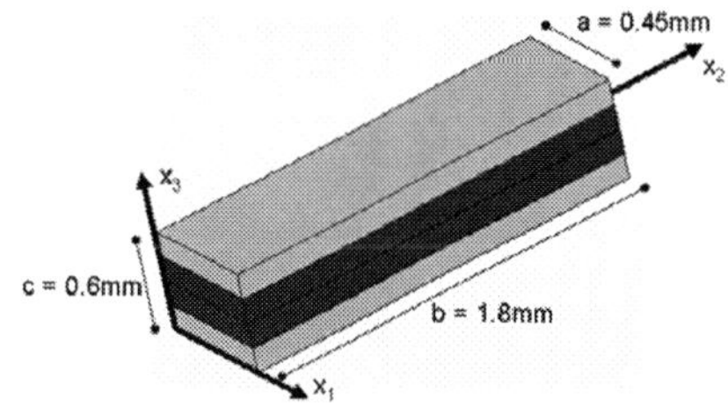

Figure 4. Four-ply plain weave symmetric comparable tape laminate.

considered ±45 degree tape laminas and a plain weave mat. The hybrid laminate was also cut at the two locations specified in Figure 3.

Figure 6 shows the ¼ unit cell hybrid laminate. Also, since free edge effects are expected to propagate further into the interior, the hybrid laminate free edge model is significantly wider than the woven laminate free edge model. This model with waviness ratio of 1/3 has dimensions *a*, *b*, and *c* of 0.45mm, 6.3mm, and 0.6mm respectively.

The comparable tape laminate model is identical in geometry to the tape laminate in Figure 4. However, instead of two lamina orientations there are now four lamina orientations.

Figure 7 presents the mesh refinement for the hybrid free edge model. The mesh for Cut 1 contains 72,576 twenty node quadratic elements and 306,505 nodes. All models compared to the hybrid model have comparable mesh refinement. The hybrid model element dimensions are approximately twice that of the plain weave model element dimensions.

To study the effect of waviness ratio, hybrid laminate models were created with waviness ratios of 1/3 and 1/9. The model with waviness ratio of 1/9 has dimensions *a*, *b*, and *c* of 1.35mm, 5.4mm, and 0.6mm respectively. Both high and low waviness ratios models have comparable mesh refinement.

Boundary Conditions

The boundary conditions for the various conditions were based on two considerations: periodicity and equivalent coordinate systems. Periodicity of the displacements, for example is imposed by using the conditions

$$u_i\left(x_\alpha + d_\alpha\right) = u_i(x_\alpha) + \left\langle \frac{\partial u_i}{\partial x_\beta} \right\rangle d_\beta \tag{1}$$

where d_α is the vector of periodicity. If the configuration (geometry, material properties, and loading) exhibits symmetries, then there are multiple equivalent coordinate systems. For displacements the equivalence can be expressed as

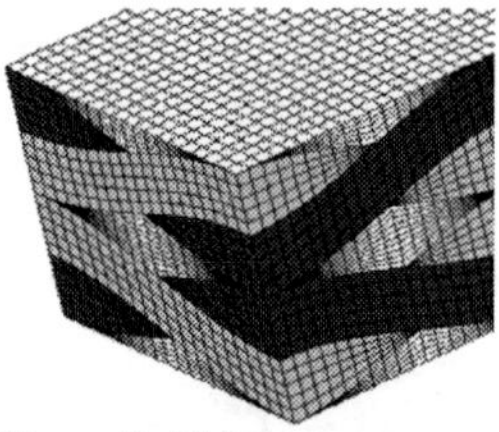

Figure 5. Plain weave symmetric model mesh refinement.

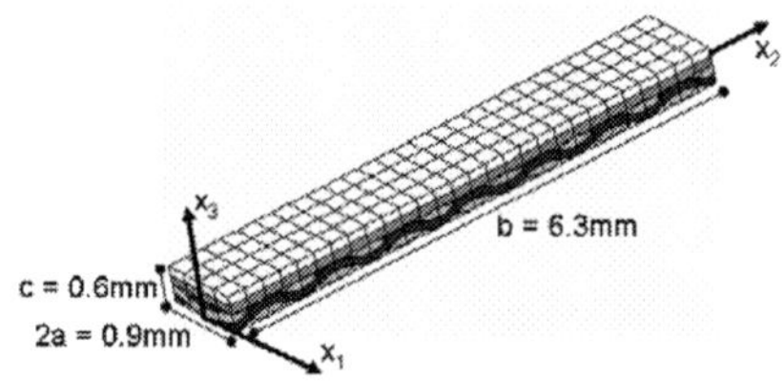

Figure 6. Four-ply symmetric hybrid laminate model WR=1/3 (1/4 unit cell).

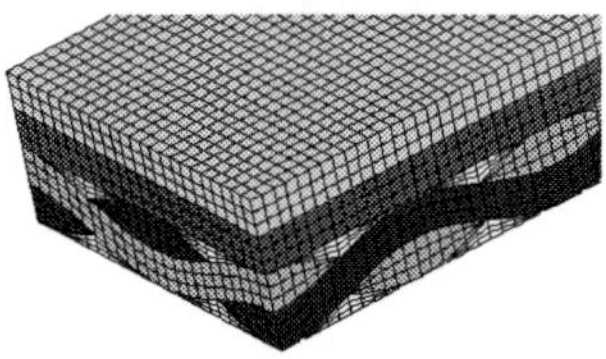

Figure 7. Hybrid symmetric model mesh refinement.

$$u_i\left(x_\alpha\right)=u_i^{'}\left(x_\alpha^{'}\right) \tag{2}$$

There is insufficient space in this paper to describe the derivation of the boundary conditions, so they are simply tabulated below. Details of the strategy can be found in [5].

All configurations were subjected to a uniaxial load in the x_1 direction that resulted in a volume averaged axial strain ($\langle \partial u_1/\partial x_1 \rangle$) of .01. The following tables present the boundary conditions. Table I and Figure 2 specify the boundary conditions for the 1/8 unit cell plain weave model and comparable tape laminate model. Table II and Figure 1 specify boundary conditions for the periodic plain weave model. Table III and Figure 6 specify boundary conditions for the hybrid model and comparable tape laminate.

TABLE I. BOUNDARY CONDITIONS FOR 1/8 UNIT CELL PLAIN WEAVE MODEL.

Plane	Boundary Conditions	Plane	Boundary Conditions
$x_1 = 0$	$u_1=0,\ t_2=t_3=0$	$x_2 = b$	$t_1 = t_2 = t_3 = 0$
$x_1 = a$	$u_1 = \left\langle \frac{\partial u_1}{\partial x_1} \right\rangle a,\ t_2 = t_3 = 0$	$x_3 = 0$	$u_3 = 0,\ t_1 = t_2 = 0$
$x_2 = 0$	$u_2 = 0,\ t_1 = t_3 = 0$	$x_3 = c$	$t_1 = t_2 = t_3 = 0$

TABLE II. BOUNDARY CONDITIONS FOR PLAIN WEAVE PERIODIC MODEL.

Fully Periodic		**In-Plane Periodic**	
Planes	**Boundary Conditions**	**Planes**	**Boundary Conditions**
$x_1 = -a,\ x_1 = a$	$u_i\left(x_1=a\right)=u_i(x_1=-a)+\left\langle \frac{\partial u_i}{\partial x_1} \right\rangle 2a$	$x_1 = -a,\ x_1 = a$	$u_i\left(x_1=a\right)=u_i(x_1=-a)+\left\langle \frac{\partial u_i}{\partial x_1} \right\rangle 2a$
$x_2 = 0,\ x_2 = b$	$u_i\left(x_2=b\right)=u_i(x_2=0)+\left\langle \frac{\partial u_i}{\partial x_2} \right\rangle b$	$x_2 = 0,\ x_2 = b$	$u_i\left(x_2=b\right)=u_i(x_2=0)+\left\langle \frac{\partial u_i}{\partial x_2} \right\rangle b$
$x_3 = 0,\ x_3 = c$	$u_i\left(x_3=c\right)=u_i(x_3=0)+\left\langle \frac{\partial u_i}{\partial x_3} \right\rangle c$	$x_3 = 0$	$u_3 = 0,\ t_1 = 0,\ t_2 = 0$
$x_3 = 0$ and $x_3 = c$ remain planar after deformation		$x_3 = c$	$t_1 = t_2 = t_3 = 0$

TABLE III. BOUNDARY CONDITIONS FOR ¼ UNIT CELL HYBRID MODEL.

Plane(s)	**Boundary Conditions**	**Plane(s)**	**Boundary Conditions**
$x_1 = -a,\ x_1 = a$	$u_i(x_1 = a) = u_i(x_1 = -a) + \left\langle \frac{\partial u_i}{\partial x_1} \right\rangle 2a$	$x_2 = b$	$t_1 = t_2 = t_3 = 0$
$x_2 = 0$	$u_1(x_1,0,x_3) = -u_1(-x_1,0,x_3)$ $u_2(x_1,0,x_3) = -u_2(-x_1,0,x_3)$ $u_3(x_1,0,x_3) = u_3(-x_1,0,x_3)$	$x_3 = 0$	$u_3 = 0,\ t_1 = t_2 = 0$
		$x_3 = c$	$t_1 = t_2 = t_3 = 0$

Material Properties

All of the configurations used the same carbon-epoxy constituents [6]. The fiber volume fraction of the tow was 0.69. See Table IV for a listing of material properties. The hybrid model and comparable tape laminate models use the carbon-epoxy tow properties with varying orientation angles. Comparable tape laminates for the plain weave models have a stacking sequence of $[0/90/90/0]_s$ and $[90/0/0/90]_s$. Comparable tape laminates for the hybrid model have a stacking sequence of $[\pm 45/90/0]_s$ and $[\pm 45/0/90]_s$.

TABLE IV. CARBON-EPOXY MATERIAL PROPERTIES

Carbon-Epoxy Tow ($V_f = 0.69$)	**Neat Epoxy 411-350**
$E_{11} = 157.95$ GPa $E_{22} = E_{33} = 9.027$GPa	$E_{11} = E_{22} = E_{33} = 3.1$GPa
$\nu_{12} = \nu_{13} = 0.2412$ $\nu_{23} = 0.3749$	$\nu_{12} = \nu_{23} = \nu_{13} = 0.35$
$G_{12} = G_{13} = 5.12$ GPa $G_{23} = 3.34$ GPa	$G_{12} = G_{23} = G_{13} = 1.15$ GPa

RESULTS

Finite element analysis was performed on the models specified in the configurations section. Because of limited space, only stresses along certain lines will be shown herein. These lines of interest are lines A $(0,x_2,0)$, B $(a,x_2,0)$, C $(0,x_2,c/2)$, and D $(a,x_2,c/2)$. Figures 1 and 2 illustrate comparable lines for woven free edge and periodic models. The same lines will be used for comparisons to tape laminates. Free edge stress distributions will be compared to both periodic models and tape laminates. Only the predicted σ_{33} stress component will be presented due to space limitations.

Plain Weave Symmetric Results

The stress distributions along the lines A, B, C, and D were examined for all plain weave models with a waviness ratio of 1/3 and the results were very similar in all cases. Figures 8 and 9 present stress distributions for Cuts 1 and 2 along line A. The coordinate x_2 has been normalized by tow thickness (h_t). There appears to be little to no thickness effect on the stress distributions. The cut location does cause a phase shift in the stress distribution which does affect the magnitude of stress at the free edge. However, the stress magnitude at the free edge is less than or about the same as the peak magnitudes of stress in the interior of the model. Therefore, it appears that the weave architecture has a much more significant effect on the σ_{33} than free edge effects. Comparison of σ_{33} to comparable tape laminates shows that the interior stress magnitude of a high waviness ratio plain weave (i.e. the 1/3 waviness ratio) is much greater than stress magnitudes due to a free edge in a comparable tape laminate.

Hybrid Symmetric Results

The hybrid laminate model has a larger free edge effect than that seen for the plain weave laminate due to the use of angle plies stacked with a plain weave mat. The addition of angle plies created a larger region containing free edge effects. Therefore, the hybrid laminate models are much wider than the plain weave laminate. Although the interlaminar stress components σ_{23} and σ_{13} were more significant than those seen in the plain weave laminate these stress components were still negligible (along the lines examined) relative to the magnitudes of the interlaminar σ_{33} stress component. Again, only σ_{33} stress distributions will be presented.

Stress distributions along the same lines as earlier were examined. The model was sufficiently wide to capture in-plane periodic behavior of the hybrid laminate and a separate periodic model was unnecessary. Results showed that there is a slight variation of stress magnitude with respect to thickness. Line B produced the largest stress magnitudes in both hybrid and tape laminates. Figures 10 and 11 show the σ_{33} stress distributions for line B for both cuts (WR = 1/3).

Comparing the interlaminar σ_{33} stress distribution at the free edge with that of the interior hybrid laminate shows approximately the same if not lower stress magnitude as interior peak stresses. Tape laminate σ_{33} is more significant than the purely cross-ply laminates, but the interior stress of the hybrid laminate is still significantly larger. Again, the cut location causes a phase shift in the stress distribution which does affect the magnitude of the stress at the free edge. However, there appears to be no significant difference in overall peak stresses between the two cuts.

In contrast, results of the low waviness ratio (WR = 1/9) model are much more comparable to the tape laminate results. Figures of these results are not shown herein due to space limitations, but will be covered in the talk.

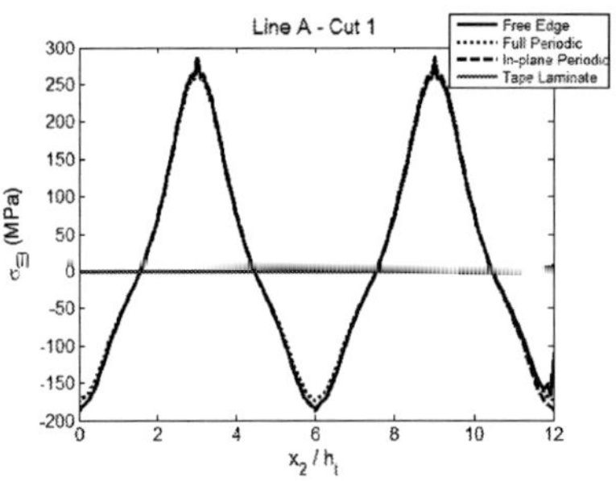

Figure 8. σ_{33} stress distribution for line A, Cut 1.

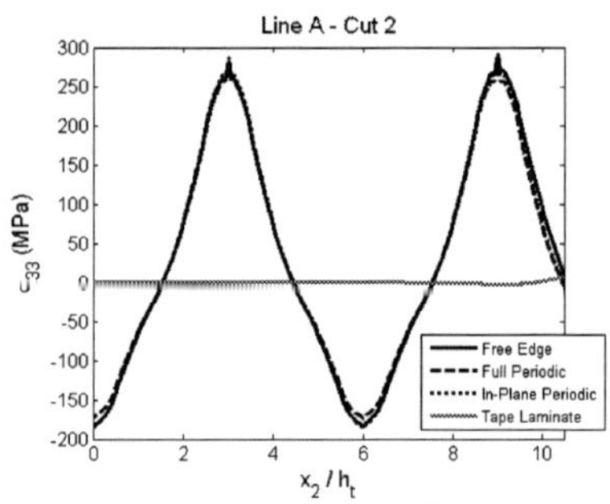

Figure 9. σ_{33} stress distribution for line A, Cut 2.

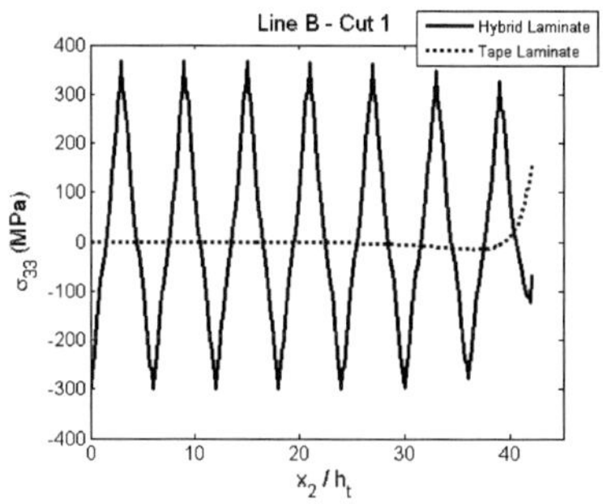

Figure 10. σ_{33} stress distribution for line B, Cut 1.

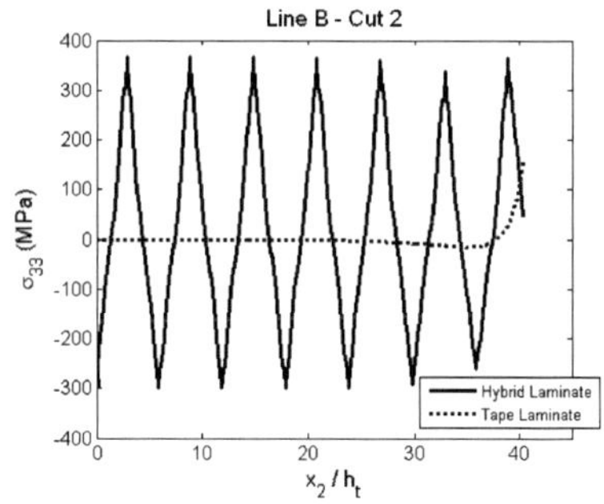

Figure 11. σ_{33} stress distribution for line B, Cut 2.

CONCLUSIONS

Various carbon-epoxy plain weave symmetric laminates and related tape laminates were studied to investigate the effects of free edges on stresses in woven composites. The study suggests that for high waviness ratio configurations (WR = 1/3) the weave architecture dominates the σ_{33} interlaminar stress distribution, and any free edge effects are negligible in comparison. It was also seen that the interlaminar σ_{33} stress for the woven laminate was much greater than the stress magnitudes of a comparable tape laminate subjected to free edge effects.

Analysis of carbon-epoxy laminates with ±45 degree angle plies showed larger free edge effects for the tape laminate, but for the hybrid laminate the weave architecture dominated the interlaminar stresses. The hybrid laminate with a lower waviness ratio (WR = 1/9) significantly reduced weave architecture effects. As the waviness ratio decreases, a woven laminate behaves more like a tape laminate, and free edge effects eventually dominate the interlaminar stress distribution instead of weave architecture.

REFERENCES

1. Pipes, R.B. and N.J. Pagano. 1970. "Interlaminar Stresses in Composite Laminates Under Uniform Axial Extension," *Journal of Composite Materials*. 4:538–548.
2. Wang, J.T. and J.N. Dickson. 1978, "Interlaminar Stresses in Symmetric Composite Laminates," *Journal of Composite Materials*, 12:390–402.
3. Whitcomb, J.D. and I.S. Raju. 1985. "Analysis of Interlaminar Stresses in Thick Composite Laminates With and Without Edge Delamination," *Delamination and Debonding of Materials, ASTM STP 876*, W.S. Johnson, ed. Philadelphia: American Society for Testing and Materials, pp. 69–94.
4. Whitcomb, J.D. 1991. "Three-Dimensional Stress Analysis of Plain Weave Composites," *ASTM STP 1110.*
5. Whitcomb, J.D., C.D. Chapman, and X. Tang. 2000. "Derivation of Boundary Conditions for Micromechanics Analyses of Plain and Satin Weave Composites," *Journal of Composite Materials*, 34(9):724–747.
6. Goyal, D., X. Tang, J.D. Whitcomb, and A. Kelkar, 2005. "Effect of Various Parameters on Effective Engineering Properties of 2 x 2 Braided Composites," *Mechanics of Advanced Materials and Structures*, 12:113–128.

Mechanical Modelling of Monofilament Technical Textile Components

V. CARVELLI

ABSTRACT

A homogenization numerical approach for the prediction of the mechanical properties of technical textile structures is described in this paper. The numerical analyses deal with modelling at two scale levels. The first scale aims to predict the mechanical behaviour of the textile by the numerical analysis of a Representative Volume. The second scale modelling evaluates the mechanical response of the entire textile component employing the homogenized textile mechanical behaviour obtained in the first step. The numerical approach is used to predict the stress-strain state of a polyester monofilament textile component employed in a narrow industrial application, screen printing or serigraphy.

INTRODUCTION

The mechanical behaviour of textiles is the most important factor in several industrial applications. This is the case for textile reinforcement used in composites or for technical textiles [1]. The forming of the textile preform in a composite component or the correct ink deposition in a screen printing application depend on the tensile response in the principal directions and on the shear behaviour of the fabric. Therefore, the knowledge of the textile mechanical features is indispensable to predict the mechanical response of textile structural components.

In the last decades, the prediction of the mechanical properties of textiles has been item of several researches by different approaches (see e.g. [1]). In particular the investigations can be collected in three main methodologies: experimental (see e.g. [2, 3]), analytical (see e.g. [4, 5]) and numerical (see e.g. [6, 7]).

This paper presents a homogenization numerical approach based on two scales modelling to predict the mechanical response of textile structural components.

Department of Structural Engineering, Politecnico di Milano, Piazza Leonardo da Vinci 32, 20133 Milano (Italy)

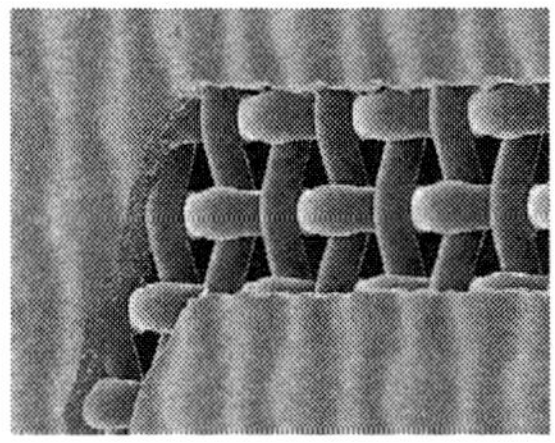

Figure 1. Dry and filled textile in the structural component.

In the first scale the mechanical behaviour of a monofilament textile is predicted by the numerical analysis of a Representative Volume (RV size of order 10^{-5}m) assuming a regular distribution of the fibres in the warp and weft directions. The accuracy of the first scale numerical model is demonstrated by comparison with experimental results from both uniaxial and biaxial tensile tests [8].

The second scale modelling is dedicated to the numerical analysis of the global behaviour of the textile component (size of the order 1m). In this context the fictitious homogeneous material having the mechanical properties obtained in the first scale analyses is employed.

The numerical procedure is applied to predict the mechanical response of a components used in a screen printing technique. The component consists of a monofilament polyester textile fixed on a metallic frame.

In the printing process the ink penetrates only in the textile parts representing the shape of the final picture. This task is obtained filling the textile area not included in the picture shape (negative picture) by a stencil (Figure 1). Therefore in the component the textile has two different mechanical behaviour: as a dry textile (where the ink penetrates) and as a filled textile that is a textile composite material. The quality of the printing mainly depends on the deformed shape of the textile i.e. on both these two mechanical behaviours.

In the following a short description of the theoretical background applied, in the first scale modelling, to the representative volumes of the dry and filled textile are presented. The numerical macroscopic results of the first scale analyses are detailed and compared (only for the dry textile) to the experimental data. Finally, the global response of a textile component for an academic screen printing application is numerically evaluated assuming the first scale results.

FIRST SCALE MODELLING

Theoretical Background

The first scale deals with the prediction of the mechanical behaviour of plane weave monofilament dry and filled textile assuming a regular distribution of the fibres in the warp and weft directions (see Figure 2a). This main assumption allows to investigate the textile mechanical features by the numerical analysis of the Representative Volume (RV size of order 10^{-5}m) following the approach described in [9] for textile composite materials.

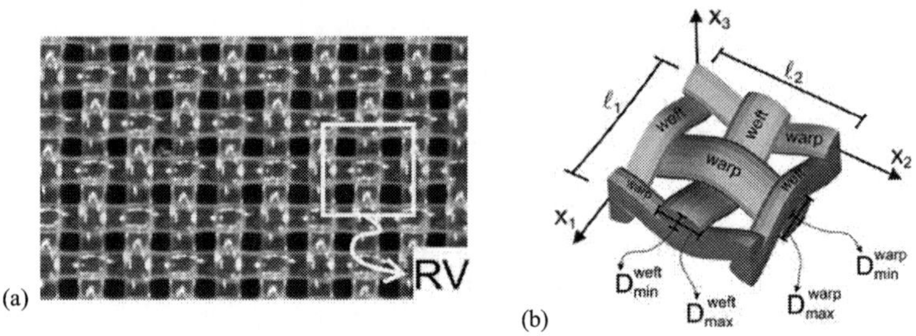

Figure 2. (a) Representative Volume (RV) in the plane weave textile; (b) some geometrical parameters of the RV.

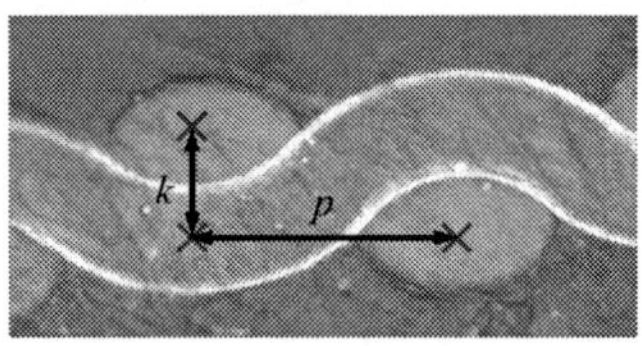

Figure 3. Microscopic observation of a plane weave textile cross-section.

From the numerical results of the RV analysis, the macroscopic (global) properties of the mechanical equivalent material are evaluated by averaging the quantities on the RV volume [10].

The three-dimensional numerical models includes the geometry of the fibres and the morphology of the textile obtained by microscopic observations [8] (Figure 3).

The numerical analysis of textiles was carried out enforcing kinematic boundary conditions on the RV boundary to simulate the periodicity (hypothesis of regular distribution of the fibres in the textile) [9].

These boundary conditions are written in term of the displacement vector $\underline{u}$, as:

$$\underline{u}_A - \underline{u}_B = \underline{\underline{E}}(\underline{x}_A - \underline{x}_B) \tag{1a}$$

$$\underline{u}_C - \underline{u}_D = \underline{\underline{E}}(\underline{x}_C - \underline{x}_D) \tag{1b}$$

In eqn. (1) A and B are two points of a weft fibre on the cross section $x_1=0$ and $x_1=\ell_1$ corresponding in the periodicity; C and D are two points of a warp fibre on the cross section $x_2=0$ and $x_2=\ell_2$ corresponding in the periodicity (see Figure 2b).

Additional kinematic conditions are enforced to avoid penetration between fibres and to consider the fibre-fibre friction contact by Coulomb's model. The numerical analyses shows that the friction contact has negligible influence on the uniaxial and biaxial tensile response of the textiles.

A three dimensional numerical simulation of a monofilament textile requires the constitutive behaviour of the fibres to be as close as possible to the real one. Therefore, the macroscopic (or global) behaviour up to failure of the plane weave textiles is investigated assuming a Ramberg-Hosgood nonlinear constitutive behaviour of the fibres. In the uniaxial state, the Ramberg-Hosgood's equation is:

$$E\varepsilon = \sigma\left(1 + \alpha\left(\frac{|\sigma|}{\sigma_0}\right)^{n-1}\right) \tag{2}$$

In eqn. 2, E is initial Young's modulus, α the "yield" offset, σ_0 "yield" stress and n is the exponent for the nonlinear term.

The parameters in eqn. 2 were determined fitting the experimental data obtained by tensile tests of fibres extracted from the textiles [8].

Dry Textile: Numerical Results

The polyester monofilament technical textile employed in the screen printing application, introduced above, has a typical plain weave geometry formed interlacing polyester fibres with the same geometry in two perpendicular directions. Some geometric features are collected in Table I, being k and p defined in Figure 3, D^{nom} the nominal diameter of the fibres, D^{max} and D^{min} the maximum and minimum diameters of the elliptic cross sections measured with an optical microscope [8].

The RV geometry was reproduced in a finite element model maintaining a constant elliptic cross section of the fibres, although the ovalization is more pronounced in the vicinity of the crossovers (see Figure 2a). The mesh is assembled with 5330 nodes and 19200 four nodes tetrahedral elements C3D4 [11].

The parameters in eqn. 2 fitting the experimental results of the fibres tensile tests, are reported in Table II.

The numerical predictions obtained with the representative volume are compared to the experimental observations up to failure. The ultimate failure is obtained in the numerical model when the maximum average Mises's equivalent stress in a fibre cross-section is grater than the experimental ultimate tensile stress of the fibre.

TABLE I. THE GEOMETRIC FEATURES OF THE MONOFILAMENT TECHNICAL TEXTILE.

Direction	Number of fibres per cm	k [μm]	p [μm]	D^{nom} [μm]	D^{max} [μm]	D^{min} [μm]
warp	150	24.3	67	34	44.92	34.01
weft	150	28.5	67	34	42.67	37.97

TABLE II. FIBRES MATERIAL PARAMENTERS FOR RAMBERG-HOSGOOD CONSTITUTIVE EQUATION.

	Weft	Warp
E [MPa]	4600	3000
α	0.544	0.132
n	3.254	5.405
σ_0 [MPa]	697	681

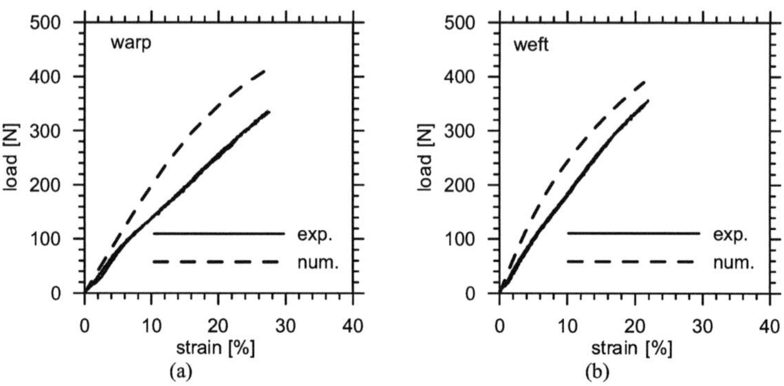

Figure 4. Dry textile: uniaxial tensile tests in (a) warp and (b) weft direction.

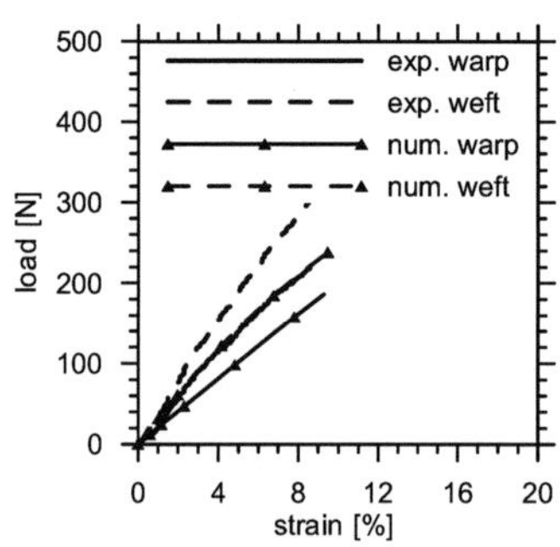

Figure 5. Dry textile: biaxial tensile test.

The numerical macroscopic mechanical responses of the RV subject to uniaxial traction in warp and weft directions are depicted in

Figure 4 and compared to experimental data [8]. The numerical results of the biaxial traction are reported in Figure 5.

Filled Textile: Numerical Results

The polyester monofilament technical textile when filled with a stencil could be considered as a composite textile material. The homogenization numerical approach, used above for the dry textile, is now applied to the RV of the filled textile in order to obtain its macroscopic mechanical features.

The RV geometry contains the same solid model of the RV dry monofilament textile employed above. Embedding the plane weave textile in a parallelepiped matrix box, the 3D geometry of the filled textile is rebuilt with a mesh having 68524 nodes and 246912 four nodes tetrahedral elements C3D4 [11].

In a screen printing application the textile component works in the elastic regime of the fibres. Therefore, in the following, the numerical analysis of the

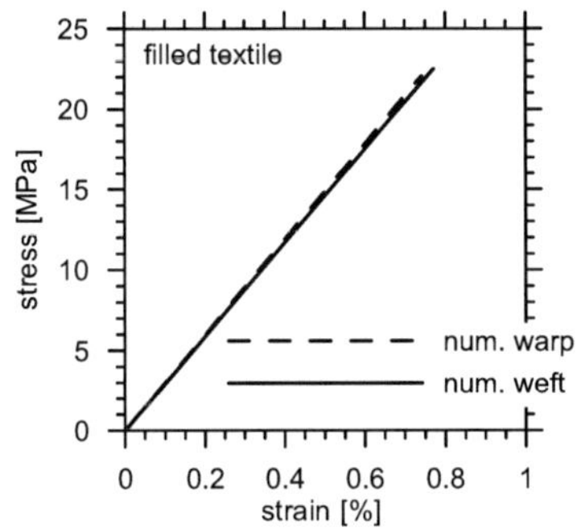

Figure 6. Filled textile: biaxial tensile test.

filled textile RV takes into consideration the elastic features of the fibres and matrix.

The fibres have the same elastic properties used in the analyses of the dry textile (see Table II), while the matrix is considered isotropic with elastic modulus and Poisson's ratio equal to 2000MPa and 0.32 respectively.

Figure 6 shows the macroscopic mechanical response during an biaxial traction of the filled textile RV. Being the stiffness of the materials in the RV similar, the elastic response of the composite textile is almost the same in both warp and weft directions.

SECOND SCALE MODELLING

The second scale analysis deals with a textile structure, made with the monofilament polyester textile previously presented (Figure 7a), intended for printing the picture 'P'. The application is only for academic purpose, it does not simulate a real screen printing component.

The geometry of the textile structure is represented in Figure 7b. It consists of a metallic square ring on which the textile is glued. In Figure 7b the grey surface represent the filled with stencil textile while the white surface is the dry textile in which the ink drops.

The finite element model of the textile component was prepared with 5720 nodes and 1814 8-nodes quadrilateral elements [11]. To reproduce the biaxial load condition before fixing the textile on the metallic ring, a displacement of 6mm orthogonal to the boundary is applied (working stain level ≈2%).

(a)

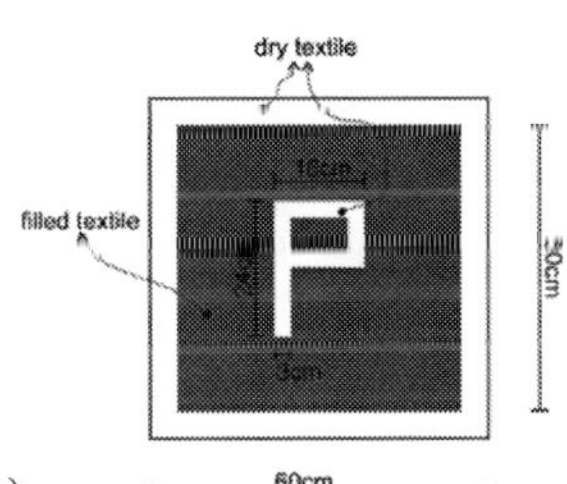

(b)

Figure 7. (a) Textile structural component; (b) geometry of the considered academic application.

The homogenized mechanical properties introduced in the numerical second scale model were obtained by the numerical first scale analyses of dry and filled textiles. They are listed in Table III, being 1 and 2 the warp and weft directions respectively; ν the Poisson's ratio and G the shear elastic modulus obtained from experimental tests for the dry textile (see [8]) and empirically set for the filled textile.

The numerical results are detailed in term of the Mises's stress in Figure 8 and strain components in the warp and weft directions in Figure 9.

The stress distribution (Figure 8) shows the low working level of the textile if compared to the strength of a polyester fibre (more than 650MPa [8]). The strain distribution (Figure 9) highlights the quasi-uniform strain components in the dry textiles (E_{11} in the range 1.6÷1.8% and E_{22} in the range 2÷2.4%). A uniform deformation of the textile in which the ink penetrates is the objective in any screen printing technique because more uniform is the deformation of the textile, more uniform is the shape of the ink drops, than increase the quality of the printing.

TABLE III. ELASTIC MECHANICAL PROPERTIES OF THE HOMOGENIZED TEXTILES.

	E_1 [MPa]	E_2 [MPa]	ν_{12}	ν_{21}	G_{12} [MPa]
Dry textile	4661	3083	0.14	0.22	113
Filled textile	2973	2918	0.32	0.31	1126

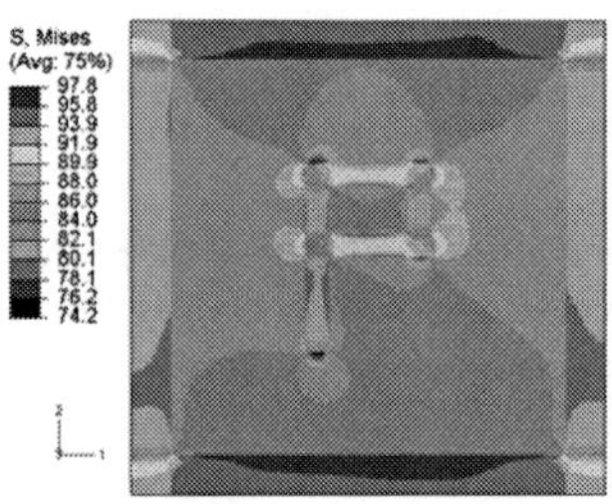

Figure 8. Contour plot of the von Mises's stress.

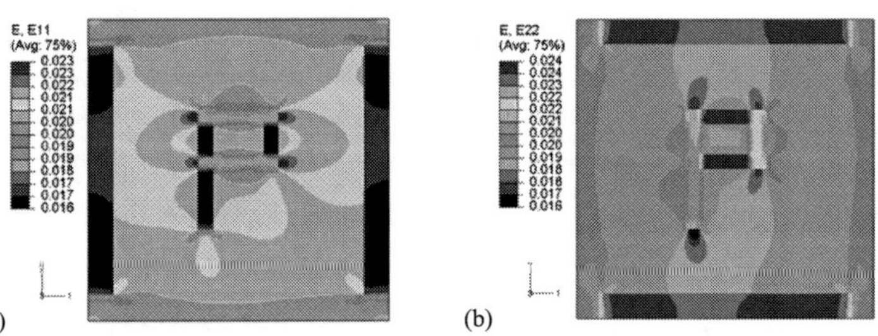

(a) (b)

Figure 9. Contour plot of the strain components: (a) E_{11}; (b) E_{22}.

CONCLUSIONS

A two scale numerical approach for the analysis of textile structural components was detailed.

In the first scale analysis the mechanical behaviour of a dry and filled with matrix monofilament textile was investigated assuming the periodic distribution of the fibres in the weft and warp directions. The numerical second scale modeling analyzed, at the structure level, the global mechanical response of a textile component employed in a screen printing technique. The predicted deformed shape shows a uniform strain distribution in the dry textile (i.e. the textile contained in the area of the picture to be printed). This is the expected results in any applications of this kind.

Therefore the numerical two scale procedure presented in this paper could be very useful for industrial textile applications to understand the influence of some geometrical features and material mechanical parameters on the overall behaviour of the textile and to design optimized textile structures for specific applications.

ACKNOWLEDGEMENT

The author thanks Pietro Camapagna for his help in the numerical analyses.

REFERENCES

1. Long A. C. 2005. Design and manufacture of textile composites. Woodhead Publishing Ltd.
2. Gasser A., P. Boisse, S. Hanklar. 2000. "Mechanical behaviour of dry fabric reinforcements. 3D simulations versus biaxial tests", Comp. Mat. Science, 12:7-20.
3. Luo Y., I. Verpoest. 2002. "Biaxial tension and ultimate deformation of knitted fabric reinforcements", Composite Part B, 33:197-203.
4. Lomov S. V., G. Huysmans, I Verpoest. 2001. "Hierarchy of textile structures and architecture of fabric geometric models", Textile Research Journal, 71:534-543.
5. Liu L., J. Chen, X. Li, J. Sherwood. 2005. "Two-dimensional macro-mechanics shear model of woven fabric", Composites Part A, 36:105-114.
6. Tarfaoui M., J. Y. Drean. 2001. "Predicting the stress-strain behaviour of woven fabrics using the finite element method", Textile Research Journal, 71:790-795.
7. Sagar T.V., P. Potluri, J.W.S. Hearle. 2003. "Mesoscale modelling of interlaced fibre assemblies using energy method", Computational Materials Science, 28:49-62.
8. Carvelli V., C. Corazza, C. Poggi. 2008, "Mechanical modelling of monofilament technical textiles", Computational Materials Science, 42:679-691.
9. Carvelli V., C. Poggi. 2001. "A homogenization procedure for the numerical analysis of woven fabric composites", Composite Part A, 32:1425-1432.
10. Nemat-Nasser, Hori M. 1999. Micromechanics: Overall Properties of Heterogeneous Solids. Elsevier Science Publishers.
11. ABAQUS Standard User's Manual, rel. 6.6, Hibbit, Karlsson & Sorensen Inc, 2003.

FEA of Textiles and Textile Composites: A Gallery

S. V. LOMOV, D. S. IVANOV, V. KOISSIN, I. VERPOEST,
M. ZAKO, T. KURASHIKI and H. NAKAI

ABSTRACT

Meso-scale finite element modelling of textiles and textile composites (scale of the unit cell of the textile structure) is a powerful tool for homogenisation of mechanical properties, study of stress-strain fields inside the unit cell, determination of damage initiation conditions and sites and simulation of damage development and associated deterioration of the homogenised mechanical properties of the composite.

The paper presents a gallery of cases of such a modelling, discussing issues of building finite element models of textile structures (geometry, meshing, boundary conditions), of interpretation of the results (homogenization, damage initiation criterion, damage propagation) and of verification of the models.

The following cases are considered:

- Composites reinforced by 2D woven fabrics: validation of FE modeling using full-field strain measurements
- 3-axial braided composites: building of an adequate geometry and mesh, boundary conditions for ¼ of the unit cell, damage initiation and propagation
- Structurally stitched thick woven composite: interaction of the stitching with the "body" of the plate
- Knitted fabrics made of shape memory alloy wires: hyperelastic material behaviour, contacts and slipping of the wires

All the cases are supported by comparison with experimental data.

Stepan V. Lomov, Dmitry S. Ivanov, Vitaly Koissin, Ignaas Verpoest, Department MTM, Katholieke Universiteit Leuven Kasteelpark Arenberg 44 B-3001 Leuven Belgium
Masaru Zako, Tetsusei Kurashiki, Hiroaki Nakai, Department of Management of Industry and Technology, Osaka University, 2-1, Yamadaoka, Suita, Osaka, 565-0871, Japan

INTRODUCTION

Finite element (FE) meso-modelling of textile composites and textiles themselves, apart from being a popular research topic, is more and more considered as a tool for industrially relevant calculations, especially for prediction of damage threshold and progressive damage. The term "meso" refers to the scale level of unit cell of the textile reinforcement. FE simulation can provide detailed information on the stress-strain fields inside the unit cell of textile composites, for elastic or non-elastic behaviour of fibres and matrix, leading to prediction of damage initiation and further development of damage. Homogenised stiffness of composite can be as well predicted using FE models, but their computational expense is much higher then simpler methods, which accurately predict homogenised stiffness (e.g. method of inclusions [1, 2]). An overview of the state-of-the-art can be found in the proceedings of a micro-symposium "FE modelling of textiles and textile composites" (St.-Petersburg, 2007). An initiative of creating "The Textile Composite Archives" [3], to be hosted on a server at Texas A&M University is a follow-up of discussions during that meeting. The "gallery" presented in this paper lies in the streamline of the proposal of this data bank.

The tools (as enumerated in conclusion of the review paper [4]) used in the examples of FE analysis given in the present paper, are shown in Table I.

TABLE I. SOFTWARE TOOLS.

	2D woven	*3-axial braided*	*Structurally stitched*	*Knitted*
Geometric modeller	WiseTex			
Geometry corrector of yarn interpenetration	MeshTex	intermediate analysis in ANSYS	mesh superposition	contact pairs in ANSYS
Meshing	MeshTex	ANSYS	MeshTex	ANSYS
Material properties	MeshTex	ANSYS macros	MeshTex	ANSYS macros
Boundary conditions	MeshTex	ANSYS macros	MeshTex	ANSYS macros
FE solver, post-processor	SACOM	ANSYS	SACOM	ANSYS
Homogenisation	MeshTex	ANSYS macros	MeshTex	ANSYS macros
Damage	SACOM	ANSYS macros	SACOM	n/a

References: WiseTex [2], MeshTex [4], SACOM [5], mesh superposition [6], intermediate FE analysis for correction of geometry [7], damage analysis [4, 5, 7]

2D WOVEN COMPOSITE: HOMOGENISED PROPERTIES AND ELASTIC STRAIN FIELDS

The work reported in this section is a part of a wider modelling effort, done in collaboration with A.E. Bogdanovich and D. Mungalov [8], which also includes modelling of 3D woven composites, performed using MOSAIC 3D model [9].

The plain weave was made of 2275 tex PPG Hybon 2022 E-glass roving in both warp and fill directions: 1.95 ends/cm and 1.60 picks/cm. Note the difference between fibre amounts in warp and fill directions, which asked to alternate fabric layer orientation in the 4-layer preform. Composite material samples were fabricated in a vacuum bag with Dow Derakane 8084 Epoxy-Vinyl Ester resin as a stack of four woven layers of 815 g/m^2 with orientation 0°/90°/90°/0°.

The geometrical model (Figure 1a) is based on the following data obtainable without cross-sectioning of the composite plate:

– Thickness of the composite plate, which was measured to be 2.45±0.08 mm. This value gives the fibre volume fraction of the plate 52.4% and the average thickness of the layers 0.612 mm, not accounting for the nesting of the yarns.

– Spacing of the warp and weft yarns – directly used in WiseTex model.

– Thickness of the warp and weft yarns was assumed to be 0.3 mm, which, together with the crimp of the yarn calculated using WiseTex, produced the thickness of one layer equal to the measured value 0.612 mm. The areal density of the fabric in WiseTex model is 809 g/m^2 and the overall fibre volume fraction is 52.0%, which correspond to the specified and measured values.

– Width of the warp and weft yarns was estimated using the images of the plate as 5.1 mm, which gives fibre volume fraction inside the yarns of 75%.

– The shape of the warp and fill yarn cross section was assumed to be elliptical

TABLE II. EFFECTIVE PROPERTIES OF 2D WOVEN LAMINATE.

	experiment	OA	M-T	FE
E_1, GPa	**26.0**±1.5	25.1	25.4	25.2
E_2, GPa	**26.0**±1.5	25.1	25.4	25.2
E_3, GPa	n/a	12.2	8.63	8.55
G_{12}, GPa	n/a	4.4	4.2	4.5
ν_{12}	**0.264**±0.148	0.118	0.117	0.128
$\nu_{13} = \nu_{23}$	n/a	0.372	0.396	0.402
$E_{45°}$, GPa	**12.2**±0.4	13.6	13.0	n/a
$G_{45°}$, GPa	n/a	11.2	11.4	n/a
$\nu_{45°}$	**0.610**±0.148	0.524	0.549	n/a

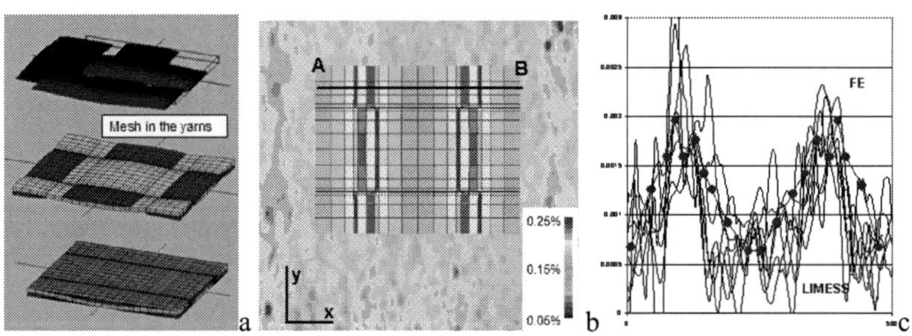

Figure 1. 2D woven laminate model: (a) WiseTex model of a ply; (b) ε_x field calculated in FE overlapped with the optically registered field, applied strain $\langle\varepsilon_x\rangle$=0.1%; (c) profile of ε_x along AB.

FE model for unit cell of the 2D plain weave composite (Figure 1a) was built and solved using *MeshTex/SACOM* software, which directly imports geometrical model of one layer of the fabric from *WiseTex*. In order to mesh the unit cell, *MeshTex* inserts thin (0.005 mm) layers of matrix in between the warp and fill yarns and on the surface of the unit cell. The resulting V_f of the unit cell is hence slightly decreased to 47.2% (52.0% in *WiseTex* model). The unit cell is unbalanced, hence it is difficult to impose correct boundary conditions on the laminate model. Because of these two factors, the following calculation route was adopted:

1. Perform FE homogenisation for a woven ply using periodic boundary conditions in all the directions
2. Scale elastic properties to fibre volume fraction 52.0%
3. Calculate homogenised laminate properties using Classical Laminate Theory with properties of lamina calculated as above

Use of periodic boundary conditions in thickness direction may be questioned as the plate has only four plies and free surfaces are not far from the middle of the plate. However, the overestimation of the stiffness, connected with that, is believed to compensate the underestimation of the stiffness resulting from the neglecting of the nesting of the layers.

Table II shows effective elastic properties of the 2D weave composites calculated using iso-strain Orientation Averaging (OA), Mori-Tanaka (M-T), FEA and experimental results. The OA, M-T and FE analyses give results very close to the experimental data for all the considered engineering constants. All theoretical values of Poisson's ratio ν_{12} are consistently lower (about two times) than the average experimental value, but they are near the edge of the experimental scatter, which is very large for this characteristic. FE gives slightly better agreement for ν_{12}. Figure 1b compares the ε_x field (loading in x direction) calculated in FE overlapped with the optically registered field (LIMESS system was used). Comparison of the two should be done bearing in mind, hat the FE calculation were done for periodicity of the fields in z-direction, whilst the experimental strains are measured on the free surface of the sample [10]. FE modelling provides good agreement with the experimental data both for homogenised properties of the composite and for the strain fields inside the unit cell.

THREE-AXIAL BRAIDED COMPOSITE: INTERPENETRATION OF THE YARN VOLUMES, ¼ UNIT CELL, DAMAGE MODELLING

The reinforcement (Figure 2a) is a triaxial braid made of carbon rovings (linear density 1600 tex), areal density of 600 g/m^2. The four-layer braided composite has interlacing yarns in three directions: braiding yarns (±45°) and inlay yarns. Five yarns (two in each biaxial direction and one inlay yarn) are interlaced at a relatively small distance, which leads to a complex shape of the yarn mid-lines and cross sections. The unit cell has a large open inner space, which is not occupied by the reinforcement and which causes nesting of the layers. There is variation of fibre volume fraction across the yarns: at the edges the fibre volume fraction is 15% lower than in the middle zone. The composite is impregnated with epoxy matrix

(Epicote 828 LV/Epicure DX 6514, mixing ratio 100/17). The composite samples were tested in tension in three directions (indicated in Figure 2a) with acoustic emission (AE) registration. Typical diagrams of cumulative energy of events are shown in Figure 2d together with diagrams of degradation of stiffness. These diagrams are used to determine the strain corresponding to onset of damage ε_1 and a second stage of the damage accumulation (called ε_2) as a "knee" on AE curves, which is close to a slope change in the tensile (stiffness degradation) diagrams. More details of the experiments can be found in [7].

Periodic boundary conditions (based on the translation symmetry of the unit cell as a whole) are used. In order to decrease the computational efforts, following [11] the internal symmetry of the unit cell is accounted for (Figure 2b). The yarn shapes were adjusted to avoid interpenetration of the volumes, using “separate and compress” algorithm, as described in [4]. This resulted in a mesh (Figure 2c) with complex non-symmetrical shapes of the yarns.

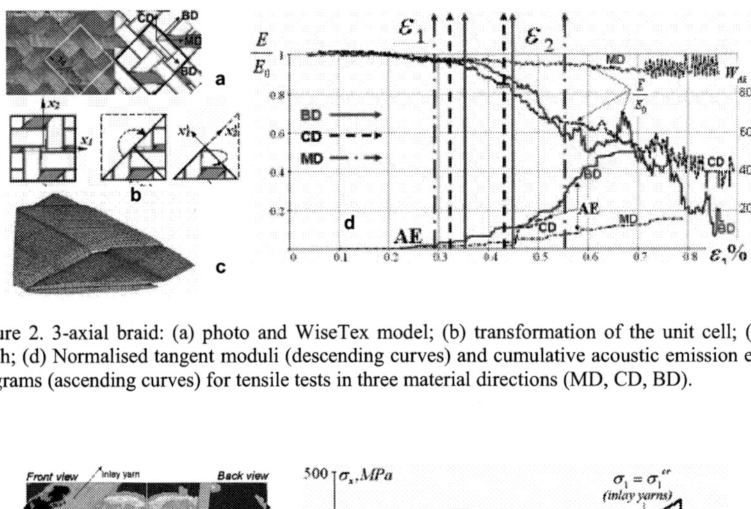

Figure 2. 3-axial braid: (a) photo and WiseTex model; (b) transformation of the unit cell; (c) FE mesh; (d) Normalised tangent moduli (descending curves) and cumulative acoustic emission energy diagrams (ascending curves) for tensile tests in three material directions (MD, CD, BD).

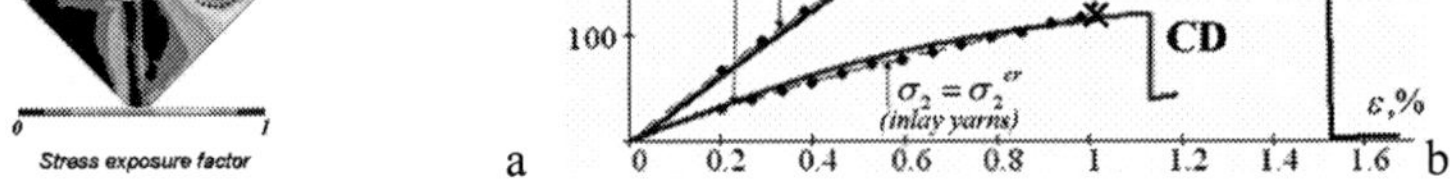

Figure 3. FE analysis of 3-axial braided composite: (a) damage initiation modelling using Puck criterion; (b) stiffness degradation modelling.

Stiffness constants for loading in MD and CD directions were well predicted by FE calculations (inside the experimental scatter). Damage initiation (ε_1 threshold) was predicted in FE calculations using Puck criterion with values of the transversal and shear strength of the UD carbon/epoxy composite 28 GPa and 74 GPa respectively. This results in a good prediction of the experimentally observed damage initiation thresholds (calculated/experimental): MD: 0.35% / 0.29±0.09%; CD: 0.24% / 0.33±0.07%, as well as the damage mode (transverse cracking) and location of the damage initiation. The stiffness degradation model was based on the damage mechanics of Ladeveze for UD composites, representing the impregnated yarns (the details can be found in [7]). Figure 3b depicts the results of the calculations. The analysis corresponds well with the experiments.

STRUCTURALLY STITCHED COMPOSITE: EARLY DAMAGE ONSET

Quasi-UD hybrid woven fabric (warp: 24K carbon tows alternated with thin polysulfone yarns; weft: the same polysulfone yarns) with the total areal weight of 226 g/m² is used as the raw material. The reinforcement was composed of 28 plies having symmetric stacking sequence [90/45/0/0/-45/90/-45/0/0/45/0/-45/0/45]s, where 90° corresponds to the warp (carbon fibre) direction in the surface plies. For the structural stitching, 1K carbon yarn and the tufting method are employed. The machine direction coincides with 0° direction of the preform. The piercing pattern is square 5x5 mm; the backside loop height is also about 5 mm. Composite plates (stitched and non-stitched) are manufactured using the liquid resin (toughened epoxy) infusion. The final thickness is 5.32 mm that gives the average fibre volume fraction of 58%. The material (Figure 4) was supplied by Dassault Aviation.

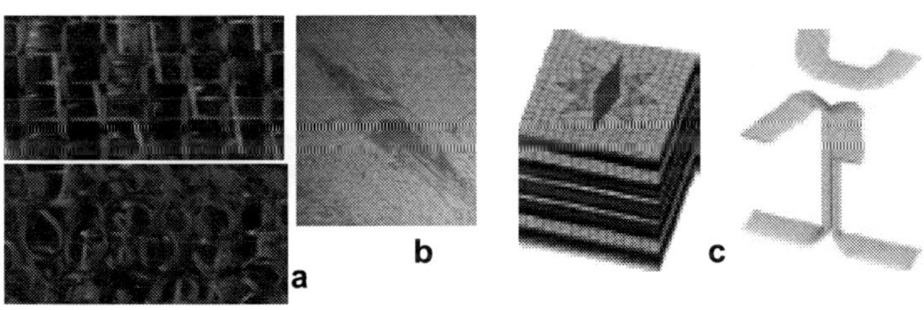

Figure 4. Structurally stitched woven composite: (a) face and back side of the preform; (b) opening inside a ply created by the structural stitching; (c) FE model: plies and the stitching yarn.

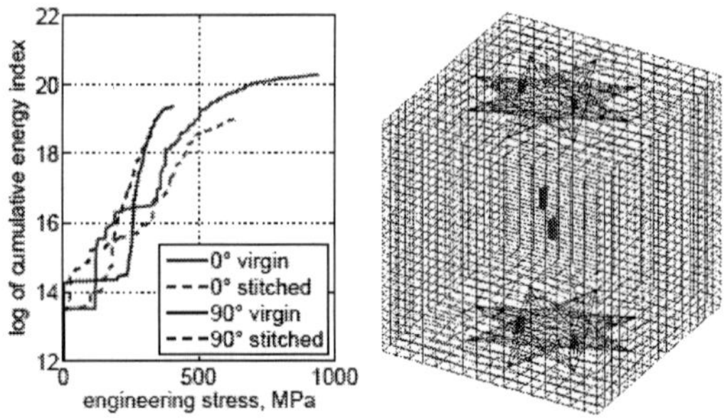

Damage initiation stress, MPa

	test	*FE*
no stitching	247	250
stitching	120	140

Figure 5. Damage in the structurally stitched laminate, loading in 90° direction: (a) AE diagrams; (b) damage initiation locations, FE model (only plies are shown); (c) damage initiation stress, MPa.

The composite samples were tested in tension in 0° and 90° direction, with registration of damage initiation and development using acoustic emission. An important observed phenomenon was earlier initiation of damage in the stitched samples loaded in 90° direction (Figure 5). X-ray observation has shown that this preliminary damage starts inside the stitching yarns. The early damage onset has been reproduced in FE simulation. Figure 4c shows the FE model, consisting of 28 layers of UD laminate (polysulfone properties are close to the epoxy, and carbon crimp is negligible). The stitching site is representing by a resin-filled opening. The direction of the fibres around the opening are deviated, following “streamlines” around the opening. The stitching yarn itself is modelled as superimposed mesh; the shape of the stitching yarn and its compressed dimensions were measured on the composite samples. Damage initiation is determined using Hoffmann criterion. Figure 5 demonstrates that the FE model predicts the early onset of damage and the location of the initial damage.

KNITTED SHAPE MEMORY ALLOY WIRES

The studied plain knitted fabric is knit with 0.15 mm diameter NiTi wires. Figure 6 shows the fabric, WiseTex model and tensile diagram of the wire. Biaxial tests were performed for two conditions: deformation in the course direction, width fixed in the wale direction, and vice versa.

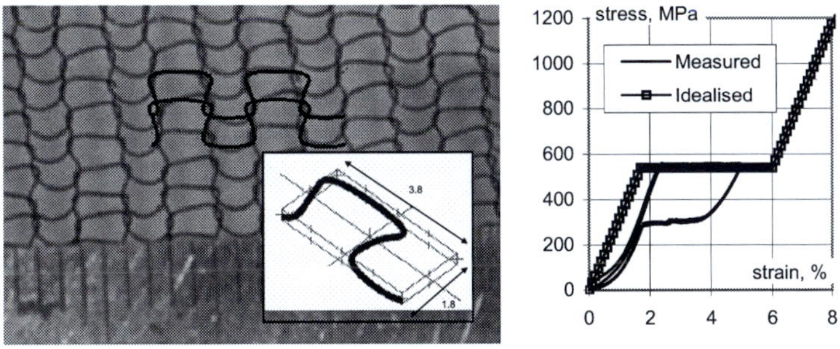

Figure 6. Knitted NiTi wires and tensile diagram of the wire.

The geometrical model of the unit cell of the fabric was first built in WeftKnit software (part of WiseTex suite), then exported in ANSYS and meshed. Interpenetrations of the wires were handled using contact algorithms in ANSYS. Figure 7 shows the results of the calculations, together with configurations of the unit cell on the different stages of deformation (note the "locking" of the loops) and distribution of the three phases of NiTi in the wires.

When compared with experiment (Figure 7), the FE simulation shows higher stiffness of the fabric starting from about 20% of the applied strain, which is not observed in the experiments. The transition from the low stiffness to the high stiffness regime is defined by the locking of the loop, which depends on the initial gap between the wires in the loop contact zone. These gaps in the real fabric have a certain statistical distribution. The model represents the average of it. When the real fabric is deformed, all the gaps should be transformed into the locking state before the fabric will start resisting by the wire tension rather than by the slippage and decrimping.

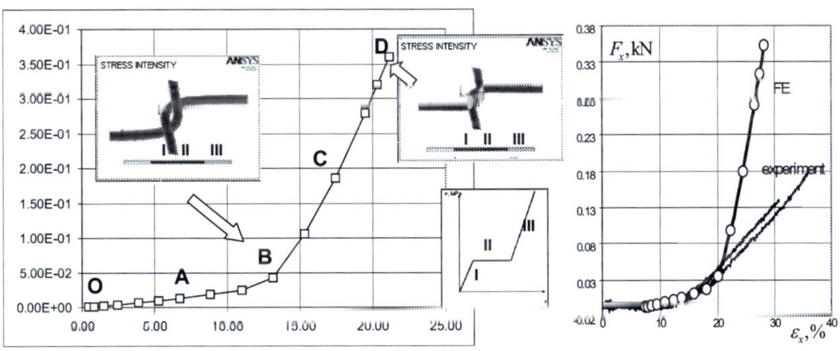

Figure 7. Computed tensile diagrams (course direction) of knitted NiTi and comparison with experiment.

CONCLUSION

The presented "gallery" of FE simulations of textiles and textile composites demonstrate that the state-of-the-art models are capable to adequately simulate deformation and damage behaviour for a wide range of materials.

ACKNOWLEGEMENTS

The work reported here was supported by EU funded ITOOL project, and KULeuven PhD grant for D.Ivanov.

REFERENCES

1. Lomov, S.V., G. Huysmans, Y. Luo, R. Parnas, A. Prodromou, I. Verpoest, and F.R. Phelan, Textile Composites: Modelling Strategies, *Composites part A*, 2001, 32(10): 1379–1394.
2. Verpoest, I. and S.V. Lomov, Virtual textile composites software Wisetex: integration with micro-mechanical, permeability and structural analysis, *Composites Science and Technology*, 2005, 65(15–16): 2563–2574.
3. Boisse, P., J. Crookston, D.S. Ivanov, S.V. Lomov, A.C. Long, I. Verpoest, J. Whitcomb, and M. Zako. Data bank for validation of finite element analysis of textiles and textile composites: a proposal. in *13th European Conference on Composite Materials (ECCM-13)*. 2008, Stockholm.
4. Lomov, S.V., D.S. Ivanov, I. Verpoest, M. Zako, T. Kurashiki, H. Nakai, and S. Hirosawa Meso-FE modelling of textile composites: Road map, data flow and algorithms, *Composites Science and Technology*, 2007, 67: 1870–1891.
5. Zako, M., Y. Uetsuji, and T. Kurashiki, Finite element analysis of damaged woven fabric composite materials, *Composites Science and Technology*, 2003, 63: 507–516.
6. Nakai, H., T. Kurashiki, and M. Zako, Individual modeling of composite materials with mesh superposition method under periodic boundary condition, in *Proceedings of the 16th International Conference on Composite Materials (ICCM-16)*. 2007: Kyoto, CD edition.
7. Ivanov, D.S., S.V. Lomov, F. Baudry, H. Xie, B. Van Den Broucke, and I. Verpoest, Failure analysis of triaxial braided composite, *Composites Science and Technology*, in print.
8. Lomov, S.V., D.S. Ivanov, I. Verpoest, A.E. Bogdanovich, D. Mungalov, M. Zako, T. Kurashiki, and H. Nakai. Predictive analyses and experimental validations of effective elastic properties of 2D and 3D woven composites. in *13th European Conference on Composite Materials (ECCM-13)*. 2008, Stockholm.
9. Bogdanovich, A.E., Multi-scale modeling, stress and failure analyses of 3-D woven composites, *Journal of Materials Science*, 2006, 41(20): 6547–6590.
10. Lomov, S.V., D.S. Ivanov, I. Verpoest, M. Zako, T. Kurashiki, H. Nakai, J. Molimard, and A. Vautrin, Full field strain measurements for validation of meso-FE analysis of textile composites, *Composites part A*, in print.
11. Whitcomb, J., C.D. Chapman, and X. Tang, Derivation of boundary conditions for micromechanics analyses of plain and satin woven composites, *Journal of Composite Materials*, 2000, 34(9): 724–747.

Effect of Voids on Elastic Properties of Real C/C Composites

M. KOSEK, P. SEJAK and P. VOZKOVA[1]

ABSTRACT

Effect of real structure defects, voids, on elastic properties of textile C/C composite was studied. The microphotographs of the structure were used for automatically finding of voids in specimen. Then the void shape was approximated either by ellipse or Lissajouse curves. In the second case, the accuracy can be specified. Applying Finite Element Method from MATLAB to a specimen containing the real void (approximated by ellipse) all the elastic parameters, local and global, was calculated, using own scripts. The void changes strain and stress in its immediate environment maximally by 15%. The influence of the void on average elastic parameters, Young modulus, for instance, is negligible, however.

INTRUDUCTION

The modeling of textile structures was started by Pierces model [1] almost one century ago. Considerably later the massive use of computers enabled the study of relations between structure and physical properties of textiles. A leading role and a lot of work in this area were made by Lomov and his co-workers, see, for instance Ref. [2] among many other excellent references. The goal of their work was the system WiseTex [7] that performs full analysis of all existing textiles and can be used for the design of textiles exhibiting specified parameters. The above mentioned and, may be, other less known approaches consider ideal periodic structure that can be very complicated, of course. This method was also applied to the efficient study of textile composites, again considering ideal structure.

However, the textile composite, due to the technological processing, exhibits structure that can considerably differ from regular one. One of type of difference is the irregularity of objects. The longitudinal yarns, for instance, cannot be modeled by

[1]Technical University of Liberec, Studentska 2, CZ-46117 Liberec 1, Czech Republic

simple sinus function [3]. From the practical point of view structure defects in the form of empty voids plays can affect the composite parameters. However, no complete study was found in literature. This is the reason why this paper deals with void finding, approximation and evaluation using Finite Element Method (FEM).

THEORY

Three basic topics are considered: void finding, void shape approximation and void shape effects on elastic properties. The void finding is fully automated, uses sophisticated algorithms and was described in detail elsewhere [4]. Therefore we will focus to remaining two topics.

Void Shape Approximation

Three different 2D void shape descriptions are used: approximation by ellipse, approximation by Lissajouse curves and smoothed (or real) polygon connecting scanned points. Approximation by ellipse is the simplest case; Lissajouse curves allow setting the level of approximation. In both the cases the analytical formula is obtained.

The approximation by ellipse was made either by geometrical procedure taken from free program on web or as a physical problem of the superposition of two mutually perpendicular vibrations of the same frequency, but different phase shift and amplitudes, of course.

The method of Lissajouse curves should be described a little in detail. The Lissajouse curves is created by the superposition of two perpendicular vibrations of n-th and m-th harmonics of amplitudes a_n a a_m, respectively, with frequencies $f_n=n{\cdot}f_o$, $f_m=m{\cdot}f_o$ (f_o is the frequency of basic harmonics) and by phase constants φ_n and φ_m. The coordinates x and y of these vibrations are given by formulae

$$\begin{aligned} x &= a_n \cdot \cos(2\pi f_n t + \varphi_n) \\ y &= a_m \cdot \cos(2\pi f_m t + \varphi_m) \end{aligned} \qquad (1)$$

The parameter t has a meaning of time for vibrations. In general $t \in <0,1>$. It is evident that ellipse of semi axes a and b is a special case of Lissajouse curve for $m=n=1$, $a_n=a$, $a_m=b$, $\varphi_n=0$, $\varphi_m=\pi/2$.

If we replace the pure harmonic vibrations in (1) by periodic ones and apply the Fourier series using of K harmonics, we get the parametric formula for a closed curve of relatively general shape. The new formula contains the sum of K equations (1). The void shape approximation starts by the application of Fast Fourier transform to the x_i and y_i coordinates of scanned points in order to get all the parameters in the sum of equations (1). Then several of the most important harmonics are selected, others are replaced by zeros in the sum of (1). The closed curve is formed by points $[x,y]$ for $t \in <0,1>$. It should be compared with scanned points $[x_i, y_i]$ in order to judge the level of void shape approximation. The degree of approximation can be set by the number of selected harmonics. In general, the higher number of the most important harmonics, the better approximation.

Elastic Parameters of Composite

In general, a composite is a heterogeneous anisotropic medium of a complicated structure. Therefore its elastic properties should be described completely only by the use of tensor physics. There are a lot of notations in different literature dealing with elasticity applications. In this paper we use the notation recommended for piezoelectricity. Furthermore all formulae use Einstein notation, the summation from 1 to 3 is made for each index that appears two times in the product. More information can be found in monographs, [5], for instance.

The elastic medium subjected any external action exhibits the shift $\vec{u} = (u_1, u_2, u_3)$ of all its points. The shift components are functions of medium coordinates in initial state, $u_i = u_i(x_1, x_2, x_3)$, $i = 1, 2, 3$.

Due to the external action, each elements of the medium performs three operations: shift as a rigid element, rotation as a rigid body and deformation, strain, typical for elastic body. The shift of element is given simply given by the vector $\vec{u}$. The elements of rotation A_{ij} and strain S_{ij} are given by formulae

$$A_{ij} = \frac{1}{2}\left(\frac{\delta u_i}{\delta x_j} - \frac{\delta u_j}{\delta x_i}\right),\ S_{ij} = \left(\frac{\delta u_i}{\delta x_j} - \frac{\delta u_j}{\delta x_j}\right), \tag{2}$$

respectively. The rotation A_{ij} is an anti-symmetric second rank tensor, with zeros on main diagonal and three independent components. Each of them gives the rotation along one of the coordinate axes. The strain S_{ij} is a symmetric second rank tensor, containing six independent components in general. The diagonal components have the meaning of elongation, while the non-diagonal components give the shear strain in the plain given by indices *i*, *j* with respect to the axis, index of which is not in indices of the strain component.

The stress is given by the symmetric second rank tensor T_{ij}. Diagonal components are normal stresses, while the other components are the tangential ones. The first index gives the normal to area on which the force takes effect; the second index is the component of the force.

The relation between the stress and strain is given by Hook's Law

$$T_{ij} = c_{ijkl} S_{kl}, \tag{3}$$

where c_{ijkl} are components of fourth rank tensor of elastic module, stiffness. In reverse form of Hook's Law the elastic compliances s_{ijkl} are used. Only 21 components of elastic tensors from total 81 ones are independent in the most general case of triclinic crystallographic class.

In the static case the basic differential equation of stress equilibrium has the from

$$\frac{\delta T_{ij}}{\delta x_j} = 0. \tag{4}$$

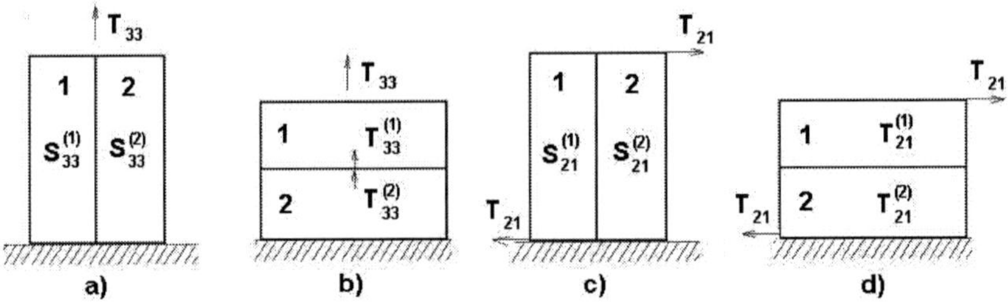

Figure 1. Boundary conditions of different types.

By substitution of strain components (2) into Hook's Law (3) and then into the equilibrium differential equation (4) we get the partial differential equation for elastic shift components $u_i=u_i\ (x_1,\ x_2,\ x_3),\ i=1, 2, 3$. Such solution must satisfy boundary conditions for elastic shift, strain and stress.

We give a quite general form of boundary conditions for boundary between two elastic media, see Fig. 1. There are two types of boundary conditions:

1. Dirichlet boundary condition requires that elastic shift changes continuously when passing the boundary.
2. Neumann boundary conditions are for strain and stress. The type depends on the orientation of boundary, Fig. 1. If the normal stress is parallel to the boundary, the strain is continuous, Fig. 1a. If the normal stress is perpendicular to boundary, the strain changes continuously, Fig. 1b. For tangential stress the conditions are opposite. The tangential stress perpendicular to boundary leads to continuous strain, Fig. 1c. If the tangential stress is parallel to boundary, it changes continuously, when passing the boundary, Fig. 1d. In the case of general angle, the stress must be decomposed into normal and tangential components.

Simple and sufficient boundary conditions on the body surface are either the shift or the stress. Usually they have zero or constant value.

RESULTS

We studied actual structure on samples of C/C composite prepared by standard technology. Sample dimension were approximately 30 x 30 x 1.5 mm. The sample surface was scanned by optical microscopy of high resolution on planes perpendicular to thickness. In order to get volume information, the layers of thickness of 0.2 mm were removed in successive steps and a new surface scanned.

After the photograph processing by standard image procedures original programs realized in MATLAB were applied to image data. In the first step the void boundaries were found. Then the boundaries were approximated by analytical curves. In the last step Finite Element Method (FEM) was applied to approximated void and the output data was processed by special programs in order to get all possible results. Typical and illustrative results are shown in this chapter.

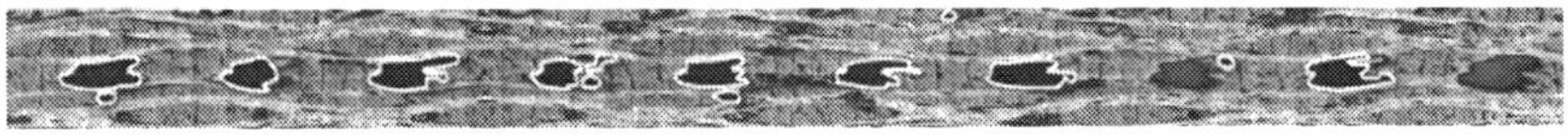

Figure 2. Automatically found voids.

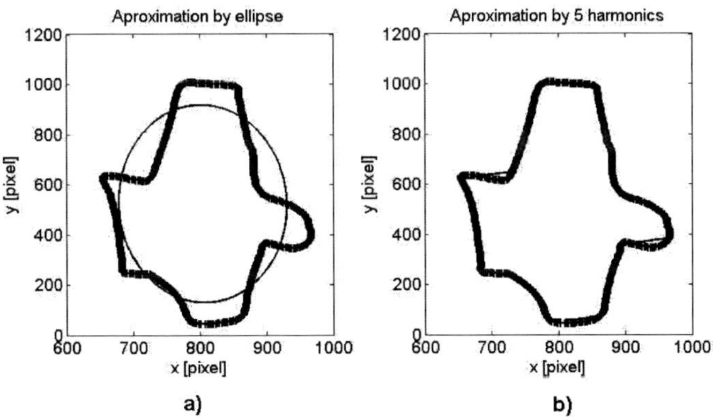

Figure 3. Comparison of approximations by ellipse (a) and Lissajouse curves (b).

Automated Finding of Voids

Typical result from the system for automated void boundary finding, described in Ref. [4] is shown in Fig. 2. All important voids were found. Also it is evident from Fig. 2 that system distinguishes between empty voids and areas filled with matrix. The last and the last but two objects in Fig. 2 are voids filled by matrix and therefore they were not found. By the suitable set of system parameters, only the voids containing matrix can be found.

Void Shape Approximation

The 2D cut of the void can be approximated by the ellipse or by Lissajouse curves. Both the approximations are shown in Fig. 3. Since the scales on axes are different, actual shape is not shown. Nevertheless, the approximation by ellipse is satisfactory for practical use. Lissajouse curves make possible very fine approximation; in Fig. 3 only five the most significant harmonics were used.

The 3D void shape can be reconstructed using photographs from plan-parallel cuts. The 3D approximation of the void by three axis ellipsoid is illustrated in Fig. 4, ellipsoid three main cuts in three planes are shown. Different scales are in different parts and axes in order to evaluate the level of approximation and data resolution. In reality the ellipsoid has different semi-axes. Only in basic plane XY there are points sufficient in numbers to allow any approximation. In other two planes the numbers of points are limited by number of removed layers in scanning procedure. Therefore, only the simplest approximation by ellipse is possible. Nevertheless, this approximation is acceptable for praxis.

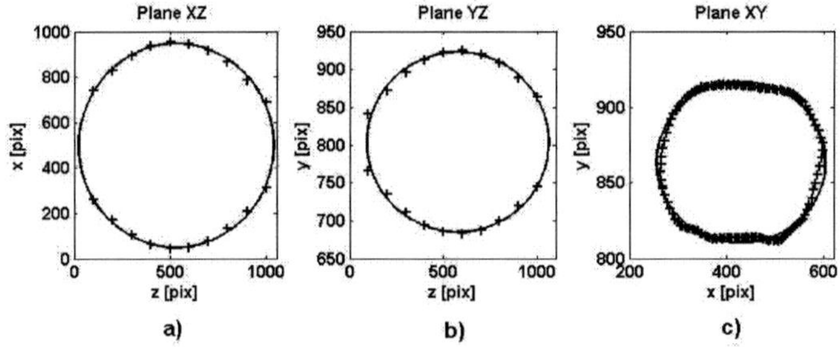

Figure 4. Void approximation by an ellipsoid, three main cuts.

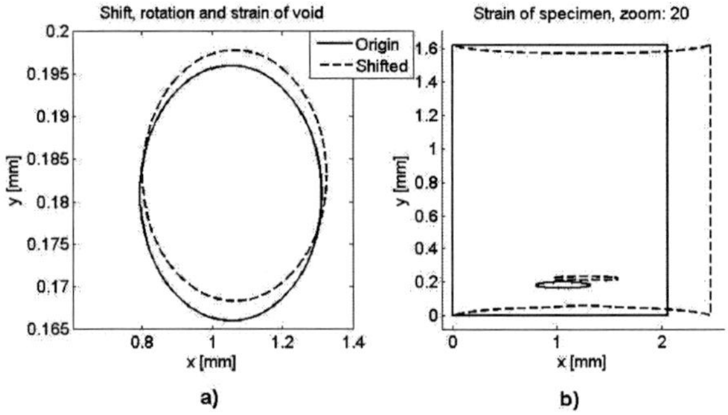

Figure 5. Shift, rotation and deformation of void, real dimensions of void (a) and in the specimen position with zooming of 20 (b).

Application of Finite Element Method

In order to evaluate the possibility of the method and from the reason of lack of correct data and other information, only 2D case of composite cut containing one elliptic void in an isotropic medium was considered. In calculations the 2D specimen has length of 2 mm in direction of X axis and height of 1.6 mm along Y axis. The calculations were made in MATLAB using the PDE (Partial Differential Equation) toolbox. The composite has Young modulus of 50 GPa and Poisson ratio of 0.3, which are the material parameters in basic differential equation (4). The boundary conditions are the following. The uniform uni-axial strain of 1 % was applied to the right hand side of specimen, while the left hand side was clamped, zero shift. Other sides are free, e.g. the stress is zero. Data was send to system partially from script (geometry), partially form work space (boundary conditions, grid, solution, etc.).

The FEM results were saved in file and later processed by original scripts in order to get all the possible information in numeric and graphical form. The following elastics quantities were calculated and visualized: shift, all its derivatives, rotation (using formula (2)), strain, stress, main values and main vectors of both the strain and stress (special procedure) and Young modulus. These quantities are either in a local form or as an average values in the plane normal to force action. Only a few of typical results are shown here.

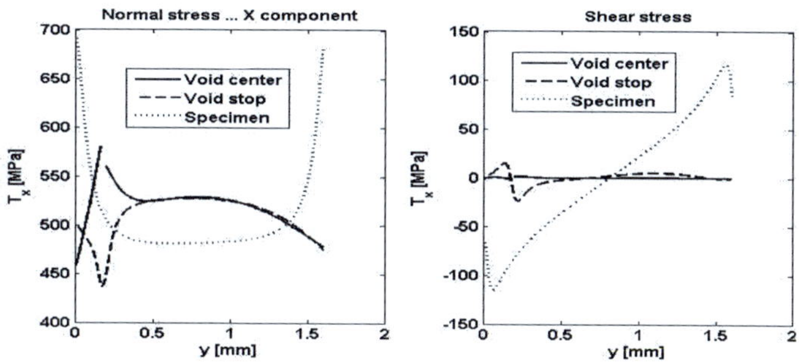

Figure 6. Local stress in vicinity of void; normal component (a) and shear stress (b).

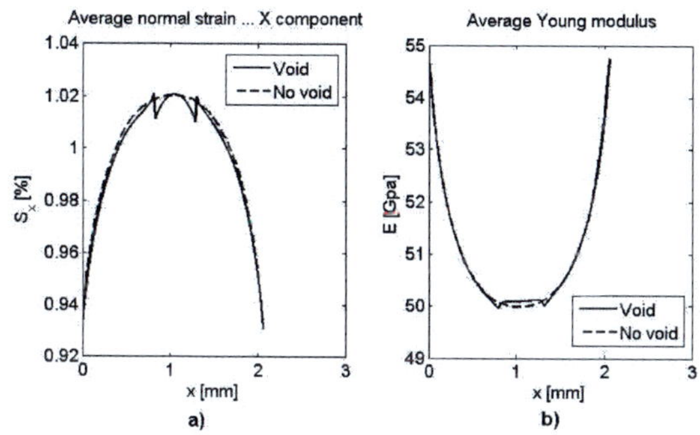

Figure 7. Average values calculated in plane perpendicular to the acting force: normal strain (a) and Young modulus (b).

The calculated shift is used to visualize total shift, rotation and deformation of the void, which is in detail in Fig. 5a. The Fig. 5b contains deformation of both the specimen and void zoomed 20 times. The upper and lower specimen parts are deformed as it is presented in basic literature. This confirms the reliability of calculations.

Local values as derivatives of shift are derived by MATLAB functions. An example of local stress in three interesting directions perpendicular to force action is in Fig. 6. The Fig. 6a shows the X component of normal stress, e.g. the component in the direction of force. In Fig. 6b there is the shear stress. The solid curve going through the void center is interrupted in the void, which again confirms the correct functioning of FEM. The dashed curve shows that the stress at void right end is different from that in the void center. The dotted curve shows the stress near the end of specimen far from the void. It follows from comparison with this curve that void changes the elastic values, but the effect of the void is local and relatively low, about 15 % with respect to specimen value very near to the void.

Local values can be verified only logically, but the average values can be compared with experiment. The average values were calculated in planes both parallel and perpendicular to force action. Average values of normal strain and Young modulus are shown in Fig. 7 and they are compared with specimen without void. Young modulus is defined as the ratio of average normal stress and strain plane. The void little changes the value of average strain and Young modulus only in planes that intersect the void or are very near to the void. The change is small, less than 1 %.

Near the specimen center the calculated strain is very near the input value of 1 % and the Young modulus is also very near to input value of 50 GPa. At the specimen ends the values are different from input one; this is due to the small dimensions of specimen. These end values are one that could be measured. Since both the curves in Fig. 7 practically do not differ at the specimen ends, the measurement of the void effect is not possible on standard apparatus.

DISCUSSION AND CONCLUSION

Physical meaning and practical consequences of main results were already discussed during their presentation. The paper focuses to study of the effect of voids to elastic properties in real structure of composite. For simplicity only one void was considered, the future work will deal with all the voids in the specimen according to Fig. 2. Because of lack of data, only 2D case was considered. On the other hand the effect of void 2D approximation was considered in detail. Extensive calculations [6] revealed that the more exact approximation (Lissajouse curves, smoothed polygon or real one) has negligible effect on FEM results. Simple void approximation by ellipse is practically acceptable. For 3D case is means that derived ellipsoid can be used to get reasonable results.

Calculations revealed that the effect of voids to average elastic parameters is negligible. The void influences elastic stress only in its immediate environment. However, the change is up to 15% maximally for typical void dimensions. The void local effects can take place only in extreme composite treating, which would lead to its damage. Since the voids are distributed in composite volume almost regularly according to Fig. 2, similar conclusions will be derived from calculations with all the voids. The only problem is a group of very near large voids; this problem should be simulated, since this case was not fond in experiments.

Main simplification of reality was the assumption of isotropic composite medium. In reality the composite is heterogeneous material, it contains areas with matrix, longitudinal and transversal yarns. Especially yarn areas contain uniform anisotropic material due to the presence of fibers. The problem of heterogeneity and anisotropy can be studied either by a simple equivalent model of composite or by considering the real structure. The only problem is that we have not found an automated procedure for yarn finding. Routine, time consuming hand search must be used. Yarn axis and yarn cross section approximation of high quality is possible as well as for voids [6]. Also the theory for general FEM application exists. It was sketched in chapter Theory of the paper, especially the boundary conditions. More sophisticated system, MultiPhysics, for instance, should be used in this complicated case.

In principle, the study of the effect of all the components of real composite structure on its physical properties is in preparation and can be realized in the future. The only limitation is the realization of suitable method for at least semi-automated finding of yarn boundaries.

REFERENCES

1. Hearle, J. W. S., Grosberg, P., Backer, S. 1969. *Structural Mechanics of Fibres, Yarns, and Fabrics,* Wiley - Interscience, New York (USA).
2. Lomov, S.V., G. Huysmans and I. Verpoest. 2001. "Hierarchy of textile structures and architecture of fabric geometric models," *Textile Research Journal,* 71(6):534–543.
3. Košek, M., Doležalová, E. 2001 "Visualization of Spatial Structure of 2D Woven Real and Virtual Composites," Eight Annual International Conference on Composites Engineering (ICCE/8), D. Hii, editor, Tenerife, Spain, pp. 487–488.
4. Košek, M. Seják, P. 2008 "Visualization of Voids in Actual C/C Woven Composite Structure," will be published in Composites Science and Technology.
5. Nye, J. F. 1964. *Physical properties of Crystals*, Clarendon Press, Oxford.
6. Sejak, P. 2008. "Geometric Models Of Textile Composite Structure Focused To The Study Ofstructure Effect Impact On Resulting Properties," PhD thesis, TU Liberec, Czech Republic (in Czech).
7. http://www.mtm.kuleuven.ac.be/Research/C2/poly/research/textile_modelling/textile_geometry_modelling_paper_sl.pdf.

TOPIC V

TEXTILE COMPOSITES IN BALLISTICS

Modeling the Material and Failure Response of Continuous Filament Fabrics for Use in Impact Applications

G. NILAKANTAN, M. KEEFE, J. W. GILLESPIE, JR.
and T. A. BOGETTI

ABSTRACT

Continuous filament yarns comprised of organic fibers are widely used in fabrics as flexible protective materials because of their superior specific tensile properties. Numerical simulations of the impact of these fabrics using finite element codes provide valuable insight into the mechanisms of deformation, interactions, and energy absorption. However in much of the numerical simulation work to date, yarn failure is based on a single deterministic strength rather than the statistical distribution of strength based on the presence of defects. In this study, a first attempt is made to implement statistical based criteria for the prediction of yarn failure in a finite element analysis. A Weibull distribution based on the experimental testing of yarns is used to characterize the strength. This distribution is then mapped onto the finite element model of the fabric. When applied as a Monte Carlo simulation of the impact of fabrics, the statistical variation of penetration resistance as a function of constituent yarn properties can be studied. The dynamic finite element code LS-DYNA® is used for the numerical simulations.

INTRODUCTION

Based on quasi-static testing, it is known that the yarn strength is actually a statistical distribution based on the presence of defects. Furthermore, the statistical distribution is a function of the gage length of the test specimens and shifts to lower strength levels with increasing gage lengths. Weibull statistics have been used to describe these effects [1]. The *length scale* effect consequently controls the strength of a yarn. This length scale effect is felt in high rate deformation events.

During the impact of fabrics, faster moving longitudinal waves and slower moving transverse waves propagate outward in the yarns involved at the impact site. The length of the yarns involved in the waves increases with time and is

Gaurav Nilakantan, Center for Composite Materials, University of Delaware, DE, USA
Michael Keefe, Center for Composite Materials, University of Delaware, DE, USA
John W. Gillespie Jr., Center for Composite Materials, University of Delaware, DE, USA
Travis A. Bogetti, Army Research Laboratory, Aberdeen Proving Grounds, MD, USA

subjected to various stress levels and strain rates. However the yarns ahead of the waves are not subjected to any stresses. Yarns are comprised of hundreds of filaments and display a progressive type of failure with individual filaments redistributing loads amongst themselves until complete yarn failure by filament fracture or pullout. In addition, the degree of twist of the constituent filaments can influence the yarn strength. Organic fibers have been shown to exhibit non-linearity and strain rate dependence of tensile properties. They stiffen at high strain rates with an increase in strength and usually a reduction in strain to failure. This viscoelasticity also affects the Weibull strength distribution [2]. Thus the behavior of fabrics under quasi static loading is considerably different than when under high strain rate loading. The statistical strength distribution can be further affected by weaving [3] and environmental conditions that degrade the yarn strength. Therefore it is important to be able to accurately model the behavior of these fabrics at high rates of deformation while considering all of the factors mentioned above.

At present, much of the numerical simulation of the impact of these fabrics typically uses only material data obtained from quasi static testing where a single deterministic strength is used rather than the statistical distribution. No length scale effects are incorporated and if high strain rate material data is used in the modeling of large dimensioned systems, it is usually based on experimental data of very small dimensioned samples obtained through split-Hopkinson bar testing. The focus of this research is to provide a framework for incorporating a material and failure model into a numerical simulation that addresses the above concerns.

For this preliminary study, a Weibull distribution is used to characterize the yarn strength distribution while the yarn constitutive behavior is assumed to be linearly orthotropic elastic till failure. No length scale or viscoelastic effects have been incorporated and the yarn is modeled as a homogenous continuum neglecting the filament level architecture. A Monte Carlo approach is used to study the effect of statistical variation in yarn strength on the fabric response to impact. This new material and failure model is incorporated as a user defined routine in the dynamic finite element code LS-DYNA, which is used to simulate the fabric impact event.

CHARACTERIZING THE STRENGTH OF KEVLAR YARNS

A Weibull distribution is often used to characterize the strength of Kevlar yarns. For quasi static data, a two parameter Weibull distribution is sufficient while for high strain rate data, a bimodal Weibull distribution is preferred because of the two competing modes of failure as well as the peculiar skin-core structure of Kevlar [2]. At low strain rates, a shear mode of failure through the breakage of secondary bonds is dominant while at high strain rates, failure is due to the breakage of primary bonds. The cumulative distribution function (CDF) of a bimodal Weibull distribution is given by Ref. [1] as

$$F(\sigma) = 1 - \exp\left(-\frac{L}{L_0}\left(\frac{\sigma}{\sigma_{01}}\right)^{m1} - \frac{L}{L_0}\left(\frac{\sigma}{\sigma_{02}}\right)^{m2} \right) \tag{1}$$

where σ is the stress level of interest, σ_{01}, σ_{02} are the scale parameters and m_1, m_2 are the shape parameters. The effect of gage length is incorporated through the '(L/L_0)' factor or the length-scale term where 'L' is the length of yarn and 'L_0' is gage length at which the experimental data was generated. A longer specimen gage length involves a larger number of defects and reflects a greater probability of a critical defect being present. Consequently the strength must decrease with length. It is important to capture this '*length-scale effect*' since in an impact event, only the yarn material behind the front of the outward propagating longitudinal strain wave is involved in the energy dissipation. The material ahead of the wave does not have any information about the impact event and does not participate. Studies have shown that the weaving process can degrade the strength of yarns and therefore shifts the Weibull probability distribution to the left [3]. It is more realistic therefore to characterize the strength of yarns extracted from a woven fabric rather than yarns unraveled from a spool. The constitutive tensile response based on the bimodal Weibull distribution is given by

$$\sigma = E\varepsilon \exp\left(-\frac{L}{L_0}\left(\frac{E\varepsilon}{\sigma_{01}}\right)^{m1} - \frac{L}{L_0}\left(\frac{E\varepsilon}{\sigma_{02}}\right)^{m2} \right) \tag{2}$$

where 'σ' is the stress, 'ε' is the strain, and 'E' is the elastic modulus. Kevlar yarns also exhibit viscoelasticity. These yarns stiffen at high strain rates with an increase in the strength. For each strain rate and gage length, there are a corresponding set of Weibull parameters that need to be determined. Consequently the term 'E' in equation (2) actually represents a rate-dependent modulus. For high rate tensile testing of Kevlar yarns using the split-Hopkinson bar, the gage length of the specimen is usually restricted to a few millimeters in order to ensure stress equilibrium within the specimen which is required by the testing method. Thus while it is possible to determine Weibull parameters for a small gage length, the length scale term must be used to extrapolate this data for longer gage lengths unless another experimental method is available. Table (1) displays the scale and shape parameters of a 2 parameter Weibull distribution, obtained from the quasi static experimental tensile testing of single Kevlar KM2 filaments of 25 mm gage length, from Ref. [4]. The corresponding set of scale and shape parameters when using a bimodal Weibull distribution is also displayed in Table (1).

TABLE 1. WEIBULL PARAMETERS FOR KEVLAR KM2 FILAMENT.

	σ_0 (MPa)		**m**		**E (GPa)**
2 parameter	3411.9		14.7960		68.7
	σ_{01} (MPa)	**σ_{02} (MPa)**	**m_1**	**m_2**	**E (GPa)**
Bimodal	3323.0	3700.4	18.0023	24.8803	68.7

Figure (1a) displays the cumulative distribution function (CDF) based on equation (1) for the data reported in Table (1) Figure (1b) displays the corresponding stress-strain curve based on equation (2). The plots have also been extrapolated from the experimental gage length of 25 mm to different gage lengths ranging from 5 mm to 200 mm using the length scale term in equations (1) and (2) Figure (1c) displays the variation of strength with gage length.

IMPLEMENTATION OF THE WEIBULL STRENGTH DISTRIBUTION AND MONTE CARLO APPROACH IN THE YARN FE MODEL

The FE model of a yarn can be of two types (*i*) using filament-level architecture as seen in figure (2a), and (*ii*) a homogenous continuum neglecting the filament architecture as seen in figure (2b).

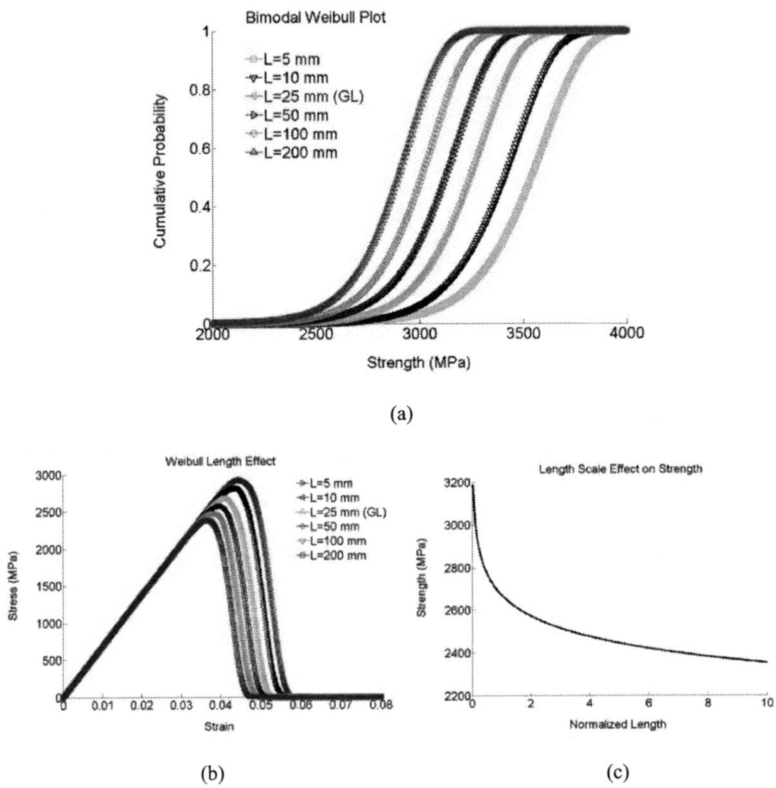

Figure 1. Weibull strength distribution for a Kevlar KM2 filament (a) CDF (b) Constitutive response (c) Effect of gage length on strength.

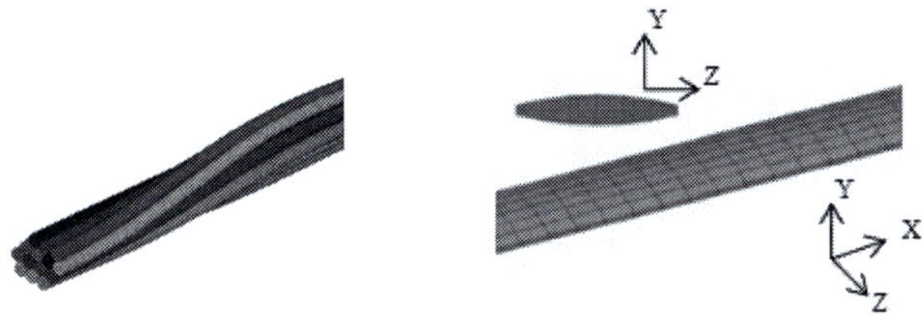

(a) (b)
Figure 2. FE model of a yarn (a) Filament level resolution (b) Homogenous continuum.

For a Monte Carlo approach in a yarn with a filament level architecture, each filament is randomly assigned to a particular material model with a unique strength, failure strain, and elastic modulus. The distribution of the material models to each filament will follow the Weibull distribution obtained experimentally from individual testing of filaments, such as in Ref. [4]. However this model with filament architecture is computationally very intensive and not widely used when modeling fabrics. For the purpose of our study, we adopted the widely used homogenized yarn model, and at this preliminary stage we further assume that the strength remains uniform throughout the yarn length.

IMPLEMENTATION OF THE WEIBULL STRENGTH DISTRIBUTION AND MONTE CARLO APPROACH IN THE FABRIC FE MODEL

As a first step to map the statistical strength distribution onto the fabric FE model for the purpose of a Monte Carlo approach, each yarn within the fabric is assigned to a particular material model that has a unique strength, wherein all elements within the same yarn have the same strength. The assigned moduli and strengths remain static since viscoelastic and length scale effects are not considered here. The distribution of material models will follow the Weibull strength distribution obtained experimentally from individual testing of yarns. Consider the case of a quarter symmetric FE model of a plain weave Kevlar S706 fabric, 4 in. long and 2 in. wide, with a count of 34 yarns per inch. For computational reasons, only a quarter of the fabric has been modeled and so there are a total of 102 yarns in the model. Figure (3) displays a histogram of the probability density function (PDF) and cumulative distribution function (CDF) for the 2 parameter Weibull strength distribution obtained from Ref. [4]. Figure (4) displays the corresponding FE model with the contours of strength that are initially assigned to each yarn. A total of 50 material models have been used since Ref. [4] used 50 samples for the experimental testing. The strengths vary from 2630 MPa to 3880 MPa. For each separate run of the Monte Carlo approach, the strength assignments of each yarn will vary, but will still follow the distributions in figure (3).

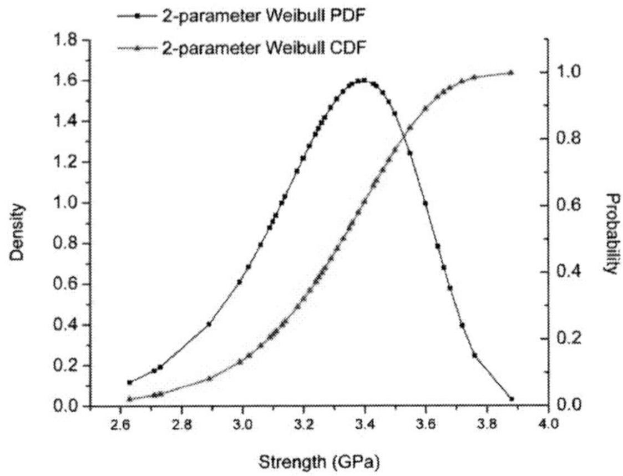

Figure 3. PDF and CDF for the 2 parameter Weibull strength distribution of the Kevlar KM2 filaments.

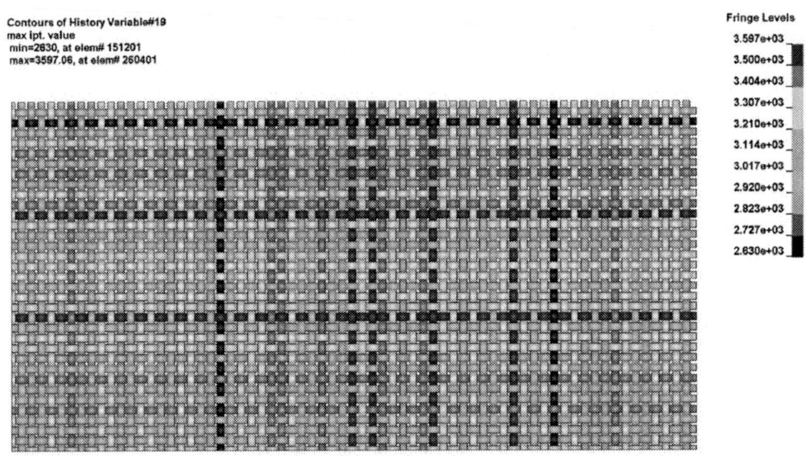

Figure 4. Fabric FE model showing contours of assigned yarn strength.

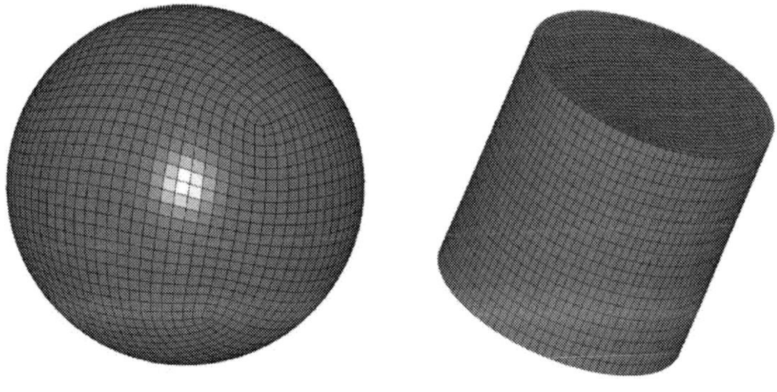

Figure 5. FE mesh of the projectile (a) Case #1, spherical (b) Case #2, cylindrical.

RESULTS AND DISCUSSION

The fabric is transversely impacted at the center by a 0.63 gm rigid projectile moving along the *y*-direction with a velocity of 100 m/s. The fabric is gripped on all four sides. It is comprised of Kevlar KM2 yarns with an areal density of 180 g/m^2. The effective material properties of the yarn are assigned as follows: longitudinal elastic modulus (E_x) of 62 GPa, transverse elastic moduli (E_y) and (E_z) of 6200 MPa, shear modulus (G) of 3280 MPa, zero Poisson ratios, and density (ρ) of 1310 kg/m^3. To keep the simulation run times small in this preliminary study, only one quarter of the fabric has been modeled, although it is more realistic to use a full scale model. A static frictional coefficient of 0.18 is specified for both the yarn to yarn and projectile to yarn contact algorithms. Two cases are studied. In Case #1, the projectile is spherical with a of diameter 5.55 mm, while in Case #2, the projectile is a right circular cylindrical with a diameter and height of 5.55 mm. Both projectiles have the same mass, and differ only in the presented area and shape of the impact face. The projectile in the second case will engage more yarns upon the first instant of impact. Figure (5) displays the FE meshes of the two projectiles. Figure (6) displays the quarter symmetric fabric FE model setup for Case #1. The undulating warp and fill yarns can be seen in the fabric. Figure (7) displays the residual projectile velocity for 10 runs of the impact simulation in Case #1. Three baseline results are also presented wherein all yarns within the model have been assigned the same strength. This allows a study of the effect of incorporating a statistical strength versus a single deterministic strength. The three strengths chosen for the baselines are 2630 MPa, 3255 MPa, and 3880 MPa, and these correspond respectively to the minimum, mean, and maximum strengths from the experimental data. The average strength of a 2 parameter Weibull distribution is given by equation (3a), and is determined to be 3293 MPa. This is different from the mean strength of 3255 MPa obtained experimentally from the 50 samples, given by equation (3b).

$$\bar{\sigma} = \sigma_0 \Gamma\left(1 + \frac{1}{m}\right) \tag{3a}$$

$$\bar{\sigma}_{exp} = \frac{1}{N} \sum_{i=1}^{N} \sigma_i \tag{3b}$$

where σ is the strength, σ_0 is the scale parameter, m is the shape parameter, Γ is the gamma function, and N is the number of samples.

The mean residual projectile velocity from the 10 runs is 70.635 m/s and the standard deviation is 3.312 m/s. Figure (8) displays the residual projectile velocity for 10 runs of the impact simulation in Case #2, along with the three baseline results. For all the 10 runs, the same fabric FE model with the exact same strength distribution as Case #1 was used. This enables a direct comparison on a run by run basis to study the effect of the projectile's geometry on the fabric response. The mean residual projectile velocity from the 10 runs is 81.150 m/s and the standard deviation is 2.323 m/s.

Since this is a preliminary study, the results presented herein are not definitive and therefore have not been analyzed further. Their purpose is to be illustrative and demonstrate what the effect is, of incorporating a statistical strength distribution on the fabric response such as on the residual projectile velocity.

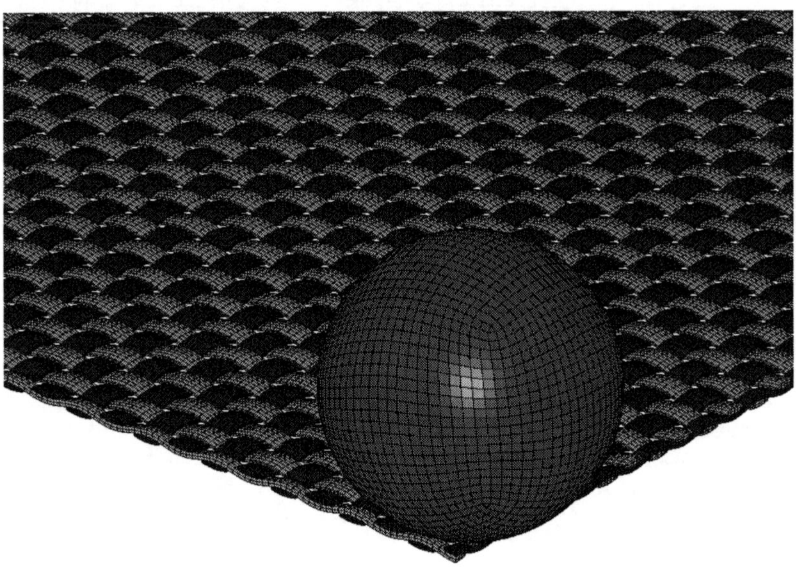

Figure 6. FE model setup for Case #1.

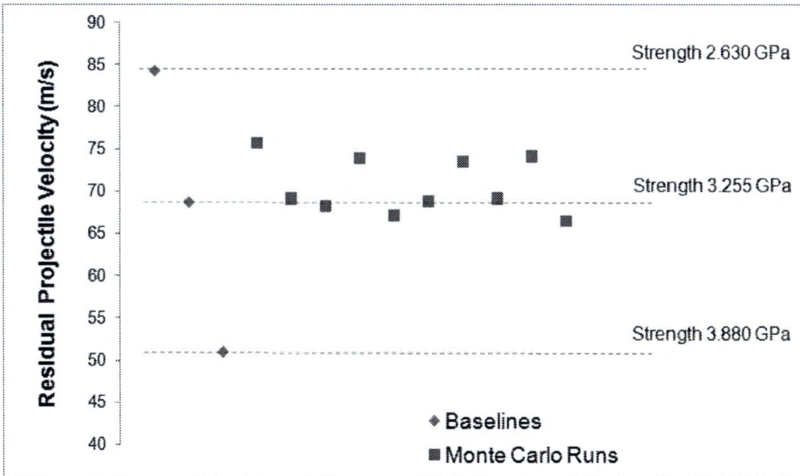

Figure 7. Residual projectile velocities from the Monte Carlo runs of Case #1, spherical projectile.

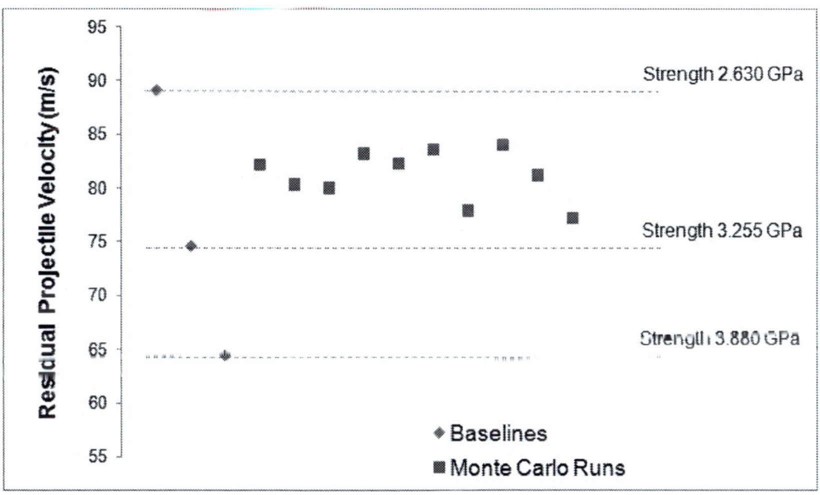

Figure 8. Residual projectile velocities from the Monte Carlo runs of Case #2, cylindrical projectile.

Future work will involve incorporating both rate sensitive effects and length scale effects into the model. This will require the tracking of the wave propagations through the fabric and determining the strain rates at every time instant. We also will study the difference between using a filament level resolution for the yarn model versus a homogenized model with all elements assigned to randomized strengths. All this will require a lot of experimental work to generate data that can be fed into the numerical model. The next step then would be to experimentally study the various effects such as weaving damages and environmental degradations to generate further data for the model input. Finally the simulations will need to be run using full scale numerical models instead of quarter symmetric models.

CONCLUSIONS

A basic framework to incorporate the statistical distribution of yarn strength into a finite element analysis of the impact of fabrics has been presented. This framework will provide a platform for developing advanced material and failure models that incorporate a wide range of coupled effects previously neglected that may have significant bearings on the predictive capabilities of a numerical impact simulation. This is accomplished by determining the parameters of the experimentally obtained statistical property distribution for each effect and then feeding them into the FE model. The *property* may include yarn moduli and strength; and the *effects* include length-scaling, viscoelasticity, environmental degradation, weaving damage, damage due to previous impacts, and frictional temperature effects to name a few. A lot of experimental testing at quasi static and high rates is necessary in order to establish a reliable source of data which will serve as the model input. The resultant material and failure model is more realistic than those currently used in fabric FE models. It is envisioned that this material model framework will enable an in depth study of the effect of the statistical variation of constituent yarn properties on the penetration resistance of a fabric system. The obtained scatter in residual projectile velocities will also enable the construction of master V_0 to V_{100} curves. This will also aid in process control during the manufacturing of the yarns and fabrics used in energy dissipating applications.

ACKNOWLEDGEMENTS

The financial support of the US Army Research Laboratory at the Aberdeen Proving Grounds, MD, USA, and the Center for Composite Materials at the University of Delaware, DE, USA is gratefully acknowledged.

REFERENCES

1. Goda, K., and Fukunaga, H., "The evaluation of the strength distribution of silicon carbide and alumina fibers by a multi-modal Weibull distribution," *J. Mater. Sci.,* Vol. 21, 1986, pp. 4475-4480.
2. Wang, Y., and Xia, Y., "Experimental and theoretical study on the strain rate and temperature dependence of mechanical behavior of Kevlar fiber," *Composites, Part A,* Vol. 30, 1999, pp. 1251-1257.
3. Lee, L., Rudov-Clark, S., Mouritz, A.P., "Effect of weaving damage on the tensile properties of three-dimensional woven composites," *Composite Structures,* Vol. 57, 2002, pp. 405-413.
4. Leal, A.A., Deitzel, J.M., and Gillespie, John W. Jr., "Assessment of compressive properties of high performance organic fibers," *Composites Science and Technology,* Vol. 67, 2007, pp. 2786–2794.

Prospects and Risks of Tailored Preforms for Advanced Composite Applications

P. SCHIEBEL, T. BLOCK and A. S. HERRMANN

ABSTRACT

Recent developments of preform technologies like Tailored Fibre Placement (TFP), Fibre Patch Preforming (FPP) or Dry Fibre Placement (DFP) have shown strong potentials of reducing structural weight by optimized mechanical utilization [1], [2]. The paper discusses modern preform technologies and intends to provide the reader an overview of the current state of the technology. Both research on the characterization of textile structures from manufacturing and the mechanical point of view as well as computation and optimization of fiber paths for complex geometries are presented.

INTRODUCTION

In aviation state of the art are preimpregnated tapes (prepregs) with a high performance for large, two dimensional shaped parts such as the skins of the wings or the vertical tail plane. The manufacturing process is automated leading to good laminar mechanical properties. Major drawbacks are the material costs, the energy consumption of the autoclave curing-process and the manual processing effort. Furthermore the manufacturing of three dimensional parts with a complex laminate lay-up is inefficient. As confirmed by studies CFRP parts manufactured from textile preforms in a quick infusion process are a very promising technology in order to overcome these problems and lead to reduced costs and a higher quality [3], [4].

High flexibility of textile manufacturing processes allow the production of dry preforms in a near net shape for two and three dimensional parts. These in combination with automated resin infusion technologies make it possible to employ CFRP in mass production.

Patrick Schiebel, Faserinstitut Bremen e.V. (FIBRE), Am Biologischen Garten 2, 28359 Bremen, Germany, E-Mail: Schiebel@faserinstitut.de
Tim Block, Faserinstitut Bremen e.V. (FIBRE), Am Biologischen Garten 2, 28359 Bremen, Germany
Prof. Axel S. Herrmann, Composite Technology Center Stade GmbH, Airbusstraße 1, 21684 Stade, Germany

TFP Technology, which is a technique to transfer effectively the results of a stress and strain calculation into a textile reinforcement structure, was developed by the Institute of Polymer Research (IPF), Dresden, Germany. Dr.–Ing. Rothe registered the patent of this technology for IPF in 1992. TFP and also DFP and FPP technologies make it possible to place rovings or dry fiber patches in any desired direction. These

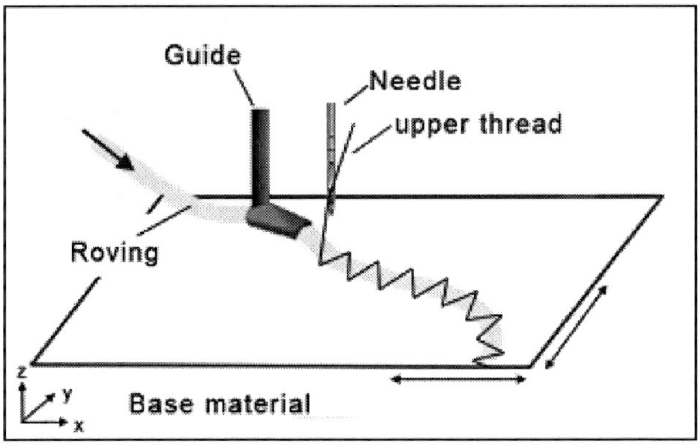

Figure 1. Principles of roving lay-up with TFP [5].

high automated manufacturing processes enable fabrication of preforms for advanced composites applications.

Attachment of the dry roving on a base material is done on a modified embroidery machine with a zigzag stitch using an upper and lower thread as it is displayed in

Figure 1. In textile preforms variable fiber orientations can be produced in each layer. Without unclamping the base material, the rovings can be laid-up one by one. The result is a near net shape preform which is easy to handle. The technology can be used for local reinforcements of big parts or complete preforms for small parts. The machine is numerically controlled and works with a high reproducibility. In order to optimize productivity a group of knitting heads can work parallel to each other on the same base material. The consolidation of TFP preforms to composites can be done with standard methods such as Resin Transfer Molding (RTM), Resin Film Infusion (RFI) or vacuum infusion etc.

MATERIAL DESCRIPTION

The main improvement of TFP is that the material properties can be designed by adapting the fiber orientation in the direction of the stress flow inside the part. To orientate the roving on defined curves the production properties have to be adapted locally. A roving path with curves requires more fixing through stitching yarn than a linear path. The manufacturing speed goes inversely proportional to the density of the stitches. An optimized TFP production has to use different stitching densities that vary along the fiber path.

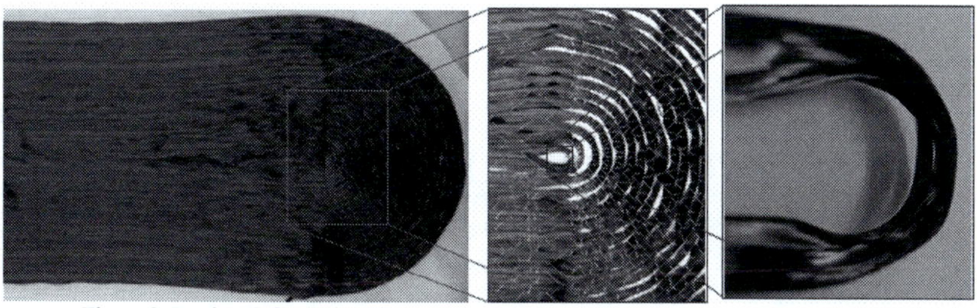

Figure 2. Roving deformations in small radii and provoke inhomogeneities in laminates.

To enlarge the applicability the state of knowledge has to be enhanced by characterization and computation of these structures. Knowledge about the local properties is required in order to describe the mechanical properties of tailored textile structures. On the mechanical point of view a lower performance in fiber direction of textiles compared to unidirectional prepreg is affected by the undulations of the reinforcements caused by the manufacturing and handling of the preforms. Furthermore compared to prepregs, less capable resin systems lead to reduced mechanical properties. This topic is the object of different research projects intending to compensate the mechanical drawbacks of the textile process chain at the FIBRE and other research institutes in the next years [6].

Different kinds of roving deformations in fiber paths such as small curves, sharp turning points, starting and end points have to be described geometrically in order to analyze their impact on the mechanical properties. Through multiscale simulations the local characteristics of fabrics, knit fabrics or interlaced yarns can be transferred into global part simulations. The software WiseText, developed at the Catholic University Leuven, Belgium, allows multiscale simulations of the manufacturing process chain of textile composites [7]. Also WiseText can describe the draping of the lay-up during the preforming and determine the mechanical characteristics of the cured part [8]. The basis of the computations is a unit-cell model which requires a sufficiently exact description of the mechanical and geometrical properties [9].

OPTIMIZATION OF FIBER ORIENTATION

In order to fully exploit the anisotropic properties of fiber reinforced composites, an optimization of fiber orientations in textile preforms is necessary. By doing so an increase of the strength to weight properties of the composites can be achieved. Different methods exist for optimization of fiber paths in the preforms [10]. The most common ones are genetic algorithms, load path methods and principal stress methods [11].

Generic algorithms are very powerful tools and therefore have a broad area of application. similar to the biological evolution a slow constant optimization is pursued by constant mutation, increase and selection [12]. Generic algorithms cannot determine an absolute optimum fiber direction. This has to be done manually.

For the determination of the “inside“ and “outside“ shape a topology optimization can be used. Within a given area a material rearrangement takes place. Iterative the local density is adapted to the strains or deformations. If the density is under a given limited value, the appropriate element will be eliminated, resulting in a new contour of the part. The result is a framework structure, which can be converted into paths for the rovings.

The material optimization method Computer Aided Internal Optimization (CAIO), developed by Professor Mattheck, orients the fibers locally toward the main stresses [13]. Through line integration of the individual main stress directions continuous fiber directions are generated. Within a few iterations a minimization of the shearing force is reached. At the same time the failure mode changes from matrix cracking to fiber breakage and thus leads to a higher material utilization.

An implementation of CAIO into the FE-programs MSC PATRAN and NASTRAN was realized under the name TACO by the DLR in Braunschweig [14], [15]. A combination of topology optimization and material optimization is investigated under the name AOPS in the project BIOTEX using the FE-program ANSYS [16].

Research at the FIBRE on fiber direction determination is based likewise upon the CAIO method. In order to reach a true optimum the integration of manufacturing restrictions into the optimization process is required. The external geometry comes from a normal CAD program and is transferred into a FEM model (MSC-NASTRAN). Loads and starting material properties have to be defined. From the result of local stresses a first tailored roving path is specified in the fibre-program taking the force flow into account. The modified material properties have to be transfer back into the FE-simulation. Now the first loop is closed and the second begins with new stresses inside the material, due to new fiber directions (see Figure 3).

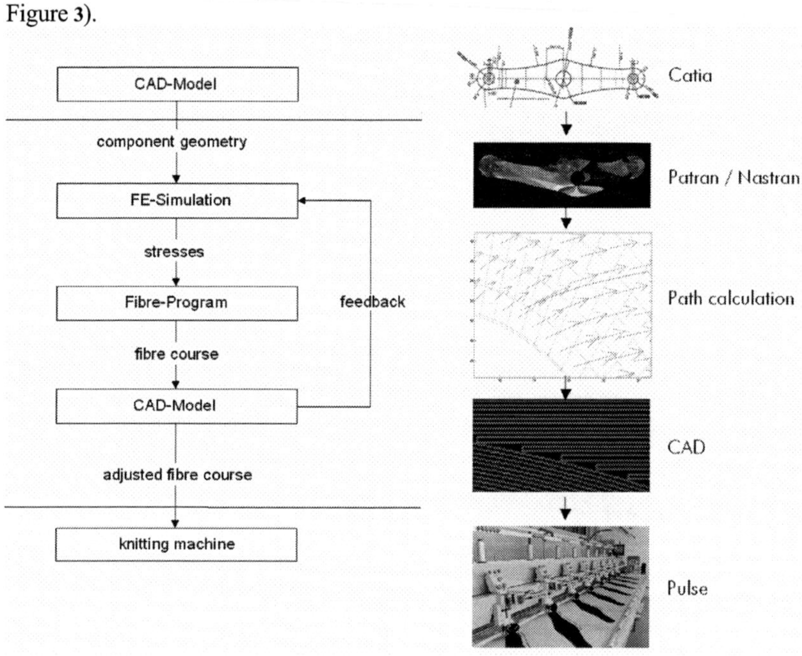

Figure 3. FE-Analysis of optimization fiber direction for TFP.

The strength and stiffness of the material is locally adapted and again one determines the local tensions and deformations. This cycle repeats itself, until an optimum is reached. If these trajectories are converted directly into fiber paths, a high maximum stress of the composite construction part is expected. This however is not reached as many influences of the production process as well as the material behavior are not considered in the optimization process. This can lead to a strong variation in the fiber volume content in the part.

Different programs exist that help the design engineer to draw rowing paths next to the main stress directions. The requirements for a TFP path however are so complex that the actual programs practically ignore the manufacturing requirements of TFP such as:

- the number of starting and end points of roving paths in a preform shall be as small as possible
- rovings shall be placed parallel
- no gaps between rovings
- avoid small radii and sharp turning points
- the weight of the preform shall be homogenously distributed
- a minimum angle between local fiber paths and optimal mechanical directions

Only an integration of the manufacturing and material constraints into the optimization process allows to fully exploit the material potential. An integration of all boundary conditions could be realized, but due to their complexity only at large research expenditures. This would lead to a large reduction of the manual work, which is required to generate fiber paths and thus to significant reduction of unit construction and development time.

Optimized fiber alignments have to be computed iteratively due to the coupling of principle stress directions and fiber orientations. Shear stresses and corresponding failures are reduced. To satisfy manufacturing restrictions an optimized fiber orientation needs some adjustments resulting in local deviations. A feedback method is necessary to project the real manufactured fiber orientations on appropriate finite element meshes. Material properties are locally adjusted in order to include effects of interior notches such as roving turns, loops, starting and end points of the fiber paths.

CONCLUSION

For an increased employment of textile technologies for lightweight construction a number of fundamental investigations are necessary. It is required to predict the properties of the finished part based on the characteristics of the dry fiber semi finished part. In particular the deformation behavior of the reinforcement fibers is not sufficiently well-known at present.

The most important aspect of new research projects is the mathematical description of the textile manufacturing conditions for their consideration in the material optimization. For this computation algorithms must be developed, which describe the placing and deformation behavior of the rovings. The description of the roving behavior during the textile processing process using FEM is the subject of current research activities at the FIBRE. A micro mechanical based approach is selected, which describes roving behavior on basis of the material behavior of single filaments and its physical interactions such as fiber fiber adhesion, friction between the filaments, etc. The simulation of the roving is performed using a unit cell model. The description of the material behavior is a basic module of each stress analysis and thus an important precondition for the generation of fiber paths for load adapted construction of parts.

REFERENCES

1. Weimer, C., T. Preller,T. Mitschang, P., Drechsler, K. "'Advanced Composite Materials: Technologies, Performance and Modelling Part I", 'Composites: Part A 31 (2000) 1261–1268.
2. Ogale, A, Mitschang, P.: Tailoring of Textile Preforms for Fibre-reinforced Polymer Composites, Journal of Industrial Textiles, Vol. 34, No. 2—October 2004 77.
3. Schiebel, P., Herrmann, A. S., Eberth, U.: Hochleistungsverbundbauteile auf Basis textiler Halbzeuge - Herausforderungen und Entwicklungen., 33. Aachener Textiltagung, 2006.
4. Brandt, J., Drechsler, K., Arendts, F. J.: Mechanical performance of composites based on various three-dimensional woven-fibre preforms, Composites Science and Technology, Volume 56, Issue 3, 1996, Pages 381–386.
5. Hightex Verstärkungsstrukturen GmbH, "Explanation of the TFP Technology", http://www. hightex- dresden.de/, 10.2006.
6. Herrmann, A., Barth, E. Pabsch, A.: OPTIMAX—Optimisation of multi-axial non-crimp fabrics for composites in lightweight structures. Materials Week, International Congress on Advanced Materials, their Processes and Applications, München, 25.-28.9.00, DGM, DKG; VDI, BMBF.
7. Lomov, L., Verpoest, I.: Modelling of the international structure and deformability of textile reinforcements: WiseTex software, ECCM, Europ. Conf. on Composite Materials, 2005.
8. Verpoest, I., Lomov, S.: Virtual textile composites software WiseTex: Integration with micro-mechanical, permeability and structural analysis, Composites Science and Technology 65 (2005) 2563–2574.
9. Middendorf, P, Drechsler, K., van den Broucke, B., Heimbs, S.: Advanced Composite Materials: Technologies, Performance and Modelling—Part 2, Course on Emerging Techniques for Damage Prediction and Failure Analysis of Laminated Composite Structures, Cachan.
10. Tosh, M. W., Kelly, D. W.: Fibre steering for a composite C-Beam, Composite Structures 53 (2001) 133–141.
11. Li, R., D. Kelly, A. Crosky, "Strength Improvement by Fibre Steering around a Pin Loaded Hole", Composite Structures, 57(1-4), 377–383, 2002.
12. Legrand, X., Kelly, D., Crosky, A., Crepin, D.: "Optimization of Fibre Steering in Composite Laminates Using a Genetic Algorithm", Composite Structures 75, Juni 2006.
13. Mattheck, C., Baumgartner, A., Kriechbaum, R.,Walther, F.: "Computational methods for the understanding of biological optimization", Computational Materials Science, 1(3), 302-312, 1993.
14. Aschenbrenner, A.: Tailored Fibre Placement Technology—Optimisation and computation of CFRP structures, ESAComp Users' Meeting, April 2007.
15. Schiebel, P., Aschenbrenner, L.: "Characterisation, optimisation and computation of TFP structures possessing complex stress fields", CFK-Valley Stade Convention, Juni 2007.
16. Spickenheuer, A., Gliesche, K., Heinrich, G., Schulz, M., Drechsler, K., Karb, I., Moldenhauer, H.: "Ansätze zur Auslegung und Herstellung beanspruchungsgerecht gestalteter Preforms für extrem leichte Composite-Bauteile", 11. Dresdner Leichtbausymposium, Juni 2007.

Trellised Fabric Architectures for Improving the Ballistic Performance of Soft Body Armor

R. G. EGRES and C. D. FIORELLA, JR.

ABSTRACT

In the fabrication of soft body armor, it is paramount that materials and constructions be chosen that both impede projectile penetration and minimize blunt trauma by reducing backface deformation during a ballistic event. This research demonstrates that significant reductions in ballistic backface deformation can be achieved with biaxial woven ballistic fabrics though modification of the angle between interwoven yarns within the individual plies, without compromising V_{50} performance. The results of .44 Magnum ballistic impact testing to determine V_{50} and backface performance of multi-layer targets fabricated using bias-distended, biaxial woven Kevlar® fabrics at several interwoven yarn angles and basis weights are reported. The use of these fabric architectures alone or in combination with other anti-ballistic structures could have utility in the fabrication of concealable soft body armors.

INTRODUCTION

In the United States, commercial concealable body armors must meet ballistic performance criteria established by the National Institute of Justice [1] to not only impede the penetration of handgun bullets, but to satisfactorily reduce ballistic impact backface deformation in order to minimize the risk of blunt trauma injury. In the fabrication of premium protective vest materials, it is the challenge of the vest manufacture to optimize ballistic garment comfort through weight and bulk reduction, while meeting these ballistic performance requirements for a particular ballistic threat.

Biaxial woven fabrics comprised of high modulus, high strength fiber yarns have long been used in the fabrication of soft body armors. Several factors can influence the ballistic response of multi-layer, woven fabric body armors toward a

Ronald G. Egres, DuPont Advanced Fibers Systems, 5401 Jefferson Davis Highway, Richmond, VA 23234, U.S.A.
Carlo D. Fiorella Jr., University of Delaware Department of Mechanical Engineering, 126 Spencer lab, Newark, DE 19716, U.S.A

particular ballistic threat, including the physical properties of the fiber, yarn denier, fabric weave construction and yarn finish. Numerous investigations of ballistic energy dissipation in woven fabrics have been conducted, and the findings from both ballistic experiments and simulations have lead to improvements in the mechanistic understanding of the response of woven fabric materials to ballistic impacts [2,3,4,5]. While biaxial woven fabric based body armors demonstrate favorable performance with regard to their ability to stop high energy handgun bullets, it is challenging for these woven fabrics alone to achieve the same backface deformation reduction (on an equivalent areal density basis) observed in some alternative ballistic fiber-based ply constructions, including unidirectional fiber laminates and some multi-axial fabrics.

In this paper, we demonstrate that modifying woven Kevlar® fabric constructions through altering the angles between interwoven yarns via bias distention or "trellising" can reduce backface deformation over that observed in panels constructed of traditional (0°–90°) plain woven fabrics, while maintaining a high degree of penetration resistance. Ballistic impact testing performed with .44 Magnum bullets in accordance with the NIJ Standard-0101.04 (Level IIIA) test protocol was used to determine backface deformation performance of 1 lb/ft^2 areal density panels fabricated from plain weave and trellised fabrics. Further ballistic impact testing was conducted to determine the ballistic V_{50} performance of each of the test panels. These experiments demonstrate that a 20% reduction in backface deformation can be achieved against the .44 Magnum threat, while achieving V_{50} performance as high as 200 ft/s above the backface test velocity established for NIJ Standard-0101.04, level IIIA certification (1430±30 ft/s). High speed photography of backface deformation in low areal density (0.16 lb/ft^2) fabric targets impacted at 700 ft/s provide insights as to the mechanism responsible for the backface improvements observed in the 1 lb/ft^2 panels.

EXPERIMENTAL

The plain weave fabrics used in this investigation were woven by JPS Industries, Inc. using 600 denier Kevlar® KM2 para-aramid fiber yarns, having yarn counts (warp×fill) of 17×17, 24×24, 28×28, 31×31, and 34×34 yarns per inch (ypi). The base fabric properties are provided in Table I, where cover % is a measure of the fabric's tightness with respect to the maximum fabric tightness, and locking angle was measured experimentally from bias extension tests as described by Harrison and et al. [6].

Two, 15"×15", 1 lb/ft^2 areal density ballistic test panels were fabricated from each of the base fabrics and trellised ply structures as described in Table II. To generate trellised fabric plies, diagonal strips were cut from each the base fabric rolls (along the bias direction). These bias-oriented fabric strips were then clamped at the ends in the trellising frame apparatus shown in Figure 1 and stretched uniformly, changing the angle between the interwoven yarns in order to achieve the desired trellised fabric ply structure. Trellis angles were chosen to achieve final yarn end counts and basis weights corresponding with those of the other base fabrics. Trellised fabric ballistic plies were then assembled in cross-plied fashion, with each successive ply having a trellis or acute angle orientation rotated 90 degrees relative to that of the former. This was performed using a frame having closely spaced pins around the perimeter to maintain fabrics in the desired trellised architecture during assembly and stitching. All fabric targets were stitched with a 2"×2" quilt pattern and about the perimeter 0.5" from the edge.

Ballistic impact backface deformation testing was performed using .44 Magnum bullets (Remington 44 caliber, semi jacketed hollow point, 240 grain) at velocities of 1430±30 ft/s. Ballistic panels were placed in front of clay witness boxes (Roma Plastilina clay) conditioned as described in NIJ Standard-0101.04. Two backface deformation tests were performed on each target, and the maximum depth of deformation into the clay witness was measured. The ballistic V_{50} was then determined for each panel based on three acceptable complete and partial penetration pairs. As with the backface deformation tests, V_{50} test were performed on unrestrained targets placed in front of a non-backed clay witness box.

TABLE 1. PROPERTIES OF 600 DENIER KEVLAR® KM2 BASE FABRICS.

Base Fabric Construction (ypi.)	Basis weight (oz/yd^2)	Cover %	Locking Angle (degrees)
17 × 17	2.71	49.8	23
24 × 24	3.80	66.9	31
28 × 28	4.46	75.8	38
31 × 31	4.93	82.1	43
34 × 34	5.63	88.0	52

TABLE 2. PANELS FABRICATED FOR .44 MAGNUM BALLISTIC TESTING.

Panel Base Fabric Construction (ypi)	Trellis angle (degrees)	Final ply yarn count (ypi)	No. plies	Panel Areal density (lbs/ft^2)
17 × 17	90	17 × 17	53	1.01
17 × 17	37	28 × 28	32	1.00
17 × 17	33	31 × 31	29	1.00
17 × 17	30	34 × 34	26	1.01
24 × 24	90	24 × 24	38	1.00
24 × 24	59	28 × 28	32	1.01
24 × 24	51	31 × 31	29	1.00
24 × 24	45	34 × 34	26	1.00
28 × 28	90	28 × 28	32	1.00
28 × 28	65	31 × 31	29	1.01
28 × 28	55	34 × 34	26	0.99
31 × 31	90	31 × 31	29	1.00
31 × 31	66	34 × 34	26	0.99
34 × 34	90	34 × 34	26	1.00

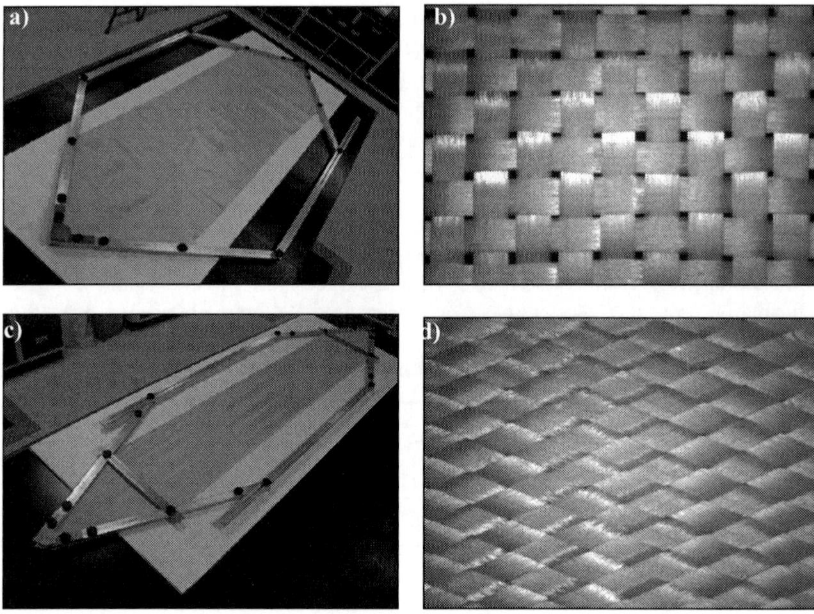

Figure 1. Trellising frame apparatus and magnified 24×24 ypi base fabric structure prior to trellising (a and b), and after trellising operation (c and d).

RESULTS

Figure 2 presents the backface deformation performance of each of the base fabric and cross-ply trellised fabric targets. The backface data is presented as the average of the four backface measurements (two from each type of panel). The best backface performance of the panels fabricated from the plain weave base fabrics was observed from the 32 layer, 28×28 ypi fabric panels, which nearly meet the NIJ backface limit of 1.73" (44 mm). The trellised fabric panels clearly demonstrate reduced backface deformation over base fabric targets of equivalent yarn count. The trellised fabric panels fabricated from the 28×28 base fabric having a 55° trellis angle and 34x34 ypi yarn count exhibited complete penetrations at the backface test velocity, and hence was not included in Figure 2.

Figure 3 presents the .44 magnum V_{50} performance of 1 lb/ft^2 ballistic panels fabricated from base fabric and trellised fabric plies. The 0°–90° plain weave base fabric panels demonstrate that increasing yarn count results in a reduction of V_{50} performance for 1 lb/ft^2 panels. Similarly, for each type of base fabric, it is evident that reduction in the angle between the interwoven yarns through trellising to achieve higher yarn count and cover also results in a reduction in V_{50} performance. This clearly demonstrates that increasing yarn count, fabric basis weight, and

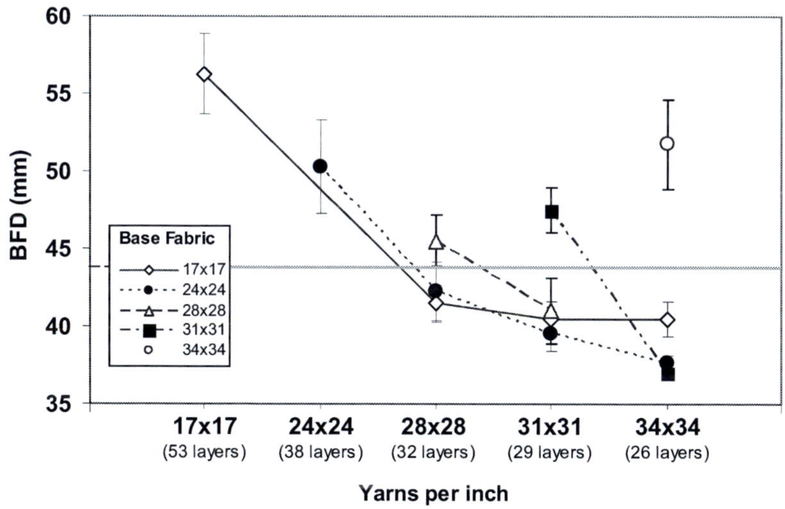

Figure 2. Backface performance of 1 lb/ft2 base fabric and trellis fabric panels (.44 Magnum, 1430±30 ft/s). The horizontal grey line represents NIJ Standard 0101.04 maximum backface deformation.

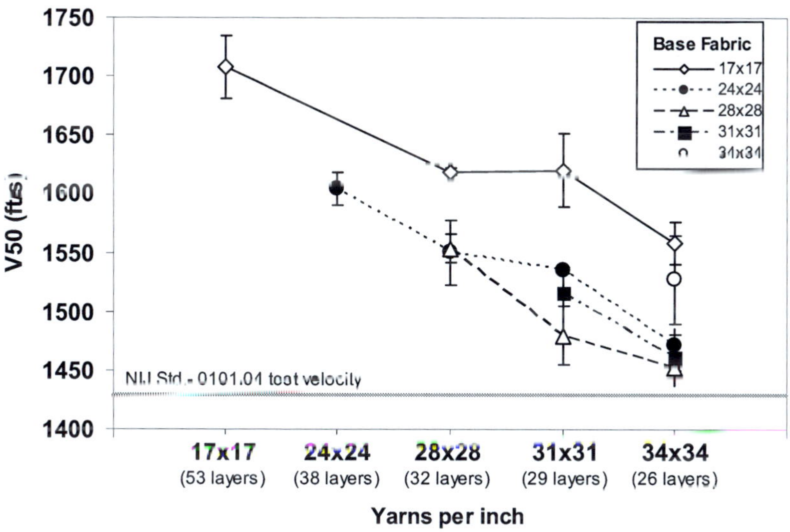

Figure 3. V_{50} performance (.44 Magnum) of 1lb/ft^2 base fabric and trellised fabric panels.

ultimately reducing the number of ballistic fabric plies reduces the V_{50} regardless of the angle between interwoven yarns.

Based on the backface deformation and V_{50} results presented in Figures 2 and 3, the best performing ballistic panels were those fabricated from 17×17 fabric trellised to 28×28 and 31×31 ypi. These panels demonstrated consistently low backface deformation, while exhibiting V_{50} performance over 1600 ft/s. While the 17×17 base fabric would be considered too unstable for vest fabrication and long term wear, trellising the 17×17 fabric to 28×28 or 31×31 ypi results in stable fabric structures that could have utility in soft body fabrication.

DISCUSSION AND CONCLUSION

During a ballistic event, traditional biaxially woven fabrics experience longitudinal wave propagation along the warp and weft yarns directly impacted by the projectile, resulting in a pyramidal transverse wave deformation out of the plane of the target, as described by Cunniff [3]. The fabric yarns not directly impacted by the projectile are driven by the cruciform strain and resulting deflection of the impacted yarns. The angle between the non-impacted, interwoven yarns within in the "facets" of the pyramidal deformation becomes less than 90 degrees as the fabric flows toward the impact point with little resistance. It was this observed trellising response due to out of plane deformation which inspired the investigation of fabrics in a pre-trellised state, where controlled resistance to further trellising of non-impacted yarns might ultimately create a more blunted hemispherical or isotopic deformation during the ballistic event.

High speed photography of ballistic impact experiments performed on pre-trellised fabric targets provides evidence in support of this hypothesis. Figure 4 presents high speed photographs of ballistic impact experiments performed on 15"× 15", 0.16 lb/ft^2 areal density targets fabricated from 6 layers of 24×24 ypi fabric, 4 layers of 34×34 ypi fabric, and 4 layers of 45° trellis angle, 34×34 ypi trellised and cross-plied fabrics. Each of the targets was stitched with the 2"×2" quilt pattern and perimeter stitching and were allowed to hang unconstrained prior to impact, being taped lightly in place along the top edge. The targets were impacted with a 0.34 inch diameter spherical steel projectile at 700 ft/s fired from a gas gun. The 24×24 ypi. plain weave base fabric and the 34×34 ypi equivalent yarn count plain weave fabric targets shown in figure 4a and figures 4b demonstrate typical pyramidal deformation at all times following impact. The impact behavior of the trellised fabric target illustrated in figure 4c demonstrates a more blunted deformation, involving a greater amount of the target at material while exhibiting an overall backface depth reduction of 17% and 21% at 0.75 ms following impact over that observed in the 6 layer, 24×24 ypi base fabric target, and the 4 layer, 34×34 ypi equivalent yarn count fabric targets, respectively.

In conclusion, we have demonstrated that stable trellised fabric architectures can potentially be used to improve backface performance in soft body armors while maintaining high V_{50} performance against .44 Magnum deformable bullets. This cursory investigation suggests that biaxial fabrics having yarns crossing at small acute trellis angles and intermediate yarn count or cover perform best against the .44 Magnum threat. Further improvements in ballistic resistance against the .44 Magnum or other handgun threats may be possible through proper selection of fabric trellis angle, yarn denier, ballistic fiber physical properties or yarn finish.

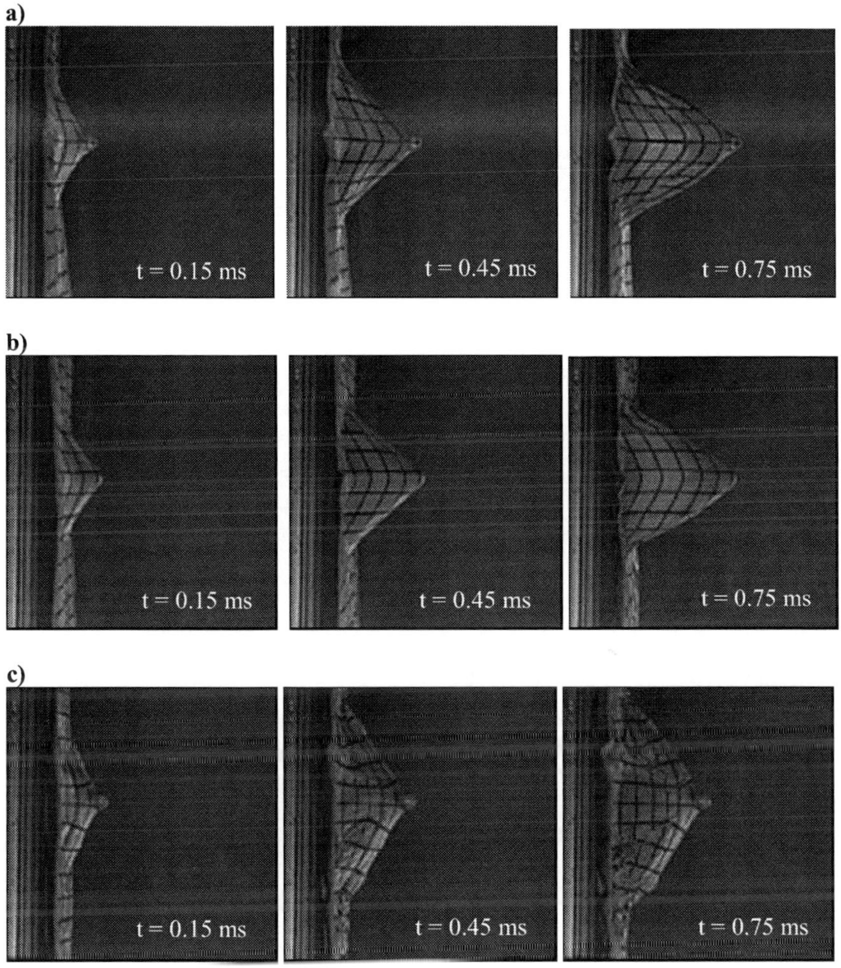

Figure 4. Ballistic impact profiles (0.34" spherical steel projectile, 700 ft/s) of a) 24×24 plain weave base fabric target, b) 34×34 ypi plain weave fabric, and c) trellised fabric plies with 45 degree trellis angle and 34×34 ypi.

REFERENCES

1. NIJ Standard-0101.04. 2001 "Ballistic Resistance of Personal Body Armor," U.S. Department of Justice, Office of Justice Programs, National Institute of Justice.
2. Figucia, F., 1975, "the Effect of Kevlar Fabric Construction on Ballistic Performance," USANRDEC TR75-103CE (AD No. B959 596L),U.S. Army Natick Research, Development and Engineering Center, Natick, MA..
3. Cunniff, P.M., 1992. "An Analysis of the System Effects in Woven Fabrics Under Ballistic Impact," *Textile Res. J.*, 62(9): 495–509.
4. Roylance, D., S.S. Wang, 1980. "Penetration Mechanics of Textile Structures," in *Ballistic Materials and Penetration Mechanics*, R. C. Liable ed. New York: Elsevier Scientific Publishing Co, pp. 273–293.
5. Cheeseman, B. A., T. A. Bogetti. 2003 "Ballistic Impact into Fabric and Compliant Composite Laminates," *Composite Structures* 61, 161–173.
6. Harrison, P., M.J. Clifford, A.C. Long, 2004, "Shear Characterization of Viscous Woven Textile Composites: A Comparison Between Picture Frame and Bias Extension Experiments," *Composites Sci. Technol.* 64: 1453-1465.

The Influence of Biaxial Stress States on the Stiffness of Glass Textile Reinforced Cementitious Composites

T. TYSMANS, H. CUYPERS, S. ADRIAENSSENS, B. BELKASSEM, O. REMY and C. RAMAULT

ABSTRACT

The application of glass textile reinforced cementitious composites (GTRC) in slender building elements such as plates or shells, can be advantageous over the widely used steel-reinforced concrete. As the glass fibre diameter is three orders of magnitude smaller than the typical steel rebar diameter, very thin and/or freeform plates and shells can be designed. In these two-dimensional elements however, complex biaxial stress states can occur under various load combinations. Up to now, only the uniaxial behaviour of GTRC is studied extensively: GTRC behaves approximately linear in compression, but is highly nonlinear in tension. In this paper, biaxial tensile tests are carried out on cruciform GTRC specimens and observed with the Digital Image Correlation technique. Experimental results are described and compared to a numerical model in the finite element program Abaqus. First results show that the uncoupled stress-strain behaviour in both principle stress directions leads to an underestimation of the biaxial stiffness of GTRC.

INTRODUCTION

Modern architecture nowadays tends towards impressive shapes with three-dimensional curvature. With this evolution, the demand for suitable materials for durable, structurally efficient and (fire)safe free form building applications rises. The most common materials used for three-dimensional forms are membranes, steel-reinforced concrete and the recently more frequently used fibre reinforced plastics. However, all three of these materials have their disadvantages for application in free form buildings.

Membranes allow for very light and freeform shapes, but exhibit problems of acoustics, fire-safety and insulation. The necessary membrane prestress requires moreover strong anchorage and increases the total weight and the amount of material used.

Tine Tysmans, Heidi Cuypers, Sigrid Adriaenssens, Bachir Belkassem, Olivier Remy, Carla Ramault, Vrije Universiteit Brussel, Faculty of Engineering Sciences, Department of Mechanics of Materials and Constructions (MeMC). Pleinlaan 2, 1050 Brussels, Belgium.

Fibre reinforced plastics also allow for free forms, applying them as faces in sandwich panels. This application leads to lightweight and stiff but relatively thick structures. For their application in buildings, the resin has to be made fire resistant, which increases the cost significantly. An underlying steel framework is usually necessitated to carry the loads to the foundations.

When properly designed, force-efficient and slender steel-reinforced concrete shells can be made to cover large spans. The form freedom of steel-reinforced concrete shells is however limited due to the practical difficulties in positioning and shaping the steel reinforcement bars. The high cost associated with placing the reinforcement makes freeform steel reinforced concrete shells an uneconomical solution. Another disadvantage of using steel reinforcement is the necessity of a concrete cover, which causes an increased dead weight and shell thickness increase (the minimal concrete cover on both sides amounts to minimally three centimeters, depending on the exposure coefficient [1]).

Textile reinforced cementitious composites (TRC) can eliminate some of the disadvantages inherent to steel-reinforced concrete, without losing the advantages. The reinforcement diameter is considerably reduced and at the same time the reinforcement ratio and thus the material's tensile capacity is retained. When using a noncorrosive reinforcement such as glass fibres, the concrete cover can be omitted and the shell self weight decreased. Consequently, thin and fire-resistant structures with strongly varying curvature can be built, which are able to carry the internal forces to foundations.

A case study on a doubly-curved, 10 m span TRC shell [2] was previously performed based on the uniaxial highly nonlinear tensile behaviour of the material. Total independence of the stress-strain behaviour in the principal directions under the occurring multiaxial stress states was assumed. This study showed that the large nonlinear tensile capacity of TRC should be incorporated in the shell design to prevent overdimensioning in the studied case. The study also showed that assuming a simplified linear behaviour could lead to underdimensioning for other geometric designs. It is thus of major importance to know and understand the non linear tensile behaviour of TRC under uniaxial as well as under multi-axial stress states. Upto now, no research has been performed on how stresses and cracks occurring in one direction affect the stiffness in another direction.

In this paper, uniaxial and biaxial (tension-tension) physical tests are performed on TRC specimens containing the same glass fibre volume fraction. Strain measurements are carried out with the Digital Image Correlation technique. The biaxial experiment is then numerically modeled in the finite element program Abaqus assuming completely uncoupled stress-strain behaviour in both directions (thus assuming no influence of cracks in both directions) and using the experimental uniaxial stress-strain data. Strain comparison from experiments and model illustrates the stiffness gain or loss with reference to the uniaxial stiffness. This analysis evaluates whether the uncoupled biaxial stiffness can be an apt approach for the design of TRC plates and shells.

TEXTILE REINFORCED CEMENTITIOUS COMPOSITE: MATERIAL CHOICES

In composite industry the choice of the most appropriate fibre material for a certain application often results from the comparison of the effective stiffness (E-modulus/density) and/or the effective strength (maximum stress/density). In the construction industry however, the material's weight often becomes less important than its cost price, and glass fibres turn out to be the more economical choice.

The disadvantage of glass fibres in combination with a cementitious matrix, is that E-glass fibres are damaged by the alkalinity of the cement. Alkali-resistant glass fibres slow down, but do not completely eliminate this process, and still undergo a strength loss [3]. Inorganic Phosphate Cement (IPC) is a cementitious matrix developed at the Vrije Universiteit Brussel which reaches a neutral pH after hardening. Research has shown that the strength loss with time of E-glass fibres embedded in the IPC-matrix is negligible [4]. Moreover, the IPC-matrix is viscous and has a very small average grain size, resulting in much higher fibre volume fractions (up to 20%) than with Ordinary Portland Cement fine tuned mortars.

In this paper, an Inorganic Phosphate cement matrix is used, in combination with randomly orientated E-glass fibre textiles (GTR-IPC). The fibre volume fraction is 20 % for all uniaxial and biaxial test specimens. This corresponds to 4 randomly orientated glass fibre mats of 300 g/m² (Chopped Strand Mat OCV M5) over a total plate thickness of 2.2 mm.

UNIAXIAL BEHAVIOUR OF GTR-IPC

The behaviour of GTR-IPC under compression is mainly determined by the cementitious matrix and is approximately linear elastic. Its behaviour in tension however is highly nonlinear. Figure 1 shows the stress-strain curve of a 500 mm x 75 mm x 5 mm, 20 fibre volume percent GTR-IPC specimen under increasing tensile load. The test is displacement-controlled (2 mm/min) and performed by an Instron 5885 universal test machine. Strain measurements are executed with a 50 mm double averaging clip-on extensometer.

Three phases can be determined in the tensile behaviour of GTR-IPC ([5]):

At the beginning of loading (pre-cracking stage I), the stiffness of the uncracked composite is determined by the "law of mixtures" for linear elastic composites. Since the volume fraction of the glass fibres is 20%, the stiffness in this first stage is mainly determined by the matrix.

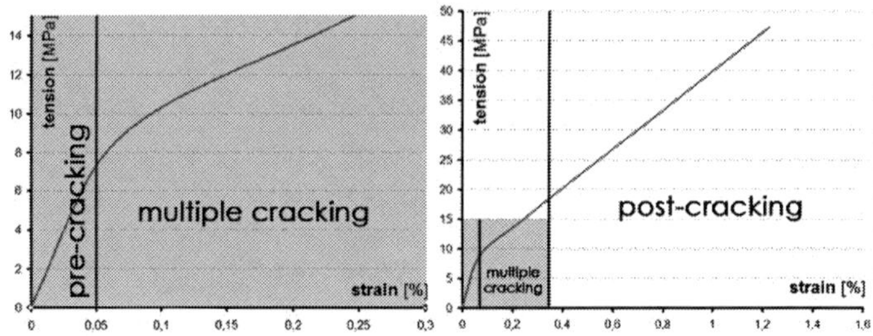

Figure 1. Experimental stress-strain behaviour of 20 vol.% GTR-IPC and fit to stochastic cracking model.

When exceeding the tensile strength of the concrete matrix, the first cracks appear. At the crack face, the whole tension force has to be carried by the reinforcement. As the amount of fibres is larger than the critical fibre volume fraction, the acting load can be carried and the composite does not fail. As the tension force increases, additional cracks occur (multiple cracking stage II): due to the frictional bond between filaments and concrete, forces are transferred in the concrete matrix until its tensile strength is reached again. The cracking distance and the crack width are determined by the properties of the reinforcement, the bond characteristics between reinforcement and concrete and the tension failure strain of the matrix.

In the stabilised crack pattern state (III), no further cracks occur. As the load increases, the filaments are theoretically strained further until their strength is reached. Failing of the specimen can however also happen as a result of fibre pull-out.

The stochastic cracking model (Cuypers et al. [6]) has been developed to describe the nonlinear tensile behaviour of GTR-IPC and has been validated before with numerous experimental data. It is based on the theory of crack formation over a certain constant stress range and describes stochastic introduction of cracking by means of a 2-parameter Weibull function. The experimental stress-strain data of the 20 volume % GTR- IPC specimens are fitted to this stochastic cracking model. Figure 1 shows that the resulting curve fits the experiments very good. This stochastic cracking model is used to implement the nonlinear tensile behaviour into the finite element program Abaqus. The advantage of using this model over the experimental data is that the model curve does not show any noise and stress is continuously increasing with strain.

BIAXIAL BEHAVIOUR OF TEXTILE REINFORCED CEMENTITIOUS COMPOSITE

Experimental Setup

The biaxial tension-tension tests are performed on cruciform specimens in a test set up at the MeMC laboratory at the Vrije Universiteit Brussel, which has been specifically designed for bi-axial tests (see figure 2, [7]). The plane biaxial test device consists of 4 independent servo-hydraulic actuators in order to avoid second order effects. The use of hydraulic actuators represents a very versatile technique for the application of the loads. When only one actuator per loading direction would be used,

Figure 2. Biaxial test set up.

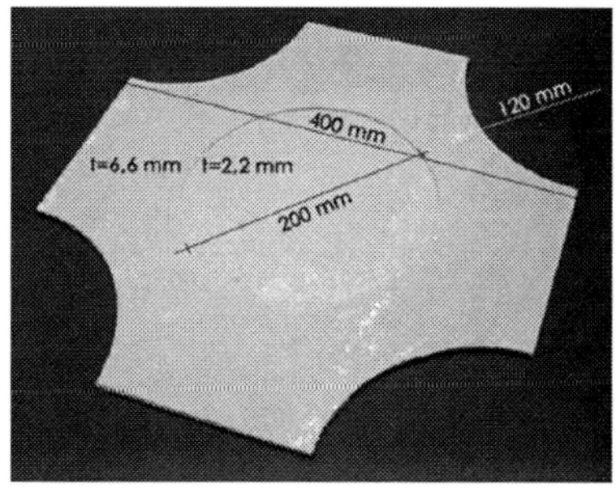

Figure 3. Geometry cruciform specimen.

the centre of the specimen would move. This would cause a side bending of the specimen, which results in undesirable asymmetric strains. The system used by the authors consists of four actuators with a close-loop servo control using the measured loads as feedback system, allowing the centre of the specimen to stand still. The device has a maximum capacity of 100 kN in both directions and is restricted to tensile loads. As no cylinders with hydrostatic bearing were used, failure or slip in one arm of the specimen would result in sudden radial forces which could seriously damage the servo-hydraulic cylinders and load cells. To prevent this, hinges were used to connect the specimen to the load cells and the servo-hydraulic cylinders to the test frame. Using four hinges in each loading direction results in an unstable situation in compression and consequently only tensile loads can be applied.

The geometry of the test specimen is shown in Figure 3. The cruciform specimen is build up out of 3 plates: one full cruciform plate, and two plates with a hole of diameter 200 mm. All three plates have a fibre volume fraction of 20 % and a thickness of 2.2 mm. The three plates are glued together with an epoxy glue (Araldite AW 136 H + hardener HY 994) to form the cruciform specimen with reduced thickness in the center. The specimen is fixed with four clamps of 5 cm depth with the same epoxy glue. Bolting is excluded as connection method as it would damage the specimen and could cause preliminary failing in the clamps. As the connection is made with glue, the force can be assumed uniformly distributed over the length of the clamp.

The specimen is loaded with a force ratio of 1/1 at a speed of 5 kN/min (load controlled). Strains in the midpoint of the cruciform specimen should thus be equal in both principal strain directions. Strain measurements are carried out with Digital Image Correlation, a full-field measuring technique which can be used to retrieve the deformation field of the specimen. A speckle pattern is applied on the top surface of the specimen. Images are taken every 1.2 seconds. By comparing the images in deformed and undeformed state, the deformation field can be obtained. Starting from this, the strain distribution is calculated over the whole specimen surface at every 0.1 kN. This surface measuring technique allows us to see a possible asymmetrical distribution of the strains before failing, contrarily to measures with strain gauges. These asymmetrical strain distributions can occur due to imperfections of the test specimen or asymmetry of the test set up.

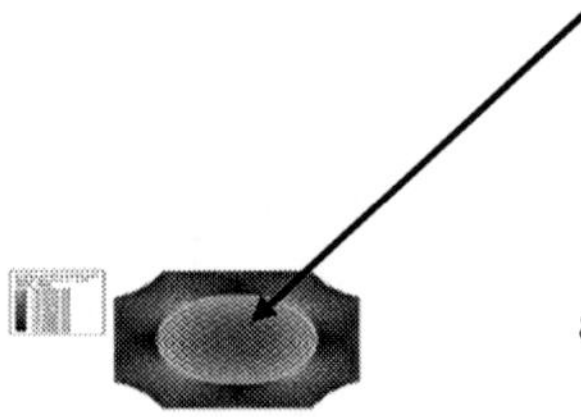

Figure 4. Maximum principal strain distribution in the cruciform specimen under biaxial load of 20 kN/direction according to finite element model.

Finite Element Model

Due to the ill definition of the load bearing area of the biaxial specimen, stresses in the biaxially loaded zone cannot be calculated from the experimentally applied loads. Therefore a finite element model is created in Abaqus (version 6.7-1), which allows comparison of experimental with the uncoupled biaxial model strains. This comparison will show the influence of stresses and thus cracks on the material stiffness in the perpendicular direction.

The geometry of the cruciform specimen is inserted as a surface with varying thickness and with 5 cm shorter arms (as this part of the arms is in the clamps during testing). Mainly quadrilateral thin shell elements with five degrees of freedom per node (three displacements and two in-surface rotations) and nine integration points per section are used. The displacement of the four edges is prevented in the vertical z-direction. The tensile force is uniformly applied at the edges as a shell edge load in steps of 2.5 kN. Selfweight is neglected in the model.

To describe the nonlinear constitutional behaviour of TRC, initially a linear elastic model is used, with an E-modulus of 12.2 GPa and a Poisson-coefficient of 0.3. For tensile stresses higher than 3 MPa, a standard plastic model is applied. The uniaxial experimental data (see Figure 1), fitted on to the stochastic cracking model ([6]), are inserted in this plastic model. In the finite element model, total uncoupled behaviour of both principal stress/strain directions is thus assumed. Comparison with experimental data will show if thus assumption is justified.

Figure 4 shows the distribution of the maximum principal strains for the modeled biaxial specimen, under a total load of 20 kN/direction. The model shows equal strains of 0.328 % in the center point of the specimen. This corresponds to principal stresses of 25 MPa. The discontinuity at the boundary of the two different plate thicknesses results in stress concentrations, which result in large strain concentrations due to the nonlinear tensile behaviour. According to the model, it is expected that the specimen will fail at this circle boundary, before achieving its failure strength in the center point of the specimen.

Experimental Results And Comparison With FEM Model

The experimental distribution of the maximum principal strain, observed with DIC under a tensile load of 20 kN per direction, is shown in Figure 5.

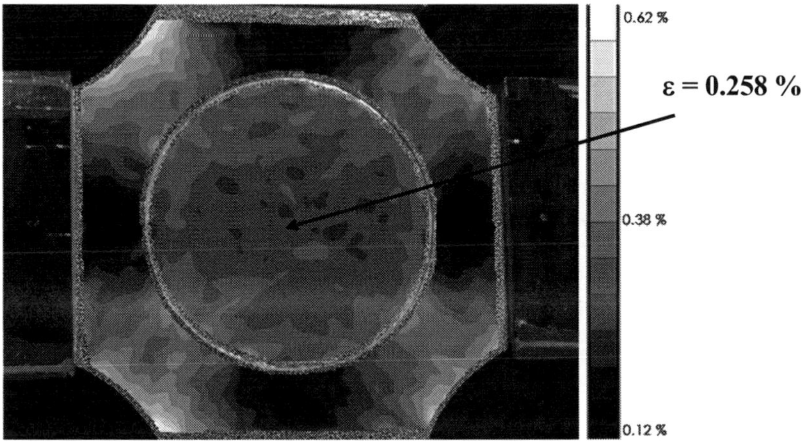

Figure 5. Experimental maximum principal strain distribution in the cruciform specimen under biaxial load of 20 kN/direction, observed with DIC.

The experimental and FEM strain distribution correspond well. With DIC, no strain values could be obtained near the edges and the thickness change boundary, so additional local measures (i.e. strain gages) have to be done to validate the model, more specifically the stress concentrations at the thickness change edge. Figure 5 also shows that the experimental strain distribution is not fully symmetric, as a result of the inherent material and geometric imperfections of the GTR-IPC specimen.

From the strain data obtained every 0.1 kN with digital image correlation, a force-strain curve is extracted for the first and second principal strain in the center point of the cruciform specimen (Figure 6). Theoretically both principal strains should be equal and this is approximately the case as can be seen in figure 6. Both curves can be compared with the force-strain datapoints that can be extracted from the finite element model. Due to the stress/strain concentrations in the FEM- model, the model curve is limited up to 35 kN, as strain failure is then reached in this concentration zone. Figure 6 clearly demonstrates that model and experimental data have initially a similar trendline, but that the experimental stiffness becomes significantly higher than the model stiffness for strains after the cracking of the specimen.

When uncoupling the stress-strain behaviour for designs in which biaxial stress states occur, the material stiffness would be underestimated according to the observations made in this paper. When designing for deflections (serviceability limit state in the Eurocode), this assumption would thus be safe. However, when designing

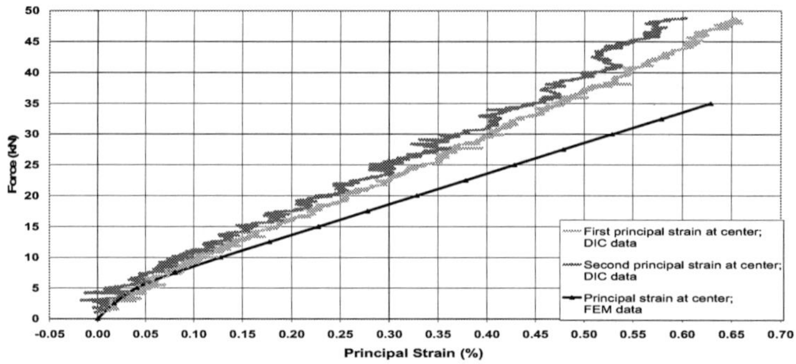

Figure 6. Force-Principal Strain diagram of biaxial test up to failure and comparison with finite element model results.

for tensile strength (ultimate limit state in the Eurocode), the in reality stiffer material could be bearing higher tensile loads than predicted by an uncoupled model, and the design could thus be unsafe. It is therefore of major importance to execute more tests in the future to validate these findings.

CONCLUSION

The uniaxial nonlinear tensile behaviour of glass textile reinforced cementitious composites is well-known. This paper described the biaxial stress-strain behaviour of GTRC based on the comparison between a physical test and a Finite Element Model. Uniaxial and biaxial tension-tension tests have been executed and experimental strains have been compared to a finite element model. The comparison of experiment and uncoupled-behaviour model shows that the biaxial stiffness is underestimated when assuming totally uncoupled biaxial behaviour of GTRC.

REFERENCES

1. European Committee for Standardisation (CEN). April 2004. "Eurocode 2: Design of Concrete Structures—Part 1-1: General rules and rules for buildings: 4: Durability and cover to reinforcement". *CEN.*
2. T. Tysmans, H. Cuypers, S. Adriaenssens, J. Wastiels and O. Remy. 2008. "Design of TRC saddle shells, part B: Influence of material nonlinearity", presented at the Challenges for Civil Construction 2008 Conference, April 16–19, 2008.
3. Orlowsky J, Raupach M, Cuypers H, Wastiels J. "Durability modelling of glass fibre reinforcement in cementitious environment". Materials and Structures 38; March 2005, pp. 155–162.
4. Cuypers H, Wastiels J, Van Itterbeeck P, De Bolster E, Orlowsky J, Raupach M. "Durability of glass fibre reinforced composites experimental methods and results". Composites Part A: applied science and manufacturing 37, 2006, pp. 207–215.
5. Brameshuber W. "Textile Reinforced Concrete: State of the art Report" of RILEM Technical Committee 201-TRC, Aachen, Germany, 2006, pp. 29-35, 135–147.
6. Cuypers H, Wastiels J. "Stochastic matrix-cracking model for textile reinforced cementitious composites under tensile loading". Materials and Structures 39; Oct 2006, pp. 777–786.
7. Smits A., Van Hemelrijck D., Philippidis T.P., Cardon A. "Design of a cruciform specimen for biaxial testing of fibre reinforced composite laminates". Composites Science and Technology 66(2006), pp. 946–975.

Experimental Investigation of Tow Pullout Behavior of Aramid Fabric

A. GAWANDI, E. THOSTENSON and J. W. GILLESPIE, JR.

ABSTRACT

Experimental investigation is performed to study tow pullout behavior for woven aramid fabric with various polymer surface morphologies. The pullout behavior is studied by conducting tow pullout tests using a pullout fixture. Various material configurations are created by applying heat and pressure to the fabric to change polymer surface morphology. The pullout tests are performed by pulling a tow from a fabric loaded with tension in the direction transverse to that of pull. The tests are performed at various loading rates to investigate the sensitivity to the loading rate. Load versus displacement data is obtained, from which the pullout energy is computed. Both, tow uncrimping and translation components of pullout energy are determined. The results are compared with baseline neat fabric as well as baseline fabric with polymer coating.

The results of the investigation show that for the aramid fabric under consideration, the neat fabric shows no loading rate dependence. On the other hand, polymer-coated fabrics with altered surface morphologies show loading rate dependence. Moreover, the heat-compressing of the fabric seems to have significant influence on tow-pullout behavior. The results of pullout tests show that by simply compressing the fabric in the presence of heat may influence the interactions at tow cross-over regions to results in higher pullout energy.

INTRODUCTION

Yarn pullout is an important phenomenon observed when a woven fabric is subjected to transverse loading. It is influenced by the friction properties at the tow cross-over points. Yarn pullout becomes one of the mechanisms by which the energy of dynamic transverse loading of the fabric is dissipated. Thus, increasing the tow pullout energy dissipation enhances the performance of the woven fabric against a

Anis Gawandi, University of Delaware Center for Composite Materials, Newark, DE 19716
Erik T. Thostenson, University of Delaware Center for Composite Materials, Newark, DE 19716
John W. Gillespie Jr, University of Delaware Center for Composite Materials, Newark, DE 19716

dynamic loading event. A class of tests reported in literature in conjunction with tow pullout behavior of woven fabrics is tow pullout tests. Increasing the tow-tow friction at the cross-over points without resulting in premature yarn tensile failure is a plausible way to increase the pullout energy. Thus, friction at the cross-over points becomes a parameter of interest. Briscoe et al. [1] studied the effect of surface friction characteristics of aramid fabrics with respect to their static deformation and ballistic performance. Various fabric weaves were studied with different levels of yarn-yarn friction achieved with addition or removal of surface lubricants. It was found that fabrics with high friction dissipated larger amount of energy relative to the fabrics with lower friction for both quasi-static and ballistic deformation processes. Bazenov [2] studied the pullout zones created by ballistic impact of laminates comprising several aramid fabric layers. He performed yarn pullout tests on plain-weave aramid fabric with varying yarn-yarn friction levels. Also, effect of water on pullout behavior was tested by immersing the fabric in water before testing. It was concluded that energy transferred to fabric increases with volume of pullout zone. Also, water reduced the penetration resistance of laminates owing to reduction of the width of pullout zone.

Shockey[3] performed pullout tests on single yarn pulled from a fabric layer while the fabric is preloaded in tension in the direction transverse to that of the yarn being pulled out. Force-displacement data was recorded. The tests were performed on Zylon (30 x 30 and 40 x 40), Kevlar (32 x 32), and Spectra (32 x 32) fabrics for various pullout rates. Both warp and fill yarns were pulled from the fabric. The results showed significant effect of pullout rate for Zylon, a smaller effect on Spectra, and a negligible effect for Kevlar. Kirkwood et al. [4] performed yarn pullout tests on Kevlar KM2 600 denier fabric preloaded in tension in transverse direction. Effects of fabric length, number of yarns pulled, arrangement of yarns, and transverse tension on the force-displacement curves were studied. A semi-empirical model was presented for predicting the yarn pullout energy as a function of pull-out distance. The model was observed to replicate the experimental data with high degree of accuracy. The energy of yarn pullout from a fabric was found to depend on length of sample, transverse fabric tension, number of yarns pulled at once, the pattern of the pull, and distance the yarn was pulled.

In this study, experimental investigation of tow-pullout behavior of plain weave aramid fabric is described. The material investigated is mainly of two types, one being the neat fabric without any coating material and the other a polymer-coated fabric with various polymer surface morphologies. Coated fabrics have applications such as described in [5]. Load-displacement curves for a single tow-pullout are generated for various material configurations at various loading rates and transverse tension. Pullout energies are subsequently obtained and compared.

MATERIALS

The materials used in the investigation are described below.

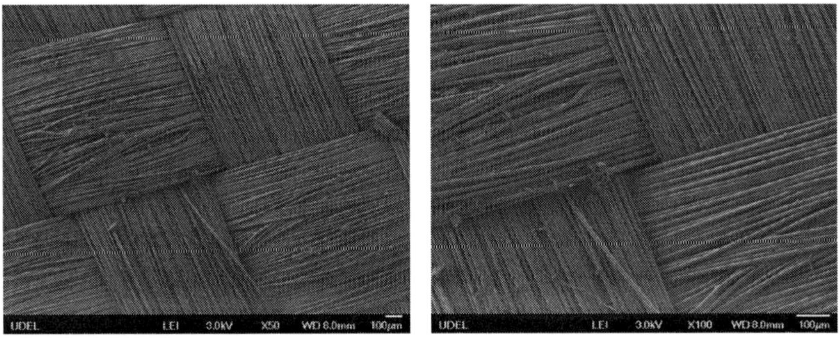

Figure 1. SEM images of baseline neat aramid fabric front and back faces.

BASELINE NEAT ARAMID FABRIC

The aramid fabric used is a plain-weave fabric. The fabric has approximately 10.5 yarns per cm (warp and fill). The uncoated version of the fabric has measured areal density of approximately 211 g/m^2. Figures 1(a) and (b) show the scanning electron microscopy images of the baseline neat fabric.

POLYMER-COATED FABRIC WITH MODIFIED MORPHOLOGY

The original surface morphology of the polymer-coated fabric is altered by compressing a single fabric layer at various temperatures. The fabric layer is heat-pressed at the temperatures of 75, 100, and 150 °C at a constant pressure of 0.86 MPa (125 psi) for 15 minutes. The layer heat-pressed at 75°C is pressed with the resin-rich side facing up, whereas the layers pressed at 100°C and 150°C are pressed with resin-rich side of the fabric facing down. This resulted in different polymer morphologies on the fabric surface. After the polymer morphology modification process, scanning electron microscopy is used to study the change in surface morphology.

The SEM analysis shows that the fabric layer pressed at 75°C seemed to drive the polymer on the resin-rich front face towards the undulations and hence results in the concentration of resin in the undulations area. Note that the fabric was heat-pressed at this temperature with resin-rich (front) face up. SEM images show that the heat-compression treatment results in a relatively larger concentration of the resin in the undulation area due to flow of polymer to the undulation area of the fabric. On the other hand, the surface modification at 100°C and 150°C seems to result in the partial impregnation of fabric tows. Note that at both these temperatures, the fabric was heat-pressed with polymer-rich front face facing down. Based on the observations from the scanning electron microscopy, the modified fabrics are classified as,

1. Coated fabric with polymer concentration in undulations area (UC-1)
2. Coated fabric with partial tow-impregnation (PTI-10)
3. Coated fabric with partial tow-impregnation (PTI-15)

EXPERIMENTAL SETUP

Figure 2 (a) shows the schematic of the pullout test. All tests are performed by pulling a single tow from a single layer of fabric while preloading the fabric in tension in the transverse direction. A fabric layer approximately 190 mm in size is used for the test. Yarns transverse to the direction of the pullout are manually removed to free up approximately 88 mm length of yarns to grip in the machine. The remaining approximately 102 mm length of the fabric is clamped in the direction of the pull. This length (L, see figure 17) is kept constant for all experiments. With this value of length, approximately 108 cross-over points are available for the tow to pull through. Shockey [3] found that multiple yarns can be pulled from a single fabric without altering the pullout behavior if at least two yarns are in between the successive yarns pulled. This observation is first verified prior to testing and although leaving two yarns in between seems to be sufficient, it is decided to leave at least three yarns between the successive tows being pulled. Considering the variation of crimping in the warp and weft directions of the fabric, pullout tests are performed in both warp and weft directions. All tests are conducted using Instron 5567 machine with a 500 N (100 lb) load cell.

Figure 2 (b) shows the fixture used for the tow pullout tests. The fixture is adapted from Shockey [3] and allows for a fabric specimen to be clamped on opposite edges using bolts. The fabric edge to be clamped is wrapped over a steel rod that fits in a trench on the fixture. A top plate is then placed firmly and tightened with bolts. Approximately 25 mm of fabric length at each end is provided for this wrapping and clamping mechanism. The fabric specimens are cut such that approximately 140 mm length of fabric remained between the two clamped edges. Similar to the fabric length in the pull direction (L), this

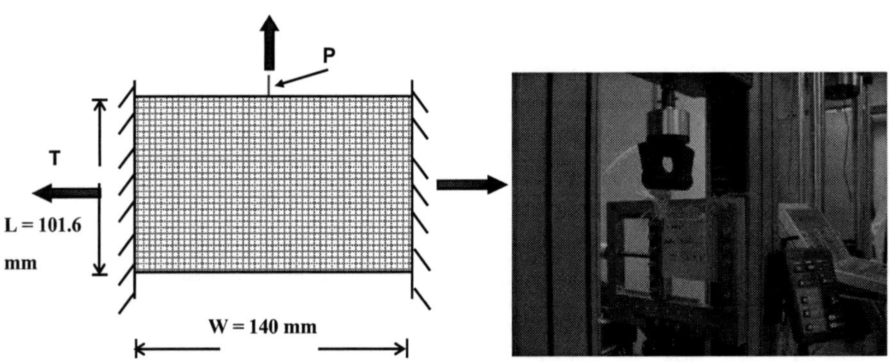

(a) (b)

Figure 2. Tow pullout test (a) schematic (b) test set up.

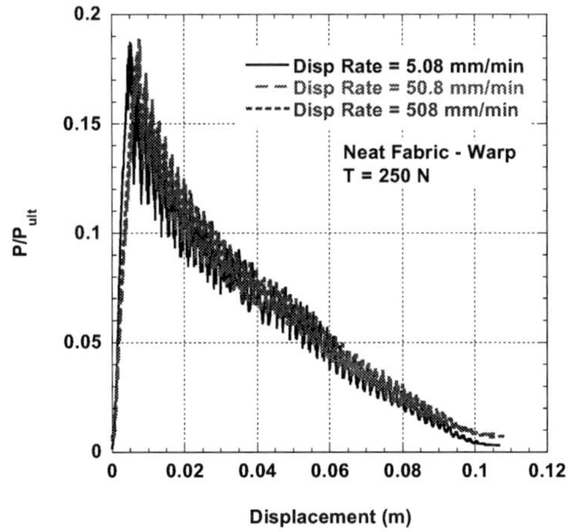

Figure 3. Load-displacement curves for neat fabric.

RESULTS AND DISCUSSION

Figure 3 shows the load-displacement curves for the baseline neat aramid fabric at various displacement rates for the transverse tension of 250 N. The load in the tow is normalized by the estimated load for the tow failure. As the figure shows, the baseline fabric without any coating shows no loading rate dependence. The curves for various loading rates almost fall on top of each other. Figure 4 shows the

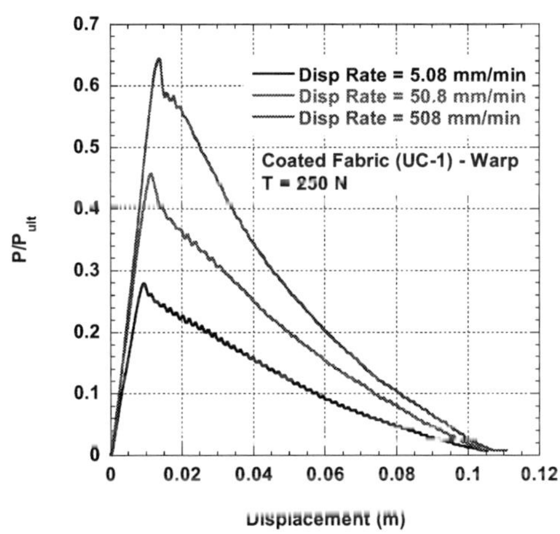

Figure 4. Load-displacement curves for polymer-coated fabric UC-1.

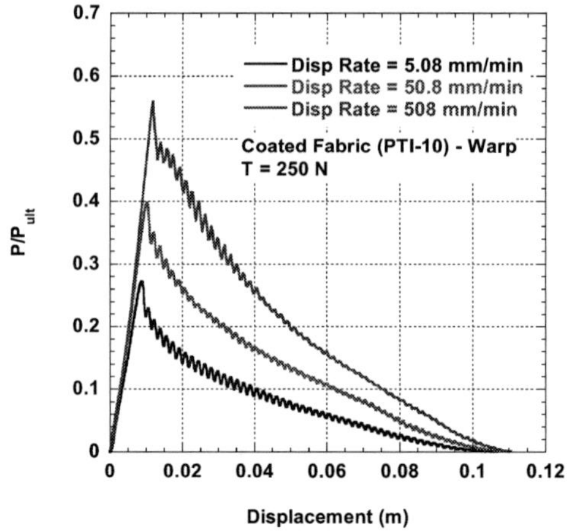

Figure 5. Load-displacement curves for modified polymer-coated fabric PTI-10.

load-displacement curves for the coated fabric with modified morphology UC-1. Now, with the presence of polymer, the fabric clearly shows loading rate dependent behavior. The peak load now increases with the increasing displacement rate. Also, the linear portion of the curve shows stiffening effect as the displacement rate is increased. A comparison with the data for the baseline neat fabric shows that the peak load at the largest displacement rate of 508 mm/min is almost 250 % larger than the

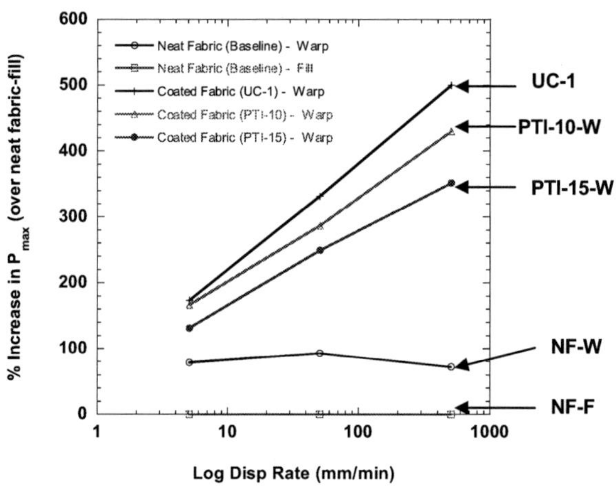

Figure 6. Peak loads as a function of displacement rates expressed as percent increase over the peak loads for neat fabric in the fill direction.

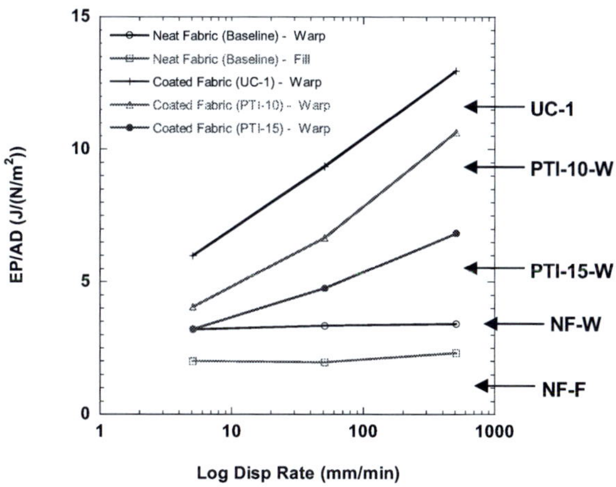

Figure 7. Pullout energies for various fabrics as a function of displacement rates.

corresponding neat fabric case. Also, the peak load for this case is approximately 65 % of the ultimate failure load. This increase obviously corresponds to an increase in the pullout energy. Figure 5 shows the load-displacement behavior for the modified polymer morphology PTI-10. The peak loads in this case are smaller than UC-1 case. The pullout tests on tows from the fabrics with modified polymer morphology (UC-1, PTI-10, and PTI-15) reveal that the pullout behavior is sensitive to the altered polymer morphology. The PTI-10 and PTI-15 fabrics show smaller peak load enhancement relative to the UC-1 case. This is shown in the figure 6 which shows the peak loads for various fabrics as a function of loading rate. The loads here are expressed as a percent increase over the smallest peak load case which is the neat fabric fill direction.

PULLOUT ENERGY

The area under the pullout curve gives the total pullout energy for the single tow pullout. The tow pullout energy is comprised of two distinct parts of the load-displacement response. In the beginning, as the load is applied on the tow, the tow starts uncrimping. This process is represented by the initial linear portion of the force-displacement curve. During the tow uncrimping, the tow being pulled experiences static friction at the cross-over points. Once the peak load is reached, the tow uncrimping stops and the tow starts to move. This is called tow translation and is represented on the force-displacement curve in the post-peak region. Summing the tow-uncrimping and tow-translation parts gives the total pullout energy.

Figure 7 shows the total tow pullout energies for various fabric material configurations plotted versus the displacement rate. The pullout energy presented in the figure has been normalized with respect to the areal density of the corresponding fabric. With the exception of the baseline neat fabric, the total pullout energy increases with the increasing loading rate. The largest pullout energy corresponds to the fabric UC-1 which has modified polymer morphology with polymer concentration in the undulations area. On the other hand, the pullout energy for the other two altered morphology cases is larger than the baseline coated fabric for the PTI-10 case and smaller than the baseline for the PTI-15 fabric. The uncrimping and translation parts of the total pullout energy also seem to follow the trends for the total pullout energy, especially at the larger displacement rates. It is clear that both uncrimping and translation components to the pullout energy are larger for the polymer-coated fabric relative to the neat fabric, with one of the altered morphology cases (UC-1) being the largest.

CONCLUSIONS

Tow pullout tests are performed on various fabric configurations comprising neat and coated fabrics with modified polymer surface morphology. A single tow is pulled out from the fabric using a pullout fixture. The tests are performed at various displacement rates and transverse fabric tension values. Force versus displacement data is acquired to compute pullout energy.

The results of the pullout tests show that modification to the polymer morphology on the fabric surface significantly influences the pullout behavior in the case of polymer-coated fabrics. The baseline neat aramid fabric considered in this study show no loading rate dependence. On the other hand, the polymer-coated fabrics with various surface morphologies obtained from heat-compressing the fabrics show strong rate dependence. The modified polymer morphology with polymer concentration in the cross-over points seems to result in the maximum pullout energy at all displacement rates used in the testing.

REFERENCES

1. Briscoe, B. J., and Motamedi, F. 1992. "The Ballistic Impact Characteristics of Aramid Fabrics: the Influence of Interface Friction," Wear, 158:, 229–247.
2. Bazenov, S. 1997. "Dissipation of Energy by Bulletproof Aramid Fabric," Journal of Materials Science, 32:4167–4173.
3. Shockey, D. A., Erlich, D. C., and Simons, J. W. 2001. "Improved Barriers to Turbine Engine Fragments: Interim Report III," DOT/FAA/AR-99/8, III.
4. Kirkwood, K. M., Kirkwood, J. E., Lee, Y. S., Egres Jr, R. G., Wagner, N. J., and Wetzel, E. D. 2004. "Yarn Pull-out as a Mechanism for Dissipating Ballistic Impact Energy in Kevlar® KM-2 Fabric: Part-II: Predicting Ballistic Performance," Textile Research Journal, 74(10): 939–948.
5. Samant, K. R., Sauer, B. B., and Minshon, J. C. 2006. "Fabrics with Strain-Responsive Viscous Liquid Polymers," United States Patent Application Pub. No: US 2006/0286882 A1.

TOPIC VI

3D TEXTILE COMPOSITES: MECHANICAL BEHAVIOR

Progressive Failure Modeling and Strength Predictions of 3-D Woven Composites

A. E. BOGDANOVICH

ABSTRACT

Modeling progressive damage and failure processes in textile composites and developing predictive strength analysis tools gain fast growing scientific interest and increasing practical importance. Having capabilities of accurately predicting failure loads by analytical tools allows one to significantly reduce or eliminate expensive and time consuming testing, facilitate design and product development, and increase reliability of resulting textile composites and structures. This paper is devoted to one specific problem—progressive damage modeling and ultimate failure strain and load predictions of 3D orthogonal woven textile composites using 3-D MOSAIC analysis approach. The essence of previously developed analysis methodology is described in first part of the paper and then illustrated on the example of textile composite reinforced with 3-D orthogonal woven E-glass preform. The analysis is performed at the level of unit cell micromechanics, with the unit cell being an assembly of 3-D resin-impregnated rectangular cross section yarns and 3-D matrix blocks between them. Each yarn is treated as a unidirectional composite with its assigned elastic characteristics and ultimate strains obtained for tensile, compressive and shear loads. Numerical results are presented for the cases of unidirectional tensile loading in the warp and fill directions. They include predicted failure initiation and ultimate failure strains and loads, as well as characteristic features of progressive failure processes. Experimental results published earlier for textile composite modeled here, are used for the model calibration and validation of predicted ultimate failure characteristics.

INTRODUCTION

Among most important and difficult theoretical problems in the area of textile composites are: modeling complex and diverse fiber architectures, 3-D stress/strain

---

Alexander E. Bogdanovich, 3TEX, Inc., 109 MacKenan Drive, Cary, NC 27511, U.S.A.

analysis and progressive damage/failure simulations, and strength predictions. Obtaining full set of ultimate failure characteristics for each practical textile composite is a big challenge on its own. In principle, it can be done by mechanical testing at a coupon level or by using predictive analysis. However, in any practical situation all available theoretical and experimental means are typically combined. The reason is not only in much higher expense and more time required for mechanical testing, but also in the fact that certain mechanical properties (through-thickness stiffness and tensile strength as an example) cannot be obtained by direct experimentation. In the above example this is due to material thickness is much smaller than the length of required experimental coupon. Therefore, predictive analysis is not only useful for saving money and time, but in many situations it is the only way to compensate for the lacking experimental data. Of course, the analysis approach and computational tool to be used for this purpose have to go through a serious verification and validation by available experimental and theoretical data.

In the context of predictive analysis role described above, the situation is much more advanced in the area of elastic properties than in the area of damage, failure and strength predictions. A lot of models and computational tools are available for elastic property predictions of various type textile composites, see recent reviews in [1], [2] and [3]. Those range from the methods relatively simple analytically and inexpensive computationally, though limited in their accuracy due to simplistic fiber geometry representation and mechanistic hypothesis (like in Orientational Averaging method, for example), to full-field 3-D specialty variational tools and commercial finite element packages. At the same time, in spite of some considerable accomplishments in the area of damage, fracture, failure and strength modeling, no universally capable and satisfying predictive analysis tool is available now. This paper describes and illustrates one practical analysis approach, 3-D MOSAIC, showing a potential to fill existing gap in failure modeling and predictive strength analysis of textile composites.

THE CONCEPT OF HIERARCHICAL 3-D MOSAIC ANALYSIS

A high complexity of the class of theoretical problems of interest here is caused by the necessity to describe three-dimensional, multi-directional reinforcement architectures, incorporate non-conventional (from the viewpoint of traditional mechanics of laminated composites) failure mechanisms and constituent material models, and consume unusually large amount of experimental and/or theoretical input material data required to run the analysis and generate results. 3-D MOSAIC approach, which is hierarchical by its nature, provides significant advantage when modeling hierarchical materials and structures. Indeed, as was illustrated in earlier publications on example of elastic property prediction, numerical results obtained from the solution of a finer structural resolution problem can be used, after appropriate averaging over the respective Representative Volumetric Element (RVE), as the necessary set of input data for a coarser structural resolution problem, see [1], [4]. Specifically, it was shown there that, one can start at the hierarchical level of individual fibers and matrix between them, select an RVE (as illustrated in Figure 1), formulate and solve appropriate boundary value problems under applied displacements or stresses, implement respective averaging procedure, and obtain complete set of effective elastic properties of the unidirectional composite of interest. The only input data required by such analysis are: elastic properties of fiber and matrix, shape of fibers and their arrangement in the unidirectional composite.

Then, the obtained effective elastic properties can be used as input data for the next hierarchical level analysis of a textile composite RVE (Unit Cell), see example in Figure 2. The procedure similar to described above is now applied to this RVE, resulting in its effective elastic properties. In order to perform this step of analysis, we only need to know shape and dimensions of each unidirectional composite, its location and orientation within the RVE. The computed effective elastic properties of the textile composite RVE (or more than one RVEs if required) can be further used as the input data for the next hierarchical level analysis, which would probably be a macro-level analysis. An excellent agreement of numerical results, obtained at three hierarchical levels, with experimental data means that this approach is not only mechanistically adequate but also converging and numerically accurate [1], [4].

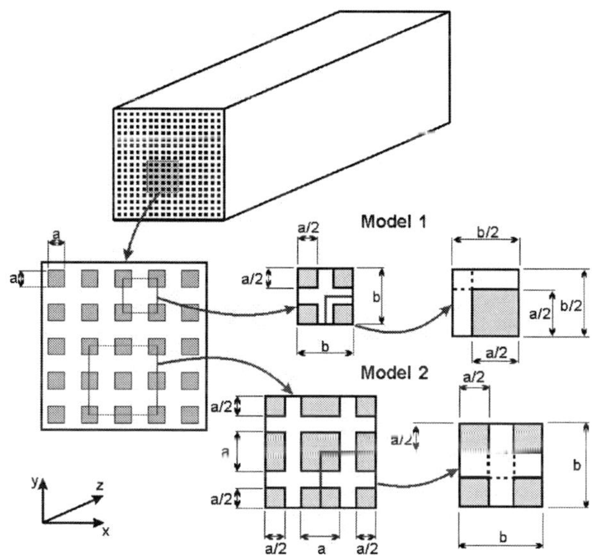

Figure 1. 3-D MOSAIC Model of a unidirectional composite [4].

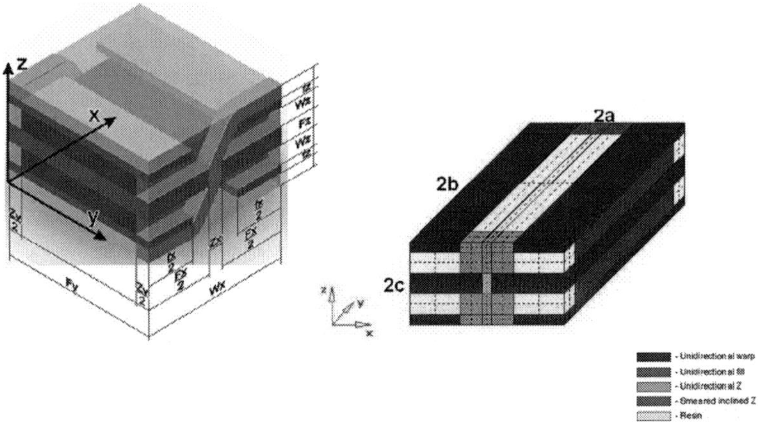

Figure 2. Model of 3-D orthogonal weave unit cell (left) and respective 3-D MOSAIC model after smearing inclined Z-yarn segments [4].

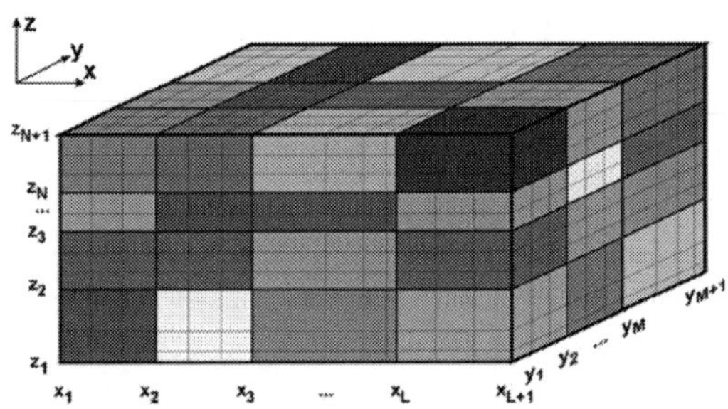

Figure 3. Generic 3-D MOSAIC model with boundaries between bricks of different materials (thick lines) and computational mesh example (thin lines).

3-D MOSAIC PROGRESSIVE FAILURE ANALYSIS

The above described general analysis concept can be used, in the same hierarchical sense, for progressive failure and strength predictions of textile composites. Of course, this type of analysis is much more complex than the one developed for predictions of effective elastic properties, and at this time only first steps have been realized. It shall be emphasized that there are several principal common features in both the analyses. The first is the representation of a material/structure to be solved in terms of a 3-D MOSAIC model, as illustrated in Figure 3.

The second principal common feature is, that all aspects of the 3-D stress/strain analysis method used in both cases are the same, including the representation of undetermined 3-D displacement field in terms of triple series with Bernstein approximation polynomial basis functions, the application of variational principles, the way how displacement continuity conditions are imposed between adjacent bricks, and the procedures of the point-wise displacement, strain and stress field evaluation directly from their approximation series, see [5]. The third common feature is that each material brick in 3-D MOSAIC model of Figure 3 is treated as 3-D anisotropic homogeneous solid, which may have its elastic and strength properties distinct from any other material brick in the model. Those properties can be obtained from experiments and/or from a different 3-D MOSAIC analysis, performed at a finer resolution hierarchical level.

Additional theoretical capabilities required for conducting progressive failure analysis and strength predictions within the framework of 3-D MOSAIC analysis approach, are provided by Continuum Damage Mechanics (CDM) methodology. This includes identification of possible failure modes depending on the nature of each brick material, selection of appropriate phenomenological failure criterion for each failure mode, and a scheme of elastic property gradual discount. One variant of such additional capabilities that were implemented in the current version of 3-D MOSAIC code, has been described in detail in [1]. The approach enables to start the analysis with local failure initiation and progression modeling within individual composite yarns, within matrix pockets, or at the yarn-matrix interfaces. The following describes the essence of this approach.

Consider again generic Mosaic parallelepiped of Figure 3. Following general concept of Continuum Damage Mechanics, each homogeneous brick is characterized by its initial, pristine elastic properties corresponding to undamaged state and a sequence of discounted elastic properties corresponding to some "damaged" states. Each of those sets of discounted properties is related to specific failure modes and to the extent of imparted gradual damage. In the analysis approach adopted here, each brick in the Mosaic structure is assumed orthotropic in its principal axes of material symmetry (denoted 1, 2 and 3), though it can belong to any of more general classes of anisotropy in the global coordinate axes x, y and z assigned to the Mosaic structure. The occurrence of each failure event is determined by a simple and mechanistically validated ultimate strain failure criterion. Accordingly, nine possible failure modes of each brick can be realized. The failure modes and failure criteria included in the analysis and the adopted elastic property discount scheme can be found in Table 1 of [1].

If necessary, any other suitable phenomenological failure criterion can be incorporated into the developed progressive failure analysis algorithm and computer code without any difficulty. However, in our opinion any additional complication of the analysis should be justified by experimental results. If intending to apply a more complex failure criterion than the ultimate strain, one has to keep in mind that a lot of additional experimental information would be needed, particularly on the "interactive term coefficients", which are contained in quadratic and other higher order failure criteria. A bulk of such data is very limited in the literature, and in order to generate new data one has to conduct biaxial loading tests which are complex, expensive and require special testing machines, fixtures, and unique specimen geometries.

Briefly, the developed progressive failure analysis is realized in the following steps.

Step 1. Run 3-D MOSAIC analysis with some arbitrary magnitude of applied load, say P_0. A set of strain rendering points $\{x,y,z\}$ is selected by choosing number of uniform intervals within each brick in *x*, *y* and *z* directions of the Mosaic structure. All strain components are evaluated at those points. This procedure can be applied to entire structure, however there are options to apply it only to one or more selected material, to one or more selected material bricks, to one or more selected strain components. Also, strains at the boundaries between material bricks may be included or not in each current strain evaluation step. All this provides sufficient flexibility to the analysis of damaged states, because repeated application of the failure criterion to those material bricks and failure modes which have been already realized and caused property discount, may result in an idle run of analysis within one cycle.

Step 2. All of the computed strains are normalized over their respective ultimate values, which are taken from the input data set for each material and transformed to the local coordinate system of each individual brick in the Mosaic structure. Of course, the normalization factors are different for distinct material bricks and even for the same material bricks having different orientation with respect to the global axes *x*, *y*, *z*. Besides, the normalization factors are generally different for same material in same direction but for cases of tensile vs. compressive strain. This step of analysis is performed for all strain rendering points chosen in Step 1.

Step 3. In the result of point-wise rendering of three tensile, three compressive and three shear strain components and their normalization over respective ultimate values, the largest magnitude among them is determined and further termed "failure strain indicator". It is associated with the following characteristics of the first failure occurrence: one integer identifying the failed material number from the menu of materials in Mosaic structure, three integers identifying location of the failed material brick within Mosaic structure, the failure mode identifiers (tension, compression or shear), three coordinates of the determined failure point, the axis index (1, 2 or 3) for tensile and compressive failures or the plane index (23, 31, or 12) for shear failure. All this together provides sufficient information for applying the elastic property discount scheme.

Step 4. The initial (arbitrarily chosen) load magnitude P_0, under which the initial analysis has been run, is divided by the largest among all failure strain indicators determined in Step 3. The obtained load magnitude P_1^F constitutes "first failure load".

Step 5. Having the information obtained in Step 3, initial elastic characteristics of the brick that failed first are discounted in accordance with the scheme of Table 1 in [1].

After that we repeat Step 1; 3-D MOSAIC analysis is run again, this time with the load magnitude P_1^F. Then Step 2 is repeated. If any of the new failure strain indicators reaches value 1, the second failure event is stated. If not, the load is increased and Step 1 is repeated. The second failure event is stated, the respective load magnitude P_2^F is determined as in Step 4. The failed material brick, failure mode and failure point coordinates are identified. Then Step 5 is realized and analysis continues in loops until the "total failure" state is reached. This can be determined by different criteria, as discussed and illustrated in [1]. One possible criterion is that some of the principal moduli (E_x, E_y, E_z, G_{xy}, G_{xz} or G_{yz}) of entire Mosaic structure becomes so small that, if applying load (tensile, compressive or shear) in respective direction, the computed strain grows beyond the allowable, thus indicating loss of load-bearing ability. The other practical criterion is that failure in fiber direction is reached for those material bricks which provide principal reinforcement of the structure. In particular case of 3D orthogonal woven composite that would be a failure of unidirectional composite bricks having warp-, fill- or Z-directional reinforcement. In the latter case, failure in the form of catastrophic delamination would be obtained, which is obviously not acceptable.

NUMERICAL EXAMPLE

One example of representative 3D woven fabric and its composite Unit Cell is shown in Figure 4. This 3.25 kg/m^2 (96 oz/yd^2) areal density E-glass fabric is one of 3TEX's major commercial products.

Mechanical properties of composites reinforced with this preform and manufactured by VARTM method using Derakane 8084 epoxy-vinyl ester resin are currently studied experimentally and theoretically. First results of experimental studies of failure initiation and progression under uniaxial tensile loading in the warp and fill directions were presented in [6]. Those results show that initial failure (detected by Acoustic Emission method) was observed in this composite at overall tensile strain level ~0.43% in warp and ~0.37% in fill direction. The respective ultimate failure strains were recorded as ~2.74% and 3.0%. The ultimate tensile failure stresses were obtained as ~429 MPa in warp and ~486 MPa in fill directions. These experimental results are valuable, because they allow us to perform initial calibration and validation of the computational model in Figure 4.

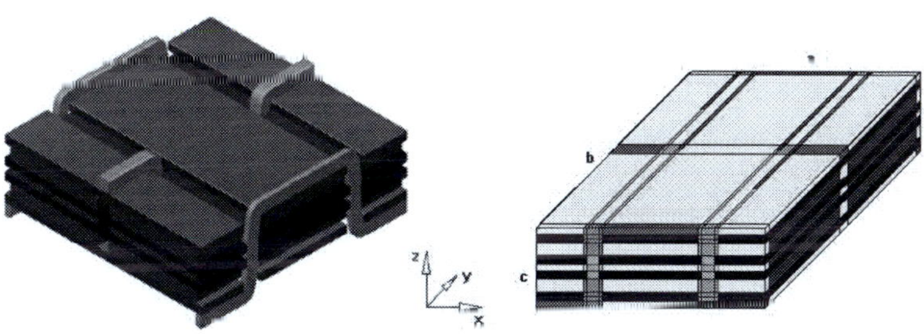

Figure 4. Full Unit Cell of 3D orthogonal woven E-glass fabric (left) and 3-D MOSAIC model of its composite Unit Cell (right) used in progressive failure analysis.

Due to a limited length of this publication, we are not able to describe here in any detail the boundary value problem formulation, geometric input parameters, material input data, computational aspects, and convergence study of the 3-D stress/strain fields used in progressive failure analysis. Note that the composite Unit Cell model in Figure 4 contains seven different materials; four of them are identical unidirectional transversely isotropic composites placed in warp, fill, Z (along z-coordinate) and Z (along x-coordinate) directions. The other two distinct (and monoclinic in the global coordinate system) materials are the inclined segments of Z-yarn smeared with surrounding matrix. The rest of the Unit Cell volume is occupied by numerous blocks of pure matrix. Total number of individual blocks, which are treated as elementary units of progressive failure process in this model, is 495. So, progressive failure analysis of this composite model is very cumbersome and requires long computer runs and a lot of analyst's time. The initial model calibration and validation showed very good agreement with experimental results from [6]. Numerical results will be presented and discussed at the conference and then published elsewhere.

REFERENCES

1. Bogdanovich, A.E. 2006. "Multi-scale Modeling, Stress and Failure Analyses of 3-D Woven Composites," *J. Materials Science*, 41(20): 6547–6590.
2. Lomov, S.V., Ivanov, D.S., Verpoest, I., Zako, M., Kurashiki, T., Nakai, H., and S. Hirosawa. 2007. "Meso-FE Modelling of Textile Composites: Road Map, Data Flow and Algorithms," *Composites Science and Technology*, 67: 1870–1891.
3. Bogdanovich, A.E. 2008. "Computational Modeling and Analysis of Textile Composites: Accomplishments and Challenges," in *CD Proc. of SAMPE'08 Conf.*, Long Beach, CA, May 18–22.
4. Bogdanovich, A.E. 2003. "Multiscale Predictive Analysis of 3-D Woven Composites," in *CD Proc. of 35th Int. SAMPE Techn. Conf.*, Dayton, OH, Sept. 28–Oct. 2.
5. Bogdanovich, A.E. 2000. "Three-Dimensional Variational Theory of Laminated Composite Plates and Its Implementation with Bernstein Basis Functions," *Computer Methods in Applied Mechanics and Engineering*, 185(2-4): 279–304.
6. Bogdanovich, A.E., Mungalov, D., Ivanov, D., Lomov, S.V., and I. Verpoest. 2008. "A Comparative Study of Mechanical Properties of 3D Woven Single-Layer and 2D Woven Multi-Layer E-Glass Composites," in *CD Proc. of SEICO 08 SAMPE Europe Int. Conf.*, Paris, March 31–April 2: 110–115.

Modeling of the Geometry and Numerical Calculation of the Elasticity Parameters of 3D Braided Shapes

Y. KYOSEV, A. BRÜCKEN, A. TILLMANNS and A. BÜSGEN

ABSTRACT

Complex braided shapes are more and more often used for different technical applications. The calculation of the parameters of braided structures, based on the machine and yarn data, has only been published for simple regular shapes yet. In this work, the yarn paths are calculated using a new simulation of the braid structure. It calculates the yarn paths, taking into account the geometry of the shape which is covered, as well as the take down speed and the yarn cross section parameters. The cover factor and braiding angle at each point of the shape are available from the simulated yarn geometry. Using additional algorithms, the elasticity modules for each base cell are calculated. Additional to the implemented calculation algorithms, an export of the yarn geometry to the FEM software ANSYS is included, which allows further calculations using the yarn axes or yarn volumes. In order to verify the results, different 3D bodies are covered with braid with a small number of carriers—from 12 up to 48 – with different take down speeds. The yarn paths in these forms are detected optically and compared with the simulated ones. The first verifications demonstrated a very good correlation between the simulated and experimental results.

INTRODUCTION

Braided structures are commonly used in different types of composites and medical applications, where complex 3D shapes have to be covered with special or high performance yarns. For the covering of mandrels, usually normal circular or special radial braiding machines are used. Here the 3D braided shapes are defined as conventional braiding performs, which cover mandrels with 3D form and not compact 3D braids, which build their cross section during the braiding without mandrel. If the

Research Institute of Textile and Clothing (FTB), Niederrhein University of Applied Sciences, Mönchengladbach, Webschulstr. 31, 41065 Moenchengladbach, Germany form of the mandrel becomes more complex, then a computer controlled take down mechanism or, for 3D forms, a computer controlled robot arm has to be used. For simple forms, like tubular and conical, there are also simple relations available which allow the calculation of the machine parameters, so that the desired braiding parameters like constant cover factor, special braiding angle can be archived. Models for the prediction of the yarn path are presented in [1], [2], the limit packing density is described in [3], and a simple implementation of such programs for tubular fabrics in [4] and [5]. These papers treat the relations for basic geometries. In [6] is presented a paper with CAD for simulation of different 3D-braided shapes, but only the problems with the yarn path and cross section modeling are reported, describing the usage of the modified Frenet-frame for the cross section orientation. Thus, the problem of modeling the yarn path of complex 3D braided shapes remains still open.

The goal of this work is to present the next step in the solution of this problem, presenting ideas for simulation of the yarn paths of braided structures around 3D mandrels with rotational symmetry. In [7] the idea of presenting the form of the mandrel as a function R=R(h) is described, where R is the radius of the concentric cross section at the length h of the mandrel. We used this idea, extending the possibility of the use of variable take down speed V=V(t).

One problem not discussed in all of the above cited papers is the slippage between the yarns and the mandrel. All these works use the yarn path mapping to the mandrel surface, which is valid only if the yarn does not slip around the surface. This condition can be true if the friction coefficient between the yarn and mandrel is high enough or if yarn path lies near to the geodesic line (shortest distance) between the crossing points. Only in this case the mapping of the yarn around the mandrel is correct. In other cases the yarn piece will slip down to more stable positions under the influence of the yarn tension. The basics of these limiting conditions are developed initially for the goal of winding packages [8], [9]. In the current work these conditions are adapted for the case of braiding and are included as additional extension of the model. In this way the model calculates the areas of possible slippage between the yarn and mandrel during the braiding.

The goal of the visualization of the structures is not only to understand them better, but also to use the geometry for future calculations of the elastic properties. The yarn paths of the braided form can be exported and converted into a set of unit cells or "basic" cells, for which the E moduli can be calculated using available theoretical models. Such unit cells of braided structures are well described in the literature [10], [11], [12], [13] etc. Because of the similarity with the woven unit cells, which are the most investigated cells in textile mechanics, there are several models for the calculation of the mechanical properties of the unit cell of braided laminated composites. As an example for the automatic generation of the unit cell geometry, the package WiseTex[13] can be used, and with the available suite of additional tools like TexComp[14], the moduli of elasticity of the cell can be calculated in different directions using the Eshelby models. At this stage in the current work, the model from [15] is implemented. For the calculation the simulated geometry is exported to the calculation module used for 3D woven structures [16], which is currently also extended in order to calculate the properties of braided ones.

METHOD

The principle of the braiding of mandrel is presented on the figure 1. The position of the braiding ring between the carriers and the mandrel determines the yarn geometry in the braiding zone and can influence the braiding process and the braiding conditions. From simple geometrical relations, the dependency between the braiding (or helix) angle α (see [1] [2], [7]), defined as half of the angle between the braiding yarns

$$\alpha = \tan^{-1}\frac{\omega R}{V}, \tag{1}$$

where ω is the angular speed of the carriers, V is the take up velocity and R is current radius of the mandrel. For conical forms, the yarn path in the parametric form is described [1] as follows

$$\begin{aligned} x &= r_0 e^{\theta \cot\alpha \tan\tau}\cos(\theta + d_i) \\ y &= \pm r_0 e^{\theta \cot\alpha \tan\tau}\sin(\theta + d_i) \\ z &= r_0 e^{\theta \cot\alpha \tan\tau}\cot\tau \end{aligned} , \tag{2}$$

where $r_0 = R(0)$ is the initial radius of the mandrel, τ the taper angle of the cone, and θ the angular coordinate of the current point in the cylindrical coordinate system (r,θ,z).

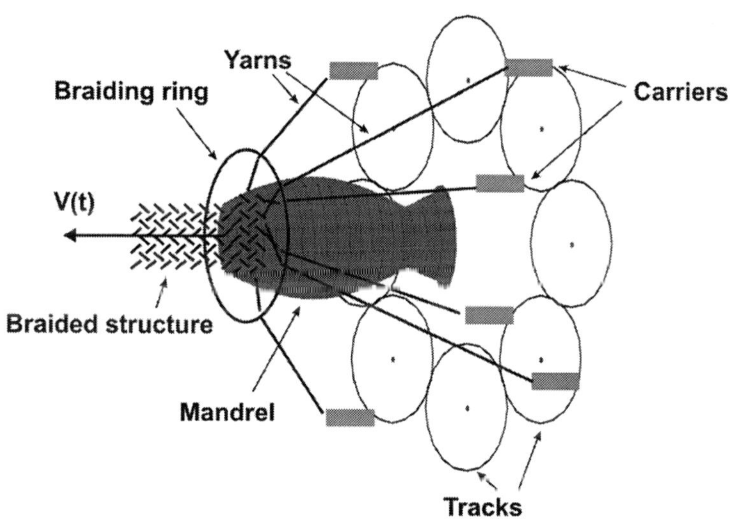

Figure 1. Working principle of braiding machine.

In the current model, we split the mandrel into set of conical parts where the relations (2) can be used. The discretization of the mandrel profile depends on the take up speed during the braiding and it is performed in such a way, that the borders between the cone parts correspond to the crossing points of the braiding yarn. At each step both of the stability conditions for the yarn path around a surface [8] are checked:

$$\tan\alpha \leq \tan\tau_G \leq \tan\varepsilon_{\max} = \mu \tag{3}$$

where τ_G and $\varepsilon_{\max}$ are the geodesic and the limit friction angles, respectively, and μ the friction coefficient between the mandrel and the yarns.

During the calculation of the yarn coordinates, first the crossing points between the yarns are calculated. They are placed on a cone surface with radius - $r_{Cone}(t)$, calculated from the current mandrel radius $R(t)$ and the yarn thickness in lateral direction $d_{y,2}$:

$$r_{Cone}(t) = R(t) + d_{y,2} \tag{4}$$

Afterwards, the yarns coordinates are calculated from these cross points taking into account the yarn thickness and the braiding pattern. Additionally the following assumptions are taken into account, in order to simplify the model:

—the yarn tension of all yarns is the same;

—yarns with rectangular, elliptical or circular cross sections can be used;

—braiding pattern 1x1, 1x2 and 2x2 could be used; other patterns are possible, but they are not important for the practical applications;

—the friction between the mandrel and the yarn is neglected as a force, but is taken into account while checking the stability conditions of the yarn parts during the simulation;

—all carriers which move in the same direction, use yarns with the same properties. Since for tubular braiding two systems of carriers are required (clockwise and anti-clockwise directions), non-crimp braid can be generated if the carriers in one directions use fine monofilaments.

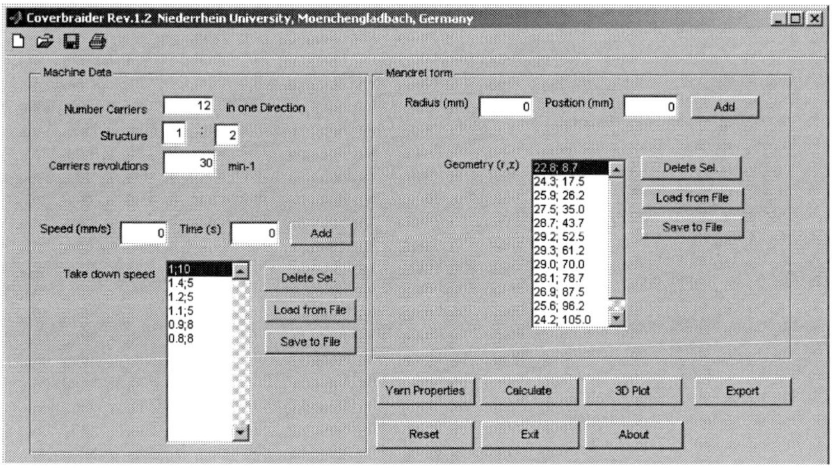

Figure 2. Graphical user interface for the input data.

IMPLEMENTATION AND NUMERICAL RESULTS

The input data for the machine parameters, braiding type (one over one, one over two, or two over two) and the yarn can be entered into a user-friendly graphical interface (Fig. 2).

The take up speed and the geometry of the mandrel can be loaded from or saved to text files as sequences. After the calculation, the yarn axes can be visualized or exported for the mechanical calculations. A simple result of the simulation of one layer of braided structure is presented in Fig. 3.

Using the information of the yarn cross section and the yarn paths, the 3D volumes of the yarns are created. An example for 2x2 braid over a mandrel is presented in Fig. 4. This geometry can be exported for ANSYS for further FEM or CFD calculation.

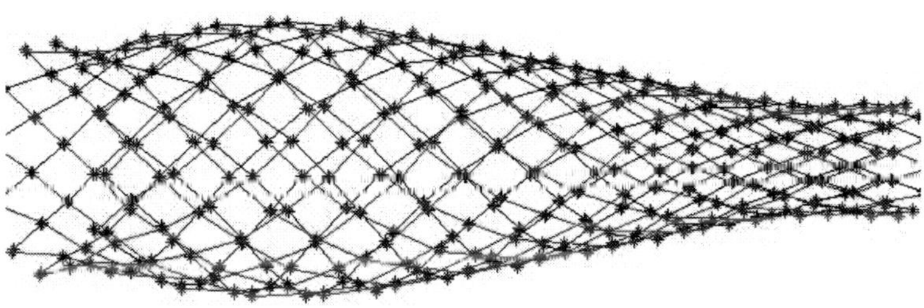

Figure 3. Simulated yarn paths of braided structure.

Figure 4. Simulated yarn paths of braided structure with constant take up speed 5.3 mm per braiding point.

Figure 5. Simulated yarn paths of braided structure with take up speed 3 mm per braiding point, calculated in order to receive a cover factor near 1.

Figure 6. Braided form with variable take up speed, in order some areas of the mandrel to be covered with higher cover factor.

PARAMETERS OF THE BRAID AND COMPOSITE LAMINATE

For the calculation of the mechanical properties, the model of Sheng [15] is used. It bases on basic cells, like the one presented in Fig. 7. Using the parameters of the basic cell, the cover factor, fiber volume fraction and elastic modulus can be calculated. For the braided structure presented in Fig. 5, the fiber volume fraction is presented in Fig. 8.

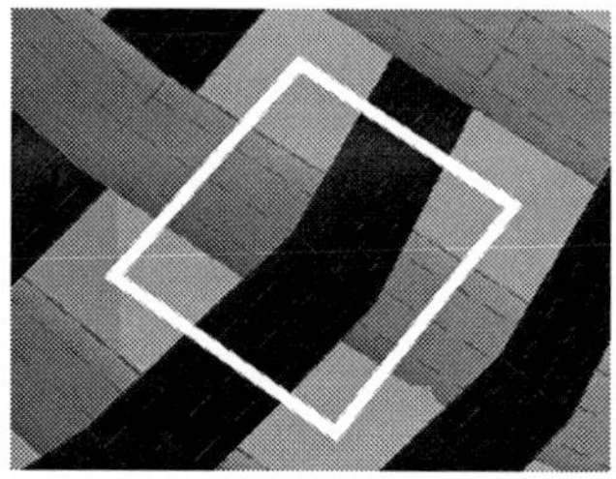

Figure 7. Basic cell, used during the calculation of the mechanical properties.

The E modulus in the axial and circumferential direction are calculated and presented in Fig. 9. All the calculations are performed for an epoxy resin with E = 3.5 GPa and ν_{LT} = 0.35 and T-300 carbon with E_L = 230 GPa, E_T = 40 GPa, G_{LT} = 24 GPa, G_{TT} = 14.3 GPa and ν_{LT} = 0.26.

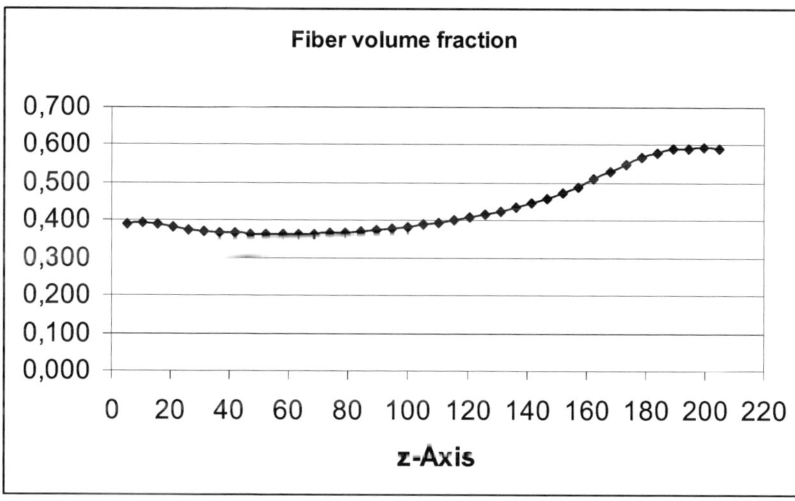

Figure 8. Fiber volume fraction, used for the calculation of the mechanical properties.

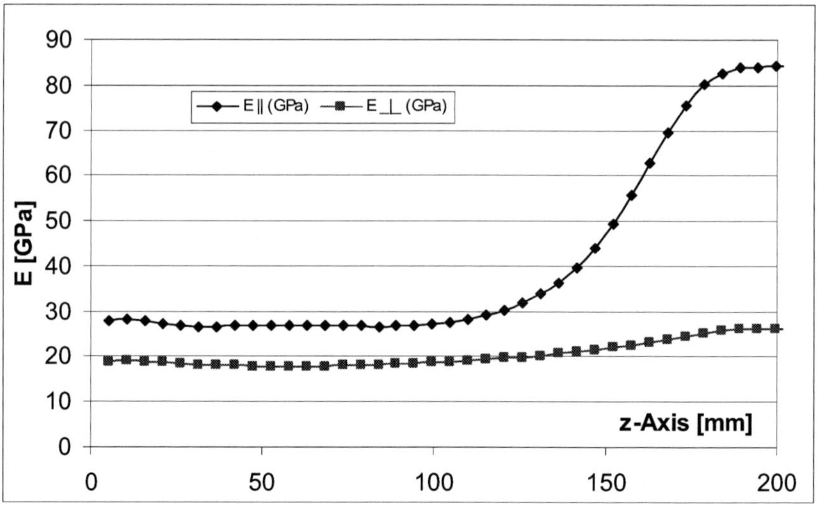

Figure 9. Calculated E modulus for the braid, presented in the Fig. 3 and 4. E|| is oriented parallel to the plane of the surface, and E _|_ is perpendicular to the surface plane.

CONCLUSIONS

A method for the simulation of braided preforms around a mandrel with rotational symmetry is presented. The mandrel is divided into a set of truncated cone where the yarn path can be analytically found. The mandrel form and the take up speed are defined as discrete functions which allow simulation of more complex relations between the braiding process and the structure. The yarn thickness is taken into account during the creation of the 3D-representation of the structure. The simulated geometry can be used for calculation of its mechanical parameters using the axis information, or it can be exported as volume for future FEM or CFD calculations. The next step in the extension of this model is including algorithms for simulation of general mandrels without rotational symmetry.

ACKNOWLEDGMENTS

The authors would like to thanks the Niederrhein University of Applied Sciences for the financial support allowing for the possibility to start this project.

REFERENCES

1. Rawal, A., Potluri, P., Steele, C., Geometrical Modeling of the yarn paths in three-dimensional braided structures, *J. of Industrial Textiles*, 2005, 35, pp. 115–135.
2. Rawal, A., Potrury, P., Steele, C., Prediction of yarn paths in braided structures formed on a square pyramid, *J. of Industrial Textiles*, 2007, 36, pp. 221–226.

3. Du, G.-W. and Ko. F., Unit Cell Geometry of 3-D Braided Structures, *J. of Reinforced Plastics and Composites,* 1993; 12; 752.
4. Naik, A., "Analysis of Woven and Braided Fabric Reinforced Composites," NASA CR 194930, June 1994.
5. Kyosev, Y., Model Generator for Tubular Braided Fabrics, Proceedings CD-ROM of *"Finite element modelling of textiles and textile composites"*, St. Petersburg, 26–28 September 2007.
6. Liao, T., Adanur, S., 3D Structural Simulation of Tubular Braided fabrics for net-shape composites, *Textile Research J.*, 2000, 70, p. 297.
7. Ko, F., Pastore, C., and Head, A., *Handbook of Industrial Braiding*, Atkins and Pearce, 1988.
8. Proshkov, A. F. Mechanism rasskladki niti, Moskva, Legpromiszdat, in Russian, orig. Title: Прошков, А. Ф. (1986) Механизмы раскладки нити (вопроы проектирования), Москва, Москва, *Легпромбытиздат.*
9. Kyosev, Y., Investigation of the internal winding of textile yarns, PhD Thesis, Technical University of Sofia, Sofia, 2002, (in Bulgarian).
10. Huang, Z.-M., Ramakrishna, S.., Predicting Elastic Constants of 2D-Braided Fiber Rigid and Elastomeric–Polymeric Matrix Composites, *J. of Reinforced Plastics and Composites,* 2004; 23; 1845.
11. Huang and Z.-M., Ramakrishna, S., Towards Automatic Designing of 2D Biaxial Woven and Braided Fabric Reinforced Composites, *J. of Composite Materials*, 2002; 36; 1541.
12. Goyal, D., Whitcomb, J., Analysis of Stress Concentrations in 2 2 Braided Composites, *J. of Composite Materials* 2006; 40; 533.
13. WiseTex http://www.mtm.kuleuven.ac.be/Research/C2/poly/software.html.
14. Huysmans, G. , Verpoest, I. , Van Houtte, P., Eshelby models applied to woven fabric composites: A benchmark study, http://www.mtm.kuleuven.ac.be/Research/C2/poly/index.htm, acc. 20.06.08.
15. Sheng, S., van Hoa S., Three Dimensional Micro-Mechanical Modeling of Woven Fabric Composites, J. of Composite Materials, Vol. 35, No. 19, 2001.
16. Birghan,A., et.al. Simulation and calculation of seamless woven 3D shells (part 1), Technische Textilien 2, 2007.

Microstructural Modeling of Three-Dimensional Woven Fiber Composites

M. A. NICOSIA, F. VINEIS, J. LAWRENCE and S. T. HOLMES

ABSTRACT

Composite materials are very attractive relative to metals for aerospace applications for various reasons, including weight savings, favorable thermal and acoustic properties, and corrosion resistance, among others. Three-dimensional woven fiber composite materials have the potential to be superior to the traditional unidirectional and laminate composites in terms both mechanical properties and cost-effectiveness in manufacturing. However, the use of these three-dimensional woven composite materials has been hampered by the absence of appropriate tools to predict stress and failure responses of composite components.

The largest challenge in this type of modeling is the disparity in relevant length scales between the unit cell dimension and size of a typical part. To address this issue, microstructural models are used, in which the unit cell geometry is used to predict average macroscopic material properties, which can then be used to perform a stress analysis on the part of interest. Prediction of mechanical failure, however, requires knowledge of the local stress—information generally not available using averaged properties.

The goal the current work is to develop a computationally feasible microstructural model for three-dimensional woven composite materials that is capable of resolving local stresses at the level of the unit cell. In general, the multi-scale model consists of two steps: (1) *Pre-processing,* in which finite element modeling of unit cells is used to estimate the effective material coefficients; and (2) *Simulation,* in which these effective material properties are used in standard finite

Mark A. Nicosia[1], Frank Vineis[2,3], Jeffrey Lawrence[3], Scott T. Holmes[3]
Departments of [1]Mechanial Engineering and [2]Chemical Engineering, Widener University, Chester, PA
[3]V System Composites, Chester, PA

element modeling of the component. As the simulation progresses, regions with high stresses are identified, and the unit-cell finite element model is utilized to predict the local stress within the fiber and matrix. This step is accomplished by projecting the macroscopic strain field onto the level of the unit cell to be used as boundary conditions. Preliminary simulations and future directions will be discussed.

INTRODUCTION

Composite materials are very attractive relative to metals for aerospace applications for various reasons, including weight savings, favorable thermal and acoustic properties, and corrosion resistance, among others. Three-dimensional woven fiber composite materials have the potential to be superior to the traditional unidirectional and laminate composites in several important areas. In terms of mechanical properties, woven composites provide more balanced stiffness and strength in all directions, which is important for general multi-axial loading states. In addition, they are easier and more cost-effective to manufacture for complex shapes using techniques such as resin-transfer molding (RTM) or vacuum assisted resin transfer molding (VARTM). However, the use of three-dimensional woven composite materials has been hampered by a lack of appropriate analytical and/or computational tools to predict stress and failure responses to be used in design.

Stress analysis of complex parts is typically performed computationally using the finite element method, which requires a geometric model of the part as well as a constitutive model describing its mechanical properties. The constitutive model for a composite material depends upon the properties and relative proportions of the constituent materials as well as their geometric arrangement. Ideally, the overall constitutive model of a material could be predicted based on knowledge of its constituents and their geometric arrangement. The largest challenge in this type of modeling is the disparity in relevant length scales. Consider a 3-dimensional unit cell resolved with a grid of 15x15x15 points. A composite part of 1 m x 1 m x 6 cm with unit cell characteristic dimension of 2 cm, would yield over 25 million nodes. Thus, for realistic geometries and unit cell structures, such models are computationally unfeasible.

To address this issue, microstructural models have been developed. These models generally consist of an averaging process, in which the unit cell geometry is used to predict average macroscopic material properties, which can then be used to perform a stress analysis on the part of interest. Early models of composite materials, referred to as *mixture theories*, were applied to unidirectional or laminate composites [1]. In these theories, the elastic modulus of the composite material was computed as a weighted average of the moduli of the constituents, with the respective volume fractions used as the weight factors. These models have been extended to consider more complex 3-dimensional fiber architectures, such as the study of Naik [2] or the work of Tabiei [3]. For example in [2], a model was presented in which the unit cell was discretized into sections perpendicular to the loading direction. Each section was further discretized into sub-elements made up of matrix or fiber. The matrix was isotropic and fiber transversely isotropic, and volume averaging was used to assign properties to each section, thus essentially treating each section as a unidirectional fiber-reinforced material. This scheme led to a fairly complex unit cell discretization and did not capture local stresses within the unit cell.

On the other end of the computational spectrum is *Direct Numerical Simulation*, in which the finite element mesh resolves the microstructural details over the entire geometry (e.g., [4]). Although possible for simulations at the level of a test coupon, current computational resources cannot be used to perform such simulations at the level of a component. Thus, for realistic component geometries and unit cell structures, such models are computationally unfeasible, necessitating the development of an intermediate strategy to project the microstructural details up to the macroscopic level for constitutive modeling as well as to project the macroscopic deformation field down to the microstructural level for estimation of local stresses. Such simulations are typically referred to as multi-scale models.

Aboudi [1] developed such an alternative micromechanical strategy referred to as the *Method of Cells*, in which the displacement field over a unit cell was approximated by linear functions and continuity between adjacent cells in displacement and traction fields led to a closed form set of equations relating the unit cell material properties and geometry to the average composite material properties. In addition to averaged composite material properties, the sub-cell displacement fields can be used to estimate the local stress within each microstructural constituent. Extensions to this theory have been developed such as the *Generalized Method of Cells* [5] and the *High-Fidelity Generalized Method of Cells* [6], which consisted of an arbitrary number of sub-cells and higher order interpolations for the displacement field, respectively. Although this method does well for the metal matrix material systems for which it has been optimized, it fails in the area of polymer matrix composites due to its lack of *shear coupling*, which is a consequence of the averaging process.

An alternative set of strategies utilize a finite element model of the unit cell geometry rather than an analytical model (such as in the *Method of Cells* family). In biomechanics, for tissue reinforced with collagen, such as heart valves, a separate finite element microstructural model was set up at each Gauss point of the macroscale finite element model [7], and a projection algorithm was developed to transfer information between the two levels. This method is computationally very expensive and has only been applied in two-dimensional cases to date.

Finite element microstructural models have also been applied to composite materials in the context of mathematical homogenization theory by the group of Fish [8]. In these methods, a set of numerical experiments is performed on a finite element model of the composite unit cell to predict the composite material properties. These properties are then used in component level simulations. During the simulation, the stress field is monitored and the unit cell finite element model used to predict local stresses in regions where macroscopic stresses become critical. This type of method combines the accuracy of a finite element unit cell model with the efficiency of using averaged material properties.

Although there has been a great deal of work in the development of modeling techniques for textile composites, there is a lack of practical application of these methods to the design of actual composite components. Specifically, many of the theoretical models have only been validated using test coupons made of unidirectional composite materials and using simple loading conditions. Whereas validation on simple geometries and loading conditions is a necessary first step for any model, ultimate application in design and analysis of large-scale components is necessary for these methods to be of practical value. With this issue in mind, the focus of the current work is to develop a multi-scale model for woven fiber composite materials that can resolve local stress distributions on the level of the unit cell while at the same time is sufficiently computationally efficient for routine use in design.

METHODS

Modeling Approach

Our approach is made up of two basic steps. First, in a pre-processing step, numerical experiments are performed using the finite element method on representative unit cells to estimate the effective material coefficients.
Once the effective material properties have been estimated, they will be used in standard finite element modeling of the component of interest. As in any stress analysis problem, geometric and loading conditions will lead to a distribution of stresses over the component. From these simulations, areas of high stress will be identified for further analysis.

In regions of high stress, the local homogenized mesh is replaced with a fully discretized model of several repeating unit cells of the material. A few unit cells away from the area of stress concentration, the model transitions back to the homogenized properties representation. The mesh density required to fully resolve the fiber architecture is much greater than needed in the homogenized regions away from the disturbance. "Tied" interfaces, which utilize a penalty method to constrain the displacement of two surfaces (or edges in 2-dimensions) with non-coincident nodes, are used to join these two regions. We hypothesize that a St. Venant-type effect will allow this method to successfully resolve the local unit cell stresses. That is, as long as the magnitude of the forces that are transmitted across the tied interface are accurate, the details of their distribution across the tied interface will not have a strong influence over the local stresses a few unit cells away from the interface. This approach has been referred to as the "mortar-element" method in the finite element literature [9] and has recently been applied to impact testing of composite materials [4]. An early presentation of an iterative variation of this method was presented by Woo and Whitcomb [10].

Numerical Example

All simulations in this work were performed using Abaqus finite element software (Dassault Systems, Providence, RI). Here, we present a simple 2-dimensional example as a proof-of-concept of the method.

A uni-directional, fiber-reinforced composite material was chosen (Figure 1). To estimate the homogenized properties associated with this unit cell architecture, a numerical test coupon was developed (60 unit cells x 60 unit cells) and subjected to simulated tension and shear tests (see Figure 1 for deformed shape of shear test; results have been scaled to illustrate the deformation). Note the overall uniform response (middle) relative to the non-uniformity across the unit cell (right). Given the unidirectional fiber architecture, an isotropic, linear elastic material model was used for the average properties. Values for fiber and matrix constituent properties were estimated based on the literature [4].

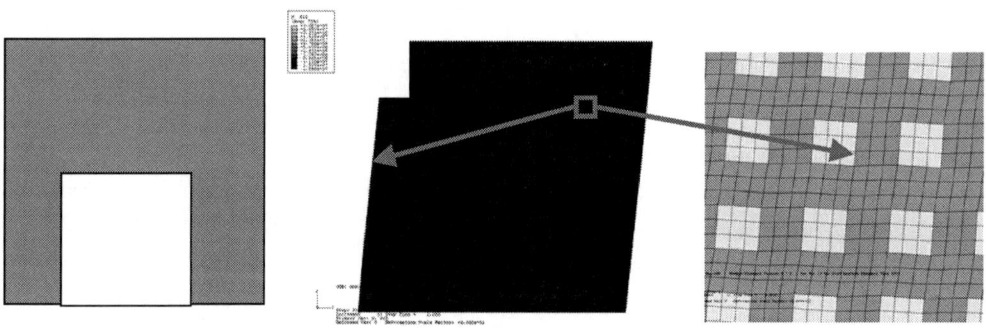

Figure 1. (left) Undeformed unit cell architecture used in current model – white indicated fibers, coming out of page and gray indicates matrix. (middle) Simulated shear test using fully resolved test coupon; note that results have been magnified to illustrate the deformation. (right) Local deformation of unit cell under shear deformation.

Once these properties associated with this unit cell were estimated (Table 1), two finite element models were generated. To provide a non-trivial stress distribution, we chose to simulate deformation of a C-channel shape under displacement boundary conditions (Figure 2). The first model was consisted of the C-Channel geometry with fully resolved fiber and matrix components. Each unit cell was discretized with a grid of 15x15 nodes for a total of just over 525,000 nodes (60 unit cells across the width). This model is referred to as a direct numerical simulation or DNS. The second model consisted of the same C-Channel, but modeled with averaged, isotropic, linear elastic properties that was locally refined in one corner with the fully discretized unit cell geometry for a total of just over 27,000 nodes.

	ν	E (MPa)
Matrix	0.3	2×10^{11}
Resin	0.3	2×10^{9}
Composite	0.35	3.86×10^{9}

Table 1. Mechanical properties of constituents as well as estimated average properties.

A numerical experiment was then performed in which the bottom edge of the C-Channel was constrained and the top edge pulled upwards. The goal was to compare the local stress distribution at the corner using our proposed approach as well as the direct numerical simulation.

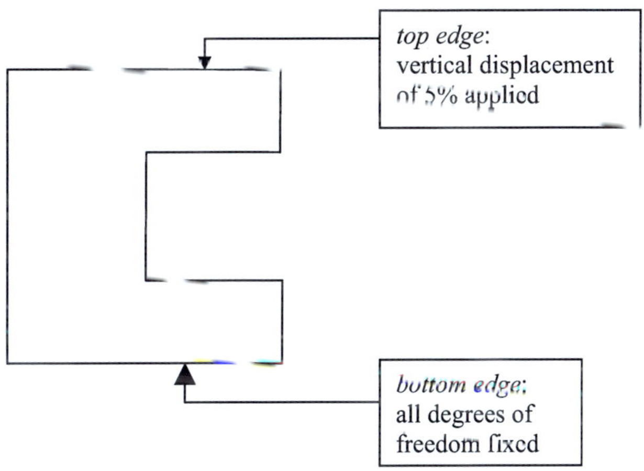

Figure 2. Geometry used for component simulation. The bottom edge was fixed while the top edge was displaced upwards by 5 %. For size reference, there are 60 unit cells across the width of the C channel, based on the channel and unit cell sizes.

RESULTS AND CONCLUSION

Figure 3 shows isocontours of von-Mises stress for the two cases run, focused on the upper, inner corner of the C-Channel. Note that all models are plotted using the same contour levels. The direct numerical simulation results are shown in 3 (bottom) and the averaged model in 3(top). Note the area outlined denotes homologous regions in the two models.

The striking similarity between the direct numerical simulation and the proposed approach, both qualitatively and quantitatively gives great encouragement that this technique will be able to resolve the local stress distributions at a reasonable computational cost. Future studies will explore its extension to fully 3-dimensional problems.

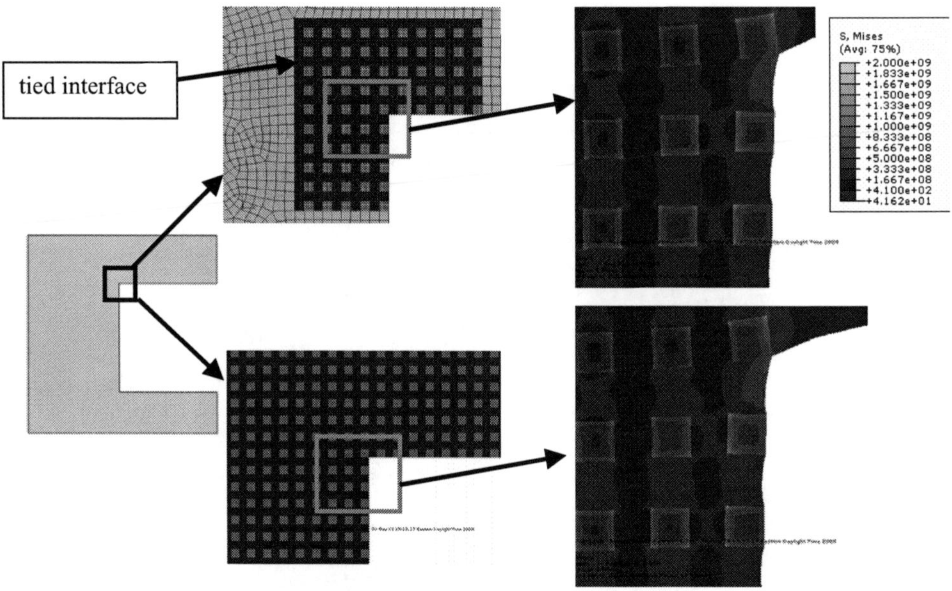

Figure 3. Results of C-Channel simulation based on the direct numerical simulation (bottom) and proposed approach (top). The same contour scale is used for both plots.

REFERENCES

1. Aboudi, J., *Micromechanics of Composite Materials: A Unified Micromechanical Approach*. 1991, Amsterdam: Elsevier.
2. Naik, N.K., et al., *Stress and failure analysis of 3D orthogonal interlock woven composites*. Journal of Reinforced Plastics and Composites, 2001. 1485–1523.

3. Tabiei, A. and Q. Chen, *Micromechanics based composite material model for crashworthiness explicit finite element simulation.* Journal of Thermoplastic Composite Materials, 2001. **14**: p. 264–289.
4. Ji, K.H. and S.J. Kim, *Dynamic direct numerical simulation of woven composites for low-velocity impact.* Journal of Composite Materials, 2007. 41: p. 175–199.
5. Paley, M. and J. Aboudi, *Micromechanical analysis of composites by the generalized cell model.* Mechanics of Materials, 1992. 14: p. 127–139.
6. Aboudi, J., M.-J. Pindera, and S.M. Arnold, *Linear thermoelastic higher-order theory for periodic multiphase materials.* Journal of Applied Mechanics, 2001. 68: p. 697–708.
7. Breuls, R.G.M., et al., *Predicting local cell deformations in engineered tissue constructs: A multilevel finite element approach.* J Biomech. Engr, 2002. 124: p. 198–207
8. Fish, J., Q. Yu, and K. Shek, *Computational damage mechanics for composite materials based on mathematical homogenization.* Intl J Num Meth Engr, 1999. 45: p. 1657–1679.
9. Baaijens, F.P.T., *A fictitious domain/mortar element method for fluid-structure interaction.* Intl J Num Meth Fluids, 2001. 35: p. 743–761.
10. Woo K., Whitcomb J., *Global/local finite element analysis for textile composites.* J. Comp. Matls. 28: 1305–1321.

Finite Element Modelling of Tow Geometry in 3D Woven Fabrics

Y. MAHADIK and S. HALLETT

ABSTRACT

A finite element model for a range of 3D woven fabrics has been developed. The aim was to produce a kinematic model of 3D woven composite preforms under compaction in order to locate areas of interest in the fabric structure, in particular the crimp of the tows which is a cause of significant reduction of composite compressive properties. The tows that make up the preform weave are represented as bundles of beam elements arranged in the cross section. Contacts are created between the elements within each tow and between the different tows. The modeled tows are arranged in a loose interlaced weave initially and a tension applied to model the loom tension and bring the fabric to the as-woven thickness. The mesh then undergoes compaction between two rigid plates moving with a fixed displacement. The model is run in the explicit finite element software LS-Dyna. Initial results show the tow cross section shape changes and tow crimp angles give good correlation when compared with computed tomography (CT) scans of a resin transfer moulded composite with the same woven architecture showing detailed views of the as-manufactured tow geometries.

INTRODUCTION

3D woven composites exhibit complex internal architectures with warp and weft tows interlaced by layers of through-thickness yarns. The weaving process that produces the fabric and the subsequent moulding compaction processes involved in creating a final composite laminate introduce distortions in the yarns that are critical to the mechanical properties of the composite.

Modelling of 3D woven fabric architectures that capture these distortions requires yarns to be flexible and have a changeable cross-section that can represent tow nesting when the fabric is compacted.

Zhou et al.[1,2] have represented each yarn as a bundle of 'digital beam chains' that are a series of small 1D beams connected by frictionless pins. 3D braiding and 3D weaving processes were modelled using this method and the results showed realistic crimping and nesting of the beam bundles.

Yusuf Mahadik and Dr. Stephen Hallett. Advanced Composites Centre for Innovation & Science, University of Bristol. Queens Building, University Walk, Bristol BS8 1TR

The work laid out in the following paper is based on this principle but used the beam modelling capabilities of the commercial finite element software, LS-DYNA. It focuses on a 3D woven fabric under compaction with the component yarns represented by bundles of beam chains.

Meshing

A finite element model of a layer-to-layer angle interlock weave 3D woven fabric was created (see Figure 1). Fabric yarns were represented as a bundle of 1D beams. The model was designed to be a kinematic representation of a fabric under compaction, and intended to show the change in crimp during compaction as well as other properties such as nesting of the yarns. The model mesh was created in MSC. Patran and solved in LS-Dyna.

The mesh was designed to be a "loose-weave" which would be tensioned to achieve a final thickness and architecture to match that of the uncompressed weave. It would then be compressed in the analysis by rigid plates to simulate the compaction of the fabric in a manufacturing mould.

The critical requirement was a mesh of beam bundles representing yarns that interlace but did not intersect each other.

The finite element mesh created in MSC. Patran was based on a repeating cell of the fabric structure. The fabric architecture properties, the warp and weft yarn spacing, tow cross sectional area and the length of the binder repeating unit were measured from the images of the fabric that were used in the sections above.

The unit cell consisted of a single repeating unit of the warp weaver with the cross sections of the perpendicular weft shifted to lie above or below the weaver section. The centre line of a weaver was represented by a cubic curve to create one repeat of the binder path (see Figure 2).

This repeating unit was duplicated along its length and then repeated across the fabric in the weft direction in accordance with the weave pattern. This produced a series of points along each weft yarn that can be joined to give a weft yarn centreline that interlaces correctly with the binder. Straight warp centre lines were then inserted in accordance with the weave pattern.

The result of this process was a mesh of one layer of 3D yarns represented by the yarn centrelines, with spacing to allow the insertion of tow bundles around this centreline. This layer could be shifted and duplicated in the thickness direction to create the complete fabric (see Figure 3).

The yarn bundles in the repeating unit were considered to lie in a circular pattern around the tow centre line in the initial mesh with the intention of them taking on their final lenticular shape upon tensioning and compacting. The cross section of each constituent of the bundle was approximately circular. Each yarn was represented by 19 circular beams in a concentric pattern around the yarn centreline with the total area of the beam cross sections equal to that of the measured yarn cross section areas.

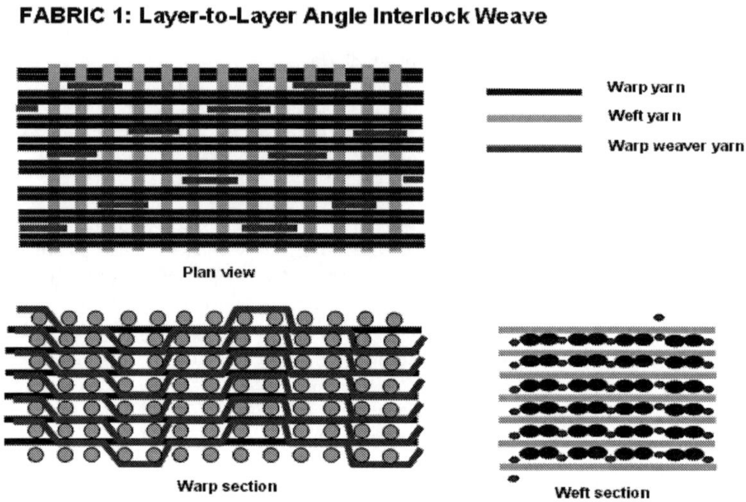

Figure 1. Schematic of Fabric 1: a layer-to-layer angle interlock fabric.

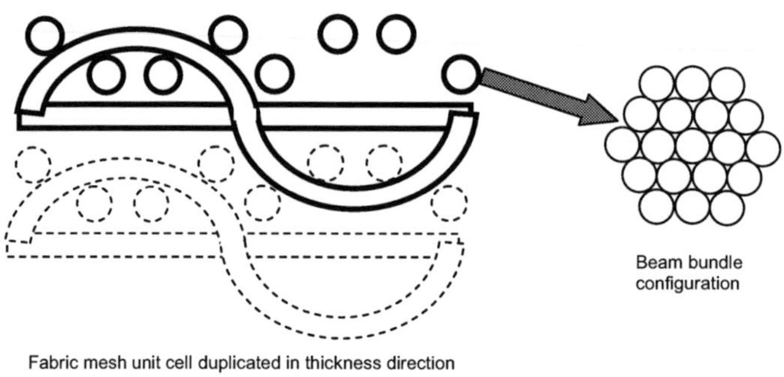

Figure 2. Schematic of the finite element mesh unit cell and tow beam bundle configuration.

The ends of the warp weavers, warp and weft yarns were extended away from the interlaced zone in order to allow boundary conditions to be applied far from the area where the fabric was to be compacted. This was chosen to represent the compaction of a small section of a large fabric sample and to minimise possible edge effects.

The compaction plates were each meshed as single 4-noded shell elements 2mm thick.

The mesh produced in MSC. Patran was exported as an LS-Dyna keyword file. Boundary conditions, loading, material properties and contact definition were all defined in the Oasys Primer pre-processing software.

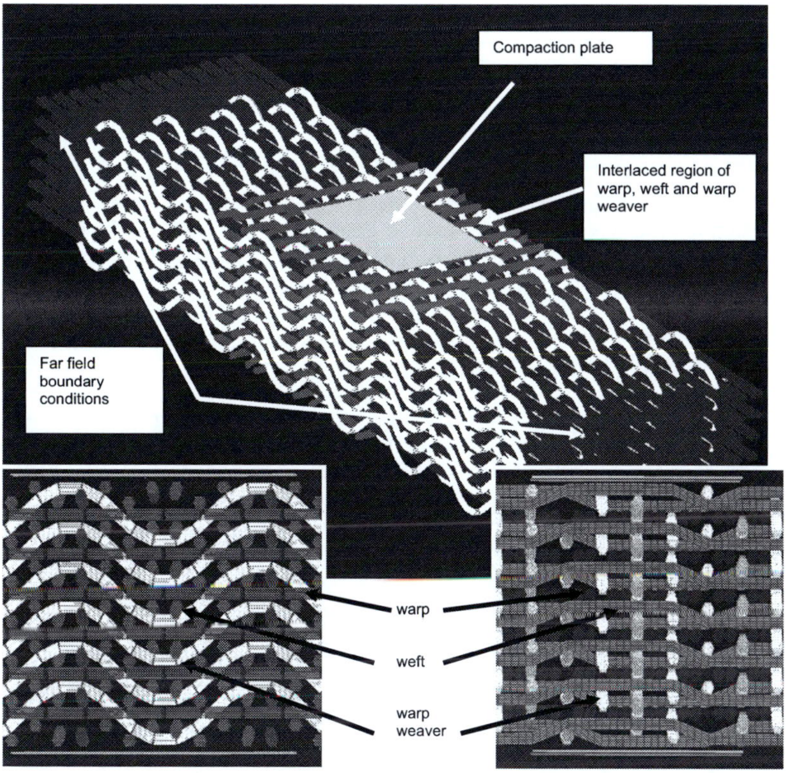

Figure 3. Final finite element mesh with sections through interlaced region inset.

MATERIAL PROPERTIES

Initial trials modelled the material of each beam bundle as elastic or elastic-plastic. It was found that using an elastic material made for unstable contacts and erratic movement of elements. Plastic materials were much more stable but showed plastic strain under large displacements which was unrealistic but minimised by the excess yarn lengths. The density, stiffness, Poisson's ratio and plastic yield stress were set to arbitrary values initially and adjusted to minimise model run time rather than to accurately model the yarn properties. The model was created to provide a kinematic representation of the fabric so accurate material properties were not paramount. Sensitivity checks for the stiffness and plastic yield stress were made by running simulations with these properties varied by orders of magnitude. No noticeable change in the final mesh geometry was seen but low plastic yield stresses produced erratic movement of elements and instabilities in the mesh. Increasing the density enabled faster processing times due to the time step being limited by the speed of the stress wave (dependent on the speed of sound through the material) moving through the structure. The material was assigned a coefficient of thermal expansion that could be adjusted via a scale factor to model the tensioning of the yarns. Once again the value chosen was optimised for model run time and stability. The material of the compaction plates were modelled as rigid.

SECTION PROPERTIES

The beams in each yarn bundle were represented as a Hughes-Liu beam section type that allowed the user to define the cross section of each beam exactly thus allowing contact algorithms to work correctly. Other sections were not compatible.

BOUNDARY CONDITIONS

The ends of each yarn were allowed to translate in the through thickness direction and allowed to rotate in all three dimension. The rigid plates were fixed in rotation and in-plane translation directions, allowing only vertical movement.

THERMAL LOADING

A linear temperature drop was applied to the mesh in order to apply a tensile force on all the yarns. This had the effect of straightening the yarns that were not already fully straight i.e. the warp weavers and weft as well as causing the collapse of the bundle cross section into something approaching the lenticular section seen in the sectioned fabric samples (see Figure 4). The straightening of the warp weavers and weft yarns caused a decrease in thickness of the fabric. Trials showed that a slow temperature drop over 10 seconds with a suitable CTE gave a stable change in thickness over a reasonable time. Once the thickness reached the measured as-woven thickness of a real fabric sample the temperature was held constant at that point until the termination of the solution.

COMPACTION

The rigid plates were then subject to a prescribed displacement regime that held them stationary until the end of the thermal loading cycle. The plates were then subjected to a high initial velocity to rapidly bring them close to the surface of the fabric and then the plate velocity was lowered to a very gradual compaction of the fabric. The slow compaction speed was chosen to increase the stability of the beams under loading.

CONTACT MODELLING

The contact between yarns was controlled by the "Automatic General" contact option in the LS-Dyna software. The contact area excluded the extended parts of the yarns away from the area compacted by the plates to reduce analysis time.

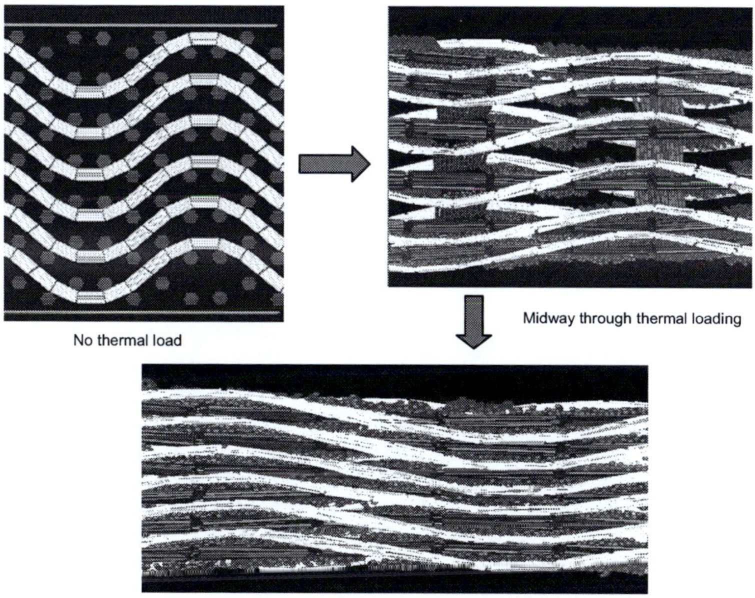

Figure 4. Effect of thermal tensioning of weavers causing decrease in thickness of the mesh.

OBSERVATIONS

Compacted samples of the modelled fabric were manufactured via resin transfer moulding (RTM) and subsequently sectioned and scanned using X-ray computed tomography (CT) which produced detailed images of the fabric internal structure. This allowed verification of the model output to be carried out by comparing the model geometry with that of the real manufactured samples.

Visualisation of the results produced by the model showed good correlation with the fabric structure before compression, with a lenticular yarn cross section being formed under thermal load. A good representation of the final compacted fabric internal structure has been achieved. A comparison of tow cross section width and height for both warp and weft showed that thinning and spreading of the yarns in the finite element model was however greater than in the moulded sample. Distinctive architectural features such as local crimp and resin rich areas at weaver/weft crossover points and flattening of the weaver under compaction as seen in the CT scan were captured well by the model (see Figure 5 and Figure 6).

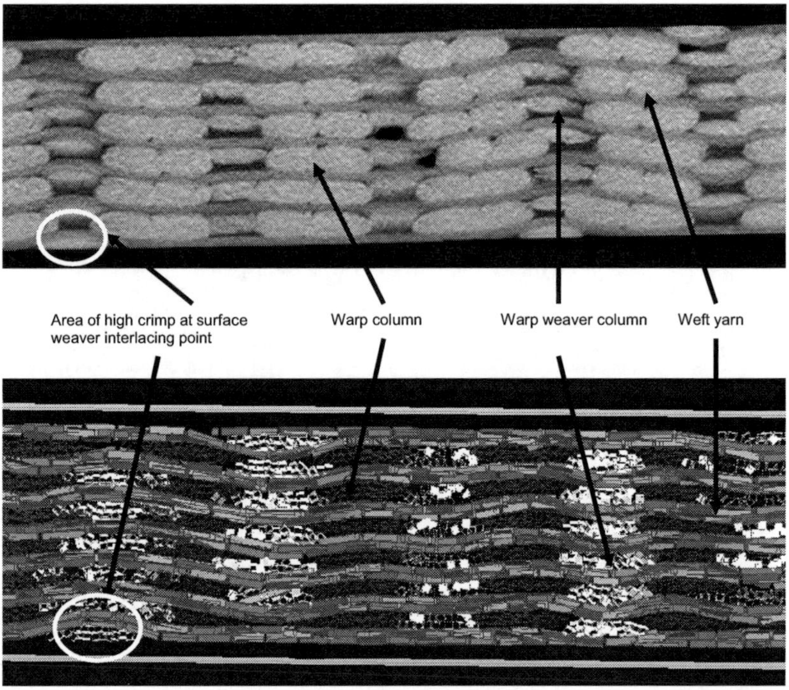

Figure 5 Weft cross section of compacted sample and fabric model.

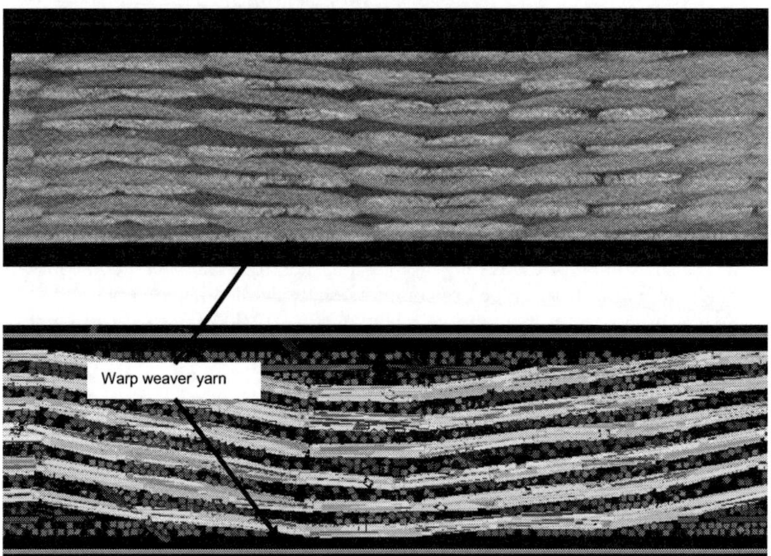

Figure 6. Warp weaver cross section of compacted sample and fabric model.

CONCLUSION

A kinematic model of a 3D woven fabric with a layer-to-layer angle interlock structure using beam elements in the finite element software LS-Dyna, was created. A tension load was applied to the fabric yarns via a thermal load in order to reduce the thickness to match that of the real fabric followed by compaction between two rigid plates to a thickness of a moulded fabric sample. Correlation between the moulded sample and the fabric model were reasonable with respect to yarn profile. For features such as local yarn crimp and resin pockets which are important for composite strength, a good level of correlation was achieved. Further work is still needed to control the yarn properties to prevent excessive tow spreading and thinning.

REFERENCES

1. Zhou G, Sun X, Wang Y, Multi-chain digital element analysis in textile mechanics, Composites Science and Technology 64 (2004) 239–244.
2. Miao, Yuyang , Zhou, Eric; Wang, Youqi; Cheeseman, Bryan A., Mechanics of textile composites: Micro-geometry, Composites Science and Technology, v 68, (2008), p 1671–1678.

Three-Dimensional Woven Composites for Automotive Applications

D. TAYLOR, A.-F. M. SEYAM and N. B. POWELL

ABSTRACT

Three-dimensional (3D) woven composite materials from high performance fibers, such as glass and carbon, have been used for automotive applications in a limited way due to the low production rate and the lack of research and development. Recently a process of manufacturing thick 3D orthogonal woven composites was developed. The process is capable of weaving thick (up to 1 inch) preforms from practically any fiber. The 3D woven preforms are conformable and can be easily molded to complex shapes and made into high performance composite materials. This capability opens numerous commercial opportunities for this specialized branch of the textile industry for automotive, marine, defense, industrial, aerospace, wind energy generator, recreational applications and other markets.

The objective of this research is to evaluate the replacement of heavy metal parts, specifically vehicle body panels, with lightweight 3D orthogonal woven composites made from glass fiber. In particular, the research examines the performance characteristics of lighter and stronger composites with the aim to improve fuel efficiency by reducing the total vehicle weight while maintaining safety standards. The performance characteristics of the 3D woven fabric composite are related to different construction parameters. Vacuum-assisted resin transfer molding (VARTM) technology is used to solidify the 3D composites. The tested properties include: tensile strength, elongation, flexural strength, impact performance, heat deteriorate temperature, and volume fraction, which is a function of areal weight, fiber density, and thickness. The recognized benefits of this research include: reduced weight of automotive parts, improved fuel efficiency, reduced cost, decreased emissions, decreased processing time, increased quality and corrosion resistance.

Dalia Taylor, Abdel-Fattah M. Seyam, Nancy B. Powell, College of Textiles, Box 8301, NC State University, Raleigh, North Carolina 27695-8301
Monsour H. Mohamed, 3TEX, Inc., 109 MacKenan Drive, Cary, North Carolina 27511

INTRODUCTION

The automotive industry is considered to be one of the fastest growing industries in the world. It is expected that global production of automotive units will grow to 70 million vehicles per year in 2010 from 55 million in 2000. Automotive industry is the most global of all industries, because of products distribution and the domination of only a few manufacturing companies [1, 2].

During the analysis of the current automotive industry it was identified that the role of the textile industry is significantly increasing in the global automotive supply chain, as new textile composite technologies are penetrating this market. Textile products are currently used for vehicle's interior, exterior and even suspension parts. Three-dimensional fiber reinforced polymer composites made by textile processes, such as weaving, braiding, stitching and knitting can not only replace metal parts of the vehicle, but also may improve performance characteristics and reduce the costs of manufacturing. These composites consist of preforms from high performance materials such as carbon, glass and aramid fibers, which are bound together within a matrix of polymer material such as epoxy, polyester or others to solidify the preforms. A liquid molding process may be used to impregnate the preform with resin to produce the final composite component of complex shapes. Lighter weight textile material parts have an advantage over metal parts in strength, stiffness, fatigue and corrosion resistance [3].

The objective of this research is to evaluate the replacement of heavy metal parts, specifically vehicle body panels, with lightweight 3D orthogonal woven composites made from glass fiber. Weight reduction is increasingly needed in automotive industry to improve fuel efficiency and to meet carbon dioxide emission requirements. Composites manufacturing helps to decrease assembly time (compared with traditional production system in a car assembly plant) and to increase quality by saving labor cost and fewer assembling errors [4].

3D WOVEN COMPOSITES ADVANTAGES

In 3D orthogonal woven composites fibers are orientated in an organized fashion in 3D space and have the greatest potential for improving mechanical performance. The fundamental of 3D orthogonal weaving is that more than one filling yarns are simultaneously inserted from one or both sides, thus building an increment of 3D fabric length with every weaving cycle of the machine, while 2D woven fabric is formatted by interlacing only two sets of yarns—one warp and one filling. In the 3D orthogonal fabric formation process, a third set of yarns, called z-yarns (see Figure 1), integrates all sets of yarns into the fabric. The three sets of yarns have no crimp, i.e. they stay straight in the fabric. It is important to mention, that the number of filling layers is always higher by one unit comparing with number of warp layers. Z-yarns take a path essentially perpendicular to the planes containing warp and filling yarns. In other words the Z-yarns are arranged in the thickness direction of the preform. In this process, warp yarns are stationary and arranged to form sheds for multi-filling insertions. Only z-yarns pass through heddles [5, 6, 7].

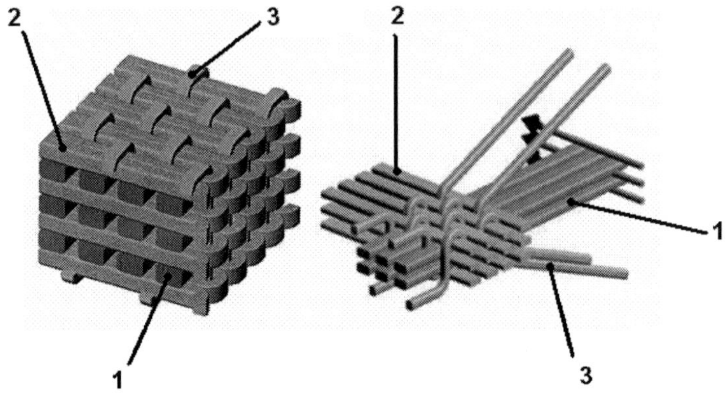

Figure 1. Model of the 3D woven fabric: 1 – warp, 2 – filling, 3 – z-yarn [6].

3D woven composites have many advantages in comparison with different composites technologies. First of all, the thickness of 3D woven preform eliminates multiple plies of 2D woven preforms and in the same time reduces labor time and decreases costs associated to build-up the desired thickness for constructing a given part. Compared with 2D woven laminated composites, 3D woven composites demonstrate higher through-thickness and interlaminar properties because of their integrated structure in the presence of orthogonal constituents [8]. The unique reinforcement geometry of 3D orthogonally woven fabric-reinforced composites (z-yarns act as capillary channels to transfer resin into preform interior from outer surfaces) creates the significantly improved resin penetration resistance during the composite. The absence of interlacing between warp and filling yarns allows 3D orthogonal woven panels to bend and internally shear easily without jamming and buckling within in-plan reinforcement that allows more easy molding process. Additionally, z-yarns, connecting different layers of warps and fillings, dramatically improve impact damage tolerance by suppressing delamination [7, 9, 10].

3D orthogonal woven preforms are highly conformable to complex mold shapes providing uniform part thickness and eliminating complicated darting and pin-wheeling. They wet out faster in both open and closed molding, improving quality, reducing molding times and facilitating migration to vacuum resin infusion. 3D woven preforms have non-crimp reinforcement with the higher fiber volume fraction (compared with fabrics with crimp) and it leads to stronger and lighter structures [10].

The reasons described above clearly explain why 3D woven composite have high potential for use in vehicle and body armor systems applications where high strength, light weights, and damage tolerance are important design parameters [9].

EXPERIMENTAL METHOD

As it was described above, this research project is testing the performance characteristics of lightweight glass fiber composites. The performance characteristics of the 3D orthogonal woven fabric composite are related to different construction parameters, such as roving density, thread densities (warp and filling), weave structure and arrangement/ratio of warp and z-yarn.

Hybon® 2022 Roving material made from electrical (E) glass fiber was used to form 3-D woven preforms for automotive application composites. Two different roving linear densities – 1102 tex (450 yd/lb) and 735 tex (675 yd/lb) - were used as filling. The warp and the z-yarns linear densities were 2275 tex (218 yd/lb) and 276 tex (1800 yd/lb) respectively. The picks were woven in 4 different linear densities 4 pick/inch (1.6 pick/cm), 5 pick/inch (2 pick/cm), 6 pick/inch (2.4 pick/cm) and 7 pick/inch (2.8 pick/cm). Three different weave structures (different number of warp and filling layers) – 4 warp/5 picks (4.5 layer fabric), 3 warp/4 picks (3.5 layer fabric) and 2 warp/3 picks (2.5 layer fabric) - were woven and analyzed. Figure 2 shows three different arrangements of warp and z-yarns in the reed for the samples woven with four layers warp. In the figure, 2a and 2b structures have the same ratio (1 warp/1 z-yarn), but different weaves – plain and basket weave; and 2c structure has lower z-yarn ratio – 2 warps/1 z-yarn. It is expected that different structures of the preforms (in the same time different fiber volume fraction and thickness of the composite) will influence different production time and different performance characteristics of the auto body parts. All 3D woven preforms were woven using machine donated by 3TEX, Inc.

Vacuum-assisted resin (DERAKANE 8084 Epoxy Vinyl Ester) transfer molding technology was used to solidify the 3-D composites. The fiber volume fraction, thickness and time of infusion of woven composite preforms will be tested. The tested final composite's properties include: tensile strength, elongation, flexural strength, impact performance, heat deteriorate temperature, and volume fraction (which is a function of areal weight, fiber density, and thickness). The testing results will be compared with required performance characteristics for metal auto body parts

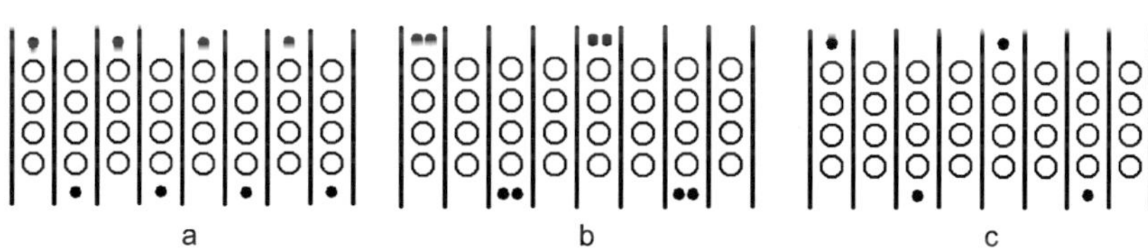

Figure 2. Different arrangements of warp and Z-yarns of 4-layer warp, a – 1z/1w (plain weave), b – 1z/1w (basket weave), c – 1z/2w (plain weave).

PRELIMINARY DATA

The properties of fiber reinforced composite greatly depend on the fiber architecture, and textile reinforcements play a major role in dominating the mechanical properties of composites. The continuity and solidity of the architecture of textile preforms becomes a main concern in advanced composites. Fiber volume fraction and thickness of the composite panel play an important role in determining structural and mechanical properties of 3D woven composites [5, 8]. Sixteen 3D woven preforms (4 warp/5 picks layers) structures for composites were examined. The linear densities of picks are 4, 5, 6 and 7 pick/inch with corresponding total densities of 20, 25, 30 and 35 pick/inch since 5 insertions are formed for each linear pick.

The thickness of the woven composite's preform panel is proportionally increasing when the filling density is increasing. Figure 3 shows thickness of 1 warp/1 z-yarn arrangement structures while Figure 4 shows thickness of 2 warp/2 z-yarn structure.

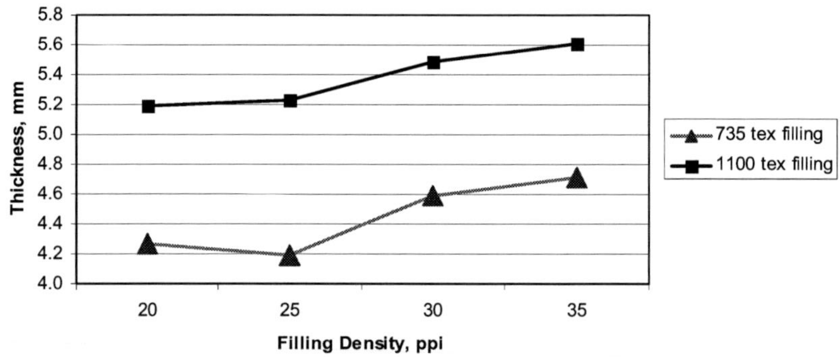

Figure 3. Thickness of the composite panel versus filling density (1z/1w woven structure).

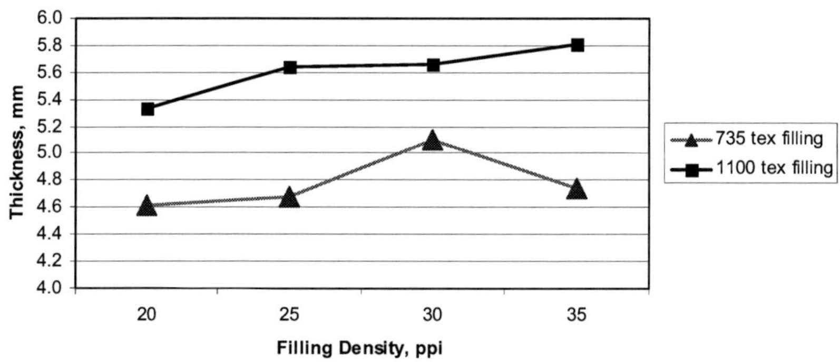

Figure 4. Thickness of the composite panel versus filling density (2z/2w woven structure).

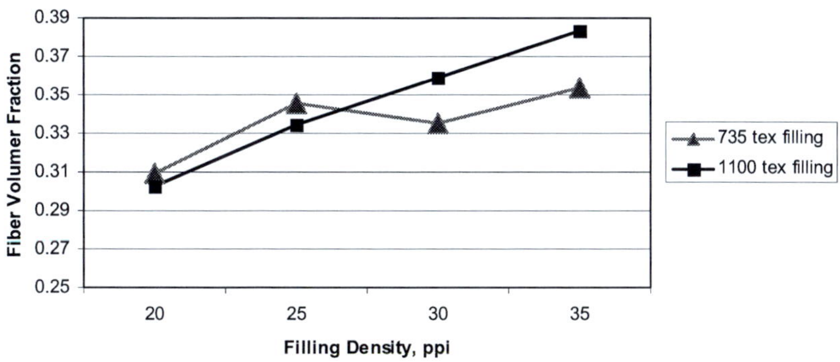

Figure 5. Fiber volume fraction of the composite versus filling density (1z/1w woven structure).

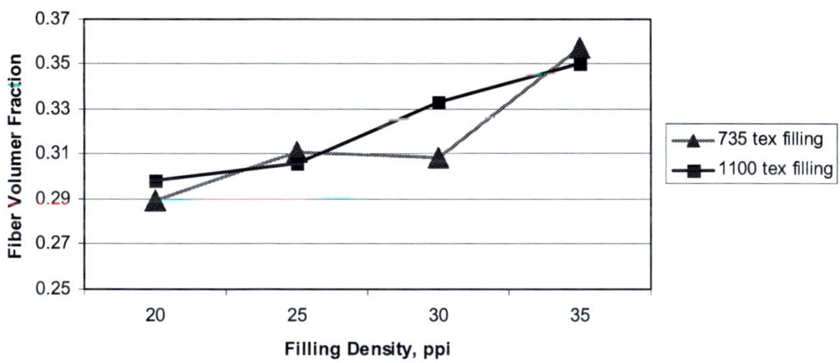

Figure 6. Fiber volume fraction of the composite versus filling density (2z/2w woven structure).

It is obvious that higher filling density will create higher fiber volume fraction of the woven preform panel. Figure 5 (1 warp/1 z-yarn arrangement structure) and Figure 6 (2 warp/2 z-yarn arrangement structure) show the fiber volume fraction increase as a result of increasing filling density.

RESULTS AND DISCUSSION

Thickness and fiber volume fraction are very important parameters of the 3D orthogonal woven composite's panel and influence not only performance characteristics of the final composite material, but also processing time, such as resin infusion and molding. The preliminary testing shows that thickness and fiber volume fraction increases while the density of filling and roving density of filling yarns were increased. However, even if performance characteristics can be increased with increase in fiber volume fraction and thickness, it does not mean that such material will be optimal for automotive application since areal weigh will also increase. It is important to find optimal thickness and fiber volume fraction of the 3D orthogonal woven preform for composite, which would have performance characteristics of the same or higher than steel or aluminum auto body parts, but in the same time would be light enough to improve fuel efficiency and to meet carbon dioxide emission requirements. 3TEX, Inc. in the past designed and developed 3D woven materials for Delphi to produce automotive Class8 composite bumper, which has resulted in 30% weight savings and assembly time from 1 day to 1 hour [4]. This project also gives a presumption that 3D woven orthogonal materials can be used for different vehicle parts, such as fender, door, roof or hood panels.

CONCLUSIONS

This paper introduces the development and testing of 3D orthogonal fiberglass woven composites for automotive applications. The preliminary testing shows how change in the linear density of filling and pick density can change thickness and fiber volume fraction of the composite's preforms, which influence production and performance characteristics. In the next steps of this research the influence of different 3D woven fabric's architecture characteristics will be tested also. One of the important characteristics—the time of resin infusion through different woven preform structures —will be determined during solidification of preforms. After solidification of preforms and producing the final composite material, performance characteristics of composite panels of the newly designed and produced material will be assessed. The tested performance characteristics include: tensile strength, elongation, flexural strength, impact performance, heat deteriorate temperature, and volume fraction. The testing results will be compared with required performance characteristics for metal auto body parts. The most important characteristic for the newly designed material is impact performance, which is critical in the automotive industry. Preliminary research and testing data gives the presumption that 3D orthogonal fiberglass woven preforms are suitable for automotive applications composites.

ACKNOWLEDGEMENT

The authors wish to thank PPG Industries, Inc. for providing the fiberglass material, Ashland, Inc. for providing resin sample and International Fiberglass for conducting the resin infusion process at their facilities, and 3TEX for the technical support on the 3D weaving machine.

REFERENCES

1. Handfield, R., Barnhardt, R., Powell, N. 2004. "Mapping the Automotive Textile Supply Chain: The Importance of Information Visibility," *Journal of Textile and Apparel, Technology and Management,* 3 (4). Retrieved March 10, 2008 http://www.tx.ncsu.edu/jtatm/volume3issue4/Articles/Handfield/handfield_full.pdf

2. Bhagwat, V. M. 2005. "Testing of Automotive Textiles," *Asian Textile Journal,* Jan-Feb: 79–82.
3. Banisster, M. K. 2004. "Development and Application of Advanced Textile Composites," *Proceedings of the Institution of Mechanical Engineers. Journal of Materials: Design and Applications*. 218 (L): 253–260.
4. Mohamed, M.H., Schartow, R.W. and Knouff, B.J. 2003. "Light weight composites for automotive applications," presented at SAMPE 2003 Conference, Long Beach, California, 11–15 May, 2003.
5. Gu, P. 1994. "Analysis of 3-D Woven Preforms and Their Composite Properties," A Dissertation for the Degree Doctor of Philosophy, North Carolina State University, Raleigh, North Carolina,.
6. Mohamed, M.H., Bogdanovich, A.E., Dickinson, L.C., Singletary, J.N., Lienhart, R.B. 2001. "New Generation of 3D Woven Fabric Preforms and Composites," *SAMPE Journal*, 37 (3): 8–17.
7. Mohamed, M., Cadogan, J. 2002. "3D Woven Carbon-Glass Hybrid Wind Turbine Blades" presented at 2002 NWTC & Industry R&D Coordination Meeting, November 5–7, 2002.
8. Chiu, C.H., Cheng, C.C. 2003. "Weaving Method of 3D Woven Preforms for Advanced Composite Materials," *Textile Research Journal*, 73 (1): 37–41.
9. Baucom, J.N., Zikry, M. A., Rajendran, A. M. 2006. "Low-Velocity Impact Damage Accumulation in Woven S2-Glass Composite Systems," *Composites Science and Technology*, 66: 1229–1238.
10. www.3tex.com, Retrieved March 21, 2008.

TOPIC VII

TEMPERATURE AND 3D EFFECTS

3D Analysis of Textile Structures with High-Resolution Computed Tomography

O. BRUNKE and F. SIEKER

ABSTRACT

Thanks to the knowledge of the 3D microstructure of a speciman by using a high resolution X-ray Computed Tomogaphy (CT) system, it is not only possible to judge the actual state of that specimen, but also predict its characteristics. By means of generating full 3D information with sub-micron resolution and moving virtually through the inner parts of the object slice by slice, complete new possibilities for analysis are available for research and quality control.

Especially for new materials like textile 3D-structures used in safety relevant or cost intensive applications, CT systems are used more and more often through the complete process of development up to the final quality check. With the new nanotom® small objects can be examined with voxel sizes <500nm (0.5μm). Further fields of applications are the analysis of metals, plastic materials and composites, ceramics and much more. Every difference within the specimen with regards to material composition, density or porosity influencing the X-ray absorption can be visualised in the 3D image. This provides exact information about the spatial distribution of different materials, density or orientation of fibres without destroying the object. By this means the CT will partially replace destructive methods like cross-sectioning – saving costs and time per sample inspected. In addition defects like cracks, voids or inclusions can not only be visualised but also their volume can be calculated.

HIGH RESOLUTION CT

For challenging x-ray inspection applications, the most remarkable results are obtainable by using high resolution computed tomography with a microfocus or

Dr. Oliver Brunke, Frank Sieker, phoenix|x-ray Systems + Services GmbH, Niels-Bohr-Str. 7, 31515 Wunstorf, Germany nanofocus® X-ray source. The achievable resolution or image sharpness is primarily influenced by the focal-spot size of the X-ray tube, which varies from a few microns for a microfocus tube to 0.8 microns for the latest nanofocus® tubes. They achieve detail detectability down to 200 nanometers (0.2 microns), using different focus modes to optimize the X-ray intensity and resolution.

CT volume generation starts with the acquisition of a series of two dimensional x-ray images while the sample is stepped in less than 1° increments through a full rotation. Each projection contains information on the position and density of X-ray absorbing object features within the sample. After image inversion and filtering a back projection is generated.

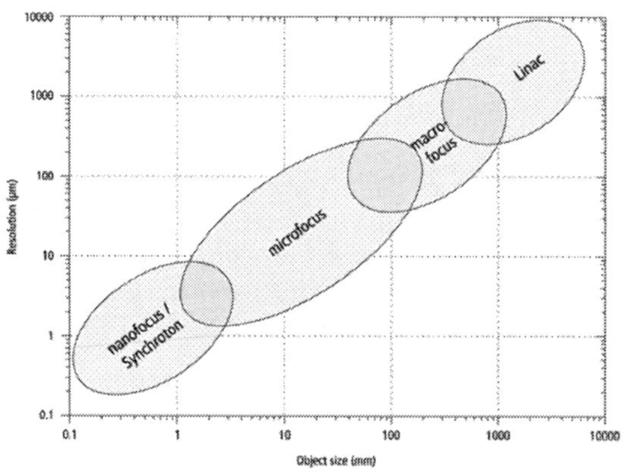

Figure 1. Microfocus CT provides state of the art for 3D analysis in a large variety of application fields with attainable voxel resolutions down to a few microns. With the use of nanofocus® tube technology, nanoCT®-systems are pushing into application fields that were exclusive to expensive synchrotron techniques.

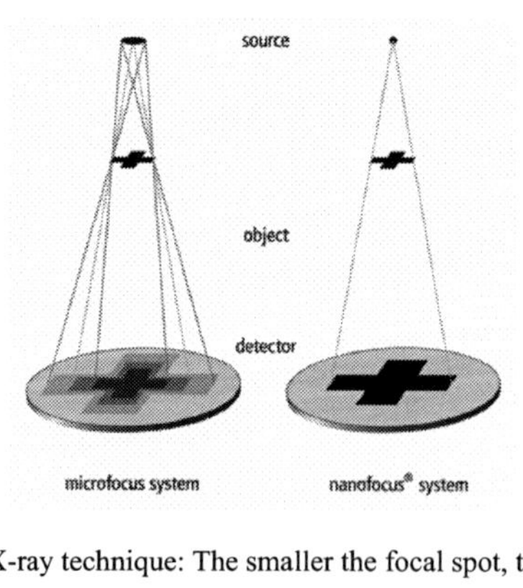

Figure 2. Nanofocus®-X-ray technique: The smaller the focal spot, the higher the sharpness.

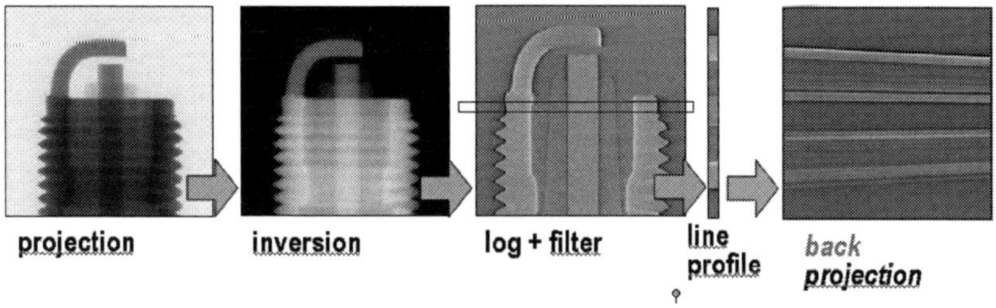

Figure 3. Data acquisition process from projection image to back projection.

This accumulation of back projection data is then used for the numerical reconstruction of the volumetric data which is visualized by slices or compiled in a three-dimensional view which can be displayed in various ways. The resolution of the scan depends on the size of the sample and the pixel size of the detector. With samples of diameters below 2mm a voxel size of 0.5μm can be reached.

By means of volume visualization software, the three-dimensional structure of the reconstructed volume is easily analyzed for pores, cracks, and density and distribution of materials with the highest magnification and image quality available

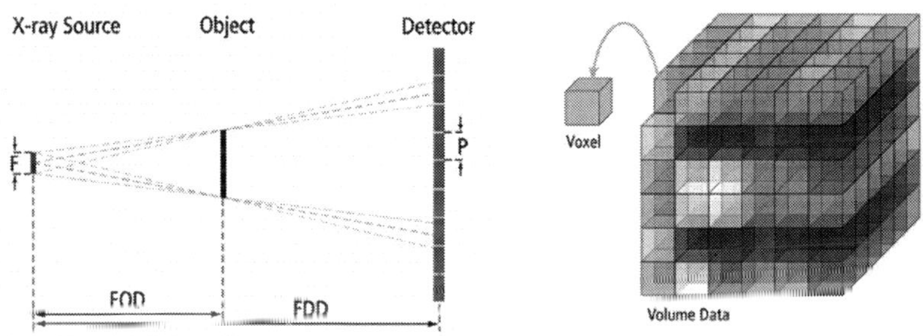

Figure 4. The voxel size directly depends on the sample size and pixel size of the detector.

Figure 5. Semitransparent CT result of a small plastic part with defect analysis. The different colour indicate differend defect sizes.

In order to reproduce an accurate reconstruction of the volumetric data, there are two conditions that must be met: First, the entire depth/diameter of the sample must remain within the field of view and cone of radiation throughout the entire 360° rotation. Since the full sample diameter must be completely displayed in each projection captured during the acquisition process, the maximum magnification is limited by the ratio of the effective detector width to the sample diameter. However, if the detector is side shifted during the acquisition for a virtual increase in effective detector width, higher ratios (magnifications) are possible. Further the use of optimized reconstruction algorithms allows an increase in resolution by scanning selected Regions-of-Interest (ROI).

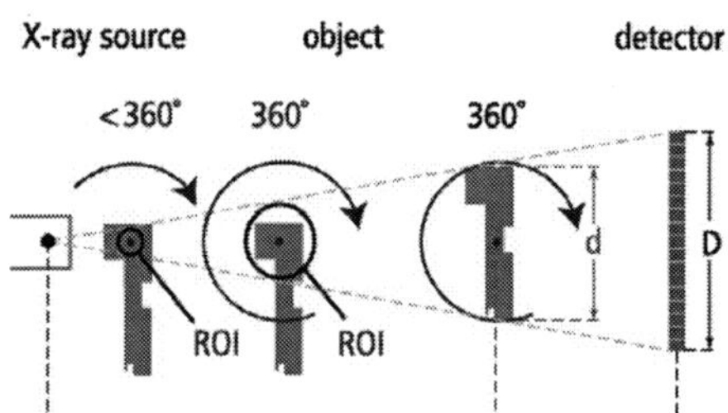

Figure 6. In general, the voxel resolution of CT scans is limited by the ratio of the effective detector width to the sample diameter because the sample should remain in the cone beam (right). However, the use of optimized reconstruction algorithms allows an increase in resolution in selected segments despite this rule (middle). The ROI|module of phoenix|x-ray even allows < 360^0 scans to achieve highest magnifications (left).

As second condition, the entire geometry of the sample, meaning every acquisition angle throughout the full 360° rotation, requires full penetration while maintaining the same X-ray energy level during each acquisition. This can be accomplished successfully even for higher density samples with a powerful x-ray source and the proper use of filters to reduce long wavelength x-rays.
There are a few physical effects influencing the CT quality, such as beam-hardening within the sample or ring-artifacts, which cannot be completely avoided. To optimize the quality of the 3D volume, advanced microfocus and nanofocus® CT systems include a variety of effective software tools to reduce ring-artifacts and correct beam hardening or drift effects which occurred during data acquisition.
After calculation the 3D volume information further software tools allow the analysis of different properties like internal defects. The size of the defects can be displayed colour coded.

APPLICATIONS WITH FIBRE MATERIALS

For scanning samples made of composite materials including glass or carbon fibres the ultra-precise high-resolution CT system nanotom® by phoenix|x-ray is a good has been used. Computed Tomography at such exceptionally high spatial resolutions as shown below requires careful design, taking into account any features which might influence the resulting resolution. These special needs for extreme precision require special manipulation systems, detectors and X-ray tubes. For example, the nanotom uses a unique 180 kV high power nanofocus® tube which can penetrate even high-absorption samples like copper or steel alloys. A 5-megapixel flat panel detector with an active area of 120 x 120 mm (2300 x 2300 pixels, 50 μm pixel size) and a 3-position virtual detector (up to 360 mm detector width) give rise to a wide variety of experimental possibilities. To avoid any negative influence of vibrations or thermal expansion, tube, detector and rotation unit are mounted on a high-precision granite-based manipulation system. Furthermore, special materials and construction techniques are used to guarantee high stability for long-term measurements. The precision of this system setup allows the high resolution and accuracy of the measurement results. Furthermore, the system includes phoenix|x ray's easy-to-use proprietary reconstruction software datos|x, which includes innovative tools for geometry calibration, detector calibration, noise and beam hardening reduction and region-of-interest-CT.

Figure 7. With its small footprint of only 163 x 143 x 74 cm, the nanotom® is suited for laboratory applications.

A few examples show the abilities of high resolution CT scans for the analysis of complex samples. Depending on the application voxel sizes between 0.5μm and 20μm have been chosen.

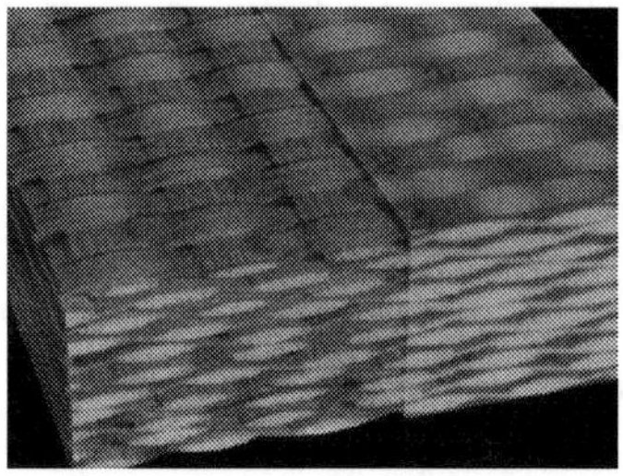

Figure 8a. Sample of glass fibre reinforces plastics showing the orientation of the fibre bundles.

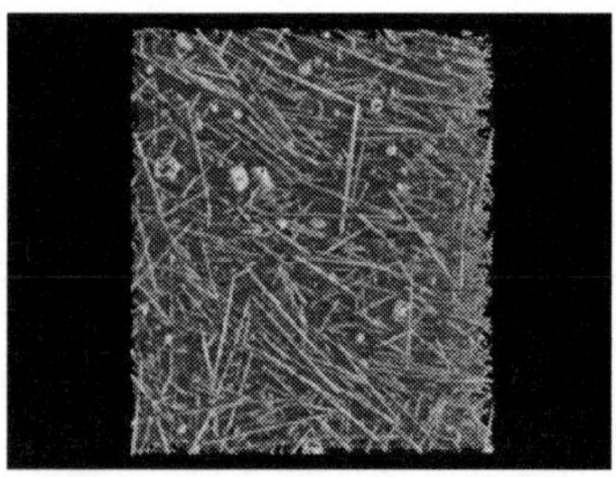

Figure 8b. Sample of glass fibre reinforces plastics showing the orientation of single fibres.

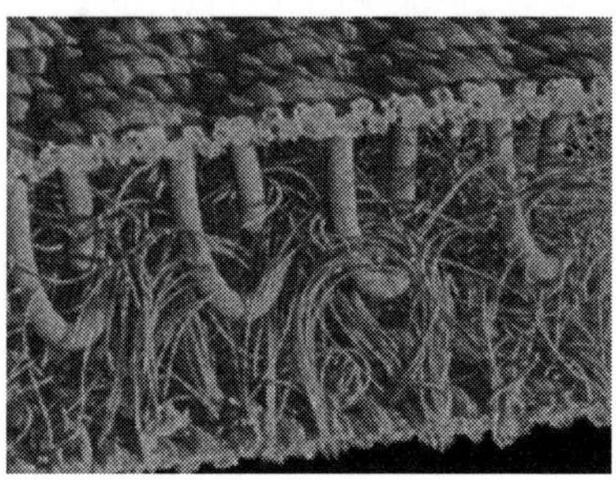

Figure 9. Technical textile. The CT reveals full information on fabric density and fibre distribution.

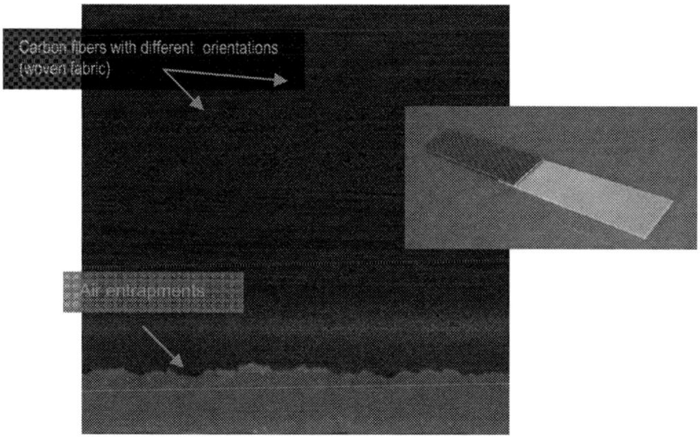

Figure 10. Composite Material welded on an aluminum plate. The fibre diameter is 7µm.

Computed Tomography at such exceptionally high spatial resolutions as shown above requires careful design, taking into account any features which might influence the resulting resolution. These special needs for extreme precision require special manipulation systems, detectors and X-ray tubes. For example, the nanotom uses a unique 180 kV high power nanofocus® tube which can penetrate even high-absorption samples like copper or steel alloys. A 5-megapixel flat panel detector with an active area of 120 x 120 mm (2300 x 2300 pixels, 50 µm pixel size) and a 3-position virtual detector (up to 360 mm detector width) give rise to a wide variety of experimental possibilities. To avoid any negative influence of vibrations or thermal expansion, tube, detector and rotation unit are mounted on a high-precision granite-based manipulation system. Furthermore, special materials and construction techniques are used to guarantee high stability for long-term measurements. The precision of this system setup allows the high resolution and accuracy of the measurement results. Furthermore, the system includes phoenix|x ray's easy-to-use proprietary reconstruction software datos|x, which includes innovative tools for geometry calibration, detector calibration, noise and beam hardening reduction and region-of-interest-CT.

CONCLUSION

High resolution CT performed with microfocus and nanofocus® technology serves a wide range of application in failure analysis and quality control for 3D analysis of a wide variety of materials. Especially for textile structures this methods leads to better knowledge of fibre distribution and therefore characteristics of the final product. Microfocus CT and nanoCT® has become a cost effective alternative to several traditional and often time consuming destructive and non destructive testing techniques.

Measurement and Modelling of In-Plane and Transverse Thermal Conductivity in Tape Prepreg, Non-Woven and Woven Textile Composites

S. HIND and F. ROBITAILLE

ABSTRACT

A summary of experimental measurements and directional thermal conductivity models for carbon/epoxy composite materials is presented. Experimental data were collected for AS4 fibre unidirectional tape prepreg and T300 fibre unidirectional non-woven and plain weave textile reinforced composites. Lamination sequences included $[0°]_n$, $[0°/90°]_n$, $[0°/90°/0°]_n$, $[0°/60°/-60°]_n$ and $[0°/+45°/-45°/90°]_n$. Manufacturing methods included hybrid hand-layup and autoclave curing, resulting in a fibre volume fraction range of 0.55 to 0.62. Classical lamination theory and models of conductivity for homogeneous assemblies of aligned fibres were used to determine thermal conductivity of tape prepreg and thermal properties of textile yarns. Unit cell simulations of textile based composites were performed to quantify the effect of changes in reinforcement geometry and configuration on directional thermal conductivity. Steady-state finite element simulations were carried out for non-woven composites, unidirectional and cross-ply, and for plain weave textiles $[0°/90°/0°]_n$ and $[0°/+45°/-45°/90°]$. For the non-woven models the effect of reinforcement configuration was quantified with a parametric study. A series of woven unit cell models were developed. The paper concludes with a discussion of the applicability of different modelling methods for the determination of thermal conductivity in composite materials.

INTRODUCTION

The National Research Council Canada - Institute for Aerospace Research has developed a conductive heating technology used for controlling the temperature of resin infusion tooling, bonded repair and secondary bonding of composite parts. Accurate heating of typical aerospace polymer matrix composite materials is challenging due to their low thermal conductivity. Non-uniform heat flux must be projected onto the surface of interest to generate a uniform temperature. To deter-

---

Simon Hind, National Research Council, 1200 Montreal Road, Bldg. M-3, Ottawa, Ontario, K1A 0R6, Canada.
Francois Robitaille, University of Ottawa, Department of Mechanical Engineering, Ottawa, Ontario, K1N 6N5, Canada.

mine the optimum configuration of heating systems, heat transfer predictions require accurate directional thermal conductivity values for the composite components.

The focus of this work is to develop a capability for the prediction of transverse and in-plane directional thermal conductivity (k_T and k_{IP}) of composite materials through collection of experimental data and evaluation of different property predicting tools including closed form equations for k_T of prepregs and yarns, classical lamination theory (CLT) and finite element (FE) models that account for textile reinforcement architecture. A number of models exist for predicting heat transfer properties arrays of aligned fibres [1,2] but analyses of textile reinforced composites are not reported extensively in the open literature.

EXPERIMENTAL

Unidirectional AS4/3501-6 carbon fibre/epoxy tape prepreg laminates were autoclave cured according to the supplier's (Hexcel) recommended cure cycle [3] giving a V_f 0.62. Non-woven textile reinforced T300 carbon fibre laminates were manufactured using unidirectional fabric style 1152 from Fabric Development Inc and room-temperature epoxy resin, mia-poxy. Cure and post-cure were conducted under a press at controlled thickness to achieve a V_f of 0.55. Woven W3T282-F263 prepreg T300 fibre laminates were manufactured according to Hexcel's recommended cure cycle giving a V_f of 0.61.

Measurements of k_T and k_{IP} were performed using the Hukseflux THASYS and THISYS apparatus. THASYS measures transverse conductivity while THISYS measures an average in-plane conductivity. THISYS was modified to permit directional in-plane conductivity measurements taken along 0°, 45° and 90° directions. All conductivity measurements were taken at room temperature. Thermal conductivity data for 11 plates are available in Table I.

MODELLING OF TAPE PREPREG LAMINATES USING CLT

CLT was used to predict the thermal conductivity of the tape prepreg laminates. Conductivity along the 0° direction, k_x, can be calculated knowing the laminate thickness $2h$, ply conductivity along and across the fibres, k_1 and k_2, ply angles, θ_i, and play thickness, t_i, using eqn (1) [4].

$$k_x = \frac{1}{2h}\sum_{i=1}^{N}(k_1(\cos\theta_i)^2 + k_2(\sin\theta_i)^2)t_i \tag{1}$$

Due to the lack of published data, fibre properties for AS4 were determined by back calculating from experimental data for Panel 1 using the rule of mixtures ROM by volume, eqn (2) and Clayton's model for aligned fibres, eqn (3) [5]. For AS4 fibre axial and radial conductivity, k_{fa} and k_{fr}, were determined to be 10.2 and 2.1 W/mK respectively. Prediction of k_{IP} for alternate tape prepreg lamination sequences was then performed. Eqn (1) predicts Panel 2 and Panel 5 k_{IP} as 3.56

Table I. Experimental Conductivity Data.

Panel	Material	Layup	THASYS k_z (W/mK) Trans. Std. Dev.	Modified THISYS k_x (W/mK) 0° Std. Dev.	k_y (W/mK) 90° Std. Dev.	k_{45} (W/mK) 45° Std. Dev.
1	UD Prepreg AS4	$[0]_6$	0.73 ±0.06	6.40 ±0.30	0.70 ±0.08	2.74 ±0.03
2	UD Prepreg AS4	$[0/90/0/90]_s$		3.87 ±0.02	3.73 ±0.06	3.89 ±0.02
3	UD Prepreg AS4	$[0/90/0]_s$	0.74 ±0.01			
4	UD Prepreg AS4	$[0/60/\text{-}60]_s$	0.71 ±0.01			
5	UD Prepreg AS4	$[0/45/\text{-}45/90]_s$	0.74 ±0.02	3.50 ±0.12	3.57 ±0.05	3.59 ±0.06
6	Hand Layup T300	$[0]_6$	0.45 ±0.00	4.77 ±0.13		
7	Hand Layup T300	$[0]_6$	0.46 ±0.00	4.95 ±0.07	0.36 ±0.00	1.68 ±0.00
8	Hand Layup T300	$[0/90]_s$	0.45 ±0.00	2.62 ±0.02	2.72 ±0.21	3.04 ±0.14
9	Plain Weave Prepreg T300	$[0]_8$	0.62 ±0.01	2.97 ±0.04	2.97 ±0.04	2.96 ±0.06
10	Plain Weave Prepreg T300	$[0]_8$	0.60 ±0.00	3.04 ±0.02	2.90 ±0.05	2.90 ±0.19
11	Plain Weave Prepreg T300	$[0/45/45/0]_s$	0.58 ±0.01	2.81 ±0.08	2.79 ±0.02	2.74 ±0.04

W/mK as compared to 3.87 and 3.50 W/mK. The 9% and 2% difference can be attributed to laminating sequence, specimen cutting angle and specimen installation angle into THISYS.

$$k_{ya} = V_{fy} \cdot k_{fa} + \left(1 - V_{fy}\right) \cdot k_r \tag{2}$$

$$k_t = \frac{k_m}{4}\left[\sqrt{(1-V_{fy})^2\left(\frac{k_{fr}}{k_m}-1\right)^2 + \frac{4k_{fr}}{k_m}} - (1-V_{fy})\left(\frac{k_{fr}}{k_m}-1\right)\right]^2 \tag{3}$$

MODELLING OF NON-WOVEN TEXTILE COMPOSITES USING FE

Thermal simulations were conducted for unit cell models of unidirectional laminates [0°] (U1x) and cross-ply laminates [0°/90°] (U2x) featuring one or two plies of non-woven unidirectional yarns with a V_f of 55%, see Figure 1. The yarn width *yw* and height *yh* were kept constant at 2 and 0.5 mm respecitvely. Yarn sections were modeled as hyper-ellipses and also kept constant. Effect of reinforcement geometry and configuration on heat transfer was quantified by varying the horizontal inter-yarn gap *gh* (0.2 and 0.4 mm) and vertical inter-yarn gap *gv* (0.02 and 0.04 mm) for each laminate. Values for *gh* are representative of reinforcement used in experiments and *gv* values are 3 to 5 carbon fibre diameters [6]. To maintain a V_f of 0.55 yarn fibre volume fraction V_{fy} was varied between models as a function of *gh* and *gv*. Typical models are illustrated in Figure 1 and non-constant parameters appear in Table II for the 8 models.

Yarn properties were calculated based on published values for Hexcel T300 carbon fibre, k_{fa} and k_{fr} are 8.4 and 0.84 W/mK, and epoxy resin, k_r of 0.20 W/mK [7,8]. Yarn axial and transverse conductivities k_{ya} and k_{yt} were calculated using eqn (2) and Nielsen's model [9], eqn (4) with A = 0.5 and $V_{fy,max}$ = 0.906. Nielsen's model produced a 5% higher k_{yt} than Clayton's and was selected as an upper limit.

$$k_{yt} = k_r \left[\frac{1 + A \cdot B \cdot V_{fy}}{1 - B \cdot \varphi \cdot V_{fy}} \right] \quad \text{where} \tag{4}$$

$$B = \frac{(k_{fr}/k_r) - 1}{(k_{fr}/k_r) + A} \qquad \varphi = 1 + \left(\frac{1 - V_{fy,\max}}{V_{fy,\max}{}^2} \right) \cdot V_{fy}$$

All geometric models were created in TexGen v2 [10] and meshed in Gambit™ using 4-noded tetrahedral elements. Steady state heat transfer simulations were performed using Fluent™ or Nastran™. Constant temperatures were imposed on upper and lower faces for k_T simulations or on opposing faces for k_{IP} simulations. All other boundaries were adiabatic. Fourier's law (eqn. 5) was used to

Table II. Reinforcement Geometry and Yarn Properties for Non-Woven Models.

			U1A, U2A	U1B, U2B	U1C, U2C	U1D, U2D
Vertical gap	gv	(mm)	0.02	0.04	0.02	0.04
Horizontal gap	gh	(mm)	0.2	0.2	0.4	0.4
Yarn fibre vol.	Vfy	(%)	0.68	0.71	0.75	0.78
Axial yarn cond.	kya	(W/mK)	5.81	6.02	6.32	6.55
Trans yarn cond.	kyt	(W/mK)	0.50	0.52	0.56	0.59

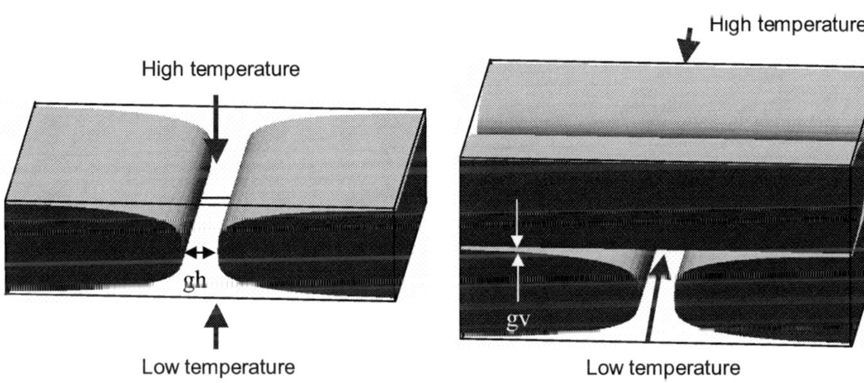

Figure 1. Typical Non-Woven Models with Geometric Parameters.

calculate thermal conductivity based on the imposed temperature gradient $\Delta T/\Delta z$ and resulting area-weighted surface heat flux q_z'' [11].

$$q_z'' = -k\frac{\Delta T}{\Delta x} \tag{5}$$

Non-woven fabric simulation results are provided in Figure 2. For the [0°] models U1A-U1D all models converged with the exception of U1C in the in-plane y direction. Compared to experimental conductivity results, average transverse, in-plane x, and in-plane y conductivity differed by 7, 2 and 16%, respectively. For the cross-ply model U2A – U2C models did not converge in the in-plane x direction. Compared to experimental values average transverse and in-plane differences are 5 and 13%, respectively.

MODELLING OF WOVEN TEXTILE COMPOSITES USING FE

Thermal simulations were also conducted for unit cell models of plain weave cross-ply $[0°/90°]_m$ and quasi-isotropic [0°/90°/+45°/-45°] laminates featuring one or two woven plies. Geometric parameters were evaluated from measurements on fabricated samples. A typical woven model is illustrated in Figure 3. Various models were created to account for the fact that in real laminates the measured total thickness is generally not equal to measured yarn heights multiplied by the number of layers, due to imperfect stacking and nesting. 4 types of models were considered: 1) a single layer with *yh* as measured; 2) 2 layers with *yh* as measured and smallest

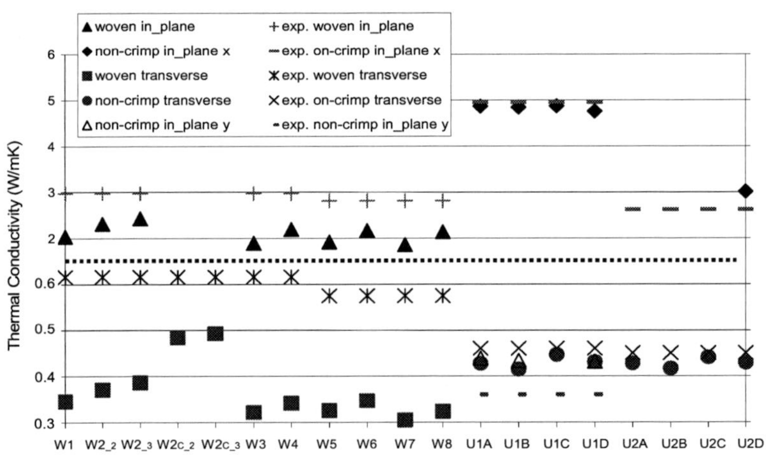

Figure 2. Heat Transfer Simulation Results for Unit Cell Models.

possible total height; 3) single layer with *yh* derived from plate thickness; 4) 2 layers with *yh* derived from plate thickness. While case 2 and 4 will allow for nesting of multiple [0°/90°] plies it is problematic to nest a [0°/90°] ply with a

[°+45/-45°] without interference. Initial woven unit cell models featured yarn element orientation. Models with more accurate V_f were generated by cutting models W2_2 and W2_3 mid-way through layers. Woven simulation results are provided in Figure 2.

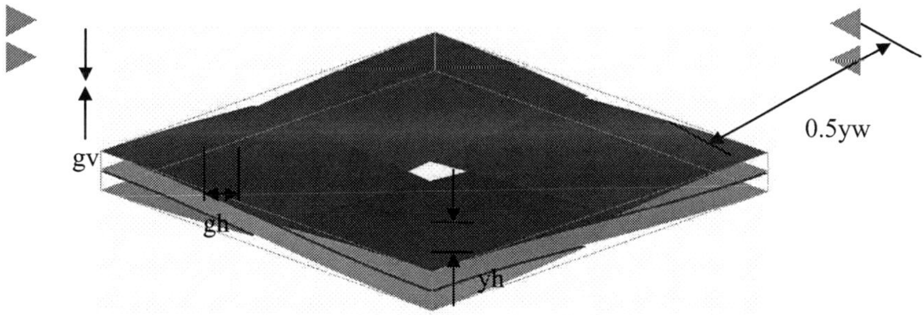

Figure 3. Typical Woven Model with Geometric Parameters.

Table III. Woven Model Plan.

Model	Orientaion	Plies	Case
W1	[0°/90°]	1	1
W2_2	[0°/90°]	2	2
W2_3	[0°/90°]	3	2
W2C_2	[0°/90°]	2	2
W2C_3	[0°/90°]	3	2
W3	[0°/90°]	1	3
W4	[0°/90°]	2	4
W5	[0°/90°/+45°/-45°]	2	1
W6	[0°/90°/+45°/-45°]	2	2
W7	[0°/90°/+45°/-45°]	2	3
W8	[0°/90°/+45°/-45°]	2	4

DISCUSSION

Extensive experimental thermal conductivity data were measured for materials made from 2 different carbon fibres, AS4 and T300, various lamination sequences, 2 manufacturing methods and 3 separate material configurations: unidirectional prepreg tape, non-woven textile and woven textile reinforcement. A range of diverse predictive models were applied to predict k_T and k_{IP} for these increasingly complex reinforcement architectures.

CLT was successfully used to predict k_{IP} of prepreg tape laminates. Used in combination with CLT, ROM and models used for the prediction of thermal conductivity in aligned assemblies of fibres, such as Clayton's and Nielsen's model, can be used to back calculate yarn properties.

FE simulations were capable of detecting variations in thermal conductivity due to subtle changes in geometry. Non-woven unidirectional and cross-ply unit cell simulation results, shown in Figure 2, demonstrate that k_T is not sensitive to reinforcement geometry or lamination sequence based on the minimal variability. There is a similar limited variability in k_T measured experimentally. These observations are supported by the widespread success of published models for the prediction of k_T, regardless of complexity [5,12-14]. It should be noted that reinforcement geometry has a much greater impact on other properties such as permeability [15]. Lamination sequence has a major effect on k_{IP} for non-woven reinforcements. Non-woven models simulated k_T values ranged from 0.41 to 0.45 W/mK compared to the average experimental value of 0.45 W/m°K. The non-woven predicted k_{IP} values for unidirectional laminates ranged from 4.76 to 4.87 W/mK along the direction of the yarns and 0.43 to 0.44 across the yarns, compared to 4.95 and 0.36 W/mK, respectively. The non-woven simulated k_{IP} value for a cross-ply laminate was 3.01 W/mK compared to average experimental measurements of 2.62 W/mK, 15% difference.

Woven unit cell models were created with realistic reinforcement geometric parameters obtained from experimental measurements of micrographs. More accurate models were developed by cutting the upper and lower faces at mid-yarn thereby increasing V_f. Woven unit cell simulation results, Figure 2 indicate that the cut models were the most accurate predictors of k_T. No in-plane results for the cut models are available. The non-cut models identified the effect of the number of plies, degree of nesting, and lamination sequence on V_f. Woven models simulated k_T values of 0.48 to 0.49 W/mK compared to experimental measurements of 0.58 W/mK.

Using FE was straightforward and reliable in predicting the directional thermal conductivities of non-woven textile reinforced composites. FE limitations and insensitivity of thermal conductivity to changes in geometric parameters suggest that simpler predictive models could be used to predict directional thermal conductivities in composites with more complex textile reinforced architectures. Future work will seek to utilise means other than FE modelling for the prediction of composite directional thermal conductivity such as CLT models adapted for textile reinforced composites.

REFERENCES

1. Zou M, Yu B, Zhang B, Ma Y. 2003. "Study on Optimization of Transverse Thermal Conductivities of Unidirectional Composites," J of Heat Trans, ASME, Vol. 125, 2003, 980–987.
2. James B.W, Wostenholm B.H, Keen G.S, McIvor S.D. 1987. "Prediction and Measurement of the Thermal Conductivity of Composite Materials," J of Physics D: Appl. Phys, 20, 1987, 261–268.
3. www.hexcel.com
4. MIL-HDBK-17-3F, Vol. 3, Chapter 5. "Design and Analysis", p. 5-27 to p. 5–40.
5. Ning Q.G, Chou T.W. 1995. "Closed-form solution of the transverse effective thermal conductivity of woven fabric composites," J of Comp Mat 1995; 29(17); 2280–2294.

6. "Construction of Fiber Reinforced Polymer (FRP) Jackets for the Protection of Pier Caps. Construction Report," Submitted by PN Balaguru, Professor of Civil engineering, Rutgers University, August 2005.
7. Gowayed Y, Hwang JC. 1995. "Thermal conductivity of composite materials made from plain weaves and 3-D weaves," Comp Eng 1995; 5(9): 1177–1186.
8. Hind S, Robitaille F, Raizenne D. 2006. "Parametric unit cell modelling of the effective transverse thermal conductivity of carbon plain weave composites," Proceedings of TexComp-8 Conference. Nottingham, October, 2006.
9. Progelf RC, Throne JL, Ruetsch RR, 1976. "Methods for predicting the thermal conductivity of composite systems: a review," Pol Eng Sci 1976; 16(9): 615–625.
10. Robitaille F, Long AC, Rudd CD. 2002. "Geometric modelling of textiles for prediction of composite processing and performance characteristics," Plast Rub Comp 2002; 31(2): 66–75.
11. Holman J.P. 1997. "Heat Transfer," McGraw Hill, New York, 1997.
12. Dasgupta A, Agarwal R.K, Bhandarkar S.M. 1996. "Three-dimensional modeling of woven-fabric composites for effective thermo-mechanical and thermal properties," Comp Sci Tech 1996; 56(3): 209–223.
13. Gowayed Y, Hwang J.C. 1995. "Thermal conductivity of composite materials made from plain weaves and 3-D weaves," Comp Eng 1995; 5(9): 1177–1186.
14. Bigaud D, Goyhénèche J.M, Hamelin P. 2001. "A global-local non-linear modelling of effective thermal conductivity tensor of textile-reinforced composites," Comp A: Appl Sci Man 2001; 32(10): 1443–1453.
15. Robitaille F. 2008. "Concurrent Predictions of Flow, Thermal and Structural Properties for Unidirectional and Bidirectional Non-Woven Textile Composites," submitted to Comp A, 2008.

Architecture Effects on Matrix Cracking for Low-Porosity 2D and 3D SiC/SiC Composites

G. N. MORSCHER

ABSTRACT

The development in SiC/SiC composites over the past few years has resulted in significant advances in high temperature performance so that now these materials are being considered for several gas turbine and reusable hypersonic applications. In this study, it will be demonstrated how important fiber architectural features can be used to maximize the critical design parameter of "matrix cracking stress" for low-porosity 2D and 3D textile-based SiC/SiC composites. Emphasis is placed on the high-performance Sylramic-iBN fiber-reinforced slurry-cast melt-infiltrated composite system.

INTRODUCTION

With the improvement in high temperature strength retention and creep resistance of SiC fibers, SiC/SiC ceramic composites are becoming more suitable for long-term hot section propulsion applications [1] as well as for nuclear applications where oxidation and/or swelling of carbon fiber reinforced composites becomes an issue [2]. Textile approaches, though a challenge due to the stiffness of polycrystalline SiC fibers, offer a range of possible solutions when attempting to fabricate near-net complex shapes.

Since at this point SiC/SiC composites are in an early stage of development, key hurdles that need to be overcome include performance optimization, reliable manufacturing, and design methodologies that account for the anisotropic nature of these composite systems. In this study, some architectural issues related to design of SiC/SiC cracking strength will be addressed by considering the effect of two key aspects of fiber architecture: the fraction of fibers in the direction of stress and the largest unbridged region containing fiber tows oriented perpendicular to the stress direction. Emphasis is placed on architectural design models for the high-performance

Gregory N. Morscher, Ohio Aerospace Institute, NASA Glenn Research Center, MS 106-5; 21000 Brookpark Rd., Cleveland, OH 44135, USA—Gregory.N.Morscher@nasa.gov

SiC fiber reinforced slurry cast melt-infiltrated (SCMI) composite system, which has shown long-term structural capability to over 1300°C. This work draws on experimental results from references 3 through 6 for SCMI composites fabricated from a variety of textile architectures and fiber volume fractions and expands upon these results to implement simple design models when first selecting the usefulness of a specific architecture for a desired component.

THE SLURRY CAST MELT-INILTRATED COMPOSITE SYSTEM

The composite materials used to generate the architecture models of this study were fabricated by General Electric Energy Ceramic Composite Products LLC (Newark, DE) using Sylramic-iBN fibers as reinforcement [1,7]. The Sylramic-iBN fibers are achieved by treating Sylramic fiber tows (originally Dow-Corning, Midland, MI and currently ATK/COIC Ceramics, San Diego, CA) with a NASA developed heat treatment [8] at a high temperature in order to stabilize the microstructure and produce a thin (~ 100 nm) BN outer layer on each fiber surface. The treatment is typically done on 3D performs consisting of stacked 2D fabric or 3D fiber architectures. Composite processing then consists of several steps as follows:

(1) Tooling of the Sylramic-iBN preform in a graphite fixture.
(2) Chemical vapor infiltration (CVI) of the tooled perform with an interphase coating consisting of a BN layer of about 0.5 μm thickness that is needed to promote sufficient debonding between fiber and matrix during matrix crack formation.
(3) CVI of a SiC layer a few μm thick which encapsulates the fiber tow and interphase to protect the fibers and interphase from subsequent infiltrations.
(4) Aqueous infiltration of a SiC particulate filled slurry into the open porosity followed by a low-temperature evaporation step.
(5) High temperature molten Si infiltration which fills in the remaining open porosity and forms a "siliconized-SiC" region of the matrix outside of the fiber tow, BN interphase, and CVI SiC perform. The composite after this step has only ~5% total porosity.

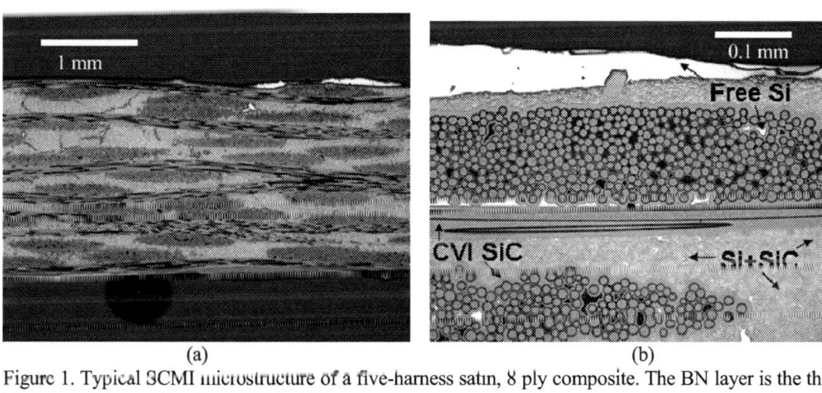

(a) (b)

Figure 1. Typical SCMI microstructure of a five-harness satin, 8 ply composite. The BN layer is the thin black layer surrounding the individual fibers.

Figure 1 shows a micrograph of a typical SCMI microstructure. An important attribute of this microstructure is that the matrix around the coated tows or minicomposites (fiber tow, BN interphase, and CVI SiC) does not contain pores or flaws that are greater in size than the minicomposites themselves. Because of this, matrix fracture or cracking is controlled by the size of the minicomposites running perpendicular to the applied stress, and thus can be modified by judicious selection of the fiber architecture.

TEST METHODOLOGY

Ultimately all pertinent elastic, inelastic, thermal, and durability properties need to be well understood for the SCMI composite system with various fiber architectures. However, it has been shown that one key design parameter is the matrix cracking stress for these materials. When matrix cracks are formed that expose load bearing fibers within a matrix crack to the oxidizing environment, strength-degradation occurs more rapidly with time [9-10]. The technique used here to determine the evolution of matrix cracks with stress is an acoustic emission (AE) method where AE is monitored during a typical tensile test and the onset of high energy AE events from the specimen gage section is considered to be the onset stress for through-thickness matrix cracking [11] (referred here simply as matrix cracking stress or MCS). The MCS typically occurs below the proportional limit stress which is the onset stress for non-linear stress-strain behavior derived from graphical methods such as an offset technique. It has been shown that stresses should be below the MCS at high temperatures in order for these composites to survive long times (>> 100 hours) [6].

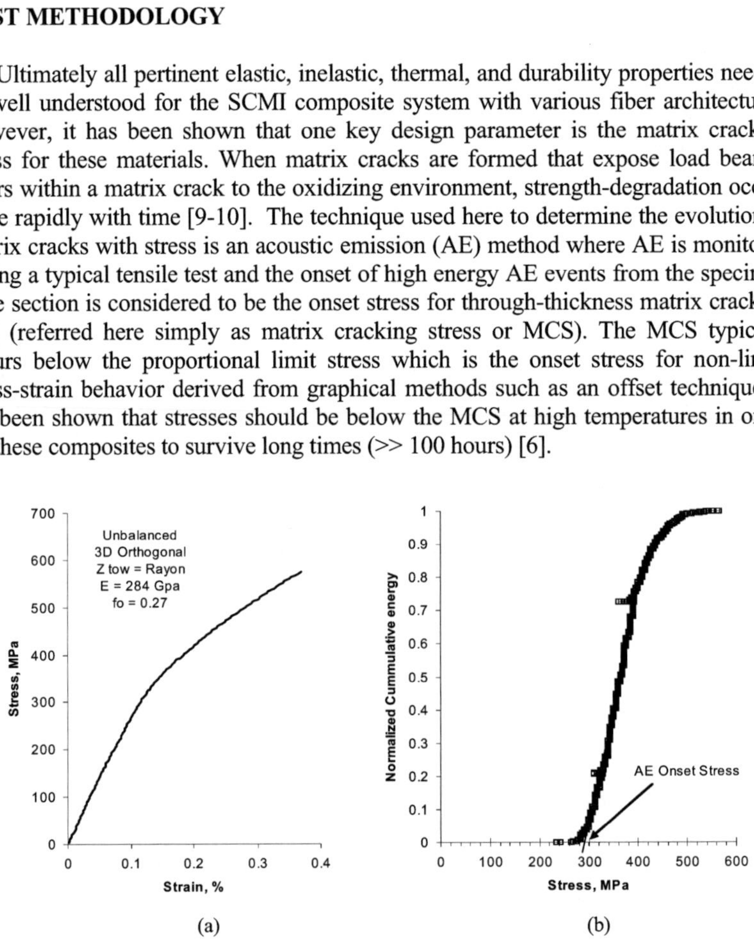

Figure 2. Stress strain curve (a) and normalized cumulative AE energy (b) from the gage section of a room temperature tensile test for an unbalanced 3D orthogonal composite tested in a high fiber content direction.

Figure 2 shows a representative room temperature stress strain curve (Figure 2a) and AE activity (Figure 2b) for a 2 mm x 10 mm dog-bone specimen machined from a thin-walled SCMI composite panel. These curves, as with all similar curves in this paper, were measured in-plane in a stress direction typically on-axis with a primary tow direction or symmetrically located between two primary tow directions. More importantly, as will be discussed, the stress was applied perpendicular to one of the primary tow directions which, due to their low transverse strength, are the source of micro or tunnel cracks that link-up and eventually initiate through-thickness cracks in the specimens. In Fig. 2b, the MCS is determined from the intercept of the steep region of the normalized cumulative AE curve down to the abscissa. The steep portion of the AE curve is dominated by high AE energy events associated with fiber-bridged matrix cracks [3, 11]. The normalized cumulative AE energy curve is determined by dividing the cumulative AE energy of each AE event from the gage section by the total cumulative AE energy from all the high energy events within the gage section for the entire test. The reduction in AE energy at the higher stresses is indicative of matrix crack saturation.

PREVIOUS RESULTS

In-plane cracking and MCS have been measured using AE for a wide variety of SCMI composites with primarily orthogonal fiber architectures, including stacked 2D woven five-harness satin fabric [3] and 2D tri-axially braided [5] plies, plus various 3D orthogonal [4,6] and angle interlock [6] preforms. Some tensile specimen cross-sections perpendicular to the loading direction are shown in Figure 3. Descriptions and some geometric and mechanical properties of the different fiber-architecture composites are listed in Table I, and stress-strain curves for some of these composites are shown in Figure 4.

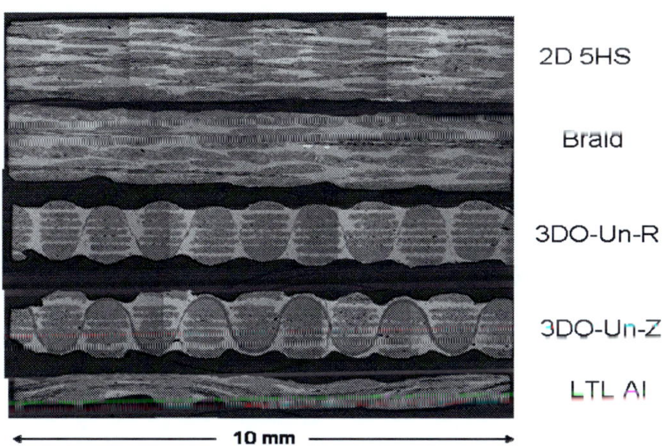

Figure 3. Some tensile specimen cross-sections of different fiber architecture composites. For the Braid composite, the cross-section is normal to the axial tows of the braid. For the 3D orthogonal composites, the cross section is normal to the "Y" direction (Table I). 3D orthogonal specimens tested in the "X" direction would be tested in the plane of the paper.

SiC/SiC Panel Type [Ref.]	Architecture Description	Fiber fraction, f_o, in load direction	E (GPa)	MCS (MPa)	UTS (MPa)	Mini. Tow height, mm
2D 5HS [3]	2D five-harness satin, balanced 0/90 lay-up (N24A)	0.19*	250	182	480	0.12
2D 5HS [3]	2D five-harness satin (double tow), balanced 0/90 lay-up	0.19	197	164	480	0.12
Braid [5]	2D Triaxial braid (double tow); cut/tooled into panel, X = axial, Y = hoop = test direction	0.26	250	213	352	2.0
3DO-Bal-R-Y [4]	3D orthogonal, Nearly balanced; Y = test direction	0.20	238	180	336	0.1
3DO-Bal-Z-Y [4]	3D orthogonal, Nearly balanced; Y = test direction	0.17	248	119	317	0.15
3DO-Bal-Z-X [4]	3D orthogonal, Nearly balanced; X = test direction	0.18	205	126	322	1.9
3DO-Un-R [6]	3D orthogonal, unbalanced, Y = test direction	0.28	275	270	>575	0.11**
3DO-Un-Z [6]	3D orthogonal, unbalanced, Y = test direction	0.27	262	245	596	0.11**
LTLAI [6]	3D Layer-to-layer Angle Interlock tested in the fill direction	0.10	125	80	204	0.18

TABLE I. SCMI COMPOSITE DESCRIPTIONS AND PROPERTIES.

* The table lists only the data for the 7.9 tow ends per cm five-harness satin architecture. Reference 6 and Figure 5 below incorporate data from 4.7 to 8.7 tow ends per cm.
** The largest tow height corresponds to the 90° Sylramic-iBN minicomposite in the X direction and not the Z-direction fiber tow minicomposite.

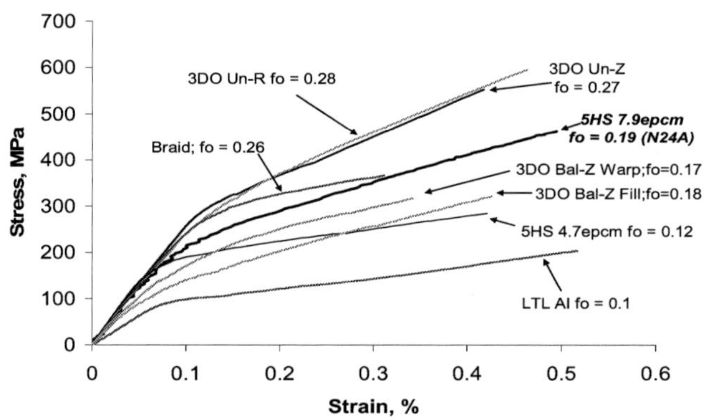

Figure 4. SCMI composite room temperature stress strain curves with different fiber architectures.

The details concerning the different fiber architecture composites can be found in references 3 through 6. The composite labeled "N24A" refers to a standard 2D d 8-ply composite [7] developed at NASA using the lay-up of balanced five-harness satin

Sylramic-iBN fabric. It is obvious that various composite mechanical properties can be improved by designing the fiber architecture in a desirable way. It should also be noted that the Braid specimen was tested in the hoop direction perpendicular to the axial tows so that the two bias direction fibers were 23° off the load direction. The f_o of 0.26 refers to all of the bias direction fibers which would bridge a transverse through-thickness matrix crack. However, since the bias fibers were off-axis to the stress direction, the ultimate stress of this composite type and orientation was less than expected if fibers were aligned in the direction of stress [5].

In references 3 and 6, it was demonstrated that the MCS was effectively linearly dependent on fiber volume fraction f_o in the loading direction with a significant amount of scatter (Figure 5a). This can be explained simply by the fact that for a given applied load, the stress on the matrix decreases as fiber fraction increases. Additionally, it was shown in reference 6 that the matrix cracking stress could be better quantified (reduced scatter) for the wide range of architectures when considering the height of the largest 90° minicomposites (fiber tow, BN interphase, and CVI SiC) tow that are normal to the applied load direction (Figure 5b). This can be explained by the fact that through-thickness matrix cracks typically emanate from regions where micro or tunnel cracks exist or are formed within these minicomposites.

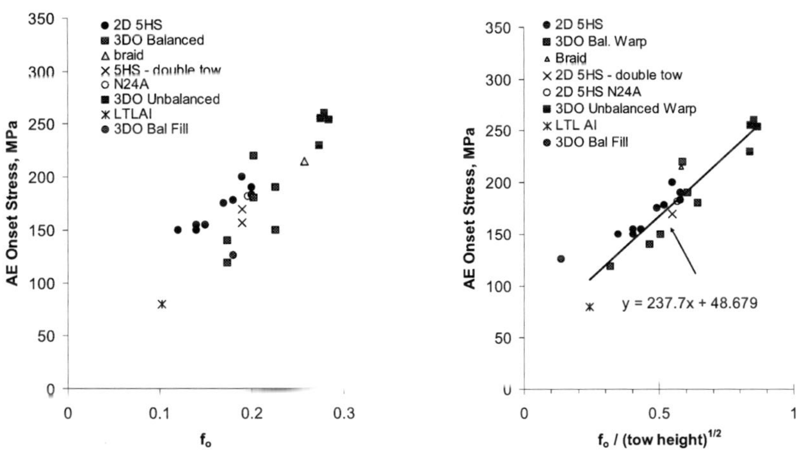

(a) (b)

Figure 5. Matrix cracking stress versus (a) f_o and (b) f_o divided by the square root of tow height.

From Figure 5b, an empirical relationship can be derived that relates MCS to fiber fraction and tow height:

$$MCS = 237.7\ f_o/h^{1/2} + 48.7 \quad (1)$$

where MCS is in MPa and h is in mm.

CONSIDERATIONS FOR FIBER ARCHITECTURE SELECTION

In the fiber architecture selection process, several factors need to be considered such as weavability or braidability of the tows, fiber volume fractions attainable, cost, etc. Once potential fiber-architecture approaches are identified which can produce the shape and dimensions of a given SCMI component, Equation 1 can be used as a first order approximation to further down-select the fiber architecture in order to maximize the matrix cracking stress where needed at different locations within the component. Equation 1 is an empirical relation for fiber fractions in the stress direction range from ~ 0.1 to 0.3 and for minicomposites 90° to stress direction which are the source of matrix crack formation.

From Equation 1 and Figure 5, it is clear that raising the fiber fraction in the primary load direction can be very beneficial for improving MCS. However, components in most cases will be loaded in multiple directions so that a high MCS architecture for one direction may actually yield an insufficient MCS for an orthogonal direction. Thus the architecture designer must first know as much as possible about the multi-directional stress requirements for the component. For the SCMI composite system, a constraint to multi-directional design is that total fiber volume fraction generally cannot exceed ~ 0.36 to 0.40.

Although Equation 1 has yet to fully established for alternate specimen dimensions and off-axis loads, it suggests that beyond simply increasing fiber fraction, MCS in a given load direction could also be increased by reducing the height h of the 90° minicomposites perpendicular to the load direction. To demonstrate, consider the typical case of an orthogonal architecture, e.g., angle interlock or 2D weave, which has a tow height of ~ 0.1 mm (a typical tow height for 800 fiber count Sylramic-iBN tows). Assuming the total fiber volume fraction to be 0.36, and considering both a single-tow and a double-tow orthogonal architecture, the estimated MCS values are shown in Figure 6 for two cases: a balanced composite ($f_o^x = f_o^y = 0.18$) and an unbalanced composite ($f_o^x = 0.26$, $f_o^y = 0.10$). Here, it is also assumed that the double tow possesses a tow height twice that of the single tow, which is probably a conservative assumption.

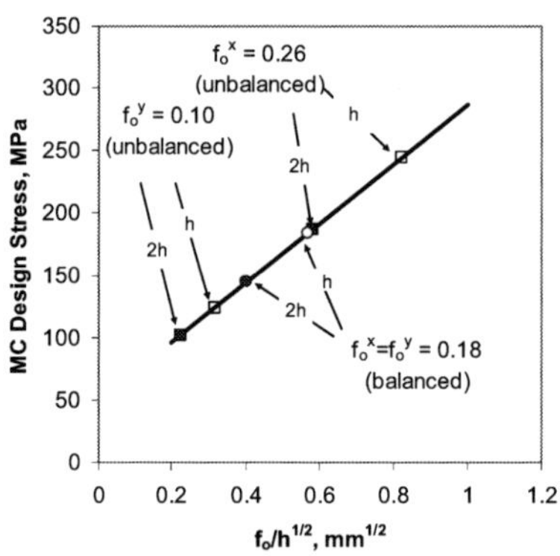

Figure 6. The effect of fiber volume fraction and tow height on the matrix cracking design stress for the case of a single tow and double tow woven composite with a single tow height of 0.1 mm.

Fig. 6 shows that the unbalanced architectures offer significantly higher MCS over the balanced architecture. However, one would need to consider the debit in the lower fiber volume fraction orthogonal direction. It is also evident that moving to double-tow architectures results in a reduced MCS over single-tow architectures on the order of 25%, something that must be considered if the weaving or braiding of single Sylramic tows is not cost effective.

SUMMARY AND CONCLUSIONS

The critical design parameter, matrix cracking stress, for the Sylramic-iBN, slurry-cast melt-infiltrated SiC/SiC composite system can be estimated using the empirical relationships developed in this paper. This in turn should allow the selection of appropriate architectures to minimize the occurrence of through-thickness matrix cracks in a given SiC/SiC component. Though not as much work has been done on other fiber-types or matrix systems, it is anticipated that similar trends and relationships can be derived for the wide range of fiber-architectures shown here as long as the 90° minicomposites are the primary source of cracks that eventually cause through-thickness matrix cracking. When the effects of fiber-type on matrix cracking stress are better understood (e.g., interfacial shear stress and residual stress effects) then a more complete comparison incorporating both fiber type, fiber architecture, and fiber off-axis load can be developed for CMC design engineers.

ACKNOWLEDGEMENTS

This work was funded under NASA's ARMD Supersonics project.

REFERENCES

1. Brewer, D. 1999. "HSR/EPM Combustor Materials Development Program" *Mater. Sci. Eng. A*, A261: 284–291.
2. Naslain, R. 2004. "Design, preparation and properties of non-oxide CMCs for application in engines and nuclear reactors: an overview" Comp. Sci. Tech., 64: 155–170.
3. Morscher, G.N. 2004. "Stress-Dependent Matrix Cracking in 2D Woven SiC-fiber Reinforced Melt-Infiltrated SiC Matrix Composites", *Comp. Sci. Tech.*, 64: 1311–1319.
4. Morscher, G.N., Yun, H.M., DiCarlo, J.A. 2005. "Matrix Cracking in 3D Orthogonal Melt-Infiltrated SiC/SiC Composites with Various Z-Fiber Types", *J. Am. Ceram. Soc.*, 88 (1): 146–153.
5. Morscher, G.N.; Yun, H.M.; DiCarlo, J.A. 2007. "In-Plane Cracking Behavior and Ultimate Strength for 2D Woven and Braided Melt-Infiltrated SiC/SiC Composites Tensile Loaded in Off-Axis Fiber Directions," *J. Am. Ceram. Soc.* 90 (10): 3185–3193.
6. Morscher, G.N., DiCarlo, J.A., Kiser, J.D. and Yun, H.M. "Effects of Fiber Architecture on In-Plane Stress-Strain Behavior and Creep in Melt-Infiltrated SiC/SiC Composite Panels," submitted to *Comp. Sci. Tech.*
7. DiCarlo, J.A., Yun, H-M., Morscher, G.N. and Bhatt, R.T. 2005. "SiC/SiC Composites for 1200°C and Above" in *Handbook of Ceramics Composites*, Chapter 4; N.P. Bansal, ed. NY: Kluwer Academic, pp. 77–98.
8. Yun, H.M., Gyekenyesi, J.Z., Chen, Y.L., Wheeler, D.R., and DiCarlo, J.A. 2001. "Tensile behavior of SiC/SiC composites reinforced by treated Sylramic SiC fibers." *Ceram. Eng. Sci. Proc.*, 22: 521–531.

9. Heredia, F.E., McNulty, J.C., Zok, F.W., and Evans, A.G. 1995. "Oxidation embrittlement probe for ceramic-matrix composites" *J. Am. Ceram. Soc.*, 78: 2097.
10. Morscher, G.N. and Cawley, J.D. 2002. "Intermediate Temperature Strength Degradation in SiC/SiC Composites," *J. European Ceramic Society*, 22(14-15): 2777-2788.
11. Morscher, G.N. 1999. "Modal Acoustic Emission of Damage Accumulation in a Woven SiC/SiC Composite," *Comp. Sci. Tech.* 59: 687-697.

Analysis of Thermal Conductivities of Three-Dimensionally Woven Fabric Composites

J. SCHUSTER, D. HEIDER, K. SHARP, G. SCHNEIDER,
C. SCHÖNBORN and M. GLOWANIA

ABSTRACT

This study investigates the effect of three-dimensional fiber reinforcement on the out-of-plane thermal conductivity of composite materials. Composite preforms with different fibers in thickness direction were fabricated. After infusion, using a Vacuum-Assisted Resin Transfer Molding process, their through thickness thermal conductivities were evaluated. The measured out-of-plane thermal conductivities showed significant increase compared to a typical laminated composite. Although the through thickness thermal conductivity of the samples increased with through thickness fiber volume fraction, the values did not match those predicted by a simple rule of mixture. Using Finite Element models to better understand the behavior of the composite material, improvements to an existing analytical model were performed to predict the effective thermal conductivity as a function of the composite material properties and in-contact thermal material properties.

INTRODUCTION

Fiber based composites offer the unique ability to tailor material properties locally. Localized matrix or fiber architecture changes can meet local mechanical, electrical, or thermal loads within a component or vehicle, without weight penalty or resort to additional external structures. However, traditional laminated uniaxial or biaxial composites have low through thickness thermal conductivities, which limit their use in a variety of aerospace applications. Leading edges in supersonic aircraft wings, inlet or exhaust areas of gas turbine engines, light weight heat exchangers, electronics packaging materials, doublers in satellite structures, and

Schuster, J., Schönborn, C., Schneider, G. University of Applied Sciences Kaiserslautern, Carl-Schurz Str. 10-16, 66953 Pirmasens, Germany
Heider, D. University of Delaware, Center for Composite Materials, Newark DE 19716, USA
Sharp, K. TEX, 109 MacKenan Drive, Cary NC 27511, USA
Glowania, M. RWTH Aachen, Institut für Textiltechnik, Eilfschornsteinstr. 18, 52062 Aachen, Germany

low matrix thermal conductivity only by a factor of 10 [1]. One alternative approach of achieving significant increase in the through-thickness thermal conductivity is to manufacture composites based on 3-D woven fiber architectures where the through-thickness fibers have high thermal conductivities. One commercially available fabric of this kind is 3WEAVE™ from 3TEX.

Electro-magnetic interference (EMI) enclosures all are subject to localized thermal loads which must be transferred through part thickness to adjacent cooler areas for subsequent radiation/convection, in part to mitigate internal stresses induced by reducing the thermal gradient across the part thickness. One common approach to increasing the through thickness conductivity of a composite, namely loading the resin with thermally conductive particles, can increase the conductivity only insignificantly. Also the use of expensive carbon nanotubes can increase the very

DESCRIPTION OF 3 WEAVE™

The 3WEAVE™ process was originally developed at the NC State University College of Textiles and further refined and enhanced at 3TEX [2]. This process places several layers of fill (y-directional) and warp (x-directional) yarns in x-y plane without any crimp (contrary to conventional 2-D weaving process), thus providing straight bi-directional in-plane reinforcement (Figure 1). The multiple layers of warp and fill yarns are interlaced by a set of "z"-yarns, which holds the fabric together. The z-yarns provide a desirable level of out-of-plane strength, so the layers of warp and fill yarns cannot separate. The amount of "z"-fibers can be varied from less than 1% to more than 30% of the total fiber amount in the preform.

Using 3WEAVE™ process, fabrics have been produced with thickness up to 2 inches (5 cm), and widths up to 72 inches (183 cm). 3WEAVE™ process is "gentle" on the warp- and fill-directional fibers, thus improving the ability to use of rather brittle fibers, such as highly thermally conductive pitch yarns over traditional 2-D. Further, it is possible to fabricate directly some complex shapes of integral 3-D woven preforms. Importantly, different fiber kinds can be hybridized in a 3-D woven integral preform. Selective placement of fiber types through the preform can make a composite material with gradually changing properties. This could be used for tailoring density, stiffness, strength, thermal expansion, thermal conductivity, and other properties.

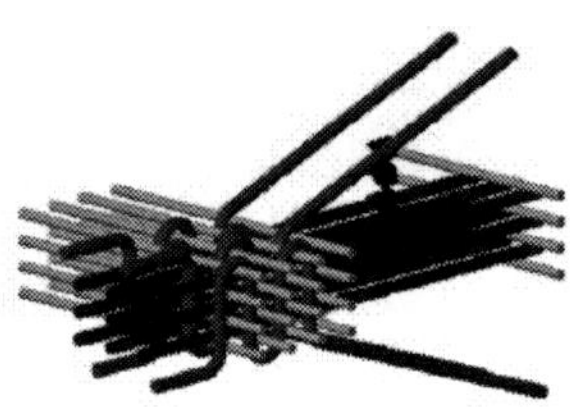

2-D weave formation 3-D (3WEAVE™) weave formation

Figure 1. Schematic of the 2-D and 3-D weaving processes.

THERMAL CONDUCTIVITY MEASUREMENTS

Thermal conductivity measurements were performed with a measuring cell built in-house at the Center for Composite Materials at the University of Delaware according to ASTM E 1225 – 04 allowing measurements of circular samples with a diameter of 2" (50.8 mm) [3, 4]. Sample thicknesses can vary from 2 until 50 mm. The cell was placed on a cool plate to provide a heat sink with a constant temperature T_{sink} beneath the bottom plate. Temperature information read by thermistors were evaluated by a LabView-based program. Conductivity paste (OT-201 from Omega with K = 2.3 W/(m K)) was used in order to facilitate coupling and to reduce interfacial thermal resistance. Prior to measurements, the thermal resistance R_{int} at the interfaces from meter bars and sample had to be determined using a stainless steel reference samples. R_{int} was determined to 533 (K mm^2)/W. The heat flux introduced by the cartridge heater was about 2000 W/m^2. Calibration measurements were performed on isotropic specimens to determine the accuracy of the measuring cell. In addition, eleven composite samples made of 3WEAVE™ by resin infusion in a VARTM-process were tested (Table 1). The thermal conductivities K of these samples were calculated according to a simple rule of mixtures [5]. It turned out that results for isotropic materials match their reference values well. However, the measured values for the composite samples with z-fiber reinforcement differed drastically from theoretically derived thermal conductivities despite the standard deviation for all material tests was less than ± 5 %.

TABLE 1. THERMAL AND CONDUCTIVITY PROPERTIES OF SAMPLE CONSTITUENTS USED.

	base material	axial thermal conductivity K_a [W/(m K)]	radial thermal conductivity K_r [W/(m K)]
Cu-fiber	Copper	390	390
E-glass-fiber	glass	1.04	1.04
CN80-fiber	pitch	320	11
YS80-fiber	pitch	320	11
T300-fiber	PAN	100	11
T700-fiber	PAN	100	11
SC-15	epoxy	0.19	0.19

This leads to the conclusion that a simple model consisting of a parallel connection between out-of-plane conductivities of in-plane fibers and z-fibers can not be applied for these materials without alteration.

FE-ANALYSIS

To get a better understanding of this result, the heat flow through a composite unit cell encompassing a single copper fiber sandwiched between two meter bars was modeled by finite elements with a strictly vertical heat flux of q" = 500 W/m^2 and T_{sink} = 20 °C. Using this volume FE model with triangled meshes, variations in the v_{fz} of the copper fibers in the composite can be coupled with variations in the thermal conductivity of the meter bar. The copper volume fraction of the composite was varied between 0 and 100% and the thermal conductivity of the meter bar was also varied in steps between 0.1 and 1000 W/(m K). Thus, the effect of volume fraction and thermal conductivity on the measurement of thermal conductivity could be determined (Fig. 2). It is obvious that the measured conductivity is influenced not only by the volume fraction but also by the connecting meter bar material. The simple approach in terms of rule of mixture used above to estimate composite thermal conductivity seems to be the limit for connecting material conductivities approaching infinity. Higher thermal conductivity in the meter bar material increases the proportion of the heat flux through the copper fiber.

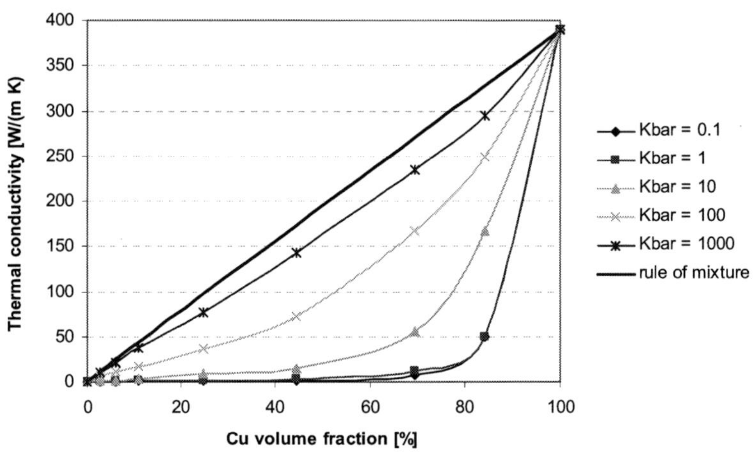

Figure 2. Thermal conductivity based on Cu-fiber volume fraction and meter bar material's thermal conductivity.

In contrast, the low conductivity bar material allows little heterogenization of the heat flow, so the still homogeneous heat flux passes primarily through the area of the composite between z-fibers. As the thermal conductivity of the meter bar increases, significant heat flow passes through a larger meter bar area near the copper, as seen in the increasing size of the "mushroom" cap near the copper (Figure 3).Therefore, it has to be stated that the thermal conductivity of anisotropic or orthotropic material such as composites is no material but a system property because its heat flux influenced by adjoining materials. In addition, FEM-modeling was performed with the same scenario as described previously but only for a copper content of 10 %. However, this content was distributed in one, four, eight, sixteen, and thirty two fibers (so called entry ports) for samples with different thicknesses (1 mm, 2 mm, 4 mm, 8 mm, 16 mm, 32 mm). The result of this computation is depicted in Fig. 4.

It is obvious that the distribution density of the sample also effects the thermal conductivity measurement. Furthermore, it turned out that contrary to isotropic materials also the thickness has an influence on the thermal conductivity (Fig. 4). This can be explained by an increasing effect of radial heat flow with depth.

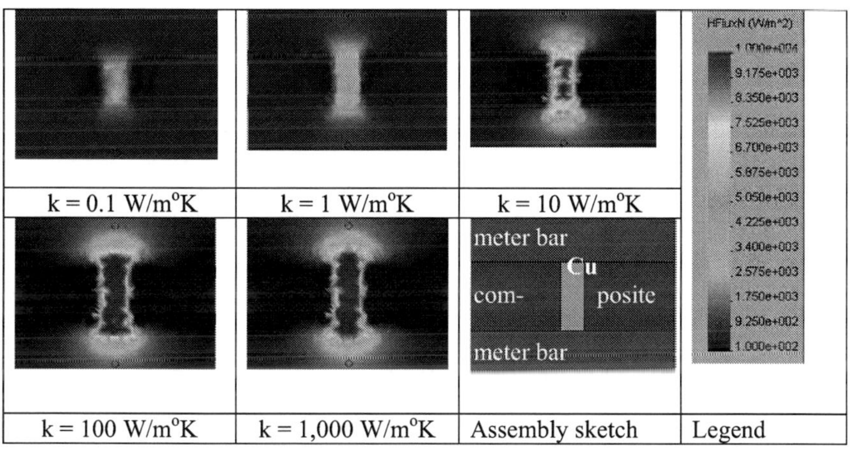

Figure 3. FEM-Results of heat flux depending on meter bar conductivity.

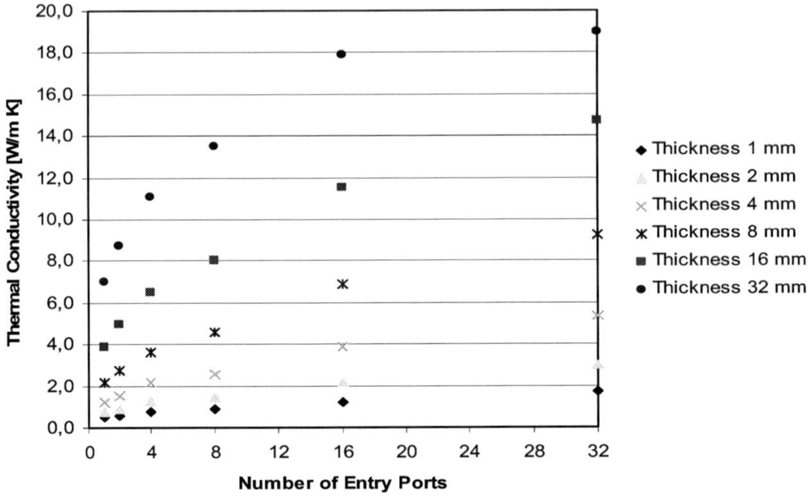

Figure 4. Effect of thickness and number of entry ports.

IMPROVEMENTS TO ANALYTIC MODEL

As described above, two effects unique to heterogeneous materials could be identified:

- Thermal conductivity of the material where the heterogenization takes place
- Distribution density of z-fibers

As common in describing properties of composites these effects can be taken into account by introducing factors to the simple rule of mixture, which can be calculated based on geometrical and material properties.

$$K_{3weave} = K_{z-fiber}\,\delta_{z-fiber}\,\eta_{KMat}\eta_{TD} + (1-\delta_{z-fiber})K_{in-planex,y} \qquad (1)$$

In this equation, $\delta_{z\text{-fiber}}$ is the fiber volume fraction of z-fiber, η_{KMat} reflects the influence of the thermal conductivity of the adjoining material and η_{TD} takes the effect of the fiber distribution density and the thickness of the material into account. $K_{z\text{-fiber}}$ and $K_{in\text{-}planex,y}$ can be calculated according to [5].

$$\eta_{TD} = \tanh\left(\frac{\sqrt{\delta_{z-fiber}}\,t_{sample}}{2\,d_{roving}}\right) \qquad (2)$$

with t_{sample} as the sample thickness and d_{roving} the diameter of the roving. The dependency of η_{KMat} on the meter bar material can also be derived from the FE-results (Fig.1). The course of the graph can be described by a vertical shifted hyperbolic tangent function. Its course is also depicted in Fig. 5:

$$\eta_{KMat} = \frac{1}{2}\left[\tanh(\log(0.1K_{Mat})) + 1\right] \qquad (3)$$

Eqn. 3 approximates empirically the effect of increase in diameter of the mushroom cap shaped heat flux area with the thermal conductivity of the meter bar material. η_{KMat} becomes e.g. 0.6 for stainless steel as a meter bar material. This approach has to be done empirically due to the two-dimensional nature of the heat flux, because such problems can only be solved for special cases. The results of the improved model are shown in Fig. 6 in comparison to the simple model results proving the accuracy of the model introduced.

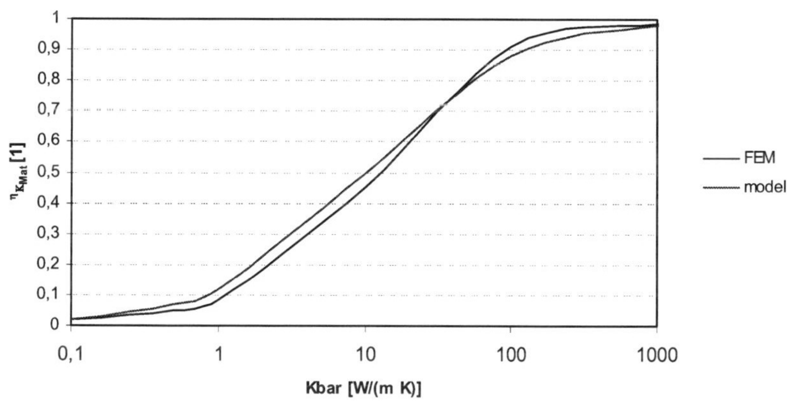

Figure 5. Dependency of η_{KMat} on meter bar material.

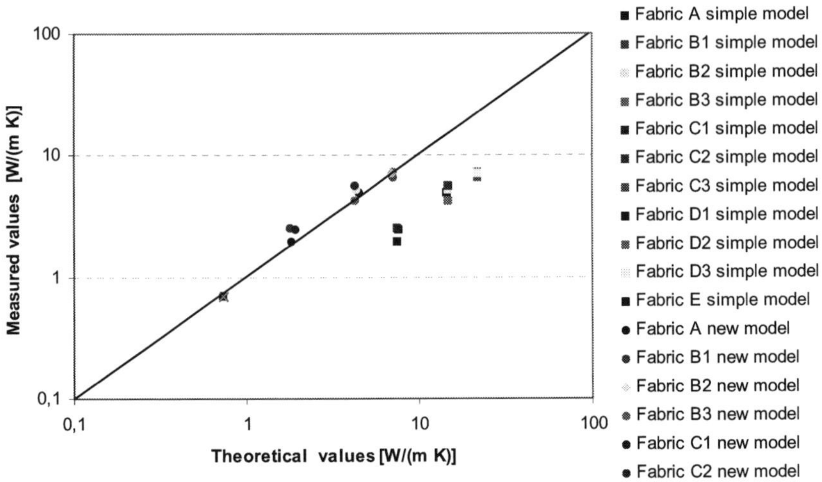

Figure 6.Comparison of simple rule of mixture values, new model and measured values.

DISCUSSION, CONCLUSIONS, AND PERSPECTIVES

The thermal conductivity measurements on three-dimensionally reinforced composite samples described previously resulted in a drastic difference between expected and measured data. One explanation of these results involves the existence of discrete thermally high conductive heat flux entry ports in the composite samples under investigation. Thus, the thermal conductivity of a 3-D composite material has to be considered as a system property and not a material property due to the heat flux distribution capability of the materials connected. Furthermore, the distances between these ports define the ability of the composite system to homogenize the heat flux and can be expressed by the distance between fibers derived via fiber diameter. This ability is affected by the thermal conductivities of the adjoined materials. Note that only isotropic neighboring materials were considered in this study. The development of a purely analytical thermal conduction model seems to very difficult if not impossible due to the two-dimensional nature of heat flux. However, the introduction of two factors to the simple rule mixture based on a parallel thermal connection model by taking into account fiber spacing and adjoining material led to reasonably accurate model predictions for the samples tested. In addition, this investigation showed that with a z-fiber volume content of about 6 % the maximum possible out-of-plane thermal conductivity can be increased by a factor of eight. This opens the possibility to design composite parts in terms of heat transfer having the thermal conductivity as design aspect to varied by fiber architecture and thus to seen equal to mechanical design criteria such as Young's modulus or strength.

ACKNOWLEDGEMENTS

The work described was conducted as part of United States Air Force contract FA8650-06-C3616..

REFERENCES

1. Bagchi A., Nomura S. 2006. "On the Effective Thermal Conductivity of Carbon N[a]
Reinforced Polymer Composites," Composites Science and Technology, 66 (11-12):
1712.
2. Mohamed M.H., Bogdanovich A.E., Dickinson L.C., Singletary J.N., Lienhart R.R. 2(
New Generation of 3D Woven Fabric Preforms and Composites," SAMPE Journal, 37
17.
3. N.N. 2004. "Standard Test Method for Thermal Conductivity of Solids by Means
Guarded-Comparative-Longitudinal Heat Flow Technique," ASTM E 1225-04.
4. Laubitz M.J. 1984. Axial Heat Flow Methods of Measuring Thermal Condu
Compendium of Thermophysical Property Measurement Methods Volume 1, Plenum Pre
York and London.
5. Schuster, J., Heider, D., Sharp, K., Glowania, M. 2008. "Thermal conductivities o
dimensionally woven fabric composites," Composites Science and Technology, 68 (9):
2091.

which is related to the inherent in-plane waviness of the fibre tows. This causes uncertainties in determination of laminate properties, which are typically addressed by introduction of a safety factor. The influence of material non uniformity on the laminate properties has been addressed in a variety of studies [1-4] by describing Young's modulus and strength by random fields.

In this study, a bottom-up approach is implemented to estimate the mechanical properties of a laminate based on a randomised local fibre configuration. The non-uniformity of textile fabrics is modelled considering local changes in the distance between fibre tows. These are described by continuous functions with random parameters, which can be related to actual textile structures. The fabric structure results in local variations in the fibre volume fraction, which determine the composite properties. The same approach has been successfully applied before to predict fabric permeabilities [5].

MODELLING

Mechanical Properties

A symmetrical laminate (E-glass fibres with a polyester resin matrix) with four uni-directional (UD) layers at orientations [45° / -45° / -45° / 45°] and a total thickness of 4 mm, representing a laminate manufactured from two layers of bi-axial non-crimp fabric reinforcement, is discussed here. Material properties for the fibres and the matrix are listed in Table I. In the absence of available data for constituent behaviour in compression and shear, it was assumed throughout the present study that tensile, compressive and shear strengths were equal, both for the E-glass fibres and the polyester matrix. The transversely isotropic elastic constants and strengths of a UD lamina are described by Chamis' micromechanics models [6]. The lamina properties depend on the mechanical properties of fibres and matrix and the fibre volume fraction, V_f, which is determined based on considerations for the representative volume of a UD lamina at a given height (Figure 1).

TABLE I. MECHANICAL PROPERTIES ASSUMED FOR E-GLASS FIBRES AND POLYESTER RESIN.

property	fibre	matrix
Young's modulus	$E_f = 73000$ MPa	$E_m = 3700$ MPa
Poisson's ratio	$\nu_f = 0.22$	$\nu_m = 0.33$
shear modulus	$G_f = 29918$ MPa	$G_m = 1391$ MPa
tensile strength	$\sigma_f = 2000$ MPa	$\sigma_m = 70$ MPa
shear strength	$\sigma_f = 2000$ MPa	$\sigma_m^S = 70$ MPa

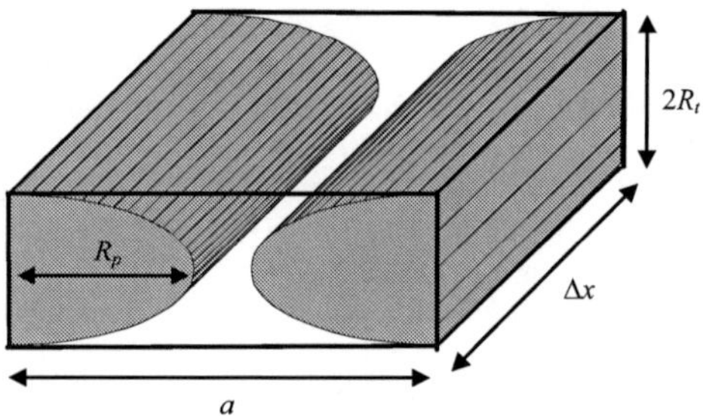

Figure 1. Representative volume of a UD lamina.

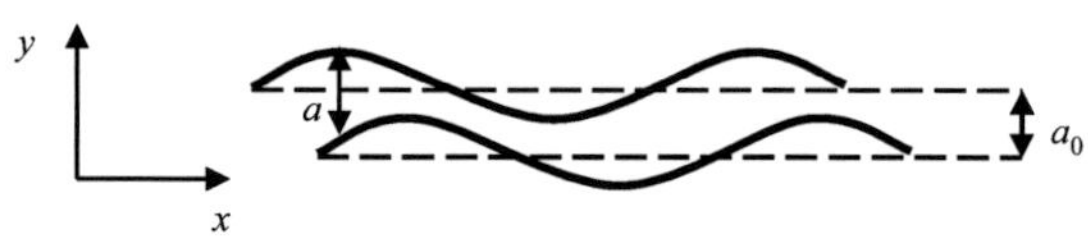

Figure 2. Variation of fibre spacing.

The fibre volume within the representative volume is given by the fibre tow cross-section, the fibre tow length and the fibre volume fraction within the fibre tows. The latter is estimated by the maximum achievable fibre volume fraction V_{fmax}. In practice, values for V_{fmax} of about 0.6 are achieved in uni-directional composite plies produced by conventional manufacturing methods [7]. The fibre tows are assumed to have an approximately elliptical cross-section with semi-major axis R_p (in the lamina plane) and semi-minor axis R_t (perpendicular to the lamina plane). The representative volume corresponds to the product of the area included by the fibre tow spacing a and the length of the segments Δx, and the height, which is identified with the value $2R_t$. Thus, the fibre volume fraction is

$$V_f = \frac{\pi R_p R_t \Delta x V_{f\max}}{2 a \Delta x R_t} = \frac{\pi R_p V_{f\max}}{2a} \quad . \tag{1}$$

Randomisation

At any point (x,y) in a lamina, the variation of the in-plane distance between adjacent fibre tows oriented along the x-direction (Figure 2) is modelled by

$$a - 2R_p = (a_0 - 2R_p)\left(1 + \frac{1}{N}\sum_{i=0}^{N-1} w_i(x)\, w_i(y)\right) \tag{2}$$

with the functions

$$w_i(x) = \sin(\omega_{xi}\, x + \varphi_{xi}) \tag{3}$$

and

$$w_i(y) = \sin(\omega_{yi}\, y + \varphi_{yi}) \quad . \tag{4}$$

The lamina is characterised by a_0, the average value of a, and the in-plane dimension of the fibre tows R_p. The values of the frequencies ω_{xi} and ω_{yi} and the phases φ_{xi} and φ_{yi}, which describe the in-plane fibre tow waviness, are determined randomly. While the phases have values between 0 and π, the maximum random frequency

$$\omega_{\max} = \frac{|da/dx|_{\max}}{a_0 - 2R_p} \tag{5}$$

is determined by the maximum value of da/dx, which can be measured based on visual observation for the actual lamina that is modelled. Since the changes in fibre distances are limited by the fibre mobility in the transverse direction, the amplitudes of the waviness are normalised by the number of terms in the expansion, N. Here, the number of terms was set to $N = 3$.

Finite Element Modelling

A rectangular laminate specimen with dimensions 175 mm × 25 mm (representing the gauge length between the jaws in a standard tensile test) was discretised into 700 square composite shell elements (Abaqus element type S4R) with an edge length of 2.5 mm. An example for the distribution of the fibre spacing a for one lamina is plotted in Figure 3. Assignment of local values of the fibre volume fraction to each finite element is based on solving Equation (2) for the co-ordinates of the element centre of mass. Here, the average distance between the fibre tows is assumed to be relatively small compared to the dimensions of the disturbances along the fibre axes. Thus, the changes in orientation of the fibre axes are neglected and the fibre tows are treated as being piecewise parallel. From the distance between parallel fibre tows, local fibre volume fractions and local values for the mechanical properties are calculated. This procedure is repeated for each of the laminae, and material properties are specified at the lamina level of the layered shell finite elements. The stiffness matrix of each element is calculated by the Abaqus/Standard FE package according to composite shell theory [8].

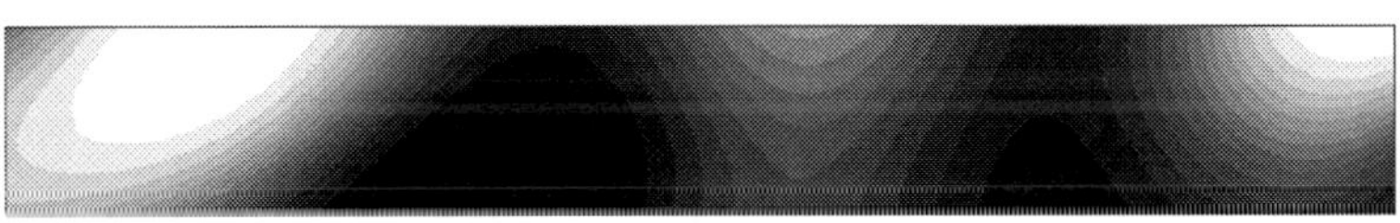

Figure 3. Distribution of fibre spacing a for the example of a lamina with $R_p/a_0 = 0.38$ and $\omega_{max} = 100/\text{m}$; white: $0.92 \times a_0$; black: $1.14 \times a_0$.

Failure Model

In order to estimate the strength of these laminates, material properties were defined to be dependent on the value of variables which evaluated terms in the maximum stress failure criterion. A brittle stiffness degradation law, suggested by Blacketter [9], was incorporated. While this neglects the non-linear post-failure behaviour exhibited by such bias tension specimens, it offers a simple, fast and purely predictive indication of the onset of failure. The failure criterion was evaluated at each numerical integration point through the material thickness and stiffness degradation was applied at the lamina level. Although other intra-ply failure mechanisms were incorporated into the analysis, the present configuration always exhibited shear failure as the active failure mechanism.

RESULTS AND DISCUSSION

Bias tensile loading along the specimen axis was simulated by constraining the displacement on one short edge of the rectangular specimen and prescribing a displacement on the other short edge. Constraints were applied to prevent rigid body motion and rotation, but the specimen was able to twist about its major axis, and transverse contraction due to Poisson's effect was also unrestrained. For each set of parameters R_p/a_0 and ω_{max} listed in Table II, a series of simulations (100 repeats) was run and statistically evaluated. The maximum frequency of the fibre waviness in the simulations is limited by the edge length of the finite elements and the dimensions of the specimen.

Figure 4 shows typical plots of transverse and out-of-plane displacement fields on the undeformed finite element mesh, while Figure 5 shows the distribution of failure index values in each of the laminae immediately prior to the onset of failure. The simulation results suggest that the finite elements, for which first failure is detected, are most likely to be located near the long specimen edges. It is not clear at this stage whether this represents a true effect or whether it is due to the method of constraining the ends of the specimen. The exaggerated deformed shape in Figure 5 suggests that there are localised in-plane bending effects due to the unsymmetric nature of the laminate, caused by its property distribution, which may be a contributing factor to this observation.

TABLE II. PARAMETERS FOR RANDOMISATION OF THE FIBRE STRUCTURE AND RESULTING PREDICTED VALUES FOR MODULUS AND STRENGTH, SHOWN WITH COMPARABLE EXPERIMENTALLY-DETERMINED MECHANICAL DATA.

R_p/a_0	V_f	ω_{max} / m^{-1}	E / MPa	σ / MPa
0.30	0.28	100	8630 ± 119 (± 1.4 %)	98 ± 5 (± 4.8 %)
0.38	0.36	-	9970	116
0.38	0.36	40	9837 ± 151 (± 1.5 %)	103 ± 3 (± 3.3 %)
0.38	0.36	100	9955 ± 92 (± 0.9 %)	100 ± 3 (± 2.6 %)
0.38	0.36	600	10001 ± 22 (± 0.2 %)	97 ± 1 (± 1.0 %)
0.45	0.42	100	11296 ± 54 (± 0.5 %)	103 ± 1 (± 1.3 %)
(exptl.)	0.36	(exptl.)	11263 ± 546 (± 4.8 %)	112 ± 4 (± 3.0 %)

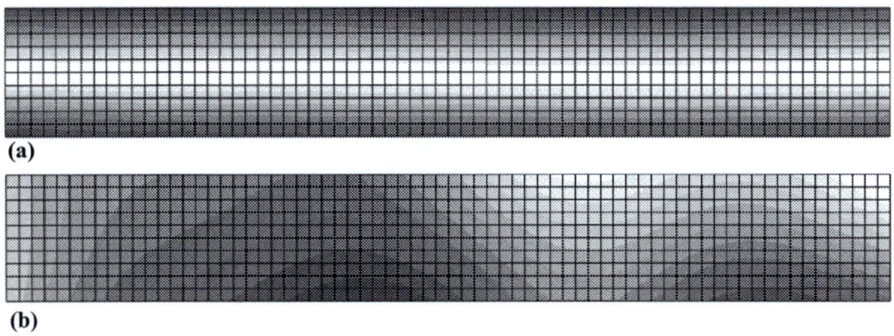

Figure 4. Typical displacement fields for bias tensile loading along the specimen axis (R_p/a_0 = 0.38, ω_{max} = 100/m); (a) transverse displacement magnitude (white: 0; black: 1.47×10^{-2} mm); (b) out-of-plane displacement (white: -1.33×10^{-2} mm; black: 2.63×10^{-2} mm); field observed within the elastic regime under applied displacement U_1 = 0.35 mm, (ε_1 = 0.2 %).

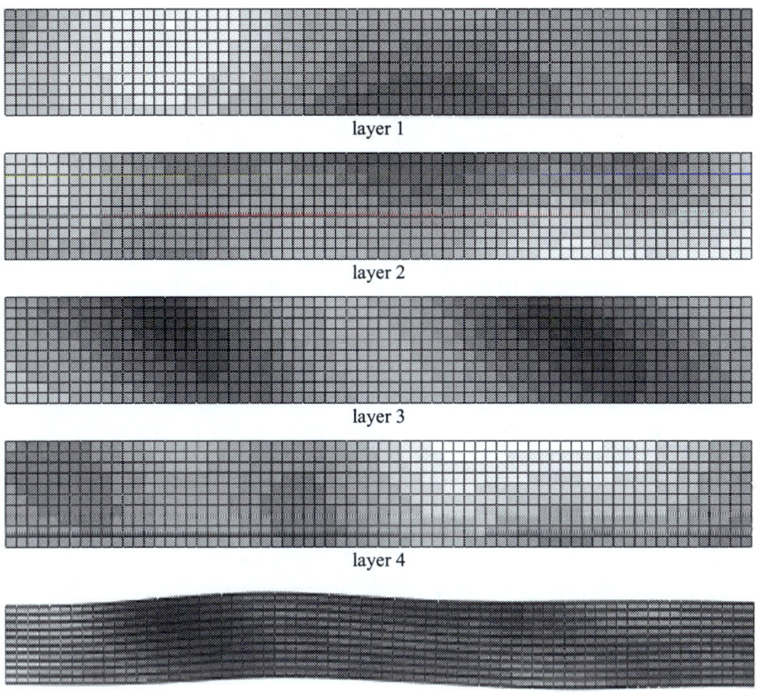

maximum value in all layers; transverse deformation scale factor 100

Figure 5. Maximum stress failure index plotted in each lamina and summarised by plotting the maximum value through the thickness (R_p/a_0 = 0.38, ω_{max} = 100/m); white: 0.70; black: 1.00. Last increment before onset of failure, U_1=1.76 mm (ε_1=1.0%).

The results for the macroscopic laminate properties (modulus E and strength σ) of the specimens are listed in Table II, and compared with experimental data for laminate specimens from four layers of non-crimp E-glass fabric reinforcement and an unsaturated polyester resin, manufactured using a liquid moulding process. Although the selection of a simple brittle material degradation law without an interlaminar damage model precluded the correlation of strain to failure and of the overall stress-strain behaviour, predicted values of stiffness and strength were within 15 % of those determined experimentally for similar materials, as indicated by the results in Table II.

Because of the effect of through-thickness averaging of the lamina properties over all four layers, the coefficients of variation for E and σ are generally small. The results suggest that:

- At given ω_{max}, E decreases significantly with decreasing V_f as expected, while its coefficient of variation (standard deviation / average value) increases. σ does not change significantly, but its coefficient of variation increases.
- At given V_f, E does not change significantly with increasing ω_{max}. For σ, a significant reduction compared to a specimen with uniform properties and a slight decrease with increasing ω_{max} can be observed. In analogy to the observations for the permeability of non-uniform textile structures [5], the coefficients of variation of E and σ increase with increasing ω_{max} (from zero for uniform laminae), reach a maximum and then decrease again, converging to a limiting value.

The global variations in mechanical properties increase with increasing ω_{max} starting from zero for a uniform fibre structure. In this range of ω_{max}, which is not considered in Table II, changes in the net fibre volume fraction may occur. On the other hand, the higher ω_{max}, i.e. the more periods of waviness there are across the specimen, the more uniform appears the lamina at the specimen-scale. Thus, the resulting global variations decrease again with increasing ω_{max}.

The decrease in global variations in mechanical properties with increasing maximum random frequency at given specimen dimensions suggests in turn that the variations also decrease with increasing specimen dimensions at given maximum frequency. The same lamina appears more uniform at the macro-scale than at the meso-scale. This scale effect implies that variations in mechanical properties, which are experimentally determined for macro-scale specimens, underestimate the actual local variations. Thus, information on local variations is required for modelling of macro-scale lamina properties. This conclusion justifies the bottom-up approach for modelling of global lamina properties based on local fibre configurations.

CONCLUSIONS

The influence of the inherent waviness of the fibre tows in fibrous reinforcements on the properties of composite laminates was discussed for the example of a symmetrical laminate with four UD layers at a total thickness of 4 mm. The local mechanical properties of a UD lamina were described by Chamis' micromechanics models, implying a dependence on the given mechanical properties of fibres and matrix and the local fibre volume fraction.

The non-uniformity in the fibre structure of a lamina was modelled considering local changes in the distance between fibre tows. The local properties are described by continuous functions with random parameters, which can be related to geometrical parameters of actual fibre structures. Based on statistical evaluation of series of non-linear finite element simulations implementing the local values of the mechanical properties, most probable failure scenarios in bias tensile loading were predicted for the laminate. The dependence of the variability in global mechanical properties (modulus E and strength σ) on geometrical parameters (average fibre volume fraction V_f and maximum frequency of fibre tow waviness ω_{max}) was discussed.

Within the parameter ranges investigated, the simulation results showed that beyond an initial local maximum, the effects of ω_{max} on average values of E and σ were small, while those due to an increase in the specimen V_f could be largely attributed to the micromechanical relations implemented in the modelling. The coefficients of variation of E and σ both decreased with increasing ω_{max}. They were also observed to decrease with increasing V_f. It was noted that local effects were less significant at the specimen level and that such a physically-based model for local variations would be particularly necessary for components in which local effects could dominate overall behaviour. Examples would include components with features such as holes and bends with small radii.

ACKNOWLEDGEMENT

The authors wish to thank Mr. Tim Baillie for programming the routines for the stochastic simulations.

REFERENCES

1. Lin, S.C. 2000. "Reliability predictions of laminated composite plates with random system parameters", *Probabilist. Eng. Mech.*, 15(4): 327–338.
2. Graham, L.L., K. Gurley and F. Masters. 2003. "Non-Gaussian simulation of local material properties based on a moving-window technique", *Probabilist. Eng. Mech.*, 18(3): 223–234.
3. Chung, D.B., M.A. Gutierrez and R. de Borst. 2005. "Object-oriented stochastic finite element analysis of fibre metal laminates", *Comput. Method. Appl. M.*, 194(12–16): 1427–1446.
4. Onkar, A.K., C.S. Upadhyay and D. Yadav. 2007. "Probabilistic failure of laminated composite plates using the stochastic finite element method", *Compos. Struct.*, 77(1): 79–91.
5. Endruweit, A., and A.C. Long. 2006. "Influence of stochastic variations in the fibre spacing on the permeability of bi-directional textile fabrics", *Compos. Part A-Appl. S.*, 37(5): 679–694.
6. Murthy, P.L.N., and C.C. Chamis. 1986. "ICAN Users and programmers manual", NASA Technical Paper, TP 2515.
7. Ashbee, K.H.G. 1986. "Bi-Dimensional Compression Moulding of Super-High Fibre Volume Fraction Composites", *J. Compos. Mater.*, 20(2): 114–124.
8. Abaqus/Standard Finite Element Package v6.7, Theory Manual. Manufacturer's web site: http://www.simulia.com (Viewed 25 June 2008).
9. Blacketter, D.M., D.E. Walrath, and A.C. Hansen, 1993. "Modeling damage in a plain weave fabric reinforced composite material", *J. Compos. Tech. Res.*, 15(2): 136–142.

Experimental Research on Bending Influence on CFRP Tensile Specimen with Different Stitching Parameter using Optical Measuring Technologies

F. F. KRUSE, D. PLEGE and T. GRIES

ABSTRACT

Within realization of complex three-dimensional textile structures in near-net-shape form, stitching is an appropriate way to join the different subpreforms. The combination of preforming and LCM-technologies leads to a cost economic method to manufacture large complex parts out of FRP. Besides all advantages, a substantial disadvantage of sewing-technologies is the fact, that the stitches lead to undulations of the fibres. Thus, reduced mechanical properties are the result, especially if the load is acting in the laminate-plane. Most remarkable are the eye-like disorientations in the laminate plane, which are the subject of numerous investigations. However, while the impregnation of the dry performs is done with vacuum-assisted resin infusion processes, also undulations normal to the laminate plane occur. Especially on thin-walled laminates, the latter have a significant influence on the fracture behaviour, on which is the focus of the following paper.

INTRODUCTION

During the manufacturing process, a textile preform goes through three important phases. The first Step is the joining of the different subpreforms in a workpiece-holder or in the female mould, if an one-side stitching technologie is used. Except for the case of tufting, the seam architecture consist of two threads, from which one part runs under the lower surface of the workpeace. The following experiments are dealing with the double-lockstitch, due to his very simple architecture. But the principle is also valid for more complex stitching technologies like the ITA-Stitch.

In the second step, the preform is put down into the mould. The lower sewing thread is thereby laying at the wall of the mould, while die dry fibrelayers runs over

---

Frank Felix Kruse, Dorothee Plege, Thomas Gries; Institut fuer Textiltechnik der RWTH Aachen, Eilfschornsteinstraße 18, D-52062 Aachen, Germany. Tel: +49-241-80-95633, eMail: felix.kruse@ita.rwth-aachen.de it. Due to gravity, some minor undulations occur already in this step. In the third step, the vacuum-set is established and the preform is compressed by the vacuum. Thus, the layers are pressed into the free spaces besides the sewing thread. In the fourth and last step, the resin is injected. After the consolidation, the now fully developed excentricity is fixed by the resin. Under tensile loads, this excentricity leads to the mentioned bending effect. The fiber-layers tends into an straight condition.

Specimen Geometry

The geometry of the specimen was selected according to the german industrie standard DIN 2747, Typ 3. Die specimen-length is 250 mm. The width of 25 mm assures an existence of minimum three stitches on the specimen. The laminat was build up from 8 unidirectional layers of weight per unit area of 125 g/m² The layersetup was [0°/0°/90°/0°]s. With an realised fibre-volume fraction of 62%, the thickness of the specimen is about 1 mm. For load introduction purposes, glassfibre-epoxy tabs are applied. The stitching-length varies from 2 – 6 mm. One of the specimen is shown in
Figure. **2**. In the further text, the mould face side of the specimen is designated as the front side.

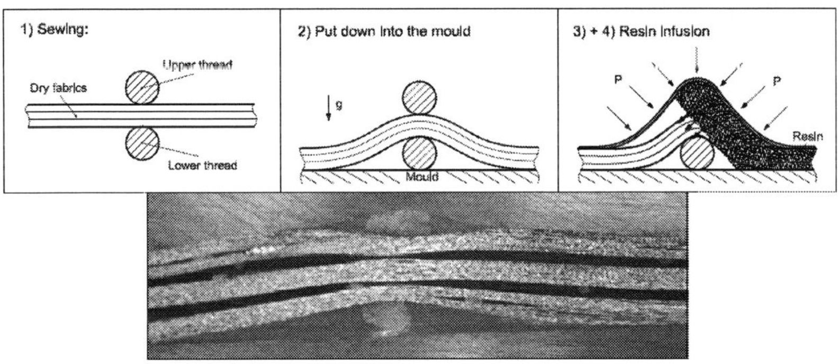

Figure 1. Development of the out-of-plane undulations

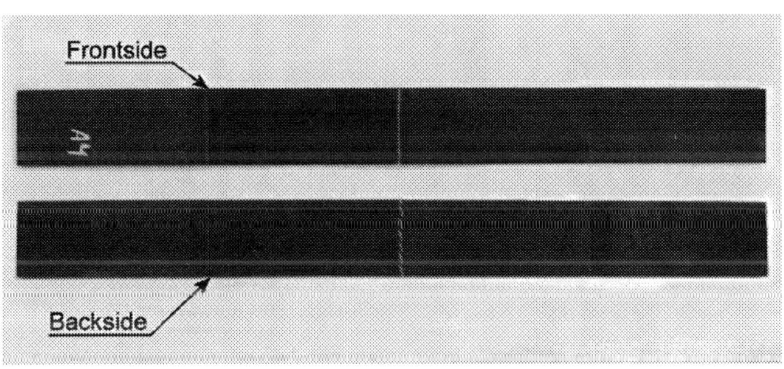

Figure. 2. CFRP-tensile specimen with 3 mm stitching-length and GFRP-tabs.

Experimental-Setup

For the experiments, an electromechanical testing-machine Zwick Z100 with hydraulic clamps was used. The measurement of strains and deformations was performend by two coupled Aramis 4M optical measurement systems. On the frontside of the specimen, a measuring-volume of 25 x 25 x 5 mm³ was applied. To measure the undisturbed strain far away from the seam, on the backside a measuring-volume of 65 x 65 x 50 mm³ was used. The experimental setup is shown in
Figure **3**.

The average drop down of the ultimate strength of the stitched specimen against the unstitched ones is about 26,5 %. A significant influence of the stitching-length on the ultimate strength his not ascertainable, as can be seen in
Figure 4.

Figure 3. Experimental setup with the two optical measurement-systems.

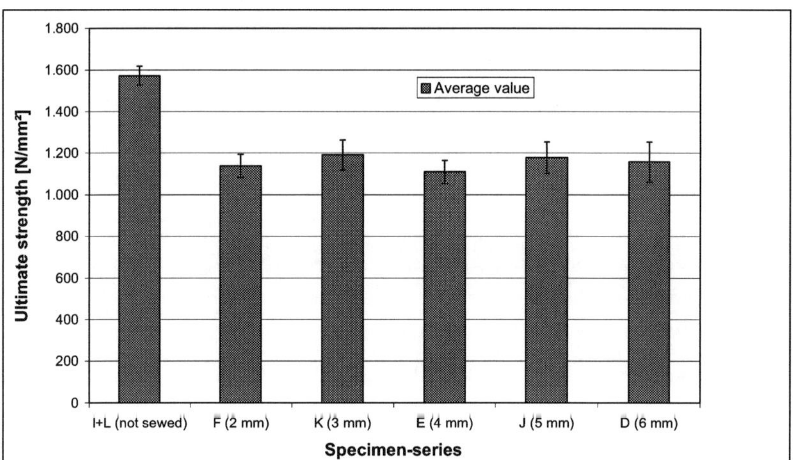

Figure 4. Ultimate Strength of the specimen series.

RESULTS

The coloured plot of the measured deformation on the front side of specimen with 4 mm stitching length is shown in
Figure **5**. In addition, a picture of the unpainted specimen is given above, to indicate the position of the stitches. The maximum values of deformation are about 0.3 mm. At the position of the stitches, the deformation is slightly lower. In
Figure **6**, the deformation in certain points on the frontside is shown over the load. From a load of roundabout 15 kN and beyond, some stepwise deformation increasing occur. At the backside of the specimen, these stepwise increasings do not occur. The deformation is increasing steady up to a value of 0.2 mm. After that the deformation stays nearly constant until the specimen collapse appears, which indicates delaminations of the fibre layers. To explain this, we keep in mind that the excentricity due the out-of plane undulations is acting in the same way like a Mode II bending-test. Thus shear forces between the layers are resulting

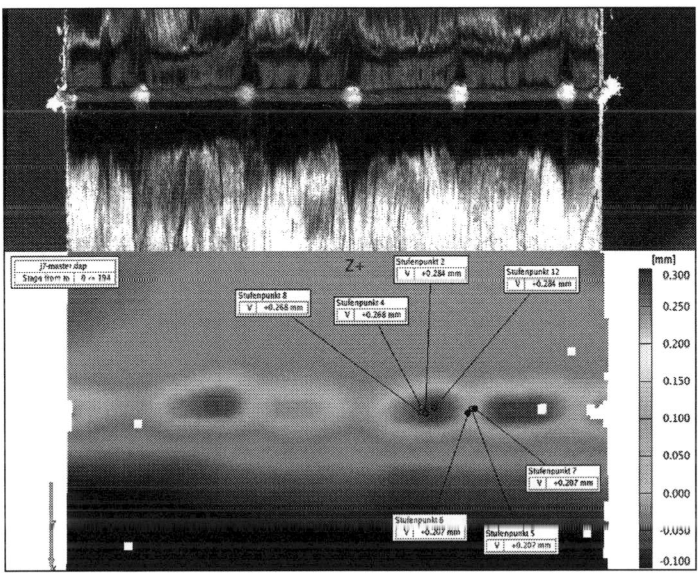

Figure 5. Measured Deformation normal to the laminate plane on the front side of the specimen shortly before the ultimate load.

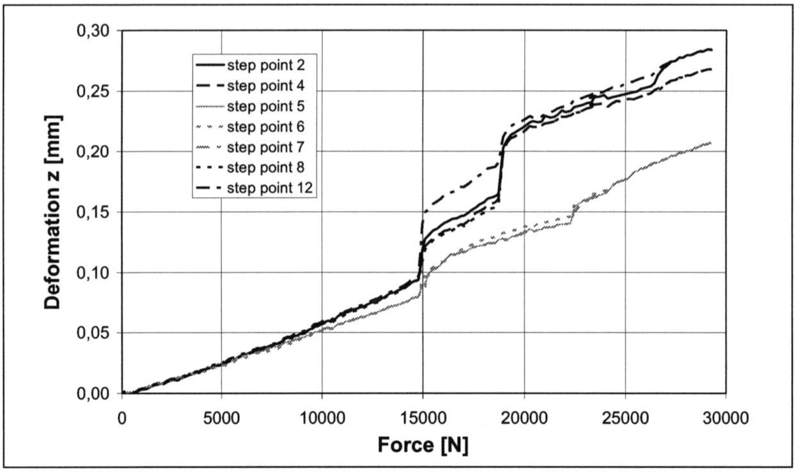

Figure 6. Deformation in z-direction on the front side of the specimen over the load.

The bending effect is clearly visible in Figure 7, in which the longitudinal strain on both sides of the specimen is drawn over the load. On the front side, the strain raises steady up to the now well known 15 kN. The resulting strain-value of roundabout 3.5 % is more than two times higher than the ultimate strain of the used HT-carbonfibres. This is due to the reason, that the optical measurement system is only detecting the strain distribution on the surface of the specimen. As visible in Figure 1 the fibres laying under the upper sewing-thread and the surrounding resin-rich area, which has an much higher ultimate strain. With the further rising of the load, some stepwise increasing occur like in the deformation plot of the front side (Figure 6). On the backside, which was titled as the slave-system in Figure 7, the strain is decreasing up to 15 kN, which is an result of the pressure forces due to the excentricity. Above this load, the strain is raising nearly to the value of the global strain (e.g. the strain far away of the seam area). It can be shown, that this is an effect, which occurs exactly at the position of the lower sewing-thread. A few tens of mm besides the sewing thread, the strain is about 0.5%.

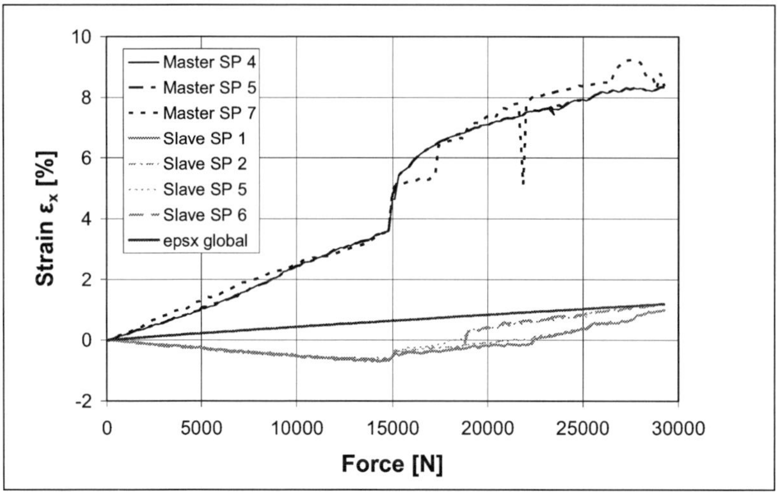

Figure 7. Longitudinal strain in discrete points on both sides of the specimen surface vs. load.

The coloured plot of the transversal strain distribution on the specimen front side shortly before the specimen collapse is shown in figure. The alternating values between transversal compression and tension are clearly visible. The undulated fibres around the stitches tends towards a straight position and therefore leading to an compression of the sewing-thread.

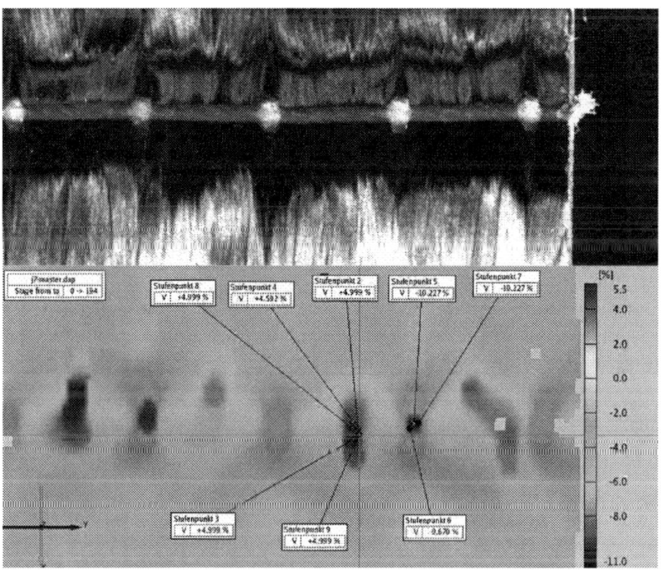

Figure 8. Transversal strain distribution on the specimen front side short before the collapse.

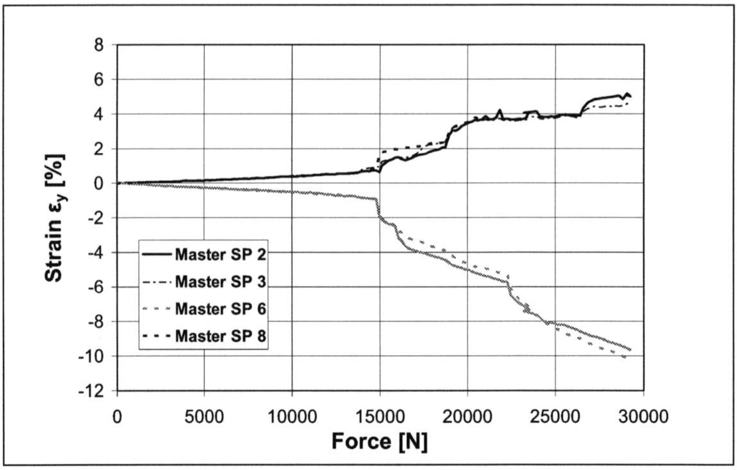

Figure 9. Transversal strain in discrete points between and over the stitches on the specimen front side vs. Load.

As a result of this effect, tension forces transversal to the fibres occur between the stitches. The values of the discrete step-points highlighted in the colour plot are drawn over the load in

Figure 9. The solid and dash-dot black line is indicating the transversal strain between the stitches, the gray ones laying at the position of the stitch. Like in the Diagram of deformation and longitudinal strain, up to 15 kN the values increasing steady. Between the stitches, some minor micro cracks could be identified at .about 13.5 kN. This is due to the unhealthy Mode I-loading case in this area. Beyond of 15 kN, the specimen is degrading successively. In the compression areas over the stitches, the highest strain values reaching nearly -10%. On the back side of the specimen, the transversal strain values are about ten times lower (max +/- 1%) which is in accordance with the mentioned bending effect.

SUMMARY & OUTLOOK

The presented experimental results indicate the existence of a bending effect due to the undulations normal to the laminate plane. Unfortunately the optical measuring system delivers only a view on the events on specimen surface. Some questions concerning the fracture behaviour inside of the specimen are remaining unanswered. Furthermore, the answer to the question of the bending effect's portion of influence on the total ultimate load can't be given. The current scope of work at the ITA is therefore to separate the out-of-plane bending effect and the effect of the in-plane eye-like imperfections with special prepared specimen. Also theoretical investigations are actually performed to determine the influence of parameters like the laminate thickness and the diameter of the sewing-thread on the bending effect.

REFERENCES

1. Heß, H.; Roth, Y.C.; Himmel, N..: 2007. "Elastic constants estimation of stitched NCF CFRP laminates based on a finite element unit-cell model". Composites science and technology 2007", Bd. 67, H.6, S. 1081–1095.
2. V. Koissin , S.V. Lomov , I. Verpoest, H. Nakai , T. Kurashiki , and M. Zako; 2008. "Structurally stitched preforms:experimental characterisation, geometrical modelling, and FE-analysis", Procedings of the ECCM 13, 2-5 June 2008, Stockholm, Sweden.

Unfolding of L-shape Specimens. Effect of the Out-of-Plane Fiber Proportion on the Strength

P. CASARI and T.-W. CHOU

ABSTRACT

This article reports an experimental study on the strength of L-shaped specimen under unfolding load. The material is described, and the focus is put on the folding process of the preform in order to ensure a good shape of the inplane tows of the material. This hands-on step is crucial for ensuring a satisfactory architecture of the staking sequence of the laminate. The results show a significant influence of the interlock yarns ratio on the inplane strength of the laminate. However some scatter is noticed, and could be due to the non-homogeneity of the material.

INTRODUCTION

Composite materials are widely used in high performance structures because they offer a high strength-to-weight ratio. Now their use turns to the replacement of metal part used as junctions like profiles in the aeronautical industry or local reinforcement patches for prototype structures. A specific family of preforms has been designed in order to fulfill the need of structural parts loaded with out-of-plane stresses [1]. Then characterization has provided some quantitative experimental data to assess the strength of such materials in damage [2], [3] or in delamination [4] for example. Modeling is also a useful tool for stress analysis [5], but the 3D composites topic remains very technological.

The goal of the present study is to assess the strength of a 3D interlock composite material by means of varying the fibers ratio used for the reinforcement across the width. At first, the material is described and the specific handling process of the preform is presented. Mechanical characterization is done by using a simple testing device on which L-shaped specimens are loaded in tension until the unfolding strength.

Pascal Casari, Université de Nantes, Nantes Atlantique Universités, Institut de Recherche en Génie Civil et Mécanique, CNRS, UMR 6183, IUT de Saint-Nazaire, 58 rue Michel Ange BP 420 44606 Saint Nazaire Cedex FRANCE. Pascal.Casari@univ-nantes.fr
Tsu-Wei Chou, Department of Mechanical Engineering, University of Delaware, Newark, DE 19716, USA. chou@UDel.Edu

DESCRIPTION OF THE MATERIAL

The main manufacturing steps are presented and explain how the dry preform is handled to ensure good mechanical properties.

Architecture of the Composite Material

The composite dry preform is called "orthogonal interlock fabric" and manufactured under the shape of a plate made of a carbon T300 tows in the warp and weft directions bound together by interlock yarns. The material can be described as a [0; 90] laminate in which some through-the-thickness yarns are interlaced and provide the main out-of-plane reinforcement (see Figure 1). Table I recalls the geometrical characteristics of the composite part.

TABLE I. GEOMETRICAL CHARACTERISTICS OF THE COMPOSITE.

	Warp direction	Weft direction	Through-the-thickness
# of plies	13	14	
Fiber volume fraction	22%	22%	1%
Thickness	Total 5.2 mm, Ply: 0.19 mm		

Folding of the Dry Preform

L-shape specimens are folded along the weft direction with an internal radius of 10mm. If the preform was just folded, the plies would be creased and their strength would be significantly affected. To avoid this, a simple device is designed and used (see Figure 2) in order to stretch the preform along the warp direction during folding and to enforce the plies to slip over each other. The ends of the preform are clamped and the dry fabric is hand stretched. The fold radius is shaped by means of two cylindrical fixed bars, ensuring that the plies remain flat and with the same length.
The shape is provided with a few efforts to overcome friction, but caution is requested in order not to damage fibers in the areas in contact to the steel bars. It is also cost-effective to shape two performs in one step because symmetry ensures to remain the plies at the same length with no effect on the clamping conditions (see the principle in Figure 2).

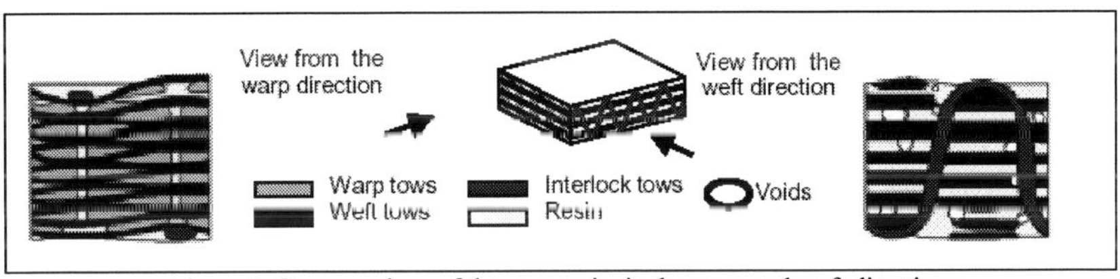

Figure 1. Cross-sections of the composite in the warp and weft directions.

Interlock Yarns Removal

In order to assess the effect of interlock yarns on the strength of the material under unfolding load, some samples are prepared with different interlock yarns ratios. The maximum amount of interlock carbon yarns is 1% of the whole dry preform and corresponds to 8 rows across the width of the specimen chosen for mechanical testing. Some yarns are then removed and provide four sets of specimen with 0.25, 0.5, 0.75 and 1% of interlock yarns reinforcement (see Figure 2 the aspect of the specimen surface).

Removing the whole interlock yarns could not be achieved because once the tows are no longer held, they move significantly and the architecture of the preform is deteriorated.

Infusion Process

Finally, the samples are manufactured by Liquid Resin Infusion at high temperature with an epoxy resin Hexcel RTM6 on a male mould. This step does not present any particularity, except that the permeability of the preform is quite high, and infusion may leave some voids in resin-rich areas of the microstructure.

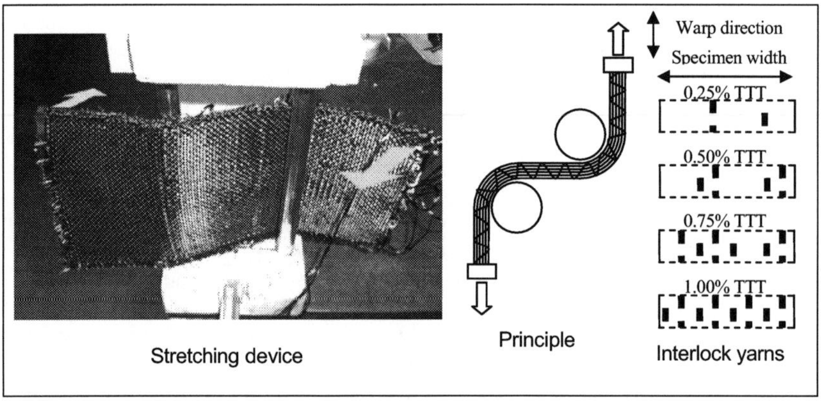

Figure 2. Forming of the dry preform and interlock yarns architecture.

MECHANICAL TESTING

In composite structures, a folded zone would mostly be loaded along its whole length, and the lowest strength would be under unfolding load. The idea is then to design a tensile test in which the specimen is loaded by a bending moment that tends to open the fold. The specimen is adapted to the test by choosing a satisfactory width according to the period of through-the-thickness reinforcement.

Testing Device

The test consists in loading an L-shaped specimen under tension on a universal testing machine (see Figure 3). The specimen is clamped to steel parts, and fastened to the machine with eye-shaped spherical bearing heads. Load and displacement are measured and recorded. The test may include large displacements, but this has not been taken into account as the result is a relative comparison among very similar specimen in terms of stiffness.

Specimens for Tensile Tests

One problem in mechanical testing is to reproduce a relevant characterization with small test coupons. In this study, the effect of interlock yarns on the strength of the material can be assessed by varying the ratio of yarns woven across the width in the warp direction. It means that a sufficient number of interlock yarns should remain in order to represent an averaged behavior. However, in order to keep a simple experimental study, only a few yarns remain in the test specimens. This choice affects the assumption of homogeneity of the material, and some artifacts may occur.

The samples are 5.2mm thick, and 22mm wide, which corresponds to 4 periods or 8 rows of interlock yarns in the weft direction. For each yarn ratio, a batch of four specimens has been tested.

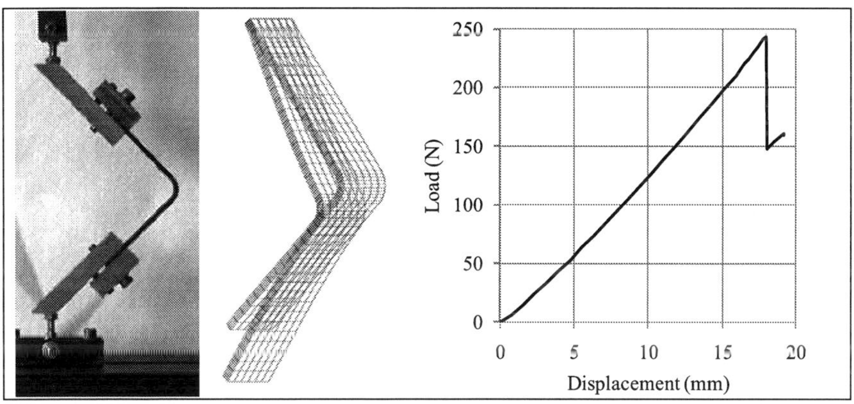

Figure 3. Testing device - Sample deformed under tension—Typical load-displacement response. Tests results are expressed as the strength of the laminate (considered as a homogeneous orthotropic material) in the central cross section of the specimen by means of the beam theory.

Results

Load-displacement curves are analyzed in terms of stiffness and maximal load. Stiffness is varying slightly among the whole set of specimens, but no correlation arises from each batch to the other (see Figure 4). The most relevant result is the strength of the L-shape specimen versus the of out-of-plane yarns ratio. Table II summarizes the tests results.

TABLE II. STRENGTH OF THE LAMINATE VS INTERLOCK YARN RATIO.

% interlock yarns	Mean	Std
0.25	174	38
0.50	174	7
0.75	262	30
1.00	306	18

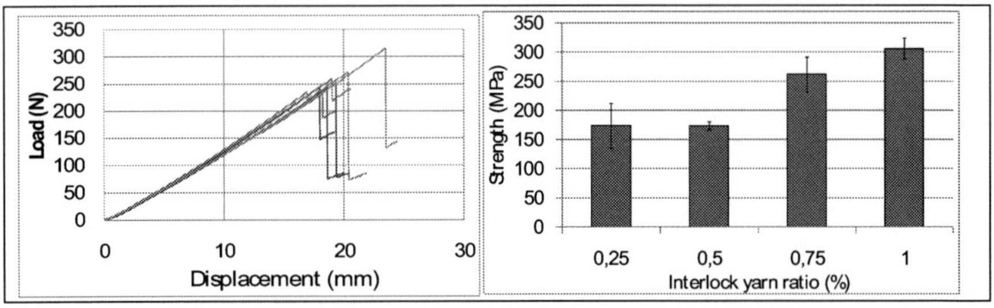

Figure 4. Typical load-displacement curves – Strength of the laminate vs interlock yarn ratio.

Firstly, scatter is significant, and recalls the remarks already expressed that the material is not homogeneous because only a few interlock yarn may be involved during specimen failure. Secondly, the inplane strength of the laminate depends on the interlock yarn ratio with a maximum for 1% of through-the-thickness reinforcement (see Figure 4). This gives a good starting point from which an optimal ratio should be sought.

CONCLUSION

This article reports an experimental study on the strength of L-shaped specimen under unfolding load. The material has been described, and the focus has been put on the folding process. This hands-on step is crucial for ensuring a satisfactory architecture of the staking sequence of the laminate. The results show a significant influence of the interlock yarns ratio on the inplane strength of the laminate. However some scatter is noticed, and could be due to the non-homogeneity of the material. The material choice with an initial yarn ratio of 1% did not allow to find a quantity of through-the-thickness reinforcement above which strength does not increase, but the direction of research is well shown, and this ratio may be increased.

Future work will focus on the modeling of the influence of interlock yarns on the homogeneity of the stress field in a composite structure made of such 3D preform. Moreover, some information about internal friction in the preform should be capitalized in order to propose a quantified mechanical approach of the preform deformation to ensure a satisfactory architecture of the composite structure.

REFERENCES

1. Chou T.-W., Microstructural design of fiber composites, Cambridge University Press, (1992).
2. P. Casari, P. Ladevèze and T. W. Chou. Damage Modeling and Characterization of a Three-Dimensional Woven Composite," Proceedings of the 12th International Conference on Composite Materials, ICCM 12, pap 717 (1999).
3. Pluvinage P., Parvizi-Majidi A., Chou T. W. "Damage characterization of two dimensional woven and three dimensional braided SiC-SiC composites" Journal of Materials Science 31 232-241. (1996).
4. Guénon V. A., Gillespie J. W., and Chou T. W., "Toughness properties of a three dimensional carbon-epoxy composite", Journal of Materials Science 24 (1989), pp. 4168–75.
5. Shenoi R. A., Wang W., "Through-thickness stresses in curved composite laminates and sandwich beams". Composites Science and Technology 61 (2001) 1501–1512.

TOPIC VIII

3D TEXTILE COMPOSITES: MANUFACTURING

Developments in Design, Manufacture and Use of 3D Woven Fabrics

X. CHEN and J. W. S. HEARLE

ABSTRACT

Most woven fabrics can be treated as planar 2D sheets. By 3D woven fabrics, which began to be important for use in composites in the 1980's, we mean those having appreciable thickness, which may be solid or hollow, or shaped 3D forms. The early developments utilised special machinery, but developments in the University of Manchester over the last 20 years have shown that fabrics can be made on commercial weaving machines with minimal adaptation. CAD-CAM software enables 3D woven fabric structures to be easily designed and linked to manufacturing instructions for electronic dobby or Jacquard weaving. The dominant market is in aerospace and military applications, but there is potential in the construction industry and other applications. A less well-known use is in reinforced hollow structures to provide light-weight protection against low-velocity impact, which may occur in civil disturbances.

INTRODUCTION

In their internal arrangement of molecules, fibers and yarns, all textiles are 3D structures, but at the coarser level of fabrics a few, such as cords, are 1D forms, a few others, such as hollow braids, are circular cylinders, but most are thin 2D sheets, albeit with a textured 3D surface. By 3D fabrics, we mean thick planar sheets or shaped solid forms with multiple layers of yarns, hollow structures and thin 3D shells. Some have been part of the long history of textiles. An example of a 3D woven fabric is a forming fabric used in paper-making with two or three interwoven layers. Velvets are made as two layer fabrics with linking threads, which are then cut to give the fabric pile. A hand-knitted sock is an old form of a

Xiaogang Chen and John W. S. Hearle, TexEng Software Ltd, Arch 30, University of Manchester, North Campus Incubator, Sackville Street, Manchester, M1 3NJ, UK.

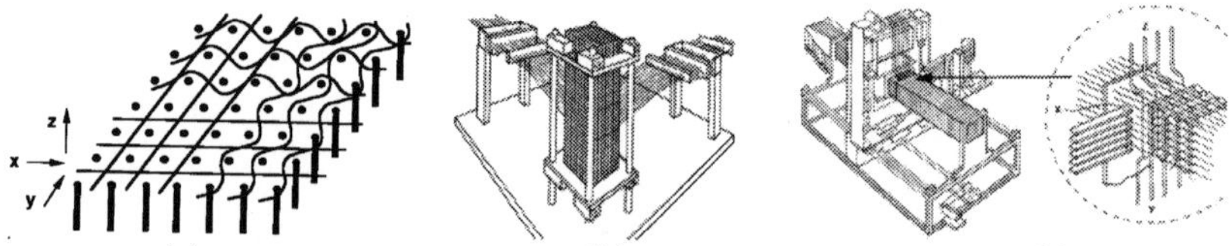

(a) (b) (c)

Figure 1. (a) Interlacing in X, Y and Z directions. From Hearle and Du [2]. (b) Replacement method.(c) Direct method. (b) and (c) From Chou and Ko [1].

3D shell structure. However these traditional uses, which are limited in number and production, do not exploit the full potential for 3D fabrics.

In the 1980s, there was a flurry of innovation resulting from the possibility of using woven, knitted and braided 3D fabrics to eliminate some of the limitations of the existing methods for making composites. 3D fabrics gave the opportunity to improve performance, particularly in through-thickness properties, and to improve performance, particularly in through-thickness properties, and to improve economics by reducing labour in both textile manufacturing and composite production. These studies were described in *Textile Structural Composites*, edited by Tsu-Wei Chou and Frank Ko [1] and in a paper on *Forming rigid fibre assemblies: the interaction of textile technology and composites engineering* Hearle and Du [2]. The economic benefit was reduced for many of the earlier methods by the need to develop special machinery, which were often labour-intensive to set up. Only a few of the early machines led to commercial production. The emphasis in this paper is on the use of commercial weaving machines, which are made for the conventional textile market and require minimal adaptation for the manufacture of 3D fabrics, particularly when advances in CAD-CAM can be exploited. The market for 3D fabrics in aerospace and military applications has stimulated a search for other uses for 3D fabrics.

In conventional weaving, the yarns lie in the X (weft) and Y (warp) directions in the plane of the fabric. The essential feature of solid 3D weaves is the presence of yarns in the Z direction, as illustrated in Figure 1. Two methods were described by Chou and Ko, 1989. The replacement method, which is unlikely to be commercially viable, is illustrated in Figure 1(b). In Fukuta's apparatus [3], it starts with an assembly of tubes in the Y-direction, which are later replaced by yarns. The direct method [4], shown in Figure 1(c), follows weaving technology and was also made in a tubular version [5]. Because of their difference from conventional woven fabrics, Ko [1] preferred to call these orthogonal nonwovens (ON 3-D). Khokar [6] adopts a more limited definition of weaving, with *noobing* as an alternative term for certain structures, but we regard all 3D structures produced on commercial weaving machines as 3D woven fabrics.

The early methods, which were pioneered by engineers in aerospace companies, began with an imagined structure and then looked for a way of assembling it. It is more productive to start with a knowledge of weaving and look for ways of adapting it. At North Carolina State University, Mansour Mohamed pioneered this approach by building a special weaving machine. Research at Manchester University has concentrated on the use of commercially available machinery, which

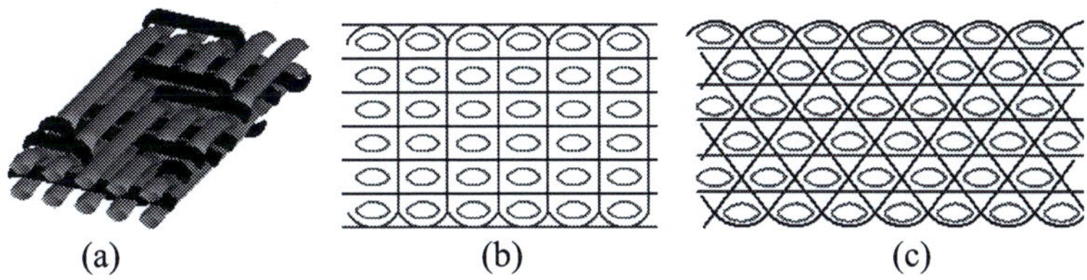

Figure 2. Cross-sectional views of 3D solid architectures: (a) multilayer; (b) orthogonal; (c) angle-interlock.

is made for the conventional textile market and needs minimal adaptation for 3D weaving.

In manufacturing, the revolutionary advance was the development of electronic Jacquards, which give on-line control of the raising and lowering of warp threads instead of mechanical control from punched cards. Weft selection can also be computer controlled. Another significant advance is in the use of textile-based, computer-aided design, which is the main function of TexEng Software Ltd. It leads to optimum forms of fabric and potentially to prediction of properties and can be linked to computer control of Jacquard or dobby weaving machines.

CLASSIFICATION OF 3D WOVEN ARCHITECTURES

3D woven architectures may be classified according to their structural features as 3D solid, 3D hollow, 3D domes and 3D nodal [7]. 3D solids refer to those woven architectures that have thick, solid cross-sections either in a broad panel or in a net shaped preform. A 3D solid architecture can be made based on the multi-layer principle, where there are neighboring layers of woven fabrics being interlaced in the Z-direction during the weaving process, Figure 2(a). The features of this type of 3D architecture are that (i) each fabric layer can be given a different weave to facilitate "hybrid" properties through the thickness of the composite; (ii) warp and weft yarns in the 3D solid textile architectures can be crimped to specified extents to suit the property requirements of the composite; (iii) required amount of vertical interlacing can be arranged between any layers of fabrics in the preform to enhance the through-the-thickness property; and (iv) the architecture can be further strengthened by adding straight wadding yarns between any adjacent fabric layers in either warp or weft or both directions. The second method for making 3D solid woven architecture is the orthogonal principle, Figure 2(b): (i) all yarns in the three principal directions are laid straight, therefore they are able to take on the load directly and most effectively; (ii) required amount of vertical yarns can be arranged by the binding weave specification; and (iii) the interlinking depth can be altered easily within the same preform, leading to variable preform thickness. The third way for making 3D architectures is the use of the angle-interlock principle, Figure 2(c), where the weft yarns are laid straight whereas the warp yarns are travelling diagonally. This structure generally leads to more flexible 3D structures. because of the reduced number of cross-over contact points. Wadding warp yarns can be

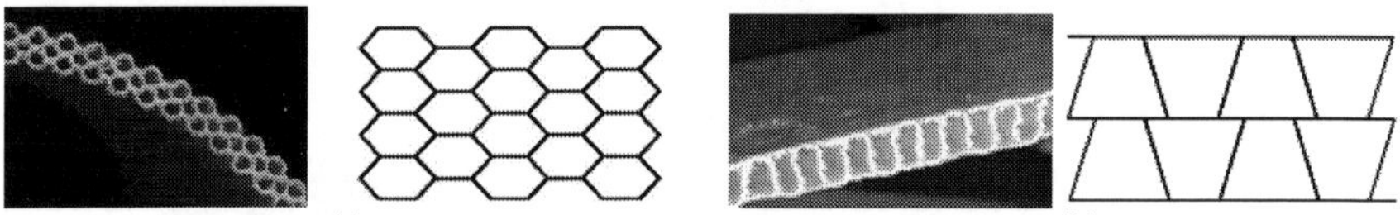

(a) (b)

Figure 3. 3D hollow structures with (a) uneven and (b) flat surfaces.

(a) (b) (c)

Figure 4. (a) Domed structure. (b) Nodal 3D architectures. (c) Woven double velour vascular graft made by Boston Scientific Company Inc. in 2004 and shown at Extreme Textiles Exhibition in New York, 2005 . From McQuaid [8].

introduced into the angle interlock fabric during weaving to increase the structural rigidity.

3D hollow architectures, also called cellular structures, have tunnels running in warp, weft, or any diagonal directions in the thickness of the 3D architecture. There are two different types of 3D hollow architectures: either with uneven surfaces, Figure 3(a), or with flat surfaces, Figure 3(b). The creation of the 3D hollow architectures is based on the multi-layer weaving principle to join and separate fabric layers in places as needed. For the flat-surface hollow fabrics, the lengths of some inner fabric layers are made longer than the surface layers. In the case of the uneven-surface hollow fabrics, the fabric layers in the same region in the cross-section are made to have the same length. The cell size determines the maximum opening thickness of the fabric.

Other complicated 3D shapes can be achieved by weave combination, discrete take-up, or molding with fabrics which have low shear rigidity. Figure 4(a) shows a domed helmet made in one piece. Nodal architecture, Figure 4(b), refers to woven tubes which are joined together. Smooth opening-up of all joining tubes must be achieved. Both of these are composites with fabrics made on commercial looms. A more complex nodal structure, not made in Manchester, is shown in Figure 4(c).

CAD/CAM FOR 3D FABRICS

Single-layer woven fabric is made from one set of warp yarns and one set of weft yarns. By arranging interlacement between warp and weft yarns, different weave patterns can be obtained and represented by a 2D binary matrix. In order for the computer to create weaves, they need to be parameterised. The parameters should be the ones that are commonly used by the weave designer and engineers.

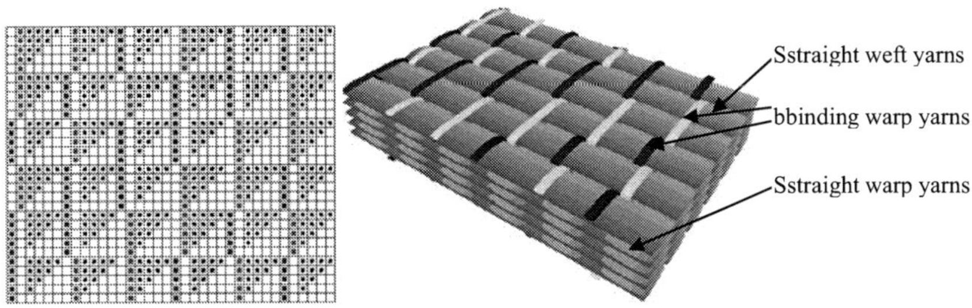

(a) (b)

Figure 5. Orthogonal weave created using Weave Engineer®.

For creating regular weaves, the input parameters are: N_f, which is the number of floats; F_i (i = 1, 2, 3, …, N_f), which gives the lengths of the ith float (if i is an odd number, F_i represents a warp-up float and, if i is an even number, it represents a warp-down float); and S, the step number of the weave [9]. A full account of the way in which these mathematical principles can be adapted to cover 3D weaves is given by Chen [10].

Based on the mathematical modeling and description of different types of 3D woven architectures, a suite of CAD software has been created to support the design

(a) weave design matrix	(b) 3D fabric
left column of each cell	black binder warp running across top of fabric
second column of each cell	white binder warp running across top of fabric
four columns on right of each cell	straight warp yarns, seen pale grey on front edge, running back across fabric

and manufacture of 3D woven fabrics. Weave Engineer® is designed to support the design and manufacture of 3D solid fabrics. Its modules include multi-layer, orthogonal, angle-interlock, backed, as well as single layer fabrics. The design of 3D fabrics is speedy and accurate. The file can be output to computer control of the weaving machine. As an example, Figure 5 shows the design for an orthogonal 3D solid fabric.

Hollow CAD® is used for the design of 3D hollow woven architectures with uneven or flat surfaces when opened up. Figure 6 illustrates a design using the software tool.

APPLICATIONS OF 3D WOVEN FABRICS

Solid Composites

The development of 3D woven fabrics is being driven by the aerospace industry. The main advantage is the ability to incorporate straight, multilayer reinforcing yarns without crimp in a consolidated solid composite. In the structures shown in Figures 2(b) and 5, the only bends are where the straight through-

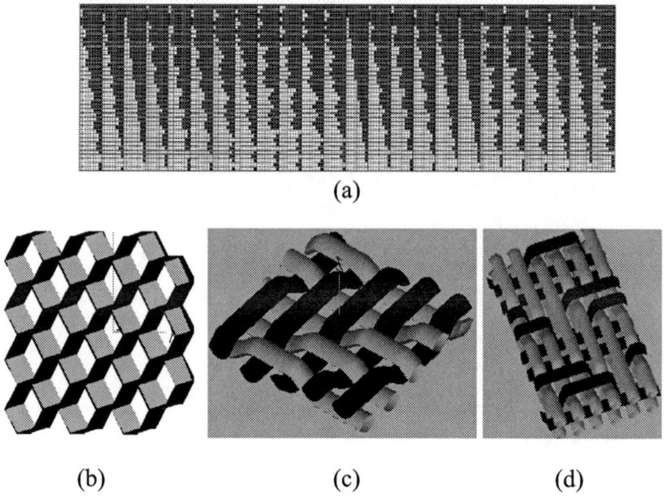

Figure 6. Design of 3D hollow architecture using Hollow CAD®. (a) Weave design matrix. (b) The cellular structure (c) The weave used for the single layer cell walls. (d) The weave used for double layers where three walls join.

thickness yarns cross over at the top and bottom of the fabric. In impregnated composites, it would not matter if the yarns broke at these points. Prichard [11] suggests that the use of 3D woven fabrics could lead to a 30% reduction in aircraft weight, with major economic and environmental benefits. Gillespie [12] has described the use of 3D fabrics in armored vehicles. Since cost is not a major controlling factor, these applications are based on high-performance carbon fibers.

There is another potential market using cheaper fibers. For example, Kuckhoff [13] describes the use of 3D knitted fabrics for reinforcing concrete walls. In general, the construction industry has not been progressive in moving from traditional materials. However, as pointed out by Hearle et al [14], it would be possible to develop a continuous weaving and consolidation line to manufacture items such as **I**-beams reinforced with glass or polyester yarns.

Cellular composites, as shown in Figures 3 and 6, could be used in aerospace and similar applications as spacers for components that need to be stiff in bending and light in weight.

A more unusual application for hollow fabrics that have been made on commercial machinery in the University of Manchester is in low speed impact protection. The motorcycle helmet shown in Figure 4(a) is one example. Another is leg protection that can be worn by riot police to guard against attack with wooden stakes or iron bars. Figure 7(a) shows the model used for finite element analysis of an impact. The forces developed, shown in figure 7(b), are much less with the cellular composite than with foam. Laboratory impact tests confirm that the energy absorption is about 20% greater with the composite than with foam. The peak force is four times less, because the impact is spread over a longer time. Figure 7(c) is an example of the geometric modeling of impact on a cellular composite.

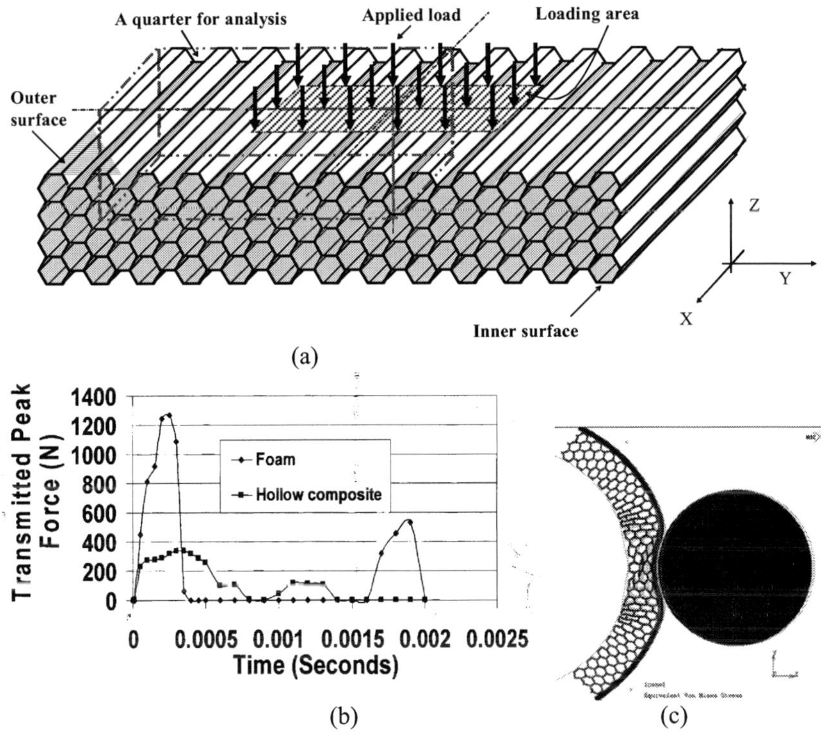

Figure 7. (a) Model for prediction of impact mechanics on a honeycomb composite. (b) FEA prediction of tension in an impact test. (c) Simulation of impact on a cellular composite.

Without any protection, a 0.4 kg wooden ball thrown at 15 m/s breaks a shin bone. In all simulations with honeycomb protection the transmitted peak forces are well below the fracture tolerance. Multi-impact simulations show the increased deformation with successive impacts. The peak force also increased. On the third impact it was twice that in the first impact, but was still below the safe limit.

CONCLUSION

The development of electronic control of the mechanical operation of weaving machines under computer input has made it possible to make a wide variety of 3D woven fabrics on commercially available machinery. Solid structures, which have a high fiber volume fraction, are made in the shape of the composite piece. Cellular composites can be flat or bent to a required shape depending on the way they are made.

The development of 3D woven fabric technology has relied heavily on the skills of experienced weaving technologists, which have developed over thousands of years and particularly since the industrial revolution two centuries ago. In contrast to the acceptance of CAD for fashion design over the last 30 years, the technical textiles industry has been slow to adopt advances in information technology for fabric design. Small companies from the traditional industry, with limited resources and great practical expertise, prefer to rely on intuition, experience and trial-and-error. This is bound to change as engineering users demand quantitative performance data and the cost of sample trials escalates. The software being developed in the University of Manchester in cooperation with TexEng Software Ltd is directed to meeting the need for CAD and predictive modelling.

3D solid fabric preforms with carbon fibers giving the highest performance have a major potential for structural composites in the aerospace industry and in some military applications. However, the use of cheaper fibers, such as glass or high-tenacity polyester, should be explored for other markets, including automotive and construction industries.

Cellular composites offer light weight and add high bending stiffness to components as spacers and protection against low speed impacts.

REFERENCES

1. Chou, T-W., and F. K. Ko. 1989. *Textile structural composites*. Amsterdam: Elsevier.
2. Hearle J. W. S., and G. W. Du. 1990. "Forming rigid fiber assemblies: the interaction of textile technology and control engineering," *J Textile Inst*, 81, 360–383.
3. Fukuta, K., R. Onooka, E. Aoki, and V. Nagatsuka. 1982. "Application of latticed structural composite materials with three dimensional fabrics to artificial bones," *Bull. Res. Inst. Polymers Textiles,* 131, 159.
4. Stover, F. K., W. C. Mark, I. Marfowitz, and W.Mueller. 1971. "Preparation of an omniwave-reinforced carbon-carbon cylinder as a candidate for evaluation in the advanced heat shield screening program," AFML, TR-20283.
5. Bruno, P. S., D. O. Keith, and A. A. Vicaro Jr. 1986. "Automatically woven three dimensional composite structures," *SAMPE Quarterly*, 17, 10–16.
6. Khokar, N. 2008. "Second Generation Woven Profiled 3D Fabrics from 3D-Weaving," First World Conference on 3D Fabrics and their Applications, Manchester, April 10–11, 2008. To be published in *Textile Res. J.*
7. Chen X. 2006. "3D Weaving and 3D Woven Fabrics", MSTI Textile Trends – Innovations and Market Insights, CD-ROM, MSTI, www.msti-aktuell.de, Berlin, Germany, February 2006.
8. McQuaid, M. 2005. *Extreme Textiles*, Princeton Architectural Press, p. 61.
9. Chen, X. 2008. "CAD/CAM of 3D Woven Fabrics for Conventional Looms", First World Conference on 3D Fabrics and their Applications, Manchester, April 10–11, 2008. To be published in *Textile Res. J.*
10. Chen, X., R. T. Knox, D. F. McKenna, and R. R. Mather. 1996. "Automatic Generation of Weaves for the CAM of 2D and 3D Woven Textile Structures," J. Text. Inst., 87, Part 1, 356–370.
11. Prichard, A. K. 2008. Presentation at First World Conference on 3D Fabrics and their Applications, Manchester, April 10–11, 2008.
12. Gillespie, J. 2008. .Presentation at First World Conference on 3D Fabrics and their Applications, Manchester, April 10–11, 2008.
13. Kuckhoff, B. 2008. "Development and industrial manufacturing of innovative reinforcement for textile reinforced concrete," Presentation at First World Conference on 3D Fabrics and their Applications, Manchester, April 10–11, 2008.
14. Hearle, J. W. S., J. T. Smith, and R. J. Day. 1995. "A technical/economic evaluation of textile composites", *Techtextil-Symposium*, Frankfurt, Lecture No. 104.

Manufacture and Characterization of Extra-light 3D Hollow Textile Composites

X. CHEN and X. GONG

ABSTRACT

This paper reports on the progress in making extra-light composite materials reinforced by 3D hollow textile architectures. The 3D hollow textiles are designed using the Weave Engineer® software tool, which allows hollow tunnels to be formed in both warp and/or weft directions. This has led to the creation of composites which are extra-light compares to the conventional composites. FE and experimental analysis show that the 3D hollow composites offer excellent energy absorption against impact, force attenuation, material strength as well as lightweight and tailorable thickness.

1. INTRODUCTION

Cellular structures are widely available in nature and engineered cellular/hollow materials have been created for many uses in various engineering fields such as automobile, aerospace, civil engineering, and so on because of the provision of special properties such as being extra lightweight, energy absorbent, and voluminous as well as being strong. Man-made polymer, metal, paper, plastic and ceramic hollow structure are now commercialised as standard products, as review by Corden and Bitzer[1], Gibson and Ashby[2], and Sato[3]. This paper aims to introduce a technical process for designing and manufacturing 3D hollow composites reinforced by cellular textiles, and to characterise the behaviours and performance of the hollow textile composites. Much work has been carried out to investigate the hollow structures and composite materials for energy absorption and crash worthiness. Evens *et al* [4] systematically overviewed the multi-functions of metallic hollow structures, examined their mechanical properties and thermal properties, and provided design analyses for prototypical

Xiaogang Chen, Xiaozhou Gong
Schools of Materials, The University of Manchester, Manchester, UK M60 1QD

systems which specify implementation opportunities relative to competing concepts. Through finite element simulations, Gibson and Andrews [5] modelled the steady state of the regular hexagonal hollow, and found that the creep rates of the structure were lower than other hollow solids.

Tan *et al* [6, 7] investigated energy absorption and deformation in hollow textile composites in relation to the structural parameters. They also pointed out that in general the density of the hollow textile composites is less that 10% of that of the same material in solid form. Theoretical analysis on the 3D hollow textile composites were also carried out by Yu and Chen [8] to simulate of trauma impact on textile reinforced hollow composites for personal protection and it was suggested that the hollow composites have advantages over other types of hollow materials in energy absorption and force attenuation. To further analysis the hollow structure composites' material property, Reid and Peng [9] investigated the dynamic crushing behavior of several wood species, including balsa wood, through the impact of free flying specimens on a rigid wall. While their tests covered a wide range of impact velocities up to approximately 300 m/s, they tested the specimens at a certain density for each species and stress strain curves were obtained. Several other uniaxial quasi-static experimental investigations have been carried out to determine the mechanical properties and deformation mechanisms of balsa wood in longitudinal and/or transverse directions by Knoell (1966), Soden and McLeish (1976), Easterling *et al* (1982) and Vural and Ravichandran (2003). Ruan *et al* [10] worked on a typical CYMAT aluminium foam, a manufactured irregular hollow material, for its strain-stress relationship and the localised cell deformation. Theoretical and experimental studies by Papka and Kyriakides [11] were carried out on an aluminum hollow with regular cell shapes crushed in the X_2 direction, leading to a good agreement between the theoretical and experimental studies.

2. DESIGN AND MANUFACTURE HOLLOW TEXTILE COMPOSITES

2.1 Design of 3D Hollow Weave

A general hollow structure is composed of an array of hexagonal which is shown in Figure 1. It is constructed by 6 free and bonded walls. The parameters that are used to describe a cell are the open angle, thickness walls, and lengths of walls. The height of the cell is $h = 2\, l_f \sin\theta$.

A woven hollow structure can be defined by specifying the structural parameters. The following general coding format is used to denote a particular hollow structure: $xL(y+z)P\theta$. In this, x is the number of fabric layers used to form the hollow structure; y is the length of bonded wall measured in the number of picks; z is the length of free wall measured in the number of picks; θ is the open angle of the hexagonal cells; L is used to denote "layer"; and P is used to denote "pick". There are situations when the lengths of free and bonded walls are the same, i.e. $y=z$. In such case, the coding format can be reduced to $xLyP\theta$; Further, when the open angle of the cell is 60°, the coding format becomes $xLyP$ an in all the above code expressions, x, y, z are integers and $x \geq 2$, $y \geq 1$, $z \geq 1$.

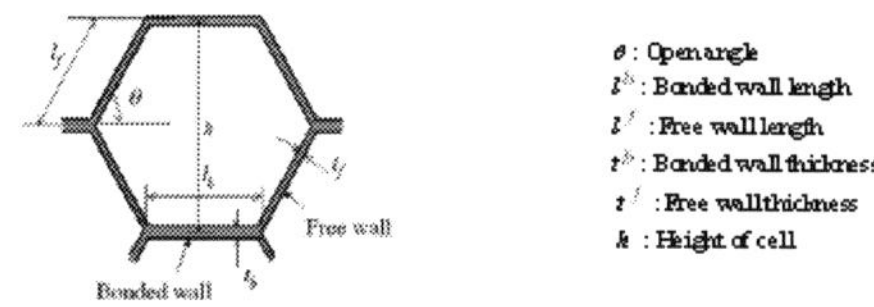

Figure 1. Parameters of a hollow cell.

2.2 Weave Creation

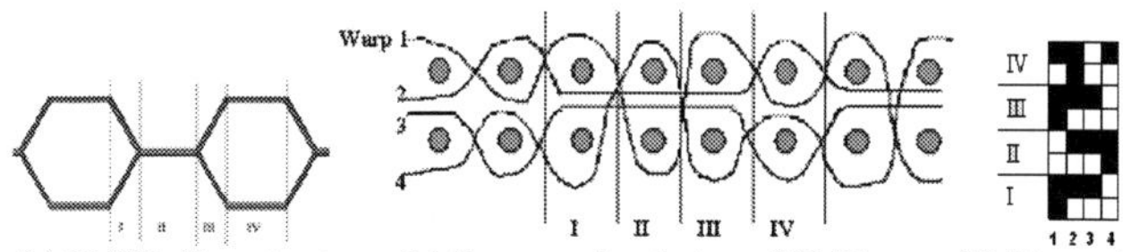

(a) 2L1P hollow structure (b) Cross-sectional view of 2L1P (c) Weave diagram

Figure 2. Hollow structure 2L1P.

A procedure for creating weaves for the hollow structure has been created and it has been implemented into software Hollow CAD© [12].

This is the simplest hollow structure where there are 2 fabric layers with 1-pick being the length for both free and bonded walls. Figure 2 shows the hollow structure when opened, the illustration of the interlacement between warp and weft yarns for this hollow structure and the complete weave.

2.3 Manufacture Of 3D Hollow Composites

In order to investigate systematically the 3D hollow composites, 10 hollow fabrics are designed and manufactured, which are 4L3P, 6L3P, 8L3P, 8L4P, 8L5P, 8L6P, 8L(4+6)P, 8L(4+3)P, 8L(3+6)P, and 8L(6+3)P. Before making the fabrics, the weft density of the overall fabric must be worked out based on the number of picks specified for each of the cell walls in the fabric specification and the actual length expected. Suppose that z is the number of picks specified for a wall of single layer fabric session (picks), l is the required length of the fabric session (cm), d_i is the weft density for this single layer (picks/cm), then

$$d_i = \frac{z}{l} \tag{1}$$

If the hollow fabric is composed of m fabric layers, then the weft density of the hollow fabric, d, will have to be set to

$$d = \sum_{i=1}^{m} d_i \tag{2}$$

In the case that all layers have the same weft density, the weft density of the hollow fabric becomes

$$d = md_i \tag{3}$$

2.4 3D Hollow Composites

To open up a hollow fabric, two sets of metal wires coated with resin were inserted to the top and bottom rows of tunnels respectively, which are pushed

(a) Opening device (b) Hollow textile composites
Figure 4. Illustrations of opening device and hollow composites.

open to the required extent by an opening device shown in Figure 4. The fabric consolidation solution was made as a mixture of resin and hardener. After the resin has been applied on the fabric, the samples are placed in the fume cupboard for quicker hardening. Figure 4(b) shows photographs of hollow textile composites.

2.5 Sample Groups

The 3D hollow composites are organised into 3 groups, according to (i) opening angles (8L6P with 30°, 45°, 60°, 75°, and 90°), (ii) cell-wall length (8L 3, 4, 5, and 6 picks respectively with opening angle being 60°), and (iii) the length ratio free wall to bonded wall (60°).

Group	Composites
Group 1: Opening angle	8L6P30, 8L6P45, 8L6P60, 8L6P75, 8L6P90
Group 2: Cell size	8L3P60, 8L4P60, 8L5P60, 8L6P60
Group 3: Length ratio	8L(3+6)P60, 8L(4+6)P60, 8L6P60, 8L3P60, 8L(4+3)P60, 8L(6+3)P60

Table 1. List of the hollow composite groups.

3. FE ANALYSIS AND IMPACT EXPERIMENT

3.1 FEA For Hollow Composites

FEA carried out to examine the impact performance of the hollow composites. The material was assumed to be isothermal and elastic-plastic, and boundary conditions were used to enforce symmetry in X-Z plane and Y-Z plane. The width×length dimension for the modelled hollow structure composite is 20×20mm.

For all the cases in this theoretical analysis work, a given impact speed of 5.5m/s (which is about 8J impact energy) were applied to the impactor. The measured Young's modulus for the textile composite material is 1250 MPa, and the Poisson ratio was assumed to be 0.3, and the density of the materials is 9.26×10^{-7} kg/mm^3. The material properties used for the impactor (tup) were that the Young's Modulus is 200GPa and the Poisson ration is 0.3.

Figure 5(b) shows transmitted force change against time. It is clear that the increase in cell wall length causes the peak transmitted force to reduce. It also shows that the hollow composite is an effective material in attenuating impact force when compared to solid materials. For the same structural parameters (8L4P60), the hollow composite demonstrated similar performance in transmitted force between FE modelling and experimental test shown in Figure 5(a).

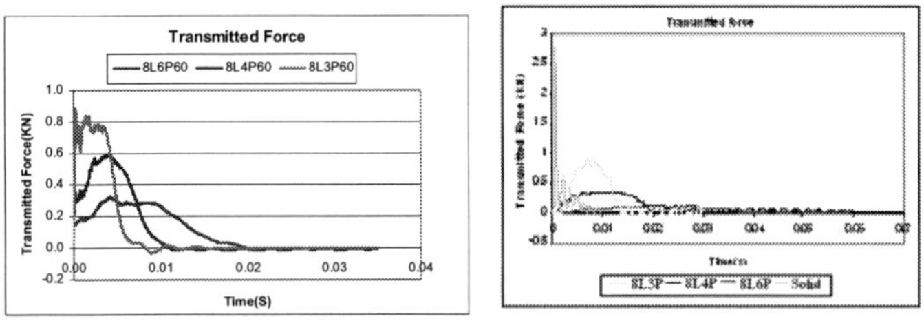

(a) Transmitted force collected by experiment (b) Transmitted force collected by FE analysis
Figure 5. Result Comparison.

Table 2 compares the energy absorption of the hollow composites in comparison to a solid material. The hollow composite with smaller cells (8L3P) is more energy absorbent than the other. Both hollow composites take in more 90% of the impact energy, while the solid material absorbs less than 17% of the impact energy.

	8L3P	8L4P	Solid
Kinetic Energy (J)	8.3954	8.30704	8.09379
Strain Energy (J)	8.10308	7.62173	1.34834
Energy Absorption	96.5%	91.7%	16.7%

Table 2. Energy absorption of various hollow structure composites.

3.2 Drop-Weight Impact Experiment

The textile based hollow composites have been impact-tested with 14 systematically designed 3D hollow textile composites. A low energy and low speed impact tester was used, which is capable of measuring the change in acceleration of the impact head and the transmitted force from beneath the specimen.

3.2 1 Influence Of Cell Opening Angle

This group involves composites 8L6P30, 8L6P45, 8L6P60, 8L6P75, and 8L6P90. Figure 6(a) shows the relationship between the impact force and the displacement of the hollow composites during the impact. Figure 6(b) depicts the influence of opening angle on the energy absorption of this group of hollow composites. The message is clear that an increase in the opening angle leads the decrease of energy absorption. This is believed to be because the hollow composites with smaller opening angles present lower resistance to the bending of its angled cell walls and therefore create larger deformation.

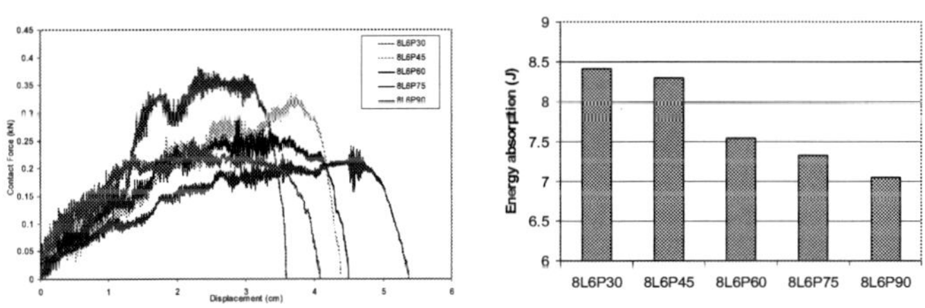

(a) Impact force –displacement curves (b) Energy absorption
Figure 6. Opening angle influence on energy absorption.

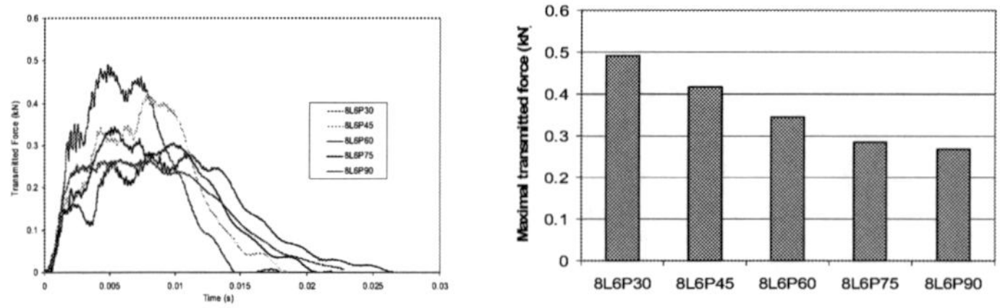

(a) Time dependency of transmitted force (b) Maximal transmitted force

Figure 7. Transmitted force with different opening angles.

Figure 7(a) shows that whilst the peak transmitted force is reduced along with the increase in composite opening angle, it is clear that the peak forces tend to be smoothed when the composites get thicker due to the increase in opening angle. It shows a clear tendency in Figure 7(b) that when the opening angle increases the maximal transmitted force gets reduced. The thickness of the composites is believed to have an important role to play in this together with the cell geometry in the cross-section of the composites.

3.2.2 Influence of Cell Size

All hollow composites in this group are made from 8 layers of fabrics involving 4 regular hexagonal cells, where the 6 walls of each cell have the same length. The opening angle of the cells in these composites is 60°. With the same weft density for all reinforcing fabric sections, the cell wall length changes from 3, 4, 5, to 6 picks, resulting in cells with increasing sizes. Figure 8(a) shows that composites with smaller cell sizes have higher impact modulus and vice versa, leading to harder and softer composite materials, respectively, with similar level of energy absorption as indicated in Figure 8(b).

Impact force attenuation is much affected by the cell size for hollow composites as demonstrated in Figure 9. It is clear that hollow composites with larger cells performs better in that the altitude of the peak transmitted force is much reduced and that the peak transmitted force gets smoothened out. This is certainly a favourable feature for materials to be used for body and limb protection against trauma impact, with composites with larger cells allowing more preparation time for the human body to react to the impact hence reducing the chances for more serious injuries.

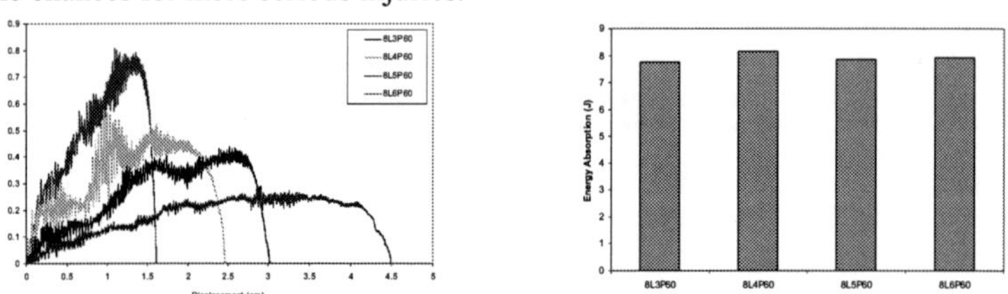

(a) Impact force - displacement curves (b) Energy absorption

Figure 8. Influence of cell size on energy absorption.

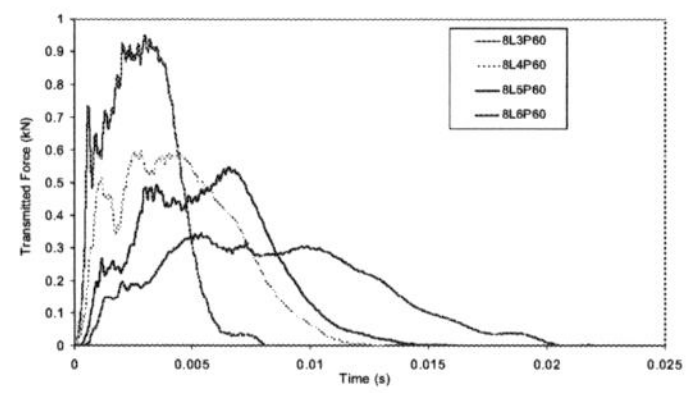

Figure 9. Influence of cell size on transmitted force.

3.2.3 Influence of Length Ratio Cell Walls

Two subgroups of hollow composites, based on the 8-layer construction, have been created for the study of influence of cell wall ratio on the impact performance. The first group involves 8L3P60, 8L(4+3)P60, and 8L(6+3)P where the free wall length is kept to be 3 picks and the bonded wall length takes the values of 3, 4, and 6 picks. In the second subgroup, the free length remains to be 6 picks while the length of the bonded walls change from 3, 6, to 6 picks. The cell opening angle for all the composites is 60°.

Figure 10(a) plots the impact force – displacement curve for these composites. It is clear that the two subgroups perform distinctively differently. The composites whose length ratio of free wall to bonded wall $\frac{l_f}{l_b} \leq 1$ demonstrated higher impact modulus than the other group. With the $\frac{l_f}{l_b} \leq 1$ subgroup, it is clear that 8L3P60 whose cell wall length ratio is 1 showed the highest impact modulus and 8L(6+3)P60, with a length ratio of 0.5, displayed the lowest impact modulus in the subgroup. For the subgroup with $\frac{l_f}{l_b} \geq 1$, a ductile behaviour was demonstrated for all three samples involved, with the 8L6P60 being the most ductile. Obviously, these two subgroups absorb impact energy in a distinctively different way. The cell wall ratio, therefore, can be used as an important parameter to influence the format the honey composites absorb impact energy. Figure 10(b) shows however the total amount impact energy absorbed by the composites of each subgroup is about the same. The subtle differences in the absorption of impact energy among the composites, especially in the $\frac{l_f}{l_b} \geq 1$ subgroup, require further investigation.

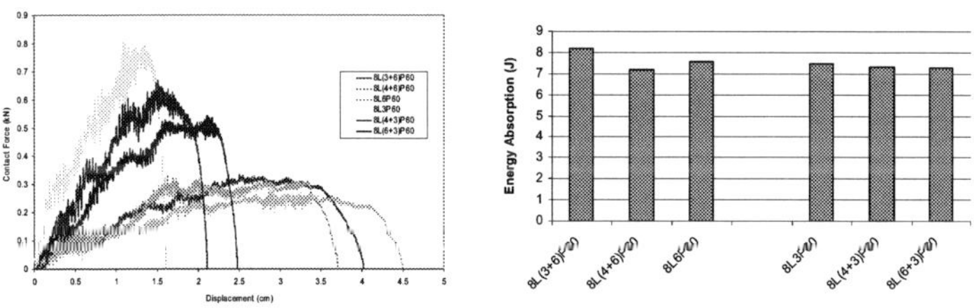

(a) Impact force – displacement curves (b) Energy absorption

Figure 10. Influence of cell wall ratio on energy absorption.

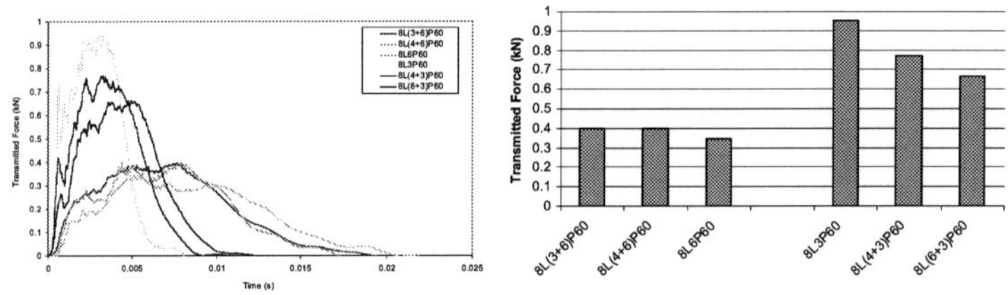

(a) Time dependency of transmitted force (b) Maximal transmitted force

Figure 11. Influence of cell wall length ratio on transmitted force.

Clear influence of the cell wall ratio on transmitted force has been demonstrated through the experiment. Figure 11(a) depicts the relationship between the transmitted force and impact time, where distinctive differences are observed between the two subgroups. The subgroup with $\frac{l_f}{l_b} \geq 1$ clearly demonstrated better capability for impact force attenuation than the subgroup with $\frac{l_f}{l_b} \leq 1$. The maximum transmitted force of all hollow composites in this group is shown in Figure 11(b). It is true for both the $\frac{l_f}{l_b} \geq 1$ and $\frac{l_f}{l_b} \leq 1$ subgroups that increase in the ratio leads to a higher transmitted force, and decrease in the ratio results in a lower transmitted force. It is also evident that $\frac{l_f}{l_b} \leq 1$ subgroup permits higher transmitted force than the other subgroup.

4. CONCLUSIONS

Experimental and theoretical study of 14 systematically designed and manufactured 3D hollow textile composites was carried out with an emphasis on the impact behaviour. The 3D hollow textile composites are created from integral 3D textile reinforcements with the required material continuity in the composites. The 3D hollow fabrics as composite reinforcements can be manufactured using the conventional textile technology, which will lead to hollow textile composites with high performance but low cost. The experimental study described in this part of the paper leads to the conclusions which provide useful information for engineering hollow textile composites against impact.

1) The opening angle of the cells in the hollow composites plays an important role in determining the properties of the hollow composites. For the same fabric, increase in the opening angle results in the hollow composites being less energy absorbent and less force attenuating, and vice versa.
2) Increase in the cell size in the hollow composites makes the composites more efficient in impact force attenuation. It also makes the peak force disappear and the maximum transmitted force delayed in time. Also, reducing the cell size can lead to the least bulky engineering material for similar impact energy absorption.
3) Manipulation of the length ratio of cell walls leads to the creation of two damping materials with different impact behaviour in absorbing impact energy. $\frac{l_f}{l_b} \geq 1$ is associated to hollow composites with high modulus and high

strength, whereas the inverse design lead to composites that has low modulus and low strength when encountered with the same level of impact. However, it was found that the total energy absorption was not affected by the length ratio of cell walls.

4) For the same volume, low density of the hollow composite leads to low contact force and low transmitted force. But, composites with too low density can be easily crushed as in the case of composite 4L6P60 in this study. So long as the composite is not destroyed by crushing, they are able to absorb similar amount of impact energy.

REFERENCES

1. Corden,J.L., and Bitzer,T.N., Hollow material and applications, Advanced Materials Technology 87, Anaheim, California,USA,6-9 Apr. 1987. Society for the Advancement of Material and Process Engineering, pp. 68–79.
2. Gibson L.J. and Ashby M.F., Hollow solids: structure and properties—2nd edition, Cambridge University Press, Cambridge, 1997.
3. Sato, T., Summary of Basic Hollow Structural Material-Kinds and Functions, Kino Zairyo(Function and Materials)(Japan), Vol.14, No.7,1994, pp.52–65.
4. Evans, A.G., Hutchinson, J.W., and Ashby, MF., Multifunctional of Hollow Metal Systems, *Progress in Materials Science*, Vol.43, 1999, pp.171–221.
5. Gibson,L.J., and Andrews, E.W., The role of hollow structure in creep of two-dimensional hollow solids, *Materials Science and Engineering A*, Vol.26, No.5 2005, pp. 424–438.
6. Tan,X., and Chen, X., Parameters affecting energy absorption and eformation in textile composite hollow structures, *Materials and Design*,Vol.26, No.5, 2005, pp. 424–438.
7. Tan,X., Chcn,X., Conway,P.P., and Yan,X.T., Effects of plies assembly on the properties of textile composite hollow structures, Materials and Design, Vol.28,2007,pp. .857–870.
8. Yu, D. K. C. and Chen X., Simulation of trauma impact on textile reinforced hollow composites for personal protection, *Proceedings to Technical Textiles for Security and Defence (TTSD)*, Leeds, UK, 2004.
9. Reid, S.R. and Peng, C., Dynamic uniaxial crashing of wood. *Int. J. Impact Engng.*, 19, 431–570, 1997.
10. Ruan, D., Liu, G., Chen, L., and Siores, E., Compressive behaviour of aluminium foams at low and medium strain rates. *Composite Structures*, 57, 331–336, 2002.
11. Papka, S.D. and Kyriakides, S., Experimental and full-scale numerical simulations of in-plane crushing of a hollow. *Acta Mat.*, 46, 2765–2776, 1998.
12. Chen, X., and Wang, H., Modelling and computer aided design of 3D hollow woven fabrics, *J. Text. Inst.*, 97, No.1, 79–87, 2006.

3-Dimensionally Carbon Fiber Reinforced Composite for High Speed Rotation

N. HIROSHIMA, H. HIROSHI, K. GOTO and H. OHKITA

ABSTRACT

Three-dimensionally carbon fiber-reinforced plastic (CFRP) was adopted to rotor disk for enhancing the burst tip speed and energy density of the disk. By placing reinforcements in the directions, r, θ, and z, the burst tip speed of the disk was numerically estimated at 1500m/sec. This newly designed disk is now under fabrication, and manufacturing process is also briefly reviewed.

INTRODUCTION

The burst tip speed of a rotating disk is dominated by specific strength. With its high specific strength, carbon fiber reinforced plastics (CFRPs) have been applied to rotating disks in order to develop high performance of flywheel for energy storage [1], In a flywheel energy storage system, kinetic energy stored in rotational body is proportional to the mass and the square of the rotation speed. Thus, the faster they rotate, the higher the reliability and the specific energy are achieved.

Until now, operational tip speeds of rotors are in the range from 500 to 650 m/s [2,3]. The purpose of this study is to enhance it over 1000m/s. In order to attain this target, burst tip speed should need to be about 1500m/s.

Under the rotation, the circumferential stress (σ_θ) is the highest among other stress components, so that fiber bundles have generally been placed to the circumferential directions in conventional flywheel energy storage studies [4,5]. Stress distribution in a rotating disk with a hole depends on the ratio of outer to inner radius (D_r); Small D_r means small hole in relatively large outer diameter.

When D_r becomes small, radial stress tends to increase and may cause radial

Noboru Hiroshima, Dept of Space science, Graduate school, The Graduate University for advanced studies (SOKEN-DAI), 3-1-1,Yoshinodai, Sagamihara-city, Kanagawa, Japan.
Ken Goto and Hatta Hiroshi, Dept of Japan aerospace Exploration Agency.
Hiroshi Ohkita, Dept of Sci. and Eng., Graduate school, Tokyo University of Science.

delamination in the circumferentially wound composites flywheel. To prevent the delamination, multi-ring method has been adopted. In this method several rings of CFRP are assembled by press-fit. In a multi-ring rotor, radial stress can be kept negative before the rotation. This suppresses the radial delamination. However, a multi-ring with small D_r less than 0.5 has never been applied for high speed rotation, because radial fiber reinforcement is indispensable for such occasion.

A CFRP disk with a hole needs a metal hub to be fitted within the hole and assembled into a rotor system. CFRP disks having small D_r are advantageous for high speed rotation, because burst tip speed of metal is smaller than that of CFRP. Relatively small outer diameter of metal hub fitted in the hole of small D_r of CFRP can be durable at the higher angular velocity. Thus, premature fracture in metal hub under high rotational velocity might be avoided.

In the present study, a new approach using 3-dimensionally carbon-fiber-reinforced composites was proposed for improvement of rotation speed of a CFRP disk. The fiber reinforcement in the radial direction can avoid the delamination. The fiber reinforcement in the axial direction enables the thickness to vary from inner to outer edge. Using these advantages, a prototype model for high speed rotation is proposed and high performance of the model is verified mainly by finite element calculation.

FLYWHEEL DESIGN

Material Design

The reinforcing fibers in the rotor are placed along three directions in accordance with the cylindrical coordinate system. Orthotropic properties of the composites with 9 independent values depend on the volume fraction of fiber reinforcement in each direction. These properties were calculated using the equivalent inclusion method High strength type PAN-based carbon fiber (TORAYCA T1000G, Toray Japan) was used in prototype design.

The upper limit of total fiber contents in 3D-fabric V_t ($=V_{FR}+V_{F\theta}+V_{FZ}$) was geometrically calculated to be 0.58, where V_{FR}, $V_{F\theta}$, and V_{FZ} denote volume fraction of the fiber in cylindrical coordinates system. In this study, calculation was repeated searching for maximum burst tip speed under the following restrictions.

1) V_{FZ} is 0.05.
2) $V_{F\theta}$ is constant within the radius.
3) The sum of V_{FR} and $V_{F\theta}$ is 0.5.

Reinforcement in the axial direction is assumed not to play a decisive roll in over all stress field of the composites disk.

Composites Rotor Design

In this study, the outer and the inner diameter of the disk was assumed to be 300mm and 40mm respectively. For attaining high strength up to the burst tip speed of 1500m/sec, continuity of fiber reinforcement from inner to outer diameter in radial direction is imperative. If a disk has constant thickness, V_{FR} geometrically decreases from inner to outer radius in proportion to its radial position. When the D_r becomes small, the difference between inner and outer radius of V_{FR} tends to be large. If V_{FR} is assumed to be 0.3 at inner radius of the disk with D_r of 0.13, it gets to be about 0.04 at outer diameter. In contrast, a disk with constant V_{FR} is obtained under the assumption that the thickness geometrically decreases from inner to outer radius. Since the flat disks have much moment of inertia than that of disks with thickness decreasing to outer diameter, the former is more advantageous for high energy density. From the practical point of view for high burst tip speeds, the disks with its thickness change was taken into consideration at the expense of energy density.

Finite Element Analysis (FEA) was adopted to obtain stress distributions in the rotating disks. Analysis was carried out using commercial software, ABAQUS Version 6.6.

Burst tip speed was evaluated using maximum stress criterion, where fracture is assumed to occur when stress reached 75% of material strength. The design and evaluation process of high burst tip speed composites rotor is as follows. First, a certain contour and V_{FR} at inner radius of the disk is given. Then, two principal stresses in the radial and circumferential directions were calculated by FEA. Second, each stress was evaluated with maximum stress criterion. This evaluation determines which stress is more likely to fracture. If the radial stress is weaker than circumferential stress, V_{FR} at inner radius of the disk is increased in the next step for further calculation. At the end of calculation, the two stresses have the same margin to strength of each direction.

Interface Design with Metal Hub

As mentioned before, the important thing of reinforcement in radial direction of 3-dimensional fabric is to adopt small size of metal hub. When the composite disk rotates at the tip speed of 1500m/s, metal hub fitted into inner diameter of Dr equal to 0.13 rotates at the speed of 195m/s (1500 m/s x 0.13= 195 m/s.). Being calculated with its density and specific strength, burst tip speed of Carbon steel (SNCM616) can be estimated about 650m/s. In this sense, SNCM616 seems to be used as metal hub. However, it is also required for the interface between the hub and composite disk that deformation of the outer surface of the metal hub always exceeds that of the inner surface of the composites disk. Several calculations showed that deformation of the inner surface of the composites exceeded that of the outer surface of hub. Judging from these calculations, conventional type of the interface cannot be used for 3-dimensionally carbon fiber reinforced composite. Another method was explored, and illustrated in Fig. 1 was shown to be promising. In this interface, the cap of the metal hub wraps the central boss of composites disk, and hence, separation of the interface can be avoided. FEA also showed that the calculated stress was allowable enough for SNCM616 to sustain the interface.

Fig.2 shows the contour of 3-dimensionally reinforced carbon fabric designed for a burst tip speed of 1500m/s. The fabric was manufactured by Shikibo Japan. The fabric has a central boss over which the cap of metal hub could be wrapped by press-fit.

Table I shows volume fraction of fiber and its deviation in 3D-fabric. FEA was done again to check the effect of the deviation in actual product shown in the Table I.

Table II shows the result of stress analysis for the composites rotor at the burst tip speed of 1500m/s. The rerun of FEM analysis confirmed the burst tip speed of 1520 m/s. Energy density of the composites rotor was also calculated to be about 130wh/kg at the burst tip speed.

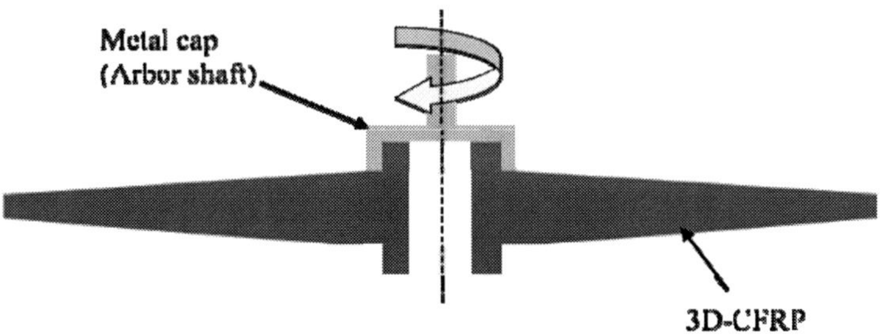

Figure 1. Construction of cap and disk.

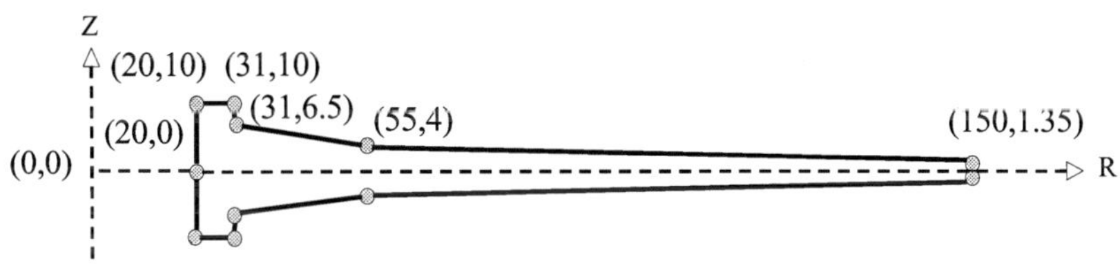

Figure2. Contour of 3-dimensionally reinforced carbon fabric (unit in mm).

Table 1. Volume fraction of the composite rotor.

	V_{FR}	$V_{F\theta}$	V_{FZ}
Average(%)	18.3	25.2	2.4
Deviation(%)	±2.0	±5.6	±4.1

Table 2. Maximum stress of composite rotor (1500m/s).

	σRmax	σθmax
Generated stress(MPa)	734	1321
Vf of position(%)	0.17	0.284
Material Strength(MPa)	855	1356
Margin	1.165	1.027
Durable tip speed(m/s)	1618	1520

Figure 3. Outlook of the specimen with the fixture.

RTM PROCESSING

RTM process was used for prototype model. The metal mold is used for maintaining dimensional precision. The fabric is packed into metal mold and vacuumed out at the level of 0.9MPa below the air. Cyanate Ester resin (EX-1545, Ten Cate, Netherland) is used for RTM processing. It took about one hour to infuse the resin into 3D-fabric within the mold. Fig.2 shows the outlook of the specimen with fixture.

CONCLUSION

The burst tip speed and energy density of CFRP rotor proposed in this study have shown to be improving burst-rotation-speed by radial reinforcement. A three-dimensionally reinforced prototype using this concept was numerically confirmed to be having burst tip speed of 1500m/s and energy density of 130wh/kg at the burst tip speed. This prototype was actually manufactured through RTM process for experimental confirmation.

REFERENCES

1. S. M. Arnold, A.F. Saleeb, and N. R. Al-Zoubi. 2002. "Deformation and Life Analysis of Composite Flywheel Disk Systems," *Composite: Part B*, 33: 433–459.
2. C. A. Mahieux. 2001. "Cost Effective Manufacturing Process of Thermoplastic Matrix Composites for the Traditional Industry: the Example of a Carbon-Fiber Reinforced Thermoplastic Flywheel," Comp. Str. 52:517–521.
3. S. Nagaya, et al. 2003. "Improvement and Enlarging of the CFRP Flywheel with Superconducting Magnetic Bearings," Physica C. 392-396:769–772.
4. Andrew C. Arvin and Charles E. Bakis. 2006. "Optimal Design of Press-fitted Filament Wound Composite Flywheel Rotors," Comp. Str. 72:47–57.
5. Sung K, HA, et al. 2006. "Design and Spin Test of a Hybrid Composite Flywheel rotor with a Split Type Hub," *J. Comp. Mater.*, 40(23):2113–2130.

An Investigation on the Effect of 3D Weaving on Carbon Fibre Tows, Fabrics and Composites

E. ARCHER, S. BUCHANAN, A. T. McILHAGGER, J. P. QUINN and M. MORGAN

ABSTRACT

3D woven textile reinforced composites allow the optimisation and tailoring of specific material properties into the final component. This paper investigates the damage imparted on load bearing carbon fibres during the 3D weaving process. 3D multi-layer woven reinforcements were manufactured on a textile loom with few mechanical modifications to produce reinforcements with fibres orientated in the warp, weft and through-the-thickness directions. Tensile tests were conducted on 3 types of commercially available carbon fibres, 12k HTA, 6k HTS and 3k HTS in an attempt to quantify the effect of fibre damage induced during the 3D weaving process on the mechanical and physical performance of the fibre tows in the woven composite. The tests were conducted on fibre tows sampled from different locations in the manufacturing process from the bobbin, through the creel and loom mechanism, to the final woven fabric.

INTRODUCTION

Woven textile reinforced composites allow the introduction of specific material properties into the final component that can provide a reduction in manufacturing cost. However, woven composites materials are susceptible to transverse impact loading which causes laminas to become delaminated [1]. Various methods have been developed to improve the impact tolerance including z-pinning or resin toughening; one method that is becoming increasingly successful is to reinforce composites with a fibre that connects the layers together running from the upper to lower surface of the laminate. It has been suggested however, that 3D woven composites generally exhibit lower in-plane properties due to degradation during manufacture. The reduced mechanical performance is believed to result from

Edward Archer, Engineering Composites Research Centre, School of Engineering, University of Ulster, Shore Road, Newtownabbey, Co. Antrim, BT37 0QB crimping and misalignment of the load-bearing fibres by insertion of the z-binder yarns during weaving. In addition, it has been discovered that damage to fibres during weaving also degrades the in-plane properties of the 3D composite [2]. During the weaving process, the fibres are bent, twisted, distorted and repeatedly abraded as they move through the creel and loom. This damage can come in two forms: fibre breakage or abrasion caused by fibres rubbing against other fibres and the loom machinery itself. It is believed to be due to the stiff and brittle nature of carbon fibre tows that there is a propensity for damage to occur during the weaving process; both the modulus and the strength of the composite are controlled substantially by the fibre so this weaving damage will affect the mechanical properties of the composite structure.

Several researchers have published work characterising damage caused to glass yarns by 3D weaving [2, 3]. The authors determined the influence of the weaving damage on the tensile stiffness, strength and failure mechanisms of fibre yarns. The results showed that the most extensive damaged occurred during the tensioning and beat-up stage, and suggested that the weaving caused a reduction in strength of dry load-bearing yarns in 3D glass preforms of approximately 30%. The same behaviour is found in the z-binder yarn having a reduction of up to 50% in tensile strength. They concluded that this greater loss in strength of the z-binder may be due to the fact that this yarn is bent more severely than the warp yarn during weaving, and also that generally the z-binder is of smaller thickness than that of the warp yarn. However, tension tests were performed on both 2D and 3D woven composites for comparison. It was discovered that the Young Modulus of the 3D woven composite was significantly lower; it was suggested that this was due to the distortion and crimp of the load-bearing yarns by the z-binders [4]. The conclusion of this work suggests that loom modifications are needed to reduce abrasion since it has been shown that fibre rubbing against the actual machinery is the biggest cause of damage to the fibre. It has been proposed that one possible modification is to cover the machinery surface in contact with the yarn with a wear-resistant material having a low co-efficient of friction, in order to reduce abrasion.

Studies have investigated the effect of fibre damage on carbon fibre tows caused by 3D weaving [5]. They tested 2 types of carbon fibre and determined that the tensile strength of the first grade was unaffected by weaving damage, and the second carbon yarns were reduced by up to 12% due to abrasion and bending. They also determined that the modulus was almost unaffected for both. However, when testing the composite, they found a large degradation in both strength and modulus caused by fibre misalignment due to crimping.

The loss in yarn strength due to weaving damage is a cause for concern, especially when these materials are to be used for structural components. More research is needed therefore, to quantify the importance of this damage, to reduce damage levels by further analysis, and to develop the 3D weaving process.

EXPERIMENTAL

3D multi-layer woven reinforcements were designed using the X-Sectional design system to provide a representation of the structure, detailing the relative positions of the yarns and also generating the lifting plan to operate a Jacquard controlled loom. Fabrics were then manufactured on a conventional textile loom with mechanical modifications to allow the addition of z-binder yarns and limit

Table I. Details of fabrics.

	Warp	Weft	% Binder	Binder	Areal weight/ gsm
Angle interlock	2 x12k HTS	12k HTS	6	6k HTA	2720
Layer to layer interlock	2 x12k HTS	12k HTS	6	3k HTA	2720

Table II. Material supplier data for filament yarn.

	Tensile strength/ MPa	Tensile modulus/ GPa
HTA	3950	238
HTS	4300	240

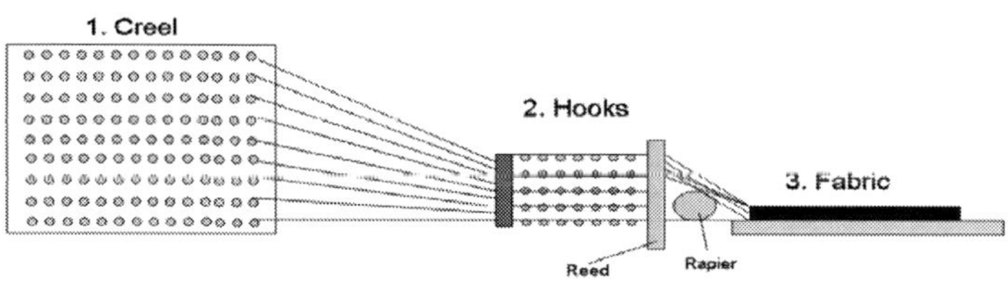

Figure 1. Weaving process.

fibre damage; reinforcements with fibres orientated in the warp, weft and through-the-thickness directions were produced. The loom used was a DATAWEAVE loom with Jacquard controller incorporating 1152 hooks. Two specific weave architectures that consist of long continuous lengths of tows for maximum strength in the 0/90° directions and have angle interlock binding configurations in-order to suppress delaminations were manufactured. Comparatively, these structures possess very low levels of crimp in the warp stuffer and weft filler tows. Both fabrics have four layers of warp and five of weft, while maintaining a 1:1 warp to weft ratio, with the binder running in the warp direction.

For the manufacture of the two fabric types, fibre was initially rewound from large bobbins supplied by the fibre manufacturer onto smaller packages for addition to the creel. A high speed single head winder was used. It was observed that some fibre damage was caused during this step. Each package was placed on the creel and the fibre was passed through ceramic guides on the creel. These guides were installed to limit entanglement. The warp ends were doubled up at the creel eye. Passive tensioning devices were added to each bobbin to apply tension in the warp direction. This was to reduce the chances of individual tows becoming entangled before the eye board, and to aid in the movement of binders past one another in the harness

Dry yarn specimens were removed from the process for assessment of tensile properties. Tensile properties for 12k, 6k and 3k were determined in several regions. This included:

—As received virgin yarn
—Rewound on creel
—After hooks
—Out of Fabric

Care was taken when removing warp and binder yarns from the woven fabrics. The fabrics were dissected using scissors, scalpel blades and tweezers.

The degree of crimp within the uncompressed fabric structure was measured for the weft and warp tows and for the binder. The crimp measurements were made in accordance with BS 2863:1984 [8]. The tows were straightened as described in the standard to determine the straight length. Crimp is calculated using the equation below:

$$C = \frac{L - L_o}{L_o} \tag{1}$$

Where:

L = mean measured length of 10 threads removed from the fabric.

L_o = length occupied by the tow in the fabric, that is the flap width.

Instron Cord Capstan grips were used for clamping dry fibres during tensile testing; each grip has a round capstan with a right-hand spiral grooved surface (Figure 2). The capstan is split axially into two halves, one half being stationary and the other movable. The specimen is wound from the outside of the capstan towards the centre before being passed through the split halves where it is clamped. The groove completes 2.5 turns around the capstan, which allows two full turns for distributing the specimen load and stress concentration. The grips self-tighten as load is applied to the specimen.

The grips were loaded onto a Zwick Z100 universal tester. A gauge length of 75mm of fibre between each clamp was used, with a test speed of 4mm/min [7].

Finally, several woven composite plaques were manufactured using the VARTM method. A caul plate was used in conjunction with a flexible membrane to consolidate the reinforcements under full vacuum, prior to the injection of resin by peripheral gating at a tool temperature of 75 °C. Araldite LY564 and Hardener HY2954 (based on bisphenol A epoxy and a cycloaliphatic amine hardener) was mixed and degassed at 2.86:1 by weight before transfer of the resin. After injection, a ramp was applied up to 100 °C. The temperature was held isothermal for 60 min; the composite plaque was then demoulded and post cured for a further 180 min at 140 °C. Fibre volume fraction was measured using the density buoyancy method in which the sample's mass in air is recorded before weighing the sample again in distilled water according to ASTM D792-91 [9]. Therefore, having obtained the specific gravity and knowing the densities of the constituent parts i.e. the fibre and resin, the percentage content of the fibres was calculated. It was found to be approximately 55% (StDv 1.60) for both composite weave architectures. Tensile specimens from both woven composite plaques were removed and tensile tested in the warp and weft direction on a Zwick Z100 universal tester. The testing was performed to CRAG standard 300 [10]. An MTS 632.85F biaxial extensometer was used to determine tensile modulus.

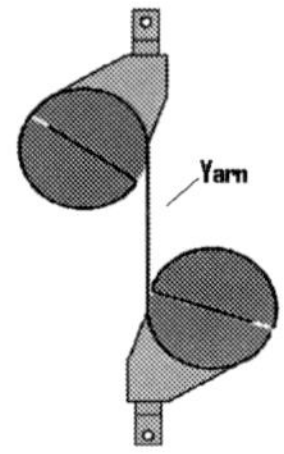

Figure 2. Instron Cord Capstan grips.

RESULTS AND DISCUSSION

Table III shows the tensile strength of dry yarns sampled from four stages in the weaving process. The results show that there is less than 3% difference in the weaving damage between the 3 fibre types. In these fabrics, the 12k stuffer follows a path with very little crimp whereas both the 6k and 3k binders follow a more crimped binding path; albeit a much less crimped path than that of the orthogonal bound fabrics reported by other authors in the area. The binder in the orthogonal fabrics go directly from top layer to bottom layer, whereas the binders in the fabrics analysed in this paper go from top layer to bottom layer at an angle. The degree of crimp within the uncompressed fabric structure was found to be less than 0.5% for the warp stuffers and weft fillers, whereas the binders in the uncompressed angle interlock and layer to layer fabrics possessed 4.2% and 3.4% of crimp respectively.

All three yarn types were found to experience a 9-10% drop in tensile strength from Virgin (as received) properties to properties found in dry yarns dissected form the fabric.

Table III shows that the dry yarn damage is induced at each stage, with an initial degradation of approximately 5% in tensile strength from virgin fibre to rewound fibre, 1.5% degradation from the rewound fibre to the fibre at the hooks, and finally 2.8% degradation in properties from the loom fibres to the fabric fibres. These results show that on average there was an overall degradation of 9.4% from the as-received virgin fibre to fibre extracted from the woven fabric.

The displacement of the crosshead was recorded to determine tensile modulus of the dry yarns. It was found that there was little variation in modulus due to weaving damage. This agrees with the findings of Lee et.al [4].

In each case the largest reduction in dry yarn tensile strength is observed at the rewinding stage (5–6%). This is a high speed process using a single head winder. The actual weaving process only imparts approximately a further 4% knock down in dry tensile properties. Prior to this work, several modifications were made to the loom to ensure limited weaving damage. For these fabrics the loom was running at a much slower rate than a conventional textile loom. The loom was run at approximately 120 picks per minute whereas a conventional textile loom would run at up to 600 picks per minute. It is expected that as the weaving production speed is

Figure 3. Weaving damage at creel eye-board.

Table III. Tensile strength of dry yarns sampled from four stages in the weaving process.

Fibre state	12K stuffer Tensile strength/ MPa	6K binder Tensile strength/ MPa	3K binder Tensile strength/ MPa
Virgin	2210	2010	1998
Re-wound on creel	2100	1915	1880
Before reed	2070	1871	1860
Woven	1985	1827	1820
Standard deviation	80	61	63
CV/ %	7	8	10

increased in terms of picks per second, greater damage will be caused as the carbon moves through the loom mechanism at a greater rate
As expected all three yarn types possessed much lower virgin (as received) properties than those values given by the material supplier.

Figure 4 shows a micrograph of the angle interlock composite in both longitudinal and transverse directions. It is clear that the warp stuffers are very straight with little crimp. However, the weft fillers appear to suffer from crimp to a greater extent. It thought that this is due to a lack of tension in the weft direction which accommodates the nesting effects caused by the binders which cross over the weft yarns at the surface layer of the fabric, bending the weft yarn down under compression from the caul plate during composite manufacture. Resin rich areas were also observed between warp stuffers in the space that is occupied by the binder on other weft layers. These resin rich areas will be difficult to eliminate with this design of fabric due to the binder pitch.

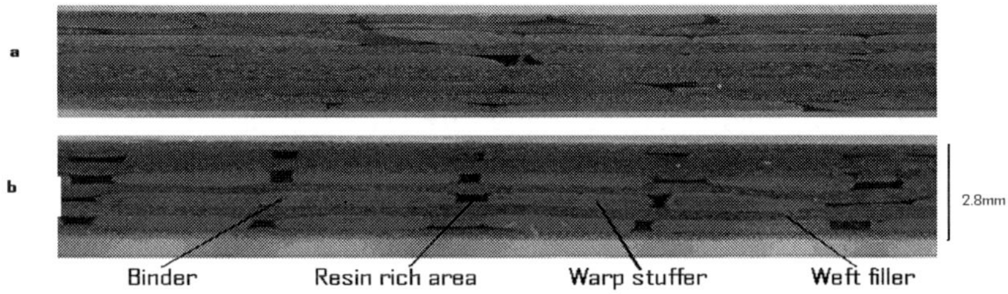

Figure 4. Longitudinal (a) and transverse (b) micrograph of angle interlock composite.

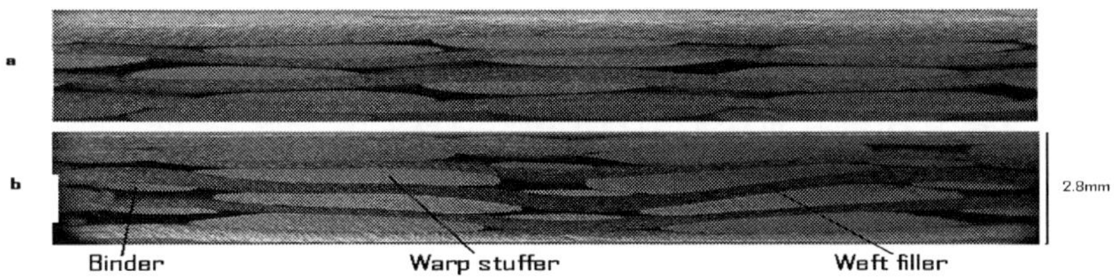

Figure 5. Longitudinal (a) and transverse (b) micrograph of layer to layer composite.

The micrographs in figure 5 for the layer to layer composite show a large degree of nesting in the warp stuffers. It is thought that this could be due to the difference in size between the large stuffer tow and the smaller 3k binder tow. There are also resin rich areas around the binder. However, as a binder is present at each layer, the resin rich area is smaller than in the angle interlock fabrics. The main observation from the layer to layer micrograph is the higher degree of crimp in the weft filler compared to the angle interlock composite.

Table IV. Mechanical properties for woven composite specimens (55% fvf).

	Modulus 0-Warp/ GPa	Modulus 90-Weft/ GPa	Standard deviation
Angle interlock	65	57	2.2
Layer to Layer	63	46	1.5

In both woven composite specimens, stiffness and strength was significantly lower in the weft direction than the warp (Table IV). The layer to layer composite was found to have a lower modulus and strength than the angle interlock composite. The layer to layer composite also exhibited a greater differential between warp and weft properties exhibiting a 26.9% reduction in modulus compared to the 12.3% exhibited by the angle interlock composite. From the micrographs (Figure 4 & 5) there is an inverse relationship between warp and weft crimp levels and modulus values.

CONCLUSIONS

It was found that the overall reduction in tensile strength for dry carbon fibre yarns attributed to weaving damage was 9–10%. This is much lower than the 30% reduction reported for glass fibre yarn and by other researchers. It is also lower than the 12 % reduction reported for carbon fibre yarns.

Surprisingly both the 12k stuffers and the 3k and 6k binders exhibited similar knock down in dry tensile strength due to weaving damage. All dry fibres were found to show virtually no reduction in stiffness throughout the weaving process.

As the results above show, the rewinding process plays a major part in the fibre damage imparting an initial degradation of approximately 5% in tensile strength from virgin fibre to rewound fibre. Further work is required therefore to investigate the mechanics of the rewinding process.

For the 3D woven composite specimens a large variation between warp and weft values were found for both tensile strength and modulus. It appears that more tension is required to maintain straighter weft fillers during fabric manufacture and consolidation for composite manufacture. This must be implemented without introducing any knock down on warp direction properties.

It is expected that as the weaving production speed is increased in terms of picks per second, greater damage will be caused as the carbon moves through the loom mechanism at a greater rate. Also, changes in material, weaving process and weave architecture will significantly affect damage caused; this must be quantified. Further work will be conducted to maintain the low levels of fibre damage, while increasing fabric production rate.

ACKNOWLEDGEMENTS

This work was supported by the 3DSIMCOMS consortium in collaboration with the Technology Strategy Board.

The Technology Strategy Board is a business-led executive non-departmental public body, established by the government. Its mission is to promote and support research into, and development and exploitation of, technology and innovation for the benefit of UK business, in order to increase economic growth and improve the quality of life. It is sponsored by the Department for Innovation, Universities and Skills (DIUS). Please visit www.innovateuk.org http://www.innovateuk.org/> for further information.

REFERENCES

1. Atas C., Liu D. 2008. 'Impact Response of woven composites with small weaving angles,' Int. J. Impact Eng: 35 (2): 80–97.
2. Rudov-Clark S., Mouritz A.P., Lee L., Bannister M.K. 2002. 'Fibre damage in the manufacture of advanced three-dimensional woven composites', Composites Part A: 34, (10): 963–970.
3. Mouritz A.P., M.K. Bannister, P.J. Falzon, Leong K.H. 1998. 'Review of applications for advanced three-dimensional fibre textile composites', Composites Part A: 30 (12): 1445-1461.

4. Lee L., Rudov-Clark S., Mouritz A.P., Bannister M.K., Herzberg I. 2002. 'Effect of weaving damage on the tensile properties of three-dimensional woven composites', Composite Structures: 57 (1–4): 405–13.
5. Lee B., Leong K.H., Herzberg I. 2001. 'Effect of Weaving on the Tensile Properties of Carbon fibre tows and Woven Composites', J. Reinforced Plastics and Composites: 20 (8): 652–670.
6. ASTM D4018-81, Standard test method for tensile properties of continuous filament carbon and graphite yarns, strands, rovings, and tows.
7. BS 2863: 1984, Textiles- Crimp of Yarn in Fabric
8. ASTM D792-91, Standard Test Methods for Density and Specific Gravity (Relative Density) of Plastics by Displacement.
9. CRAG Standard Method 302 "Method of test for the Tensile Strength and modulus of multidirectional Fibre Reinforced Plastics.

Experimental Investigation and Behavior Modeling of a 3D Interlock Woven Fabric Composite: Part 1

J. SCHNEIDER, L. MARCIN, Z. ABOURA and D. MARSAL

ABSTRACT

Recent improvements in weaving processes in composites allow realization of 3D woven fabric composites with polymer matrices. Moreover, due to the through-the-thickness reinforcement, 3D woven fabric composites offer a high delamination resistance under impact and a strong capacity to dissipate energy by damage. Thus, degradation mechanisms are very specific to this kind of materials [1] due to its complex architecture. The large dimension of the Representative Volume Element (RVE) involves difficulties about strain measurements. The aim of this article is to propose a methodology in order to define damage scenarios and perform accurate measurements for the identification of a macroscopic. On this paper we focus on a Carbon / Epoxy layer-layer angle interlock in quasi-static loadings.

1 INTRODUCTION

More and more complex structures in aeronautics involve the use of composite fabrics. According to its high delaminating resistance and impact damage tolerance, woven composite introduce a new capability. Therefore, damage comprehension was required to use fitted materials and to develop and identify phenomenological models.

So, to reduce certification costs and to design well-adapted structures we have to set up models identified on the damages occurring and their evolutions. This paper deals with experimental approach in order to identify parameters of macroscopic visco-damage model [2]. This step imposes well knowledge of damage scenarios occurred according to the nature of the tests.

The material of this study is a Carbon / Epoxy fabrics. The weaving was an unbalanced layer-layer angle interlock. The pattern generates non-common specificities like a large Representative Volume Element (greater than 15mm x 15mm)

J. Schneider[1 & 2], L. Marcin[3], Z. Aboura[1], D. Marsal[2]

[1]Université de Technologie de Compiègne, Laboratoire de Mécanique Roberval, BP 20529, F-60206 Compiègne Cedex, France.

[2]SNECMA Villaroche, F-77550 Moissy-Cramayel, France.

[3]Composites Structures and Materials Department, ONERA/DMSC ,BP 72, 29 Avenue de la Division Leclerc, F-92322 Châtillon Cedex, France.

with a complex inner architecture. Due to the complexity of this architecture the used approach needs to crosscheck different techniques: full-field analysis [3], acoustic emission, in situ and post mortem observations and classical strain measurement (strain gages and extensometer) [4].

2 MODEL PRESENTATION

For this innovative material, a phenomenological macroscopic model has been developed in order to design structures. This model was developed in the thermodynamics frame with internal variables [5]. According to the different tests realized on that interlock woven fabric, the model has to describe the degradation mechanisms and takes into account the viscoelastic behavior (See Figure 1). The proposed formulation leads to a fully 3D model allowing to describe behavior above first failure.

The material heterogeneity in woven composites made of large fiber yarns involves significant strain heterogeneities (Figure 2b). Due to mechanical contrast between composite constituents, damage is preferentially oriented by the microstructure. Therefore, scalar damage variables are used. Set of parameters have to be identified according to the experimental observation.

To identify the in-plane parameters of this phenomenological macroscopic formulation for three dimensional woven composite, classical tests has been performed in order to determinate elastic moduli. Thereby, loading-unloading tensile cycles with dwell time along the 3 main directions of the woven fabric were carried out to describe damage. Loading-unloading cycles with tensile dwell time at 45° off-axis were also realized to highlight the damaged and the viscous matrix aspect. And finally, to describe accurately the non-linear visco-elastic behavior of the polymer matrix, 45° off-axis creep test was performed. All these tensile tests were carried out in a quasi static case.

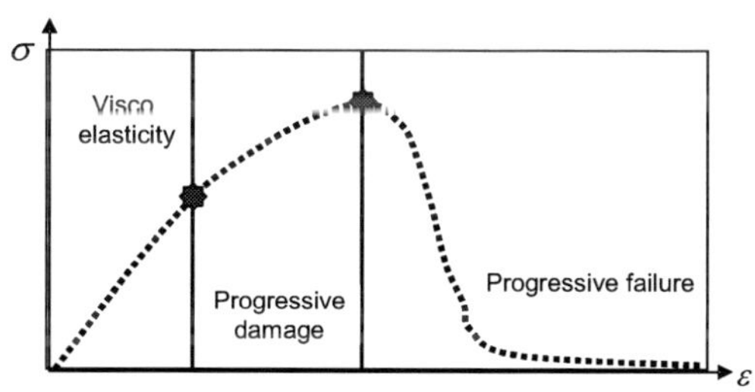

Figure 1. Different aspects of a visco-elastic damage model for 3D Polymer Matrix Composite (PMC).

3 EXPERIMENTAL PROCEDURE

3.1 Instrumentation Used Due To The VRE Specificities

First of all, keep in mind that the interlock studied includes many large yarns with significant resin areas (millimeter scale). That involves a lot of local heterogeneities.

Contrary to an ordinary laminated composite or an interlock fabric made with less significant yarns, a quite important material heterogeneity exists. Thereby, a sizeable heterogeneous strain is generated and confirmed [4] by full-field displacement/strain (see Figure 2b). That's why strain gauges with a minimum size of 20mm, or a full-field strain average on an area close to the size of the Representative Volume Element (RVE) are retained to identify the formulation (see Figure 2a).

Figure 2a. Virtual gauge area used to fulfill accurate strain measurements on the face of the specimen.

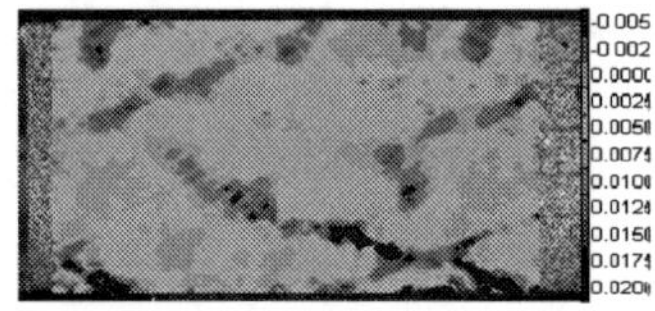

Figure 2b. Representation of the longitudinal heterogeneous strain under warp tension on the face of the specimen.

3.2 Acoustic Emission Technique

Various studies in the University of Compiègne have allowed to establish a relation (see Figure 3) between amplitude of acoustic activity and damages [6] [7] [8]. Obviously, due to the complexity of the weaving and the 3D damages occurring inside, Acoustic Emissions treatments is considered as qualitative information. This technique allows to monitor the damages phenomena during the tests and more particularly to analyze their kinetics until the ruin.

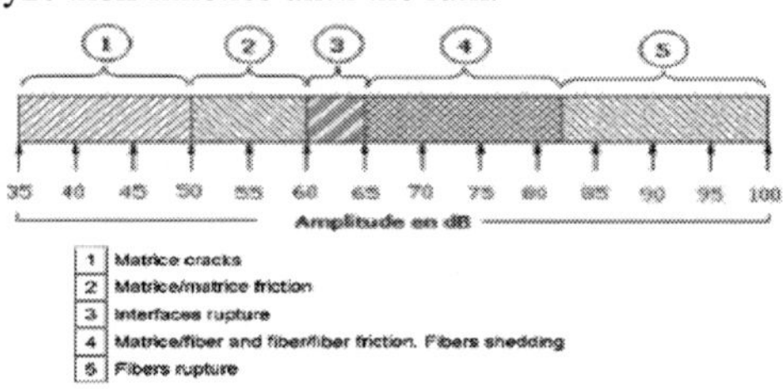

Figure 3. Amplitude ranges according to various damages.

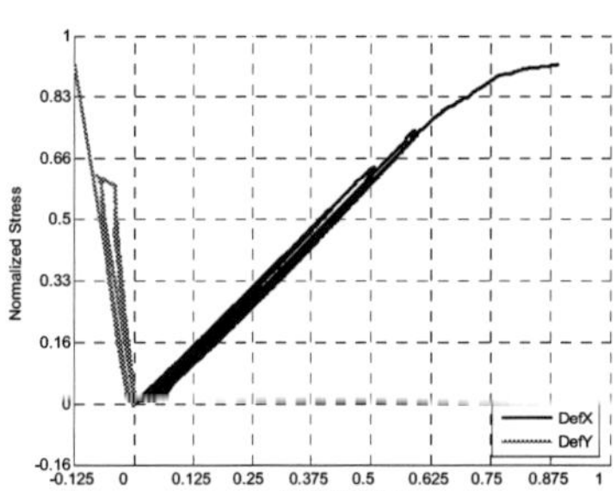

Figure 4a. Stress versus longitudinal and transversal strain under warp direction Tension.

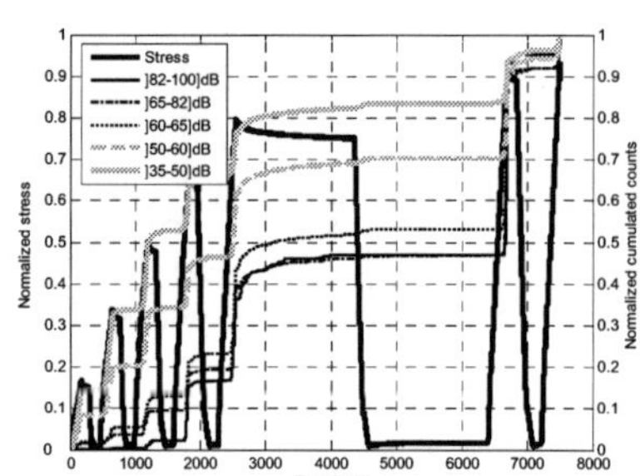

Figure 4b. Acoustic activity under warp direction tension according to the different ranges defined (cumulated counts).

4 EXPERIMENTAL RESULTS

4.1 Uniaxial Warp Direction Tensile Test

The figure 4a shows the normalized stress-strain curve. Longitudinal and traverse strain was collected for identification during the loading – unloading cycles. The behavior is linear until the last cycle. The final loading involves large damage responsible to the loss of linearity. The residual deformation is limited.

The figure 4b shows the evolution of normalized cumulated acoustic events (counts) gathered by range of amplitude (by phenomena of damage according to figure 3) as well as stress versus time. It appears that cracks of resin (range of amplitude between 35 and 50 dB) begins at the first moments of loading. That damage mechanism remains negligible and doesn't affect the mechanical behavior. It is necessary to wait the 1000th second to see an acoustic activity related to the fibers failure (higher than 82 dB). Fiber and interface breakings become significant from the 5th cycle and have doubled in the 6th. Residual deformation appears and increases from the 5th loading and confirms a high damage level.

It should also be noted that during dwell time, a stress relaxation due to the matrix viscosity appears.

4.2 Uniaxial Weft Direction Tensile Test

According to the residual stress and the no-linear to failure stress-strain behavior (see Figure 5a), a more important viscous and damaged behavior should be introduced in this direction. Compared to the warp case tension, a greater matrix section was stressed and should involve a more sizeable viscous aspect.

Nevertheless, acoustic activity characterized by fiber breakings (82-100 dB) has been observed from the beginning (see Figure 5b). Figure 6 also highlights a greater acoustic energy linked to the interface damage (60-82 dB) in the weft tension case.
Moreover, in-situ observations and full-field analysis inform about the predominant aspect of the damage in the composite behavior. Indeed, structured system of macroscopic cracks in interfaces occurs during the fifth loading. The acoustic activity gap during the fifth loading in Figure 5b also confirms this significant damage. That network of cracks initiation explains the slope modification illustrated with grey dashed lines in Figure 6a. Obviously, all these events take part in the behavior of this interlock fabric under weft tension.

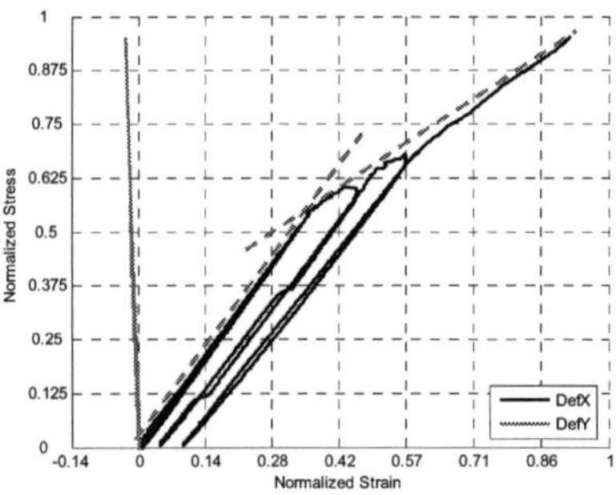

Figure 5a. Stress versus longitudinal and transversal strain under weft direction tension.

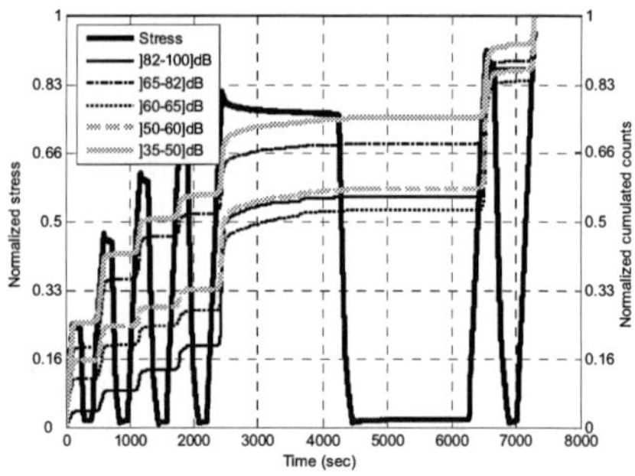

Figure 5b. Acoustic activity under weft direction tension according to the different ranges defined (cumulated counts).

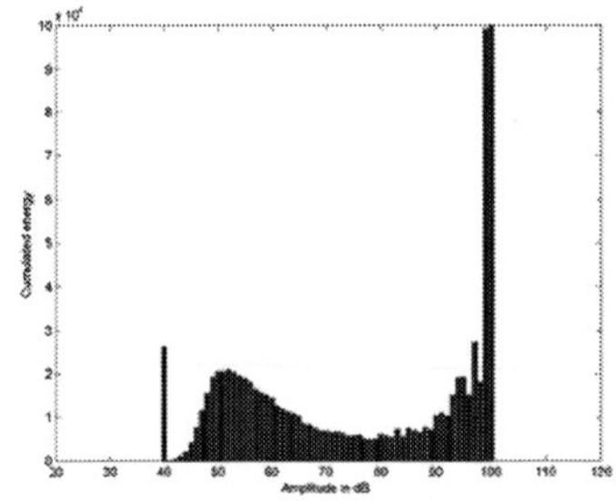

Figure 6a. Cumulated energy versus amplitude under warp direction tension.

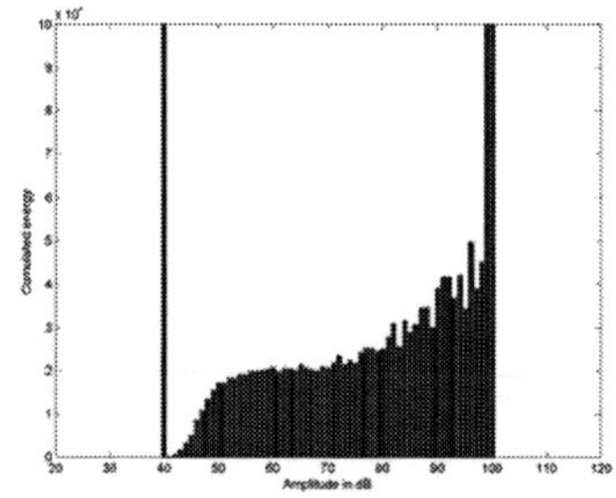

Figure 6b. Cumulated energy versus amplitude under weft direction tension.

To comprehend this structured network of cracks occurring, it's essential to understand what happens in the core of the material. The weaving principle of a layer-layer angle interlock focus on the undulation of warp yarns across the weft tows. After a sufficient damage level, the particularity of this weaving generates, for each yarn column perpendicular to the loading axis, its own behavior. Among other things, a local shear condition was generated.

The first interlock fabric investigated had a distance between weft yarn columns greater than warp yarn columns. So, according to that specificity and the weaving principle, cracks evolve variously and explain the damage mechanism divergence seen between warp and weft tensile test. For a weft tension case, the narrow distance existing between warp yarn columns make passing cracks in interfaces and in yarns (Figure 7a) and allow a "inner-relaxation" of the weaving. If inter-tow distance is greater, matrix plasticity can stop cracks growth and only meso debondings are generated (Figure 7b). Beyond a certain stress level in the weaving, matrix can no longer absorb energy, cracks grow very quickly and generate an explosive rupture by delamination and fibers wrenching.

These two damage scenarios rely on observations, acoustic emission and full-field technique conduce to the different behavior between warp and weft tensile test.

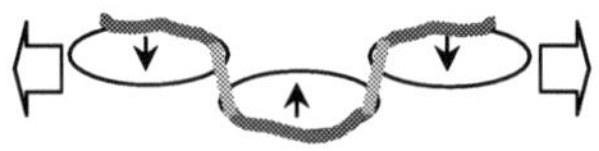

Figure 7a: Damage mechanisms under weft tension

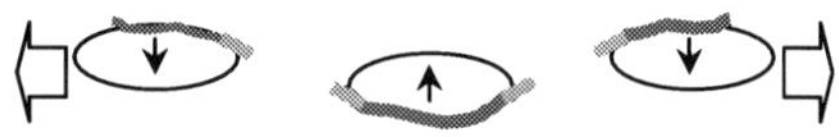

Figure 7b: Damage mechanisms under warp tension

4.3 Uniaxial Through-Thickness Tensile Test

Through-thickness testing is difficult to carry out due to the small-scale thick sections [9] and the large size of the VRE. Introduction of loading without end effects and stress concentrations have been also considered (shape optimized by FEM) [10]. To determine through-thickness tensile properties two assembly were machined with different specimen design and dimensions (Figure 8). Currently, only monotonous loadings were carried out. Loading - unloading cycles with tensile dwell time are in progress. To ensure specimen and assembly alignment, full-field displacement was controlled during the experimentation.

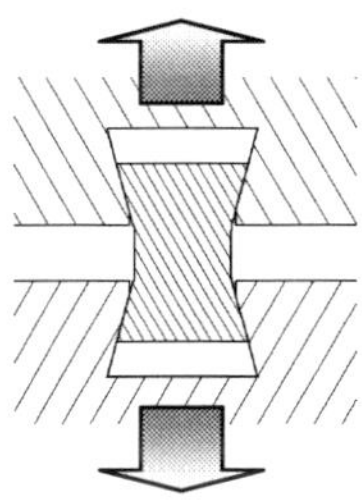

Figure 8a. Assembly based on dovetail principle.

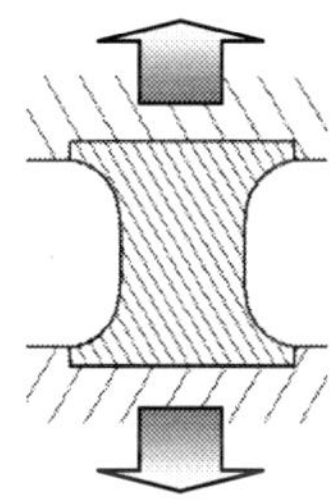

Figure 8b. Assembly based on diabolo sticked.

The Figure 9a below shows a linear to failure stress-strain behavior. Further test will allow confirming that behavior and the conceivable viscous and damaged aspect. The acoustic activity occurring during the test was mainly undertaken by low amplitude (35-60 dB). Thus, only resin and interfaces were damaged, as could be expected for that loading. Nevertheless, beyond a certain loading level, delaminating and fiber breakings appear and lead to the rupture of the specimen (Figure 9b).

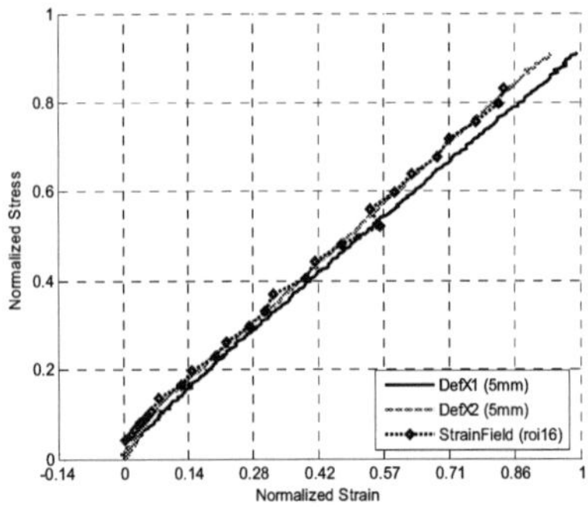

Figure 9a. Stress versus longitudinal strain (resulting to gauges and strain field analysis) for out-of-plane tensile test.

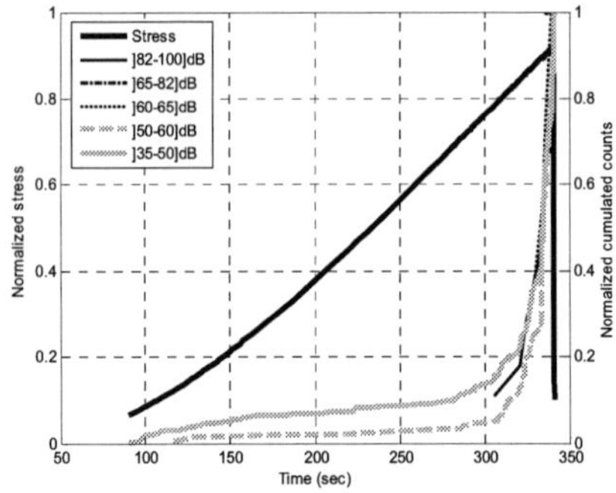

Figure 9b. Acoustic activity during through-thickness tensile according to the different ranges defined (cumulated counts).

4.4 Uniaxial Tensile Test At 45° Off-Axis

To identify in-plane shear behavior, off-axis tensile test at 45° have been chosen. With accurate strain measurements (Figure 10a) and the laminate theory applied in the interlock case, we can assess the in-plane shear stress-strain evolution.

Unlike the tests previously presented, off-axis tensile test let appears the viscous behavior. Loading - unloading cycles also generate damages. Acoustic emission treatment (Figure 10b) confirms damage mainly located in the matrix (amplitude less than 60dB) until the last loading. Obviously, interface cracks and fiber debondings take part in the damage behavior and finally produce yarn breakings.

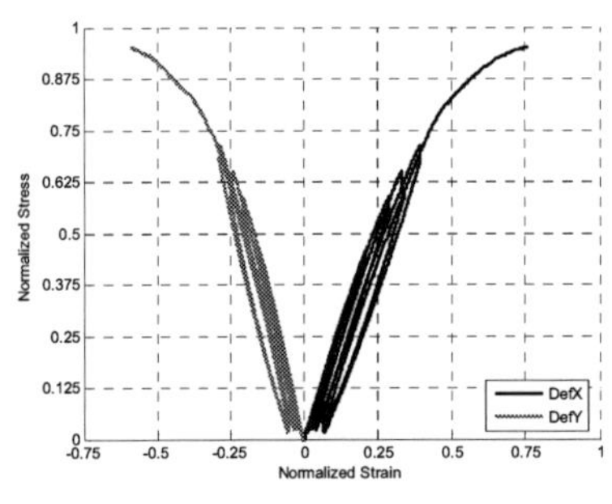

Figure 10a. Stress versus longitudinal and transversal strain for tensile test at 45° off axis.

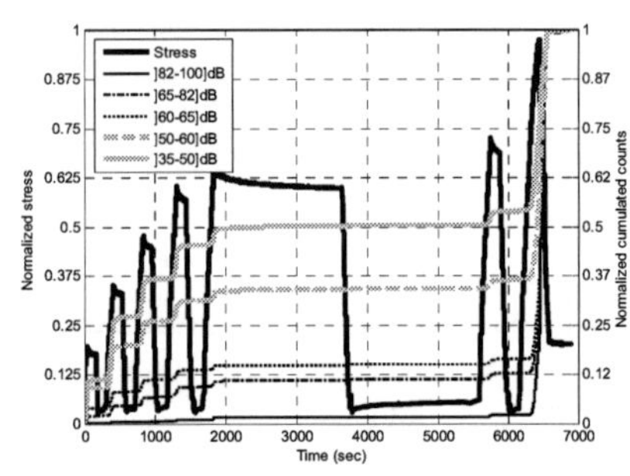

Figure 10b. Acoustic activity during tensile at 45° off-axis according to the different ranges defined (cumulated counts).

4.5 Creep Test at 45° Off-Axis

To mainly highlight the viscous behavior of this interlock fabric, a multi-level creep test at 45° off-axis was also fulfilled (submitted to a constant applied stress).

Specimen was submitted to a constant applied stress with duplicated creep levels during 1 hour (see Figure 11b). That kind of creep test allows characterization of the non-linear visco-elasticity behavior (see Figue 11a). The aim of these two kinds of 45° off-axis tensile tests was to try to separate the damage and the viscous behavior for an accurate identification of the phenomenological macro model.

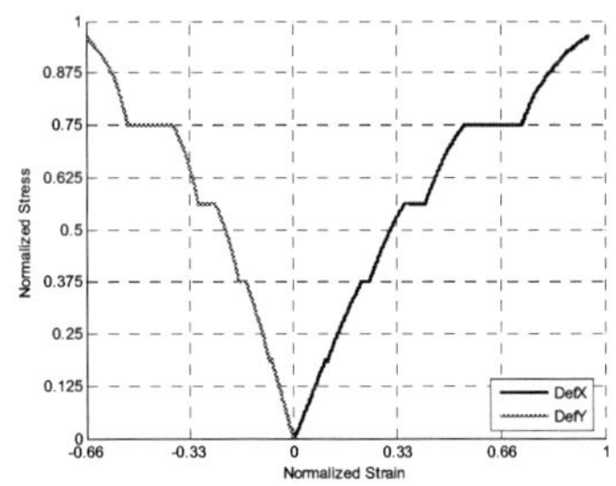

Figure 11a. Stress versus longitudinal and transversal strain for multi-level creep test at 45° off-axis.

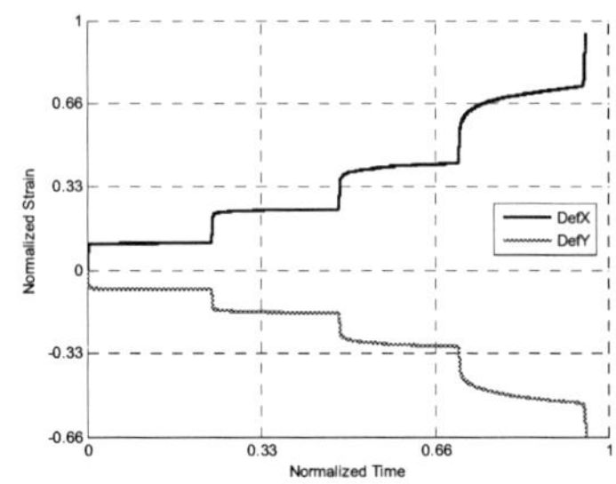

Figure 11b. Strain versus time for multi-level creep tensile test at 45° off-axis.

5 CONCLUSION

An experimental investigation to perform an accurate identification of a 3D interlock woven fabric composite has been proposed through five tensile tests.

Obviously, the use of several techniques like full-field, acoustic emissions, and observations was strenuous to set up. Nevertheless, the cross-checking of several investigation techniques allow to built a self-improving approach and enhance the confidence towards understanding and conclusion. With woven fabric composite more and more complex, all this information was required to carry out accurate damage scenarios.

This work emphasise the damage mechanisms key points for this kind of material and have allowed the development and the identification of a phenomenological macroscopic viscoelastic damage model used to design structures. To evaluate predictive character of the proposed model, a comparison with experimental data resulting of a specimen submitted to torsion was performed [11].

Further works concern the improvement of the cross checking of the several investigation techniques for off-axis and off-plane tests.

Thanks go to SNECMA for its technical and financial support. This work takes place in the framework of the MAIA mechanical research and technology program sponsored by CNRS, ONERA and SAFRAN Group.

6 REFERENCES

1. J.P. Quinn, A.T. McIlhagger, R. McIlhagger, "Examination of the 3D woven composites". *Composites Part A : Applied science and manufacturing 2008* : 273–283.
2. L. Marcin, N. Carrère, J-F. Maire, "A macroscopic viscoelastic damage model for three-dimensional woven fabric composites" presented at the European Conference on Composite Materials (ECCM13), Stockholm, Sweden, June 2–5 2008.

3. G. Besnard, F. Hild, and S. Roux, "Finite-element displacement fields analysis from digital images: Application to Portevin-Le Châtelier bands". *Experimental Techniques*, vol. 46, pp 789–803, 2006.
4. J. Schneider, Z. Aboura, K. Khellil, M.L Benzeggagh, D. Marsal. "Contribution of damage mechanisms comprehension of interlock-reinforced composite materials" presented at the European Conference on Composite Materials (ECCM13), Stockolm, Sweden, June 2–5 2008.
5. J.-F. Maire and J.-L. Chaboche, "A new formulation of Continuum Damage Mechanics (CDM) for Composite Materials". *Aerospace Science and Technology*, pp. 247–257, 1997.
6. S. Barré, M.L Benzeggagh "On the use of acoustic emission to investigate damage mechanics in glass-fibre-reinforced polypropylene", *Comp. Sci. Tech.*, 1994, 52, pp 369–376.
7. D.Scida, Z.Aboura & M.L Benzeggagh "Ageing effect on the damage events in woven composites materials under different loading conditions" *Composite Science and Technology* 2002 Vol. 62 pp 551–557.
8. C. El Hage, 2006, "Modélisation du comportement élastique endommageable de matériaux composites à renfort tridimensionnel", PhD thesis of University of Compiègne (France).
9. IM. Daniel, J-J. Luo, PM. Schubel, "Three-dimensional characterization of textile composites", *Composites: Part B : Engineering* 2008 : 13–19.
10. MJ. Lodeiro, WR. Broughton, GD. Sims. "Understanding the limitations of the through-thickness test methods". In : Proceedings of the 4th European Conference on Composites : Testing and standardization. London: IOM Communications, Ltd; 1998. p. 80–90.
11. L. Marcin, J. Schneider, M. Kaminski, J.-F. Maire, D. Marsal, "Experimental investigation and behavior modeling of a 3D interlock woven fabric composite: Part 2" presented at this conference.

Experimental Investigation and Behavior Modeling of a 3D Interlock Woven Fabric Composite: Part 2

L. MARCIN, J. SCHNEIDER, M. KAMINSKI, J.-F. MAIRE, and D. MARSAL

ABSTRACT

The aim of this study is to propose a macroscopic viscoelastic damage model for 3D interlock woven with polymer matrices. This macroscopic model is built in the thermodynamic framework with internal variables. It describes the degradation mechanisms experimentally observed and takes into account the viscous behavior due to the polymer nature. With the aim at evaluating predictive character of the proposed model, a comparison with experimental data resulting of a specimen subjected to torsion is performed.

INTRODUCTION

Composite materials are used for about thirty years in military and civil aircraft industries in order both to reduce fuel consumption and improve mechanical performance. Indeed, laminates using unidirectional plies are now used to manufacture primary aircraft structure such as A380 centre wing box. Nevertheless, for a few years, weaving processes have been used to produce 3D woven fabric composites. This kind of composite materials allows manufacturing complex shapes in few operations and so reduce manufacture costs [1]. The use of these innovative materials in aircraft structures requires development of constitutive law that take into account the non-linear behavior in order to design aircraft structures and so decrease certification costs. This work aims at proposing a macroscopic viscoelastic damage model for 3D interlock woven with polymer matrices. In order to show the efficiency of the proposed approach, a comparison is performed with experimental data from torsion tests on a bar with a rectangular cross section.

L. Marcin, M. Kaminski, J.-F. Maire, Composites Structures and Materials Department, ONERA/DMSC ,BP 72, 29 Avenue de la Division Leclerc, F-92322 Châtillon Cedex, France.
J. Schneider, Université de Technologie de Compiègne, Laboratoire de Mécanique Roberval UMR CNRS 6066, Dep. GM – Polymères et Composites, BP 20520, F-60200 Compiègne Cedex, France.
D. Marsal, SNECMA Villaroche, F-77550 Moissy-Cramayel, France.

VISCOELASTIC DAMAGE MODEL FOR 3D WOVEN COMPOSITE

The macroscopic model, developed in the thermodynamics frame with internal variables, describes the degradation mechanisms above first failure and takes into account the viscoelastic behavior due to the polymer matrix (see Figure 1). The current formulation is based on a damage approach developed for a few years at Onera [2]. The proposed formulation leads to a fully 3D model that takes into account possible couplings during multi-axial loading.

Matrix Damage Modeling

The damageable behavior is modeled in the framework of the Continuum Damage Mechanics. The damage model allows to describe the physical phenomena experimentally observed in this kind of material: initial anisotropy, effects of damage on the elastic behavior, unilateral character of damage and residual strain due to the damage. Due to mechanical contrast between yarn and matrix, damage is preferentially oriented by the microstructure. Therefore three scalar damage variables (d_1, d_2, d_3) are used.

The behavior law in the material axis is defined as following:

$$\tilde{\sigma} = \tilde{\tilde{C}}^{eff} : (\tilde{\varepsilon} - \tilde{\varepsilon}^o - \tilde{\varepsilon}^{th}) - \tilde{\tilde{C}}^o : (\tilde{\varepsilon}^s + \tilde{\varepsilon}^r - \tilde{\varepsilon}^o) \qquad (1)$$

with $\tilde{\tilde{C}}^o = (\tilde{\tilde{S}}^o)^{-1}$ $\tilde{\tilde{C}}^{eff} = (\tilde{\tilde{S}}^{eff})^{-1}$, $\tilde{\tilde{S}}^{eff} = \tilde{\tilde{S}}^o + \sum_i \eta_i d_i \tilde{\tilde{H}}_i^o$ and $\tilde{\varepsilon}^{th} = \tilde{\alpha}(T - T^o)$

where $\tilde{\varepsilon}$ is the total strain, $\tilde{\varepsilon}^{th}$ the thermal strain, $\tilde{\varepsilon}^o$ the closing strain (taking into account residual strain due to material forming), $\tilde{\varepsilon}^s$ stored strain, $\tilde{\varepsilon}^r$ residual strain (depending on damage), $\tilde{\tilde{S}}^o$ the initial compliance, $\tilde{\tilde{H}}_i^o$ a matrix damage-effect tensor on the elastic behavior, d_i the scalar damage variable and η_i the damage deactivation index.

The damage evolution follows the standard thermodynamic framework, with the limits of the non-damaging domains. Independent criteria are used for each scalar damage variables (y_i^n and y_i^t are thermodynamic forces associated with variables d_i):

$$F_i(y_i, d_i) = g_i^n(y_i^n) + g_i^t(y_i^t) - d_i \qquad (2)$$

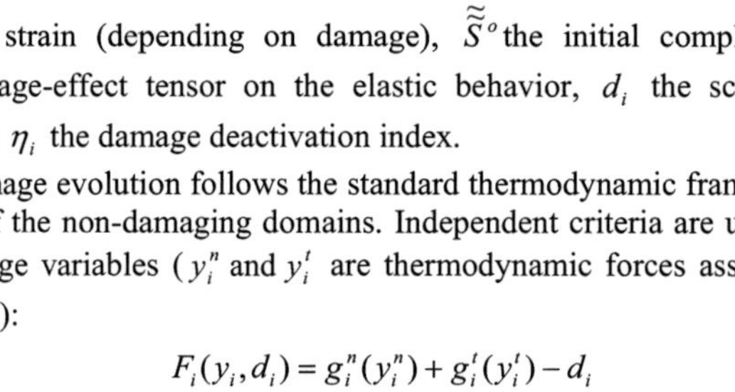

Figure 1. Different aspects of a viscoelastic damage model for 3D interlock woven.

In order to describe saturation effects, commonly observed in composites, the kinetics of the damage variables have been written in an exponential form as follow (d_c, y_o, y_c and p are parameters to determine):

$$g_i = \mathrm{d}_{c(i)} \cdot (1 - \exp - \left(\frac{< \sqrt{y_i} - \sqrt{\mathrm{y}_{o(i)}} >}{\sqrt{\mathrm{y}_{c(i)}}} \right)^{\mathrm{p}^i}) \quad i = \{j\_n; j\_t\}, j \in [1;3] \quad (3)$$

Viscous Behavior Modeling

Under shear loading, a main part of the non-linearity is due to the nonlinear behavior of the polymer matrix. In order to take into account this viscoelastic behavior a model based on a spectral formulation and using a non-linear function has been chosen [3]. The main idea of this approach consists in expanding the viscosity into a series of elementary viscous mechanisms. This viscoelastic model is able to describe creep and relaxation tests and to take also into account the influence of the loading rate on the macroscopic behavior.

The behavior law, deriving from a thermodynamic potential, is defined as follow:

$$\tilde{\sigma} = \tilde{\tilde{C}}^o : (\tilde{\varepsilon} - \tilde{\varepsilon}^{th} - \tilde{\varepsilon}^{ve}) \text{ with } \tilde{\varepsilon}^{th} = \tilde{\alpha} \cdot (T - T_o) \quad (4)$$

where $\tilde{\tilde{C}}^o$ is the elastic tensor, $\tilde{\varepsilon}^{th}$ is the thermal strain, $\tilde{\varepsilon}^{ve}$ is the anelastic strain which is determined as follow:

$$\dot{\tilde{\varepsilon}}^{ve} = g(\tilde{\sigma}) \sum_i \dot{\tilde{\xi}}_i \text{ with } \dot{\tilde{\xi}}_i = \frac{1}{\tau_i} (\mu_i g(\tilde{\sigma}) \tilde{\tilde{S}}^R : \tilde{\sigma} - \tilde{\xi}_i) \quad (5)$$

where $\tilde{\tilde{S}}^R$ is the viscous tensor (homogeneous to a compliance), $\tilde{\xi}_i$ is a family of 2nd order tensors homogeneous with a strain and corresponding to the elementary mechanisms of viscous flow. Each mechanism $\tilde{\xi}_i$ is associated with a relaxation time τ_i, weighted by μ_i. The whole weight μ_i describes a Gaussian continuous spectrum.

$$\tau_i = \exp(i) \text{ et } \overline{\mu}_i = \frac{1}{n_o \sqrt{\pi}} \exp\left(- \left(\frac{i - n_c}{n_o} \right)^2 \right) \text{avec } \mu_i = \frac{\overline{\mu}_i}{\sum_i \overline{\mu}_i} \quad (6)$$

where $g(\tilde{\sigma})$ is a non-linear function.

The coupling between matrix damage and viscosity is introduced as following. First of all, thermodynamic forces depend on total strain which allow to take into account time-effect on damage evolution. Secondly, the effect of damage on the viscoelastic behavior is introduced in the viscous tensor as follow (H_i^{ve} is a matrix damage effect tensor on the viscoelastic behavior):

$$\tilde{\tilde{S}}_{eff}^{R} = \tilde{\tilde{S}}^{R} + \sum_i \eta_i d_i \tilde{\tilde{H}}_{i\_t}^{ve} + (1-\eta_i) d_i \tilde{\tilde{H}}_{i\_c}^{ve} \tag{7}$$

Softening Behavior Modeling

For structure with high stress, gradient structural strength is generally underestimated if a first element failure criterion is used [4]. Therefore effect of failure is described so as continuum. A distinction between plan failures (such as yarn crack in warp and weft direction) and out-of-plane failure (such as meso debonding) is made.

The behavior law in the material axis is written as:

$$\sigma = \tilde{\tilde{C}}^{eff} : \tilde{\varepsilon} \text{ with } \tilde{\tilde{C}}^{eff} = (\tilde{\tilde{S}}^{eff})^{-1},\ \tilde{\tilde{S}}^{eff} = \tilde{\tilde{S}}^{o} + \sum_i \eta f_i df_i \tilde{\tilde{Hf}}_i^{o} \tag{8}$$

where $\tilde{\varepsilon}$ is the total strain , $\tilde{\tilde{S}}^{o}$ the initial compliance , $\tilde{\tilde{Hf}}_i^{o}$ softening damage-effect tensor, df_i softening scalar damage variable and ηf_i the damage deactivation index (generally equal to one).

More precisely, a scalar damage variable df_i is introduced for each mode (tensile, compression and shear) and direction (warp, weft and out-of-plane). For damage growth, independent criteria are used for each scalar damage variables:

$$F_i(yf_i, df_i) = g_i^f(yf_i) - df_i \leq 0 \tag{9}$$

In order to avoid numerical problems related to this type of modeling (mesh dependence, localization) a numerical tool such as the delay effect is used [5]. The main idea of this approach is to limit damage rate evolution:

$$\dot{df}_i = \frac{1}{\gamma_i}(df_i^{nc} - df_i) \tag{10}$$

where γ_i is the characteristic time, df_i the current softening damage, df_i^{nc} the softening damage without delay effect.

MODEL IDENTIFICATION STRATEGY AND APPLICATION

In order to identify the model, a strategy has been defined. The key point of this strategy is to be established on few tests.

Firstly, in plane damage matrix parameters are identified thanks to an incremental tensile-compression test at 0°, another at 90° and one at 45°.

Secondly, in plane viscous behaviour parameters are identified thanks to a creep test in three directions: 0°, 90°, and 45°. Thirdly parameters linked to softening behaviour may be identified thanks to tests performed on holed plates subjected to uniaxial tensile loading. It is very difficult to carry out experimental tests at the coupon level in order to identify the out-of-plane behavior of material. Nevertheless out-of-plane behaviour could be determined thanks to structural tests that induce out-of-plane stresses, such as an unfolding test on a corner L-angle.

The in-plane parameters of the proposed model have been determined thanks to test results presented in [6]. Following figures show an illustration of the model identification. More precisely, Figure 2 shows identification of the model on an loading-unloading tension test in the warp direction for a 3D interlock woven with polymer matrix. While Figure 3 shows identification performed on an off-axis creep test at 45°. These figures outline the efficiency of the proposed formulation to reproduce correctly non-linear behavior due to viscosity and degradation mechanisms

COMPARISON WITH A TORSION TEST ON BAR WITH RECTANGULAR CROSS SECTION

In order to show the efficiency of the proposed model, a comparison with tests on a bar with rectangular cross section subjected to torsion has been performed, specimen axis being parallel to the warp axis.

Experimental Results

Bars manufactured with a 3D interlock woven have been tested. In order to have contrast between plane shear and out-of-plane shear, different shapes have been subjected to torsion. Moreover with the aim at outlining viscous behavior, diverse loading rate have been chosen. Table I presents shape dimensions and loading rate.

TABLE I. SHAPE DIMENSIONS AND LOADING RATE.

Shape	Dimension (mm×mm×mm)			V (°/min)
	length	wide	thickness	
C1 (×2)	182	30	4	9
C2	185	30	4	3
C3 (×2)	185	74	10	9
C4	185	30	10	9
C5	185	74	10	5

In this paper we focus on results obtained for the shape C4. This torsion test was observed in situ with a video microscope to evaluate the damage evolution on the surface. This torsion loading generates a state of stress damageable for the matrix mainly (confirmed by Acoustic Emission), thus accentuating the viscous aspect of the material behavior. This kind of test is interesting because it induces a multi-axial stress field.

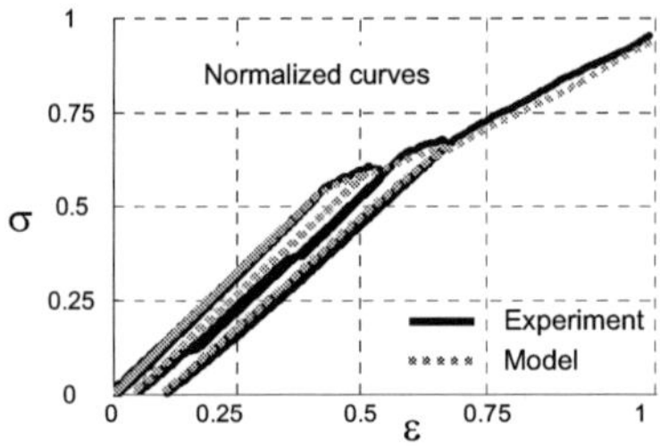

Figure 2. Incremental loading-unloading test in the warp direction.

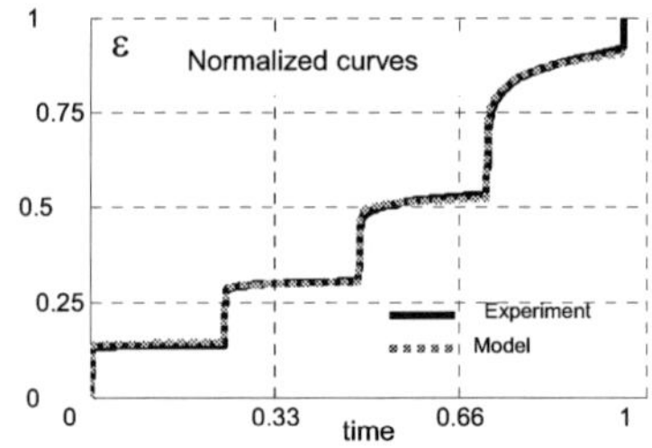

Figure 3. Multi-level creep test at 45°.

Numerical Results

In order to show the influence of non-linearities on the global response, various FE calculations have been performed (see Figure 4). Drawback of the delay effect approach is to introduce an artificial dependence on loading rate. The use of viscoelastic behavior together with this kind of approach is still problematic for static simulations. Therefore, for this current study this two non-linearities are not used together. Because of high rotation (more than 45°), calculations have been performed under high displacement assumptions. Figure 5 & 6 provides a comparison between simulation and experimental results. For each simulation different part of the proposed model are activated as follow:

- Simulation 1: elastic behavior,
- Simulation 2: non softening elastic damage behavior,
- Simulation 3: softening elastic damage behavior,
- Simulation 4: non softening viscoelastic damage behavior.

First of all, for this test, to take into account non softening behavior has no influence on the global stiffness. Secondly, to take into account viscoelastic behavior of the matrix allows to have a quite good agreement with experimental results. Indeed this test induces mainly shear stresses which outline material viscous behavior. Finally some works have to be done to improve unloading path. This is one of the drawbacks of usually viscoelastic formulation [7]. A way to correct this drawback is to introduce for instance plasticity in the formulation.

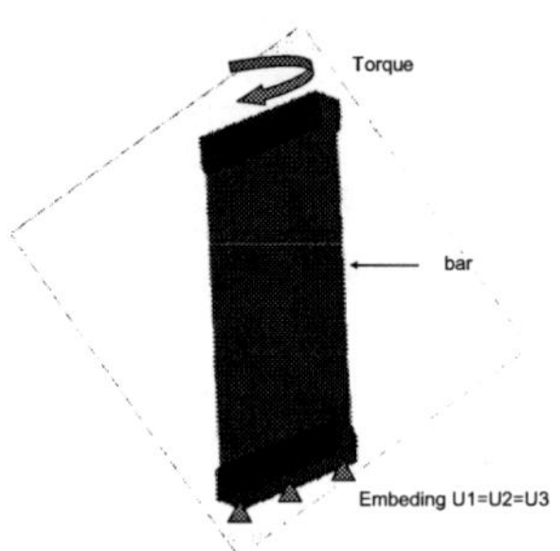

Figure 4. Mesh and boundary conditions for simulation of a Torsion test on a bar with a rectangular cross-section.

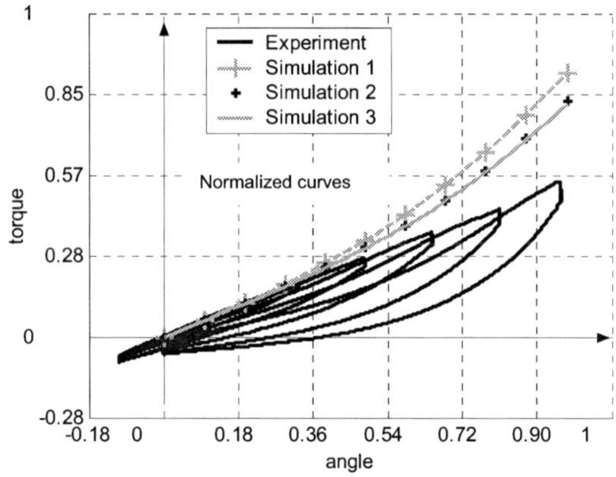

Figure 5. Torsion test on a bar with a rectangular cross-section—Global level.

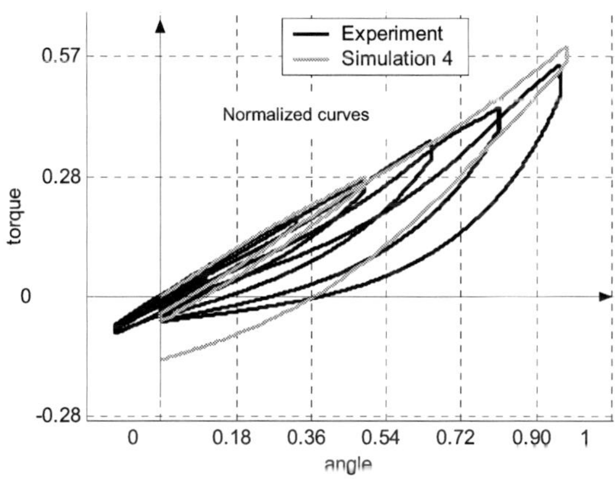

Figure 6. Torsion test on a bar with a rectangular cross-section—Global level.

The proposed model allows to have an idea of amount of matrix damage induced by this kind of test. This kind of results permits to have an idea of damage localization both on surface and in the shape in complement with experimental observations (Acoustic Emission, in-situ observation).

CONCLUSION

A macroscopic viscoelastic damage model for three-dimensional woven fabric composites with polymer matrix has been proposed. This phenomenological model is built from the knowledge of the mechanical properties of this kind of composite. Identification of the model has been performed thanks to classical carry out at the coupon level. Predictive character of the model has been tested on a bar subjected to torsion. Indeed predictions are in quite good agreement with experimental data even if improvements have to be done to have a better accuracy with experimental results.

Further works concern the improvement of results in order to improve accuracy with experimental data (in particularly for unloading) and have a better understanding of degradation mechanisms with the aim at enhancing experiment.

ACKNOWLEDGEMENT

Thanks go to SNECMA for its technical and financial support. This work takes place in the framework of the MAIA mechanical research and technology program sponsored by CNRS, ONERA and SAFRAN Group.

REFERENCES

1. A.P. Mouritz, M.K. Bannister, P.J. Falzon and K.H. Leong, 1999, "Review of applications for advanced three-dimensional fibre textile composites", *Composites Part A: Applied Science and Manufacturing*, pp.1445–1461.
2. J.-F. Maire and J.-L. Chaboche, 1997, "A new formulation of Continuum Damage Mechanics (CDM) for Composite Materials", *Aerospace Science and Technology*, pp 247–257.
3. J.-F. Maire, 1992, "Etude théorique et expérimentale du comportement de matériaux composites en contraintes planes", *PhD thesis of University of Besançon (France).*
4. J. M. Withney, R. J. Nusimer, 1976, "Strain gradient in composite laminate structure", *Journal of Composite Material*, 35:733–45.
5. P. Ladevèze, 1995, "A damage computational approach for composites: Basic aspects and micromechanical relations", Computational Mechanics, 17: 142–150.
6. J. Schneider, L. Marcin, Z. Aboura, D. Marsal, "Experimental investigation and behavior modeling of a 3D interlock woven fabric composite: Part 1", presented at the 9th International Conference on Textile Composites, October 13–15, 2008.
7. Schapery R.A, 1969, "On the characterization of non linear viscoelastic materials", *Journal of Polymer Engineering Sciences*, vol 9, pp 295–310.

TOPIC IX

GEOMETRY AND MECHANICAL BEHAVIOR OF DRY TEXTILES AND PREFORMS

General Definition of Warp Interlock Structures

F. BOUSSU, X. LEGRAND and C. BINETRUY

ABSTRACT

3D textile composites have a vast range of properties that are superior to traditional 2D laminates. Among all these 3D woven fabrics, warp interlock structure is commonly defined into three main types as layer-to-layer, angle and orthogonal. However, beyond this quick clustering, the definition of any type of warp interlock fabric appears more complex due to the different available values of the numerous parameters like, for instance, the description of all yarns and their location inside the multi-layer fabric. The objective of this paper is to give a clear and sufficient definition of any kind of warp interlock fabric and then to ensure a more realistic 3D modeling view.

INTRODUCTION

Different 3D textures can be obtained from the expertise of weaving, braiding or knitting. A classification of textile textures has been proposed by Nemoz [1] on the basis of:

- the number of axis or fiber directions in the fabric plane,
- the number of directions in space or though the perpendicular axis of the fabric plane,

Uni-dimensional texture: the simplest, is a set of fibers, filaments or parallel threads.

Two-dimensional texture or textile in surface as:

- Non-woven or mats whose fiber orientation is random,

François Boussu, Univ. Lille North of France, F-59100, ROUBAIX, ENSAIT, GEMTEX
Xavier Legrand, Univ. Lille North of France, F-59100, ROUBAIX, ENSAIT, GEMTEX
Christophe Binetruy, Polymers and Composites & Mechanical Engineering Department, Ecole des Mines de Douai, 59500 Douai, France.

- Unidirectional layers made of parallel fibers or yarns linked together by a resin or a weft yarn, soluble or volatile, of low-density
- Warp and weft fabrics with two orthogonal axes or tri-axial or multi-axial textiles.

Three-dimensional texture or textile in volume:

- Shaped fabrics of knits or double face structures,
- 3D braids in volume or in shape layer to layer,
- Double face non-woven linked by stitched yarns.

Three-dimensional preforms can be made using weaving, braiding, knitting and non-woven processes. However, each technology has its own strength and weakness as described in Zhang [15].

3D braiding composites have the following disadvantages:

- The mechanical properties of braided composites are not constant and large variation was found in tests performed by Crane and Camponeschi [2]
- The in-plane modulus of braiding structure can be significantly lower than that of unidirectional composite due to the large angle between fibers and loading direction.

3D knitting composites have the following disadvantages:

- Most conventional knitting machines cannot produce thick preforms.
- Because the stretching of the fabric during fabrication will result in a structure change, knitted composites usually contain 'soft spots' and 'hard spots' as described in Mouritz et al. [3].

Stitching application is restricted by several drawbacks:

- Most stitching machines can only handle less than 1 m wide and 5 mm thick preforms. Large capacity sewing machine is usually very costly.
- Sewing machine cannot stitch curved composites; neither could sewing machine handle complex structures.
- Stitching breaks the fiber in the needle path and degrades the in-plane properties of the composites.

Compared with 2D laminates, 3D woven composites have the following advantages:

- 3D weaving can produce complicated near-net-shape preforms. This capability greatly reduces the material cost and handling time. The integrated complex shaped structure tends to have better mechanical properties
- Through-thickness properties can be adjusted by controlling the amount of z yarns.
- Z yarn could arrest the cracks formed during impact loading, therefore, 3D woven composite has high ballistic impact damage resistance and low velocity impact damage resistance. 3D composite will also have improved post-impact mechanical properties than that of the 2D laminates [4][5]
- 3D woven composites have a higher failure strain [6][7] and mode I fracture toughness than that of 2D laminates

• Woven fabrics usually exhibit good dimensional stability in the warp and weft directions. It also has highest yarn packing density and provides higher out-of-plane strength. Thus it can carry higher secondary loads and local buckling [8]

• Woven fabrics have a very low shear rigidity that gives a very good formability [9]

The angle interlock woven fabrics are widely used in the fabrication of composite materials because of higher inter-laminar shear strength as mention in Nie et al. [10].

But what is the main definition of an angle or warp interlock fabrics? First, several definitions can be found in the literature and some papers are focused on the problem to describe these structures by some models of representation.

WARP OR ANGLE INTERLOCK STRUCTURE

Warp Interlock Structure Definitions

Two of the most common architectures are the orthogonal and layer interlock weaves.

Nie et al. [10] begin their paper by asking the problem to represent the angle-interlock fabric in a more concise and efficient way, which is a major problem for the computer-aided design (CAD) of multilayered angle-interlock woven fabric as referred by Ko [11].

Lomov et al. [12][13][14] put forward a symbolized method to characterize the construction of a 3D fabric, including 3D woven fabric and knitted fabric.

A simplified expression method is given by Nie et al. [10] using fractional formulae to represent warp interlacement. The weaving plan of angle-interlock woven fabric with a large number of layers could be easily and simply characterized by this method than any schematic diagram for any layer.

Zhang [15] describes two types of 3D weaving technologies: angle interlocking weaving and orthogonal weaving. Angle interlocking weaving can be further divided into layer-to-layer and through the thickness. Whitney et al. [16] notice that, compared with angle interlock method, orthogonal weaving generally brings much higher through-the-thickness reinforcements. This method also allows greater variability to the movement of the shuttles which could lead to structures with much more complicated cross section shape as mention in Shyong and Bing [17]. Composites made from 3D orthogonal woven preforms and epoxy resin showed superior mechanical properties to those made from other 3D woven preforms as related by Brandt et al.[18][19].

Lomov et al. [20] mention that multi-layer integrally woven reinforcements are often called 3D or warp-interlaced 3D weaves.

Multilayered composite reinforcements are termed:

• Orthogonal when Z-yarns go through the whole fabric between only two columns of weft yarns,

• Through-thickness angle interlock when Z-yarns go through the whole fabric across more than two columns of weft yarns,

• Angle interlock when Z-yarns connect separate layers of the fabric.

General Clustering Of Warp Interlock Structure

As regards the exact definition of a textile structure made by a weaving process including up to two layers of warp, the warp interlock woven fabric appears sufficient to cover all the different types of interlacing warp layers. Thus, the angle, layer to layer or orthogonal interlock clusters must be avoided as the 2,5D warp interlacing

TABLE I. CLUSTERING OF WARP INTERLOCK FABRIC.

Warp Interlocks with constant crimp for each layers	**Warp Interlocks with non constant crimp for each layers**
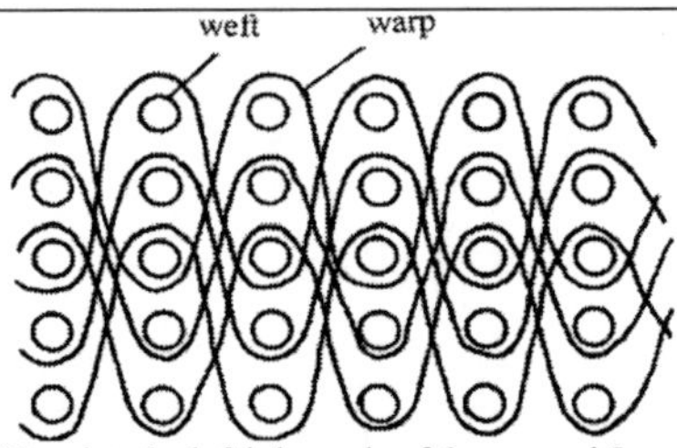 Warp interlock fabric made of 6 warp and 5 weft layers given by Gu and Zhili [21]	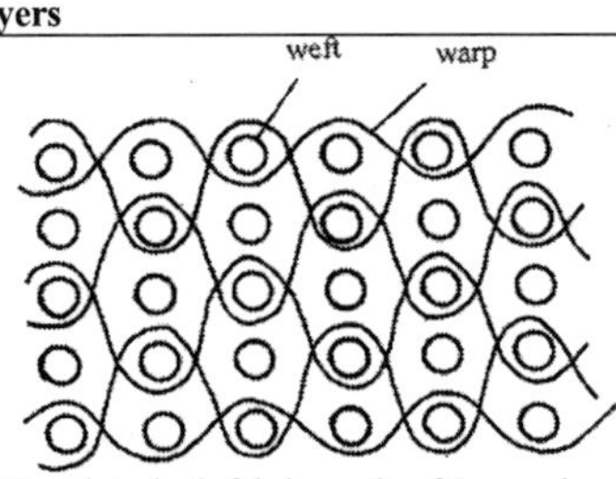 Warp interlock fabric made of 4 warp layers with one crimp value plus 2 warp layers with another crimp value and 5 weft layers
Warp interlock fabric made of 9 warp and 4 weft layers given by Hearle and Chen [22] 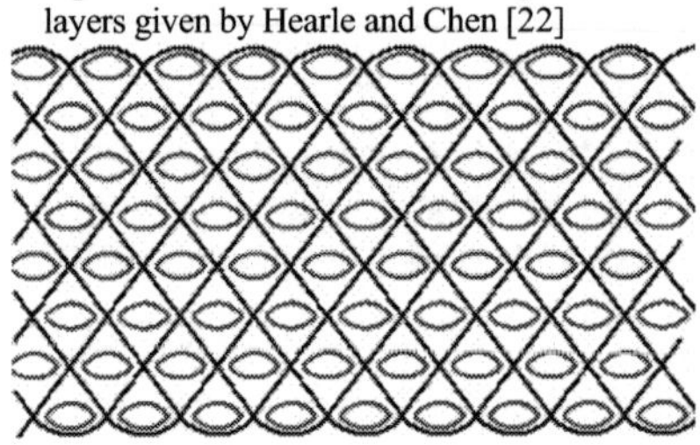	Warp interlock fabric made of 2 warp layers with one crimp value plus 4 non interlocking warp layers described both by Gu and Zhili [21], Zhang [15] and Hearle and Chen [22] as an orthogonal-Interlock structure 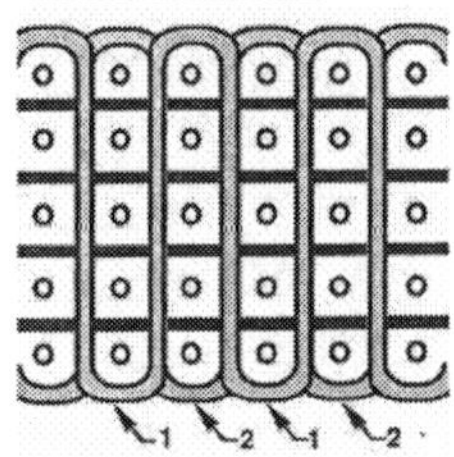

structure or 3D woven fabric which involves an insertion of the Z yarns not included in the warp direction.

A clustering only based on the crimp or warp yarn consumption inside the structure has been recently proposed [23]. This new approach leads to define two warp interlock structure which can differentiated by the constant or non constant crimp for each warp layers. To illustrate the proposed clustering, different warp interlock structures are proposed and classified by the crimp criterion in Table I.

Considering the literature review, a precise description of a three-dimensional angle-interlock woven fabric is given by Sun et al. [24] and detailed in Table II. The fiber volume fraction is about 40% and the thickness of the 3-D woven fabric composites is 3 mm.

TABLE II. WEAVE PARAMETERS OF 3-D ANGLE-INTERLOCK WOVEN FABRIC.

Layers of warp	**12**	**Layers of weft**	**11**
number of warp	Warp 816	Stuffer warp 272	Binding warp 544
Total number of warp 1632	Number of warp per weave repeat 12	Total number of weave repeat 136	
Warp density 41,73 ends/cm		Weft density 99,37 picks/cm	
Average density per layer 3,47 ends/cm		Average density per layer 9,34 picks/cm	
Warp yarn yield Tex 168		Weft yarn yield Tex 254	

It can be noticed that the pick density is higher than the warp density in order to increase the fiber volume fraction but leads to a decrease of the efficiency production rate. But, this description is not enough sufficient to describe entirely the interlock fabric. Thus, to ensure a complete representation of the warp interlock fabric, it is necessary to take into account all the parameters which can be helpful to define the 3D fabric. Thus, the main parameters to define any kind of warp interlock fabric are:

- Number of interlocking or binding warp yarns (up to two layers)
- Number of non interlocking warp yarns (between two weft yarns layers)
- Number of weft yarns layers
- Weave diagram per layer (Number of harness or frames per layer)
- Warping formulae of all yarns (including the different warp yarns yields in Tex) which corresponds to the distribution of the different warp yarns yields inside the fabric
- Weft formulae (including the different weft yarns yields in Tex) which corresponds to the distribution of the different weft yarns yields inside the fabric
- Density of interlocking yarns per layer
- Density of non interlocking yarns per layer (yarns located between no more than two weft yarns layers)
- Density of weft yarns per layer

The following figures will help us to illustrate this new definition of warp interlock fabric.

First, the layer to layer warp interlock fabric is represented in figure 1 with regular section of the different warp and weft yarns.

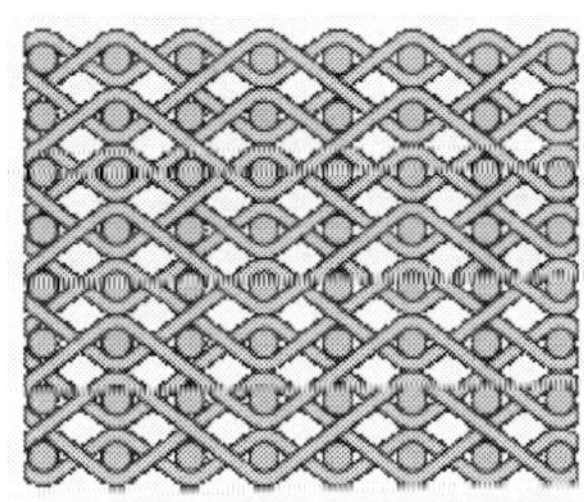

Figure 1. Schematic representations of one layer-to-layer warp interlock structure.

All the parameters given in Table III help to define precisely this warp interlock fabric. It can be noticed the different warp yarn consumption inside the fabric with respect the interlocking and non interlocking warp yarns.

TABLE III. DESCRIPTION OF THE NON CONSTANT CRIMP WARP INTERLOCK.

Number of interlocking or binding warp yarns (up to two layers)	14
Number of non interlocking warp yarns (between two weft yarns layers)	2
Number of weft yarns layers	8
Weave diagram per layer (Number of harness or frames per layer)	2
Warping formulae of all yarns (including the different warp yarns yields in Tex)	1 carbon yarn with 6k filaments (Tex 223)
Weft formulae (including the different weft yarns yields in Tex)	1 carbon yarn with 12k filaments (Tex 446)
Density of interlocking yarns per layer	6 ends/cm per layer
Density of non interlocking yarns per layer	6 ends/cm
Density of weft yarns per layer	2 picks/cm

The representation in Figure 2 of the warp interlock fabric made with different carbon fibers yields reveals the real shape and yarns distribution inside the 3D fabric.

Considering the representation of this warp interlock fabric in Figure 3, the distribution of the weft yarns layers appears non homogenous and the shape the weft yarns looks like more elliptic (or lenticular) than circular.

Figure 2. Injected layer-to-layer warp interlock structure made with different carbon fibers.

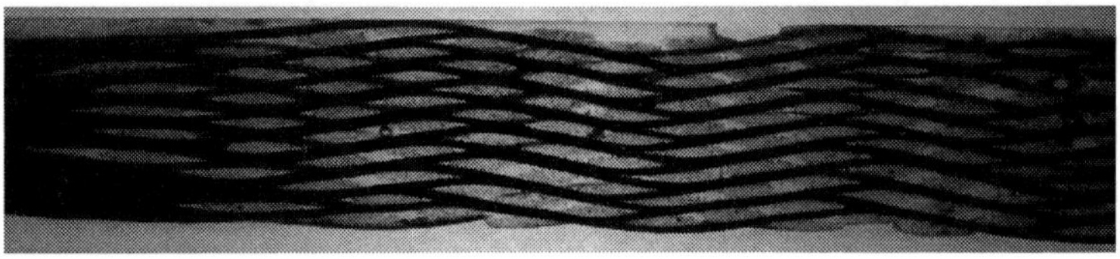

Figure 3. Slice of the layer-to-layer warp interlock fabric.

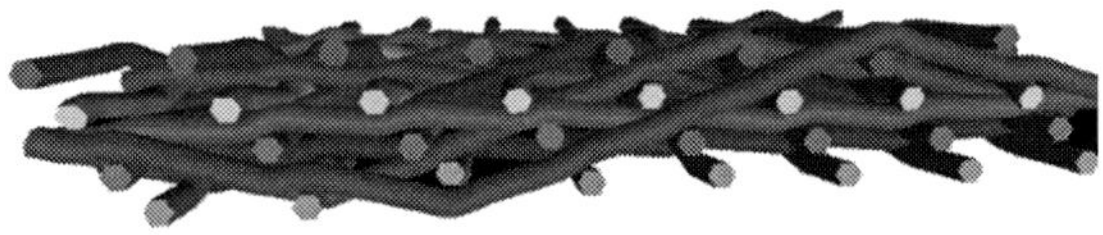

Figure 4. 3D modeling of a layer-to-layer warp interlock fabric.

Second, the 3D modeling representation given in figure 4 of this layer-to-layer warp interlock fabric including non crimp warp yarns is more precisely described with all the parameters included in Table IV.

TABLE IV. NON CONSTANT CRIMP WARP INTERLOCK (10 + 4 WARPS/ 5 WEFTS).

Number of interlocking or binding warp yarns (up to two layers)	8
Number of non interlocking warp yarns (between two weft yarns layers)	3
Number of weft yarns layers	4
Weave diagram per layer (Number of harness or frames per layer)	1
Warping formulae of all yarns (including the different warp yarns yields in Tex)	1 carbon yarn with 6k filaments (Tex 223)
Weft formulae (including the different weft yarns yields in Tex)	1 carbon yarn with 6k filaments (Tex 223)
Density of interlocking yarns per layer	3 ends/cm
Density of non interlocking yarns per layer	4 ends/cm
Density of weft yarns per layer	2 picks/cm

The final shape of the 3D modeling warp interlock fabric considering all these parameters will better fit the real view of the multi-layer fabric.

Thanks to this new definition, all the different types of warp interlock structures can be defined and help to make a finer modeling view.

CONCLUSION

Compared to the previous proposed clustering of warp interlock structures based on the criterion of crimp or warp consumption inside the fabric into two different clear groups, the proposed definition helps to better integrate all the parameters leading to a better fitted modeling view at a macro scale. A precise definition of the warp interlock fabric helps to estimate the final shape of yarns inside the fabric at a meso scale. It can be also estimated the rearrangement of filaments inside the yarn at a micro scale. Thus the interaction between all the scales, micro to meso and meso to macro, will help to better understand the distribution of the different constraints inside the 3D fabric.

Future works tend to better understand the link between a given shape of a warp interlock structure and its different mechanical properties as tensile, compression, shearing, ageing, fatigue and non laminating stress.

REFERENCES

1 Nemoz G. , octobre 2003, “Textures textiles tridimensionnelles », *Techniques de l'ingénieur, traité Plastiques et Composites*, doc. AM 5 122, pp. 1–13.

2 Crane R.M., Camponeschi Jr E.T. , 1986, “Experimental and Analytical Characterization of Multidimensional Braided Graphite/Epoxy Composites”, *Experimental Mechanics*, pp. 259–266.

3 Mouritz A.P., Bannister M. K. Falzon P. J. Leong K. H. , 1999, “Review of applications for advanced three-dimensional fibre textile composites”, *Composites Part A: Applied Science and Manufacturing*, 30 (12), pp. 1445–1461.

4 Dickinson L.C.; Farley G.L., Hinders M.K. , 1999, “Prediction of Effective Three-Dimensional Elastic Constants of Translaminar Reinforced Composites”, *Journal of Composite Materials*, 33 (11), pp. 1002–1029

5 Farley G.L.; Smith B.T., Maiden J. , 1992, “Compressive Response of Thick Layer Composite Laminates with Through-the thickness Reinforcement”, *Journal of Reinforced Plastics and Composites*; 11, pp. 787–810.

6 Cox B.N.; Dadkhah M.S.; Morris W.L., Flintoff J.G. , 1994, “Failure Mechanisms of 3D Woven Composites in Tension, Compression, and Bending”, *Acta-Metallurgica-Et-Materialia*, 42 (12), pp. 3967–3984.

7 Cox B.N.; Dadkhah M.S., Morris W.L. , 1996, “On the Tensile Failure of 3D Woven Composites”, *Composites-Part-A -Applied-Science-and-Manufacturing*, 27 (6), pp. 447–458

8 Tan P., Tong L. Steven G. P. , 1997, “Modelling for predicting the mechanical properties of textile composites - A review”, *Composites*, 28A (11), pp. 903–922.

9 Chou T.W., Ko F.K. , 1989, « Textile Structure Composites », New York, Elsevier.

10. Nie J., Lu S., Gu B. , 2006, “Fractional Formula Description of Angle-interlock Woven Fabric Construction”, *Journal of Industrial Textiles*.

11. Ko, F.K. , 1999, “3-D Textile Reinforced in Composites Materials”, *3-D Textile Reinforcements in Composite Materials*, pp. 9–40, Woodhead Publishing, CRC Press, England, ISBN 1 85573 376 5.

12. Lomov, S.V., Gusakov, A.V., Huysmans, G. , 2000, “Textile Geometry Pre-processor for Meso-mechanical Models of Woven Composites”, *Composites Science and Technology*, 60 (11), pp. 2083–2095.

13. Lomov, S.V., Huysmans, G., Luo, Y., 2001, “Textile Composites: Modelling Strategies”, *Composites (Part A)*, 32 (10), pp 1379–1394.

14. Lomov, S.V., Verpoest, I., Peeters, T. , 2003, “Nesting in Textile Laminates: Geometrical Modelling of the Laminate”, *Composites Science and Technology*, 63 (7), pp 993–1007.

15. Zhang C., 2003, “Characterization and Modelling of 3D woven composites”, a dissertation submitted to the Graduate Faculty of North Carolina State University in partial fulfilment of the requirements for the Degree of Doctor of Philosophy, Mechanical engineering & fibre and polymer science, Raleigh, USA.

16. Whitney T.J., Tsu-Wei C., 1989, “Modeling of 3D angle-interlock Textile Structural Composites”, *Journal of Composite Materials*, 23 (9), pp 890–911.

17. Shyong K.W., Bing C.K., 1999, “Processing and microstructures of 3d woven-fabric composites incorporating solid rods”, *Composites Science and Technology*, 59 (12), pp 1833–1846.

18. Brandt J., Drechsler K., Mohamed M., Gu P., March 9-12, 1992, “Manufacture and Performance of Carbon/Epoxy 3D Woven Composites”, *37th International SAMPE Symposium*, Anaheim, California.

19. Brandt, J., Drechsler K., Arendts F. J., 1996, “Mechanical Performance of Composites Based on Various Three-Dimensional Woven-fiber Preforms”, *Composite Science and Technology*, 56, pp 381–386.

20. Lomov S., Verpoest I., Robitaille F., 2005, “Manufacturing and internal geometry of textiles”, chapter 1, pp 1-60, *Design and Manufacture of textile composites*, edited by A Long, Woodhead Publishing, ISBN : 10-1-85573-744-2.

21. Gu H., Zhili Z., 2002, “Tensile behavior of 3D woven composites by using different fabric structures”, *Materials and Design* 23, pp. 671–674.

22. Hearle J.W.S, Chen X., 2004, “Orthogonal structure and Angle-interlock structure”, TexEng Software Ltd, http://www.texeng.co.uk/index.php.

23. Boussu F., Legrand X., Serret A., November 15–17, 2007, "General Clustering of Warp Interlock Structures", *1st International Conference—Intelligent Textiles and Mass Customization*, ITMC, Casablanca, Morocco, pp. 83–90, ISBN 9954-8878-0-6
24. Sun B., Gua B., Dinga X., 2005, "Compressive behaviour of 3-D angle-interlock woven fabric composites at various strain rates", *Polymer Testing*, 24, pp. 447–454.

Contribution to Formulate Bending Rigidity of Tensioned Yarn

M. K. DOLATABADI and R. KOVAR

ABSTRACT

To product a composite, reinforced with textile materials, it is necessary to know accurate behavior of reinforcement in different conditions of the production. Yarns are mostly one of the basic constructive elements of reinforcements which their properties should be considered for this purpose. The bending rigidity and deformation of yarn had been subjected to study in several investigations; however, they dealt with pure bending deformation by regards to some mechanical theorems. Where as behavior of yarns in many cases dose not follow continuum mechanic laws, some proposed method and standards are not trustworthy. The main attempt of this paper is finding a new approach for better understanding and modeling behavior of tensioned yarns at bending. Accordingly, a new method for measuring bending stiffness of a tensioned yarn is put in forward successfully. Large deflection of tensioned yarn during bending is formulated in this paper. Further more; variation of stiffness of yarn due to change bending angle is the main outcome of this paper.

INTRODUCTION

Response of yarns in structure of reinforcements play vital role in its properties and as well as final productions. Several kinds of reaction can be considered for a yarn in a structure, namely: consolidation due to tensile strain, flattening under compressive forces, bending deformation and finally, in most structure the combination of these deformations. Within these reactions, Stiffness of yarns has a great effect in mechanical properties of textile materials. For example, covering a surface, resistance to shear deformation and stiffness of final product are depended on stiffness of constitutive yarns. Here, some efforts of researchers for determining the bending properties of yarn are presented. M. L. Realff [1] developed a micromechanical model to evaluate tensile properties of woven fabric. He used a simple beam model to determine the resistance of

Mehdi Kamali Dolatabadi and Radko Kovar, Technical University of Liberec, Department of Textile Technology; Hálkova 6 Liberec 46117, Czech Republic the cross yarns for any fabric extension. For this purpose, the four point bending tester for very short length of yarn, 1150 microns, is utilized. Kawabata [2] considered the bending rigidity of a fiber for this purpose. Z. Somodi [3] tried to simulate flexibility testing of textile materials by discrimination of the beam into finite segments. Also, one method had been developed to determine the bending rigidity of yarn by using bending length of yarn [4]. Weidong Yu [5] suggested the bending evaluation system for fabric and yarn (BES-FY) on the base of three points bending at low tension of sample. Jae Ho Jung [6] tried to demonstrate that bending rigidity can effect on capstan equation. However, Backker [7], Platt [8] and J. W. Park[9,10] had been dealt with modeling of bending rigidity of yarn according to mechanical and geometrical relation between fibers in structure of yarn. Nevertheless, all researchers focused on bending rigidity of yarns as a constant quantity. Accuracies of proposed models are devisable. For instance, we tried to use J. W. Park equation for our samples but we do not find good agreement between theory and experiment values. Thus, the main attempt of this paper is find outer factors that can affect of bending rigidity of yarns except of internal factors which researchers dealt with them so far. In other word, most of researchers tried to formulate bending rigidity of yarns by regards to fibers properties and their configuration in structure of yarn. However, outer factors namely imposed tension and amount of bending angle can affect of internal structure and then bending rigidity of yarn.

DEVELOPING A METHOD FOR MEASURING BENDING RIGIDITY

Schematic concept of tensioned yarn at bending is presented in figures 1, 2. Free body diagram of this concept is shown in figure 3. From continuum mechanic and isotropic materials the Euler-Bernoulli beam equation (1) is governed. This equation reveals the relation between curvature and moment at certain point of (x,y) by considering *EI* quantities. Where, *EI* shows stiffness of materials that it consists of *E* (Young's module of material) and *I* (area moment of inertia of beam).

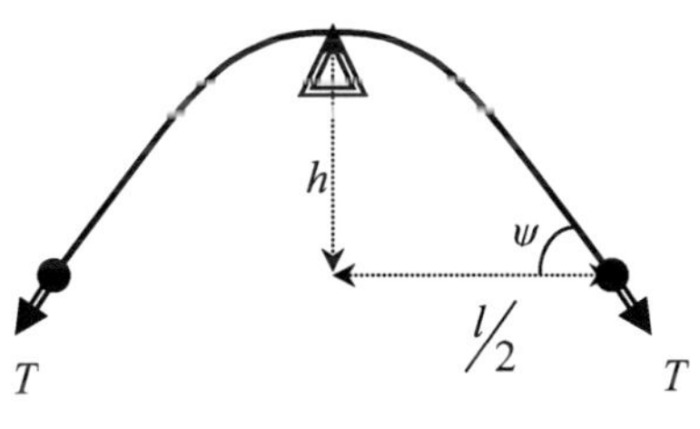

Figure 1. Schematic concept of tensioned yarn at bending.

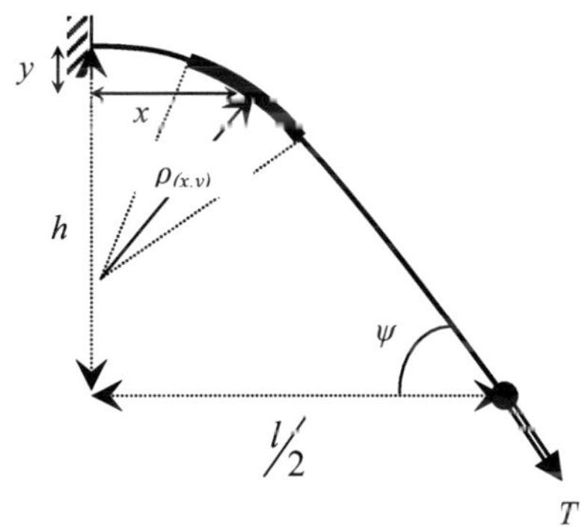

Figure 2. Curvature and cantilever bending at fixed point.

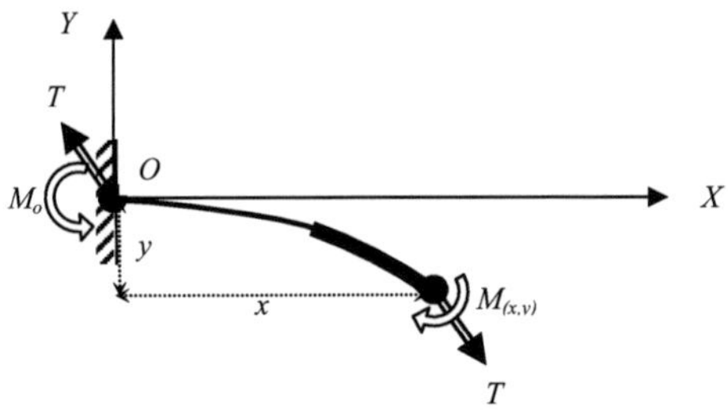

Figure 3.Schematic of free body diagram at point (x,y).

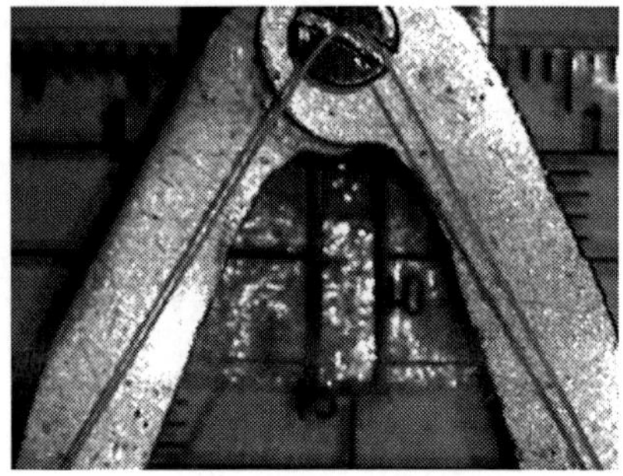

Figure 5. Real yarn at bending.

This quantity is considered as constant value for evaluating deflection of beam in continuum mechanic.

$$\frac{1}{\rho_{(x,y)}} = \frac{M_{(x,y)}}{EI} \tag{1}$$

However, there are some restrictions for applying this equation to evaluate bending stiffness of textile materials as well yarn. First of those is textile materials are not isotropic. Therefore, behavior of textile materials is not the same as continuum materials when they subjected to bending moment. For derivation equation (1) it is assumed that response of materials in compressive and tension forces are the same and the Hooke's equation is governed. It is cleared that response of yarns in axial compressive forces is not follow Hooke's equation. In other hand, when a yarn is subjected to bending moment the geometry and arrangement of fibers in structure are beginning to change. These changes are depended of many factors namely; structure of yarn, technology of spinning, imposed tension and amount of curvature. Despite of these arguments, one can logically consider that curvature of certain point (x,y) is proportional to bending moment of this point, equation (2).

$$\frac{1}{\rho_{(x,y)}} \propto M_{(x,y)} \tag{2}$$

Then, instead of EI quantity we will try to introduce a coefficient of stiffness of yarn as a function of tension and angle of bending such as equation (3). It will be demonstrated empirically that this coefficient will change with change of angle of bending. Further more the consolidation phenomenon is occurred due to increasing of tension. Thus, change in bending stiffness of yarn is expected during increasing tension. Indeed, we will try to demonstrate that this quantity is not constant during bending deformation of yarn.

$$\frac{1}{\rho_{(x,y)}} = \frac{M_{(x,y)}}{f(T,\psi)} \tag{3}$$

Components of main force T are:

$$T_1 = T\sin\psi \tag{4}$$

$$T_2 = T\cos\psi \tag{5}$$

And, moment of bending at point *O* is:

$$M_o = \frac{T_1 l}{2} + T_2 h \tag{6}$$

Also, moment of bending at certain point of (x,y) follows:

$$M_{(x,y)} = T_1 x + T_2 y - M_o \tag{7}$$

From geometry one can shows that curvature is corresponded with two times derivative of yarn deflection function:

$$\frac{1}{\rho_{(x,y)}} = \frac{d^2 y}{d^2 x} = y'' \tag{8}$$

Therefore:

$$y'' f(T,\psi) = T_1 x + T_2 y - M_o \tag{9}$$

$$y'' f(T,\psi) - T_2 y = T_1 x - (\frac{T_1 l}{2} + T_2 h) \tag{10}$$

Outcome equation is a differential equation which the general equation consists of two parts particular and homogeneous respectively. Where, each of them can be solved by regards to differential equation theorems as equations (12) and (13).

$$y = y_p(x) + y_h(x) \tag{11}$$

$$y_p = -\frac{T_1 x}{T_2} + \frac{T_1 l}{2T_2} + h \tag{12}$$

$$y_h = C_1 e^{x\sqrt{\frac{T_2}{f(T,\psi)}}} + C_2 e^{-x\sqrt{\frac{T_2}{f(T,\psi)}}} \tag{13}$$

$$y = -\frac{T_1 x}{T_2} + \frac{T_1 l}{2T_2} + h + C_1 e^{x\sqrt{\frac{T_2}{f(T,\psi)}}} + C_2 e^{-x\sqrt{\frac{T_2}{f(T,\psi)}}} \tag{14}$$

To obtain constant quantities of C_1 and C_2 one can apply boundary conditions (15) and (16):

$$\begin{cases} x = 0 \\ y = 0 \end{cases} \mapsto C_1 + C_2 + \frac{T_1 l}{2T_2} + h = 0 \tag{15}$$

$$\begin{cases} x = 0 \\ y' = 0 \end{cases} \mapsto C_1\sqrt{\frac{T_2}{f(T,\psi)}} - C_2\sqrt{\frac{T_2}{f(T,\psi)}} - \frac{T_1}{T_2} = 0 \tag{16}$$

Therefore:

$$C_1 = -\frac{T_1 l}{4T_2} - \frac{h}{2} + \frac{T_1}{2T_2}\sqrt{\frac{f(T,\psi)}{T_2}} \tag{17}$$

$$C_2 = -\frac{T_1 l}{4T_2} - \frac{h}{2} - \frac{T_1}{2T_2}\sqrt{\frac{f(T,\psi)}{T_2}} \tag{18}$$

Eventually, the function of deflection of yarn can be derived as equation (19):

$$y=(-\frac{T_1 l}{4T_2}-\frac{h}{2}+\frac{T_1}{2T_2}\sqrt{\frac{f(T,\psi)}{T_2}})e^{x\sqrt{\frac{T_2}{f(T,\psi)}}}+(-\frac{T_1 l}{4T_2}-\frac{h}{2}-\frac{T_1}{2T_2}\sqrt{\frac{f(T,\psi)}{T_2}})e^{-x\sqrt{\frac{T_2}{f(T,\psi)}}}$$
$$-\frac{T_1 x}{T_2}+\frac{T_1 l}{2T_2}+h \qquad (19)$$

EXPERIMENTAL WORK

Traditional methods for evaluating bending rigidity of yarns are on the base of cantilever method which one end of yarn is clamped by a gripper and other end of yarn is free. On the base of an ordinary method, the bending length of sample which is inclined due to its weight up to certain angle is used for evaluating bending rigidity of yarn. In other method, the curve of sample is used to compute bending rigidity. This method had been developed at TUL as internal test method [4]. Although, current methods are used in industrial and investigations there exist some defects in this method. The samples are evaluating with tension free while it is clear that the tension can effect on structure and then bending properties of materials under consideration. In addition, clamping one end of sample can change the cross section of sample and we know that this change has a great effect on the bending properties of sample. Other defect of the traditional methods is the twist of sample under test can be changed due to free end of yarn sample. Accordingly, it is tried to overcome to stated problems by improving cantilever method.

EXPLANATION OF APPARATUS

This apparatus is very simple and effective for evaluating bending stiffness of yarn sample under different tensions and different angles. The schematic of this apparatus can be observed in figure 4. The curvature of sample can be changed by adjusting position of levers symmetrically. Two low fraction rollers transfer the imposed tension to whole of the sample. It is interesting to mention that instead of simple support we use a loop from a rigid mono filament. Figure 5 shows a picture from real 30 Tex yarn which is bent up to 60 degree. One of the advantages of this method is the both ends of the sample are involved and bending point is not clamped. Further more, the response of the sample at different angle and different tension can be evaluated.

SPECIFICATION OF MATERIALS

For better understanding of response of tensioned yarns at bending deformation and evaluating $f(T,\psi)$ three different cotton yarns were used. The specifications of these yarns are represented in table 1. Two samples per each test are measured and number of repetition for each sample is four. Thus, eight measurements had been performed fro each sample. Position of sample is recorded by taking a picture after each 10 degree bending. Then, the curve for bent yarn is analyzed by image processing and angle of applied tension is measured. The best value of $f(T,\psi)$ which is satisfied equation (19) is computed by numerical solution method. The interval of bending angle in apparatus is between 10-80 degree and the imposed tension for all samples is 3.53E-3 (N). Figure 6 indicates the measured and evaluated positions of 30 Tex yarn in three different bending angles.

Table 1. Specifications of samples.

Sample code	Spinning tech.	Tex	Twist[t/m]
10Tex	Ring	10	1240
20Tex	Ring	20	882
30Tex	Ring	29.5	710

RESULTS AND DISCUSSION

Figures 7-9 represent the measured values of $f(T,\psi)$ which can satisfied equation (19) to evaluate yarn deflection at certain T and ψ. According these pictures, it is clear that any change in bending angle caused to change in coefficient stiffness at certain tension. Therefore, in structure of yarn bending rigidity or stiffness is not constant and it is strongly depended on bending angle. By increasing bending angle the bending stiffness of yarn reduced. The trend of this change is expected follows an exponential function. At low bending angle variation in results is significantly high. However, the errors of this method in high bending angle are extremely increased. Accordingly, the results in interval 20-70 degrees are trustworthy and this interval of bending angle is more used in textile applications as well as fabrics. Figure 10 illustrates the differences among three empirical equations of yarn's stiffness. It is observed that there is significant difference in evaluated stiffness of yarns. As it is expected the 30 Tex yarn has a much more bending

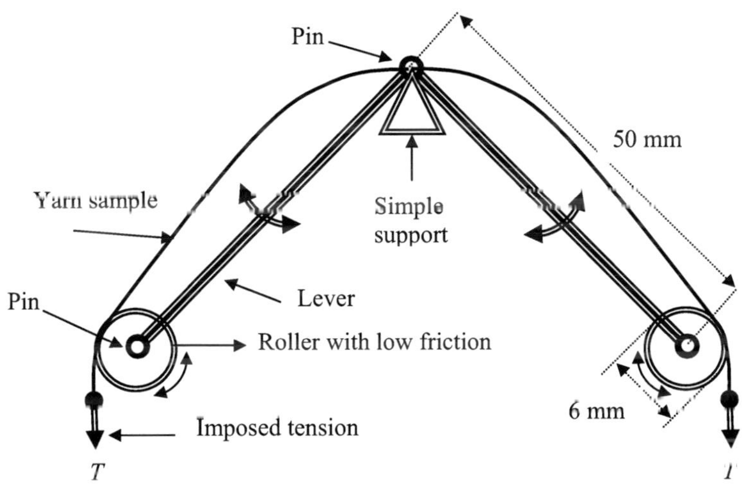

Figure 4. The fundamental principals of bending apparatus.

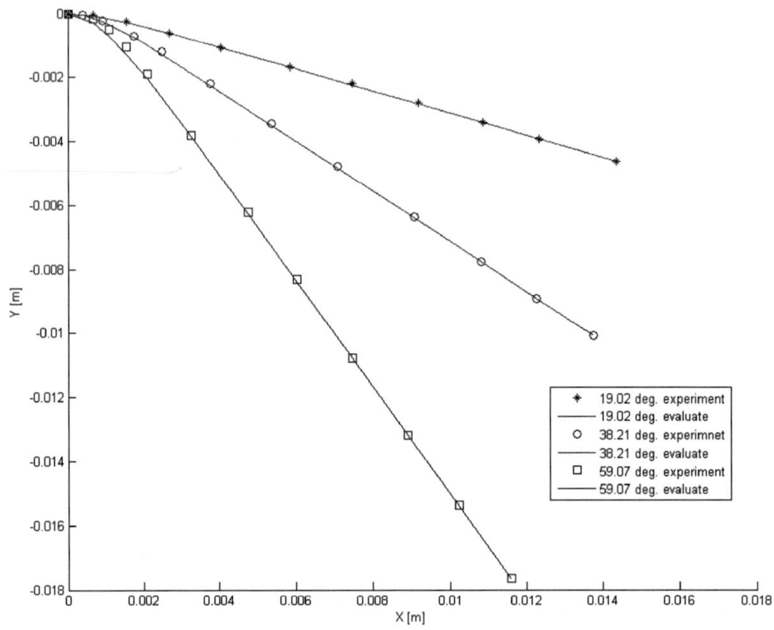

Figure 6. Position of a real 30 Tex sample and evaluated cure at three different bending angle.

stiffness in compare with finer yarns. The general trend of coefficient of stiffness shows that by increasing bending angle the resistance of yarns to bending is decreased. One source of this decrease maybe corresponded with change in cross section. Thus, it is tried to measure changes in cross section dimensions. The outcomes of this study are summarized in figures 11 and 12. These figures reveal that the trend of change in dimensions of cross section of yarns is not continuously subtractive like figure 10 by increasing bending angle. Nevertheless, probably change in configuration of yarns can contribute to these variations.

CONCLUSIONS

The apparatus and extended equation (29) are compatible to evaluate bending stiffness of a tensioned yarn at different bending angle. The coefficient of stiffness of a yarn is depended on bending angle and probably tension of yarn. However, dealing with tension was not the scope of this paper and in future the effect of tension on bending stiffness of yarns would be studied. To formulate bending stiffness of yarns it is necessary to consider bending angle and tension of yarn, due to these factors have a great effect on arrangement and configuration of yarns during bending deformation. Some of the mechanic's theorems are not compatible with textile materials.

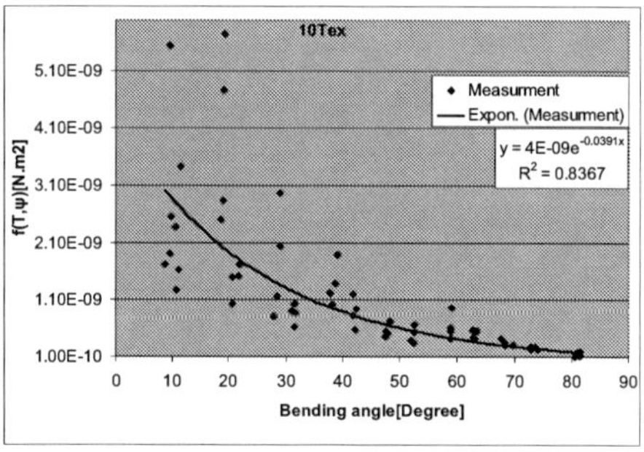

Figure 7. Summarized data of evaluated stiffness v. bending angle for 10 Tex sample.

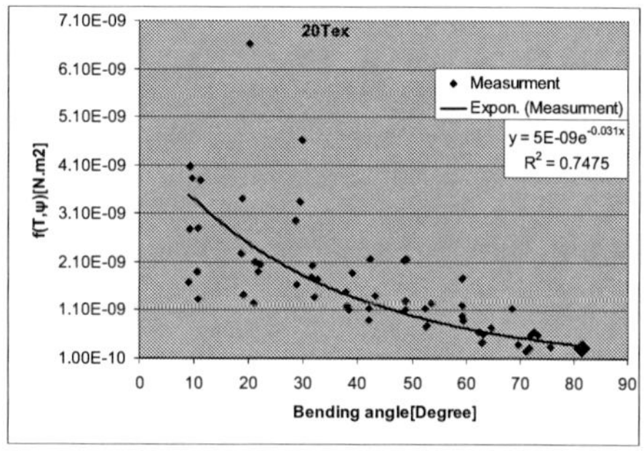

Figure 8. Summarized data of evaluated stiffness v. bending angle for 20 Tex sample.

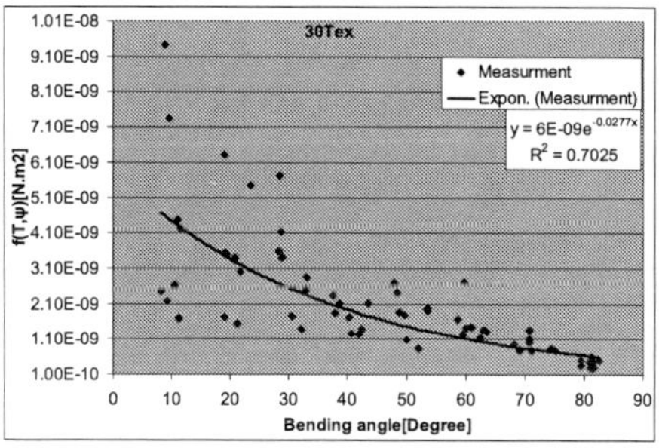

Figure 9. Summarized data of evaluated stiffness v. bending angle for 30 Tex.

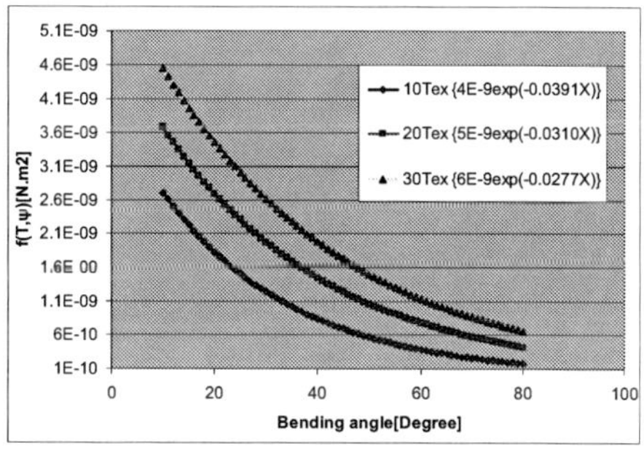

Figure 10. Empirical equations for evaluating stiffness of samples.

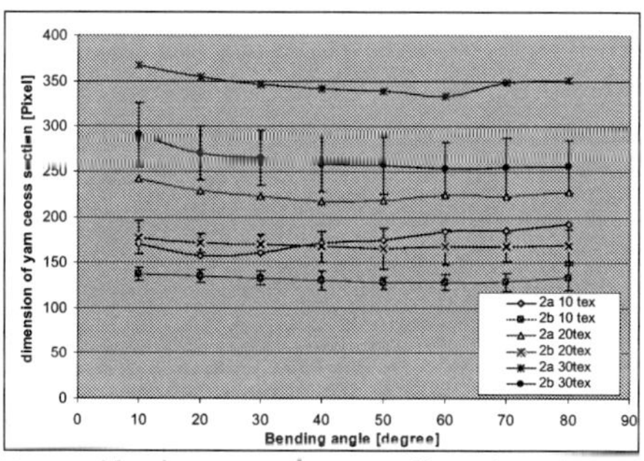

Figure 11. Average of yarns dimension during bending deformation.

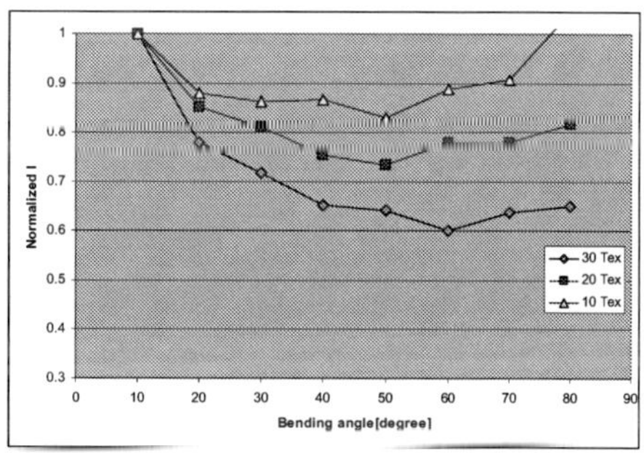

Figure 12. Normalized value of moment of inertia of samples during bending deformation.

For instance, the stiffness of yarns not only is depended on *E* and *I*, but also other conditions namely tension and bending angle which have a great effect on configuration of yarns during bending deformation.

ACKNOWLEDGMENT

This work was supported by the research project 'Research Centre Textile' of Czech Ministry of Education. Also we would like to appreciate 'Azad University, Campus Science and Research' for student funding.

REFERENCES

1. M. L. Realff, M.C. Boyce and S. Backer, A michromechanical model of the tensile behavior of woven fabric, *Textile Res. J.* 67(6), 445–459(1997).
2. Kawabata, S., Niwa, M. and Kawai, H., The finite deformation theory of plain weave fabrics, Part II. *J. Textile Inst.* 64, 21 (1973).
3. Z. Somodi, A. Hursa and D. Rogale, Numerical simulation of textile flexibility testing, *Int. J. of Clothing Sci. & Tech.* vol.15 No. 3/4, 276–283(2003)
4. Internal standard No 22-201-01/01, Bending rigidity of yarn, TU of Liberec.
5. Weidong Yu; Zhaoqun Du, Determination of the bending characteristic parameters of the bending evaluation system of fabric and yarn, *Textile Res. J.* 76, 9,702–711 (2006).
6. J. Jung, T. Kang and J. Youn, Effect of bending rigidity on the capstan equation, *Textile Res. J.* 74,12,1085-1096 (2004).
7. Backer, S. Mechanics of bent yarn. *Textile Res. J.* 22, 668 (1952).
8. Platt, M. M., Klein, W. G. And Hamburger, W. J., Mechanics of elastic performance of textile materials, Part XIV, *Textile res. J.* 28,611 (1958).
9. Jung-Whan Park, Ae-Gyeong Oh, Bending rigidity of yarns, *Textile Res. J.* 76,6,478–485 (2006).
10. Jung-Whan Park, Ae-Gyeong Oh, Bending mechanics of ply yarns, *Textile Res. J.* 73, 6, 473–479(2003).

Tow Visualisation in Woven Composites using X-Ray Computed Tomography

L. P. DJUKIC, I. HERSZBERG, G. A. SCHOEPPNER and L. A. BROWNLOW

ABSTRACT

This paper discusses a technique to visualize the tow architecture in a non-destructive manner, with the final intention of allowing tow visualisations in three dimensional woven carbon-fibre preforms. The tow architecture of plain woven fabric carbon composites, as well as glass composites, has been visualised with the use of MicroCT. Two methods of image acquisition have been discussed. The first uses traditional X-ray imaging. The second uses a newer method known as X-ray phase contrast imaging, combined with the traditional method. Because of the similar densities of carbon and polyester resin it is somewhat difficult to discriminate between tow and resin interfaces when viewed using X-rays, leading to the need for contrast enhancement. This problem is somewhat reduced when using glass in the place of carbon. Attempts have been made to improve contrast, which have included treatments of the fabric and/or resin prior to consolidation. The outcomes from various specimen production are presented and discussed with particular emphasis on the generation of three dimensional images of the tow architecture.

INTRODUCTION

Composite structures manufactured from three dimensional woven preforms have improved through thickness properties compared to those manufactured from unidirectional lamina, however, their in-plane properties are reduced[1]. In order to model the mechanical properties of such structures it is essential to determine the tow architecture and in particular any local deformations due to the interaction of adjacent

Luke P. Djukic[1,2], Israel Herszberg[2], Greg A. Schoeppner[3], Les A. Brownlow[4]
[1]The University of New South Wales, School of Mechanical and Manufacturing Engineering, Sydney, NSW, 2052, Australia.
[2]Cooperative Research Centre for Advanced Composite Structures, 506 Lorimer Street, Fishermans Bend, Victoria 3207, Australia.
[3]Air Force Research Laboratories, Wright-Patterson Air Force Base, Dayton, Ohio, 45433-7750, USA.
[4]XRT Limited, A3.0, 63 Turner Street, Port Melbourne, Victoria 3207, Australia

tows during compaction. In-situ tow visualisation has been achieved in the past by a very labour intensive technique. Images of the specimen edge are obtained using an optical or Scanning Electron microscope. Progressive removal of a small depth of edge material allow the production of a series of section images which can be reconstructed to display the tow architectures[2].

MicroCT is a non-destructive method which allows a large amount of the internal tow geometry to be determined with a single scan. This was used previously in the investigation of tow architecture in three dimensional woven glass preforms[3], and in generating three dimensional visualisations of glass fabrics[4]. There is a need for contrast enhancement when applying this method to visualisations of architecture in carbon-fibre composites due to the similarities in resin and infused tow densities. This has been discussed, along with a number of techniques used to enhance contrast. Contrast enhancement methods have include tow and fabric coatings, and resin particle additives. All methods present a change in the X-ray attenuation of the region of interest, which is achieved through a change in the density. Unmodified glass tows have also been used.

In addition to the traditional methods of MicroCT scanning[5], a phase difference contrast imaging technique has been considered[6]. These results from both methods are compared.

SAMPLE PREPARATION

A plain weave laminate with configuration $[0/90]_8$ consolidated using epoxy resin has been included for the purpose of benchmarking. A series of samples designed to enhance contrast was then manufactured. All infusions were made with polyester resin under vacuum at room temperature. Samples were trimmed to a width of 3.5mm and length of 10mm in order to allow placement into the MicroCT scanner. Thickness was dependent on the layup configuration and is discussed individually. Two methods of contrast enhancements to fabrics were attempted, including coating of the outside of the fabrics and resin particle additives. Scans were performed on a Skyscan 1072 X-ray Microtomograph, with voltage and current set to the maximum of 83kV and 104μA, respectively.

Gold Sputter Coatings

Gold coatings were applied to dry fabrics prior to consolidation. The laminate configuration was $[0/90]_8$. A coating of either 25nm or 50nm was applied using an EmiTech K550x gold sputter coater. The laminate was consolidated using vacuum bag resin infusion (VBRI), with resin flow parallel to the plies. No distribution medium was used. The application of the gold layers did not result in a significant change in the flexibility of the fabric. This method was then applied to a second sample, in which individual weft tows were coated prior to weaving. The treated fabric was then placed at the centre of a $[0/90]_7$ laminate, and infused using a wet layup with vacuum bag compaction. 50nm gold coatings were used for coating individual tows.

Iodine Contrast Agent Coatings

Iodine contrast agent, Isovue®370, was applied to two fabrics prior to consolidation. This is a contrast agent commonly used in the field of medical and clinical X-ray imaging. Coated fabrics were placed at the centre of a $[0/90]_8$ laminate, and consolidated using VBRI. Plies were soaked in a solution of 100ml distilled water: 15ml contrast agent, and 100ml distilled water: 15ml contrast agent, and removed after ten minutes. They were then dried quickly using a hot air gun. A slight reduction in the flexibility of the fabrics was noted after they had dried. In an attempt to promote softening, a small piece of fabric was heated with a hot air gun set at 200°C, and a second piece was placed in polyester resin at room temperature. No significant improvement was noted. This method was applied to a second sample, in which individual weft tows were coated prior to weaving. A solution of 100ml distilled water: 15ml contrast agent was used. Following the observation that the fabric stiffened after coating, the treated tows were wrapped around a rod with diameter of 15mm after coating. This increased the flexibility somewhat. The treated fabric was then placed at the centre of a $[0/90]_7$ laminate, and infused using a wet layup with vacuum bag compaction.

Alternating Layers Of Glass And Carbon Fabrics

A laminate of configuration $[0/90]_8$ was prepared by interleaving layers of glass and carbon fabrics. The laminate was infused using a wet lay-up with vaccum bag compaction. No additives were used. In a second sample, glass tows replaced the weft of a carbon fabric prior to consolidation. Two glass tows were required for each carbon tow due to their relative difference in size.

Aluminium Oxide Resin Additive

A 0.3μm Aluminium Oxide powder was mixed into a polyester resin with the ratio of 1:10. VBRI was used with resin flow perpendicular to the plies. In a second sample, 3μm Aluminium Oxide powder was used with resin flow parallel to the plies. A distribution medium was used to aid the spread of particles. Unlike the samples discussed previously, the fabrics were not coated in this sample.

RESULTS

Benchmark Plain Weave

The need for contrast enhancement is due to the similar densities of the infused tows and resin. Tows oriented across the page appear marginally lighter than those oriented out of the page in Figure 1. The individual tows can be discerned by the human eye without modification, however, there is some difficulty in creating three-dimensional reconstructions due to resin and tows sharing common shades of grey. Tow interfaces are visible when in contact with a resin rich region. However, when two tows are touching, particularly if they are oriented in the same direction at their contact region, the visibility of the interface is greatly reduced. Thus, there is a need for contrast enhancement.

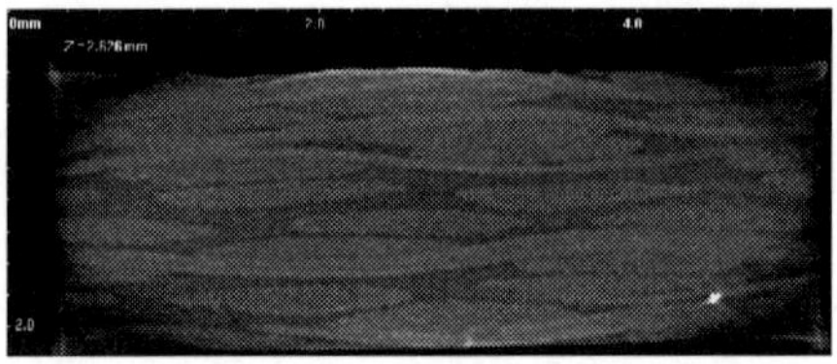

Figure 1. Benchmark plain weave specimen.

Gold Sputter Coatings

The sputter coating process was found to increase the contrast around the outside of the fabric interfaces very well. Interiors were not coated. A limitation is that the interfaces between two tows belonging to the same fabric ply or the interior of the tows cannot be coated as shown in Figure 2(a). Following this observation, the tows were coated prior to weaving. This was effective in capturing the entire tow, as shown in Figure 2(b). When viewed under an optical microscope, it was revealed that the thickness of the gold layers had been exaggerated by the MicroCT scan. There was a tendency for voids to form around the coated fabric interfaces, possibly due to changes in surface tension. Voids were not confined to the coated fabric in the second attempt, indicating that this may have been a result of the manufacturing process.

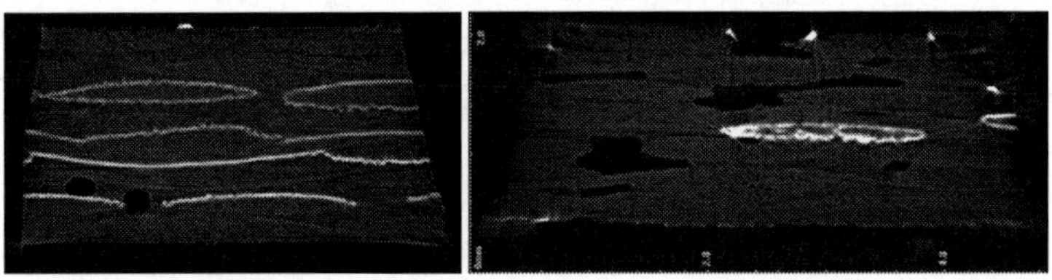

(a) (b)
Figure 2. Gold sputter coating method images; (a) fabric; (b) individual tow.

Iodine Contrast Agent

Coating of fabrics in iodine contrast agent prior to consolidation has allowed good contrast to be obtained between the coated and uncoated fabrics as shown in Figure 3(a). Unlike the gold coating of fabrics, both the inside and outside of fabrics were coated. However, interfaces between tows within the same fabric are still difficult to distinguish leading once again to the need for individual tow coating prior to weaving. Individual coatings were effective in visualising tow architecture, as shown Figure 3(b). A relatively high void content was found in both samples, however, the content was lower in the second. This was attributed to the use of a wet lay-up with vacuum bag compaction, and increased tow flexibility.

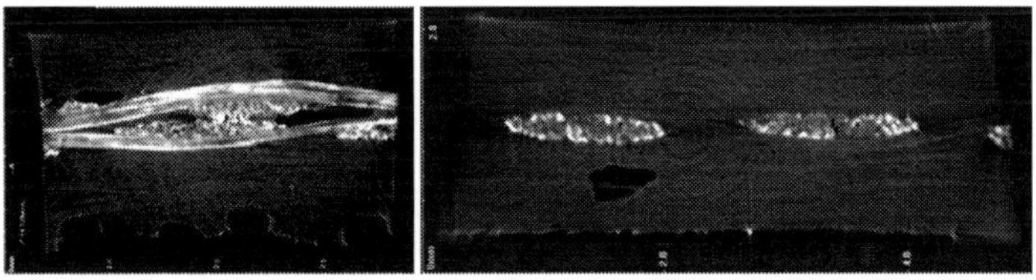

(a) (b)
Figure 3. Iodine contrast agent images; (a) fabric; (b) individual tows.

Alternating Layers of Glass and Carbon Fabrics

Glass fabrics were found to contrast very well with the infused carbon fabrics. Much like the iodine contrast agent coated fabric, the interface between the glass tows is difficult to distinguish as shown in Figure 4(a). The individual tows woven into the fabric are once again shown to allow an advantage in visualising individual tow architecture in Figure 4(b).

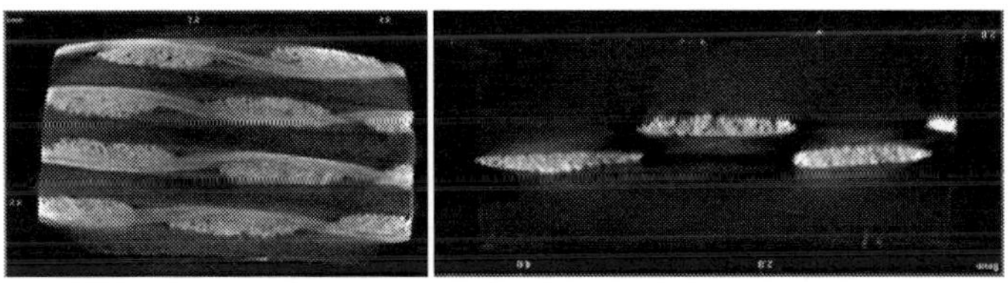

(a) (b)
Figure 4. Contrast of glass to carbon; (a) interleaved glass and carbon fabrics; (b) individual glass tows.

Aluminium Oxide Resin Additive

The effect of the Aluminium Oxide resin additive can be clearly seen in Figure 5. The interfaces have been highlighted. There is some penetration of the powder into the tows in the case of the 0.3μm powder, and the distribution of particles is quite uniform. The 3μm powder distribution was less uniform, and penetration into tows did not occur. In terms of allowing contrast enhancement of the tows, this method rates poorly compared to the fabric and tow coatings, and it was not possible to perform three dimensional reconstructions.

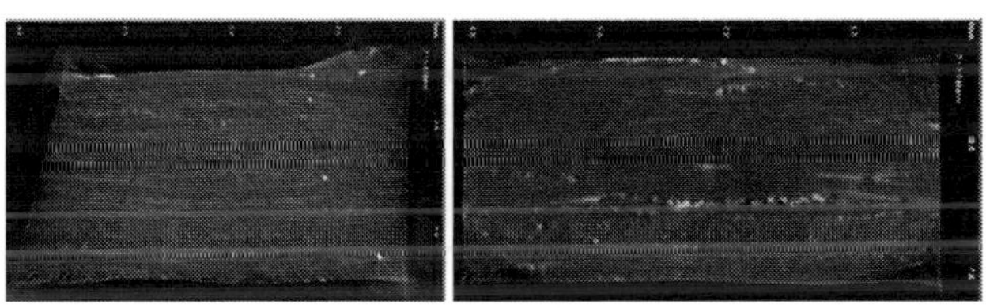

(a) (b)
Figure 5. Aluminium Oxide resin additive; (a) 0.3μm; (b) 3μm.

THREE DIMENSIONAL RECONSTRUCTIONS

Gold Sputter Coatings

The gold coatings placed on the surface of the fabrics and tows have been reconstructed in Figure 6(a) and Figure 6(b), respectively. The untreated composite has also been reconstructed and assigned a different colour set to semi-transparent for the purpose of showing its location within the laminate. A number of lines can be seen on the surface of the tows. This is an artifact of the imaging process, which helps to identify tow direction. This is especially useful in the case of the fabric. The success of this method was attributed to a very uniform coating and very little interference from image artifacts that could not be corrected with the use of software.

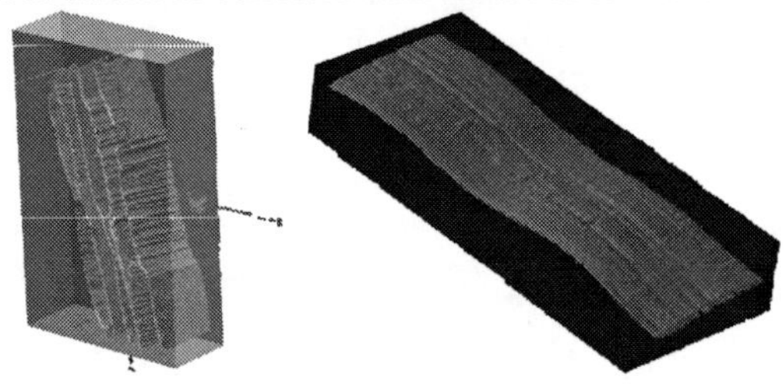

(a) (b)

Figure 6. 3D reconstruction of gold; (a) fabric; (b) individual tows.

Iodine Contrast Agent

It was possible to reconstruct both the interface and interior of the treated fabrics in fabrics and tows coated in the iodine contrast agent, as shown in Figure 7(a) and Figure 7(b), respectively. The artifact which aids in determining the tow direction is once again present. For the most part, the visualisation is very good, however, in some areas of the image the neighbouring tows oriented at perpendicular angles do not always appear to have solid boundaries. They rather appear to meld into one another. It was not possible to separate individual tows within the coated fabric, due to reconstructions being made according to density.

The three dimensional reconstruction of individual coated tows was very successful. There was minimal loss in accuracy at the free edges. Some loss of accuracy around the perimeter of the tows resulted from the need to account for a small amount of beam hardening artifact and grayscale differences due to variations in the coating thickness. The shape and direction of reconstructed tows appears to be intact.

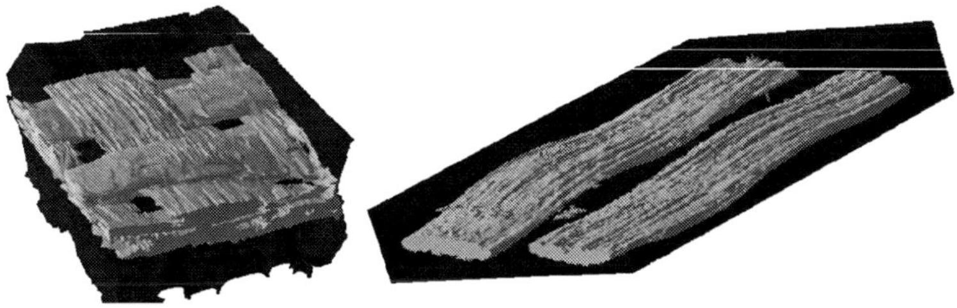

(a) (b)
Figure 7. 3D reconstruction of iodine contrast agent; (a) fabric; (b) individual tows.

Alternating Layers of Glass and Carbon Fabrics

The three-dimensional visualisation of the glass fabric plies has been made in Figure 8(a). The top surface shown in the image is not the outside of the specimen. Rather there is a layer of carbon fabric on top, which has been graphically removed for the purposes of visualisation. The separation of the layers through the thickness of the laminate is due to the interleaving of carbon fabrics. A loss of accuracy can be found at the edge of some reconstructed tows, and where the perpendicular tows cross over. This was due to the reduced apparent density of the tow in these locations. Examination of the reconstructed volume has revealed additional information regarding the architecture of the fabric. The tow located in the fabric at the centre is seen to twist along the longitudinal axis. This type of localised deformation may result in a reduction in the stiffness of this area.

A very good three dimensional reconstruction of the individual tows was possible as shown in Figure 8(b). This was owing to the difference in density between the glass tows and the surrounding areas. There were once again difficulties in the reconstruction of tow edges, resulting in them appear to be broken and discontinuous.

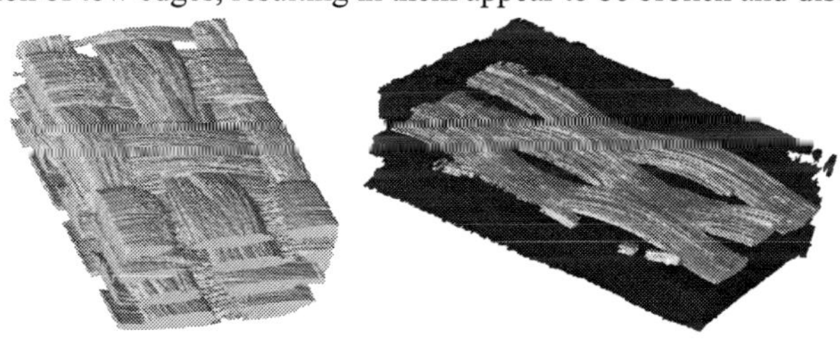

(a) (b)
Figure 8. 3D reconstruction of glass; (a) interleaved glass and carbon fabric; (b) individual glass tows.

PHASE DIFFERENCE CONTRAST IMAGING

The Phase difference contrast imaging method was developed by XRT® and combines two visualisation techniques. The first is that used in a regular MicroCT scanner, which is based on the X-ray attenuation of the sample. The second relates to the change in phase of the X-rays, which occurs due to refraction as they pass between materials of different density. This is particularly useful in visualisation of materials of low atomic weight, cracks, voids and material interfaces, although images must be interpreted correctly. These two techniques are combined to generate an image.

Figure 9(a) was taken from a scan conducted on the sample containing gold sputter coatings on fabric, whilst Figure 9(b) shows results from the 0.3μm Aluminium Oxide powder resin additive. It was possible to obtain images of higher resolution using this method compared to those generated on the MicroCT scanner used in this research. The resin particle additive method has shown a significant improvement in comparison to the regular MicroCT method.

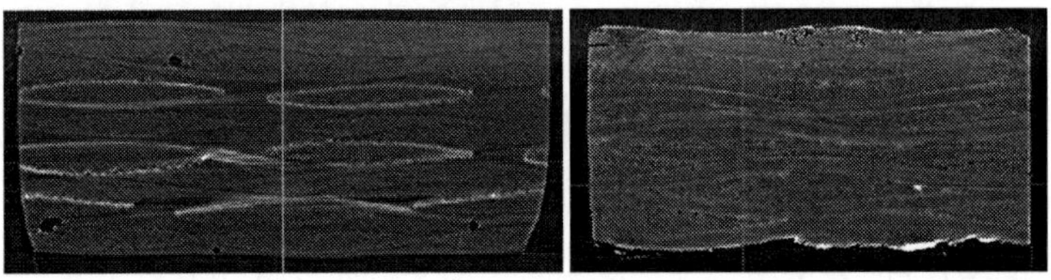

(a) (b)

Figure 9. Phase Difference Contrast Imaging Method; (a) Gold Sputter Coated Fabrics; (b) 0.3μm Aluminium Oxide Powder Resin Additive.

DISCUSSION

From these results, the most successful methods of contrast enhancement are the tow coatings. The shortcoming of the fabric coating method is that the contrast obtained between two tows within the same fabric ply is poor. In the current research tow coatings were applied to the weft only. Future work will include application of these methods to the visualisation of three dimensional performs. Coating of tows prior to weaving shows the most promise towards this end. It is likely that a combination of contrast enhancement methods will be used on different tows within the weave. If a single coating method were to be used prior to weaving of a three dimensional preform, it is expected that a similar shortcoming to the fabric coatings will occur, that is, poor visibility at neighbouring tow interfaces.

CONCLUSIONS

There is a need for contrast enhancement when visualising carbon-fibre/epoxy and carbon-fibre/polyester composites using X-ray MicroCT scanning. In order to address this issue a number of methods have been developed, including the use of fabric and tow coatings, and particle resin additives. Out of these methods, the most successful were the tow coatings, where individual tows were coated prior to weaving. Resin additive methods rated poorly. A second visualisation technique known as phase difference imaging was found to be of higher resolution than images obtained with the regular MicroCT scanner used in this research. In order to apply these methods to the visualisation of tows in 3D preforms, it is expected that tows will require treatment with materials of different density prior to weaving.

ACKNOWLEDGEMENTS

This work was supported by AFOSR/AOARD grant AOARD-07-4003 Enhancement of tow visualisation in 3-D woven composites, program manager, Dr J.P. Singh. The staff from CRC-ACS Bankstown are acknowledged for their work in manufacturing specimens and for helpful discussions.

REFERENCES

1. Mouritz, A.P., M. Bannister, P.J. Falzon, and K.H. Leong. 1999. "Review of applications for advanced three-dimensional fibre textile composites," *Compos Part A: Appl Sci Manuf.*, 30:1445–1461.
2. Bannister, M., I. Herszberg, A. Nicolaidis, F. Coman, and K.H. Leong. 1998. "The manufacture of glass/epoxy composites with multilayer woven architectures," *Compos Part A: Appl Sci Manuf.*, 29: 293–300.
3. Desplentere, F., S.V. Lomov, D.L. Woerdeman, I. Verpoest, M. Wevers and A. Bogdanovich. 2005. "Micro-CT characterization of variability in 3D textile architecture," *Compos Sci Tech.*, 65:1920–1930.
4. Schell, J.S.U., M. Renggli, G.H. van Lenthe, R. Muller, and P. Ermanni. 2006. "Micro-computed tomography determination of glass-fibre reinforced polymer meso-structure," *Compos Sci Tech.*, 66: 2016–2022.
5. Stock, S.R. 1999. "X-ray microtomography of materials," *Int Mat Rev.*, 44(4): 141–164.
6. Mayo, S.C., T. Davis, T. Gureyev, P. Miller, D. Paganin, A. Pogany, A. Stevenson and S. Wilkins. 2003. "X-ray phase-contrast microscopy and microtomography," *Optics Exp.*, 11(19): 2289–2301.

Particulate Methods for the Mechanics of Dry Textiles: Compaction of Yarn Assemblies

R. SAMADI and F. ROBITAILLE

ABSTRACT

This paper describes modeling of the mechanical behavior of assemblies of fibers as found in dry textile reinforcements using particulate methods. Examples are presented for the 2D compaction and relaxation of arrays of parallel fibers in constrained domains, and for the compaction of a similar array of fibers in a domain that is not constrained horizontally. Experimental trends are replicated including compaction stiffening, relaxation towards a stable compaction pressure, effect of compaction speed on compaction and relaxation as well as limited lateral spread and build up of compaction pressure in yarns that are not constrained laterally. The methods replicate experimental trends. Formal relations to be defined between simulation and experimental parameters are discussed.

INTRODUCTION

Designers of composite parts rely increasingly on simulation software to assist in quantifying manufacturing processes and to ensure that parts meet performance goals. Diverse physical phenomena are involved in the manufacturing of composite parts including textile draping and compaction, resin flow and preform relaxation, possible displacement of the preform in the mold, void creation and transport, heat transfer and resin cure as well as resin shrinkage, residual stresses and part warpage. Similarly, composite parts must fulfill diverse performance goals related to stiffness, strength, damping characteristics, heat transfer, surface finish and dimensional tolerances, to name a few.

Meaningful simulations of the processing and performance of composites parts require extensive material data. Generally such data is not available from suppliers of textile reinforcements. Matters are complicated by the large selection of available textile reinforcements, the limitless combinations of reinforcements that can be used is making a preform, and by the fact that reinforcement configurations change from

Department of Mechanical Engineering, University of Ottawa, 161 Louis Pasteur #A209, Ottawa, ON Canada K1N 6N5 point to point in a preform as a result of draping and compaction. Generating the required data from physical tests can be costly and time consuming. Typically, a limited number of data points will be generated for pre-selected reinforcements.

In an effort to alleviate the need for physical testing, significant research work has been directed at developing unit-cell simulation methods and software for diverse processing and performance properties of textile composites. The aim consists in predicting these properties and their dependence on reinforcement configuration, either as absolute values or as quantitative trends linking a given engineering property to different geometric parameters of the reinforcements such as the fiber volume fraction, yarn spacing, etc. Such trends may be used in complementing a limited number of physical measurements.

Unit-cell simulations require unit-cell geometric models and different groups have proposed software to create such models efficiently. However, generic models have limitations as illustrated in the following example. Figure 1a shows one such model representing 2 layers of a plain weave with [0°/90°/0°/90°] stacking sequence defined in the *x*-*y* plane. Vertical cell boundaries are normal to the 0° and 90° orientations. The thickness of vertical gaps between the yarns making each woven layer is uniform and corresponds to approximately 4 fiber diameters. In this case nesting can be approximated without creating interference between the 2 textile layers, by shifting the second layer by half the inter-yarn distance along *x* and *y* and then reducing the separation distance along *z*. As seen in Figure 1a this results in reasonable resin gaps between the 2 layers. Figure 1b shows a similar model with [0°/90°/+45°/-45°] stacking sequence. In this case the periodicity of the 2 textile layers along *x* and *y* differs and nesting cannot be induced as described above, resulting in unacceptable resin gaps between the 2 layers.

Most unit cells cannot be modeled faithfully using generic models. Therefore, computational methods that can be used in defining realistic geometric models of unit cells from first principles are needed. Related solutions based on finite element (FE) methods were proposed [1-3]. In this paper, particulate methods are used to model the compaction behavior of yarns in 2D. One algorithm is used in modeling the compaction, relaxation and spreading of yarn sections. The occurrence of compaction forces in yarns that are not restricted from spreading laterally is demonstrated. The modeling of further deformation modes in 3D and integration in unit cell modeling software for predicting full unit cell geometry constitute long-term goals of this work.

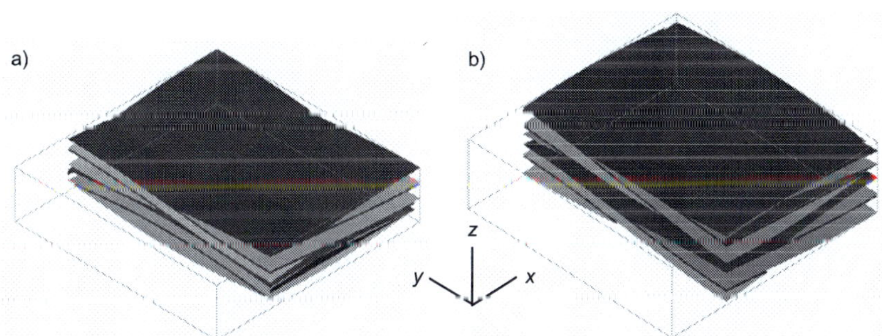

Figure 1. Geometric models of [0/90/0/90] and [0/90/-45/+45] 2-layer plain weave laminates.

PARTICULATE METHODS AND PARTICLE INTERACTIONS

Results presented here were generated using a particulate method evolved from [4]. In the basic method fiber centers in a yarn section are modeled as particles. Particles are seeded randomly in a 2D (x, z) domain. Then fibers are displaced simultaneously along x and z in the domain, in successive runs. Fiber displacement amplitudes are determined randomly within an upper limit termed *movement step*. Repulsive inter-particle forces are calculated as a function of particle separation distance derived from Gutowski's model [5]. Random displacements of fibers in the domain without change in domain dimensions are attempted for a certain *number of runs*. If the sum of inter-particle forces on a run is lower for the displaced fibers, displaced positions are maintained. During the runs the sum of inter-particle forces decreases and fiber positions evolve progressively towards a stable configuration of lower energy. Once the runs are completed the domain dimensions are reduced and the runs are repeated. Reference [4] features validation examples for 1D compaction cases. The method predicts stiffening compaction behavior as described in the literature [6]. Figure 2 shows fiber positions during 1D compaction generated using this method. The evolution of fibers to positions of lower energy and compaction pressure can be paralleled with reinforcement relaxation described in the literature [7].

The main difference between the basic particulate method and Gutowski's model lies in a more realistic relative positioning of fibers, leading to the generation of lateral inter-particle forces. This enables the modeling of compaction for yarns of irregular sections and of lateral spreading as described in the literature [8]. The ability to model relaxation within a single framework is another major characteristic.

The following 3 sections discuss the compaction of a 53 fiber assembly in a domain constrained horizontally (x) and vertically (z), the relaxation of the same fiber assembly in the same constrained domain, and the compaction of the same fiber assembly in a domain constrained in z but unconstrained in x. In the latter case pure repulsive inter-particle forces produce lateral spreading of fibers without build-up of compaction pressure along z. This is equivalent to compacting aligned rigid rods between frictionless platens: compaction force does not appear until all rods are spread out in a single layer. In practice, intertwining and imperfect alignment of fibers in yarns result in compaction pressure being generated when single yarns are compacted between platens. Fiber spreading is present but limited. Particulate methods provide a flexible framework for the inclusion of terms representing such phenomena. In this paper intertwining and imperfect alignment were replicated by blocking the relative motion of some fibers in a modified particulate method. The occurrence of compaction forces in yarns that are not restricted from spreading laterally when compacted is demonstrated in that section of the paper.

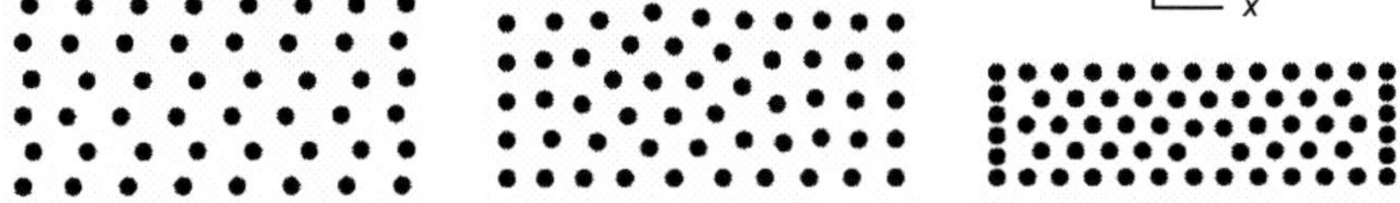

Figure 2. Steps in 1D compaction of a 64 fiber assembly in a domain constrained along x and z.

COMPACTION IN A CONSTRAINED DOMAIN

Two compaction simulations were conducted for a 53 fiber assembly in a rectangular domain constrained along *x* and *z*. Initial width (*x*) and height (*z*) of the domain were 108.0×10^{-6} m and 83.5×10^{-6} m. Final width and height of the domain were 98.75×10^{-6} m and 65.0×10^{-6} m. Fiber diameter was 10.0×10^{-6} m. Parameters used in the repulsive inter-particle force equation as defined in [4] were $L_o = 50.0 \times 10^{-6}$ m, $L_{min} = 10.0 \times 10^{-6}$ m and $\beta = 350$. These numbers correspond to an increase in fiber volume fraction v_f from 0.46 to 0.64 during compaction.

The two 2D compaction simulations were characterized by numbers of runs of 10 000 and 5 000 respectively. In the second case the fiber assembly is given twice less time to relax between *compaction steps* of constant amplitude, effectively corresponding to faster compaction.

Simulation results are presented in Figure 3 as curves of compaction pressure *P* as a function of v_f. Faster compaction results in a small increase in compaction pressures at a given v_f. At this stage a formal correspondence between the number of runs, movement step, compaction step and time has not been established. However, the pressure increase described above can be compared with similar small increases in pressure observed experimentally when commercially available reinforcements are compacted at faster rates in the range of those used in composites processes such as resin transfer molding (RTM) and vacuum assisted resin transfer molding (VARTM) [6]. Good correspondence of the amplitudes of simulated and experimental values of *P* and v_f is also obtained. Figure 4 shows fiber positions during 2D compaction generated using this method.

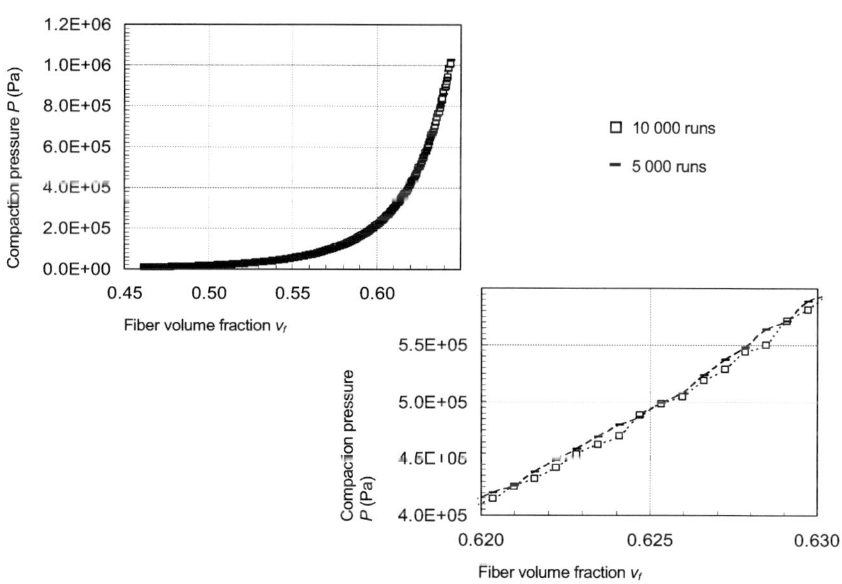

Figure 3. Compaction curves, 53 fibers, domain constrained along *x* and *z*.

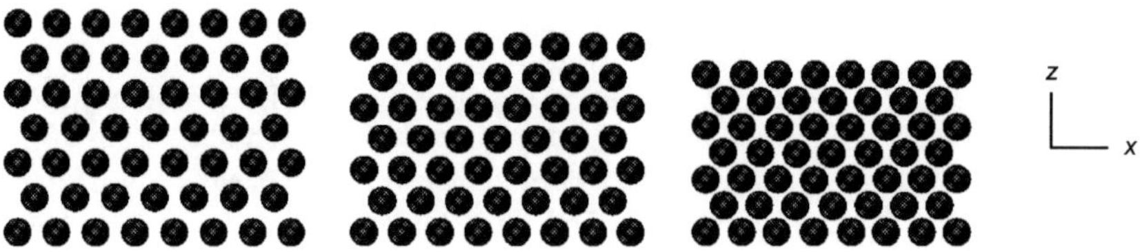

Figure 4. Steps in 2D compaction of a 53 fiber assembly in a domain constrained along x and z.

RELAXATION IN A CONSTRAINED DOMAIN

In the particulate methods used in this work, reorganization of fibers in configurations of lower energy results in progressive reduction of compaction pressure as runs progress. Relaxation is not represented by distinct functions or data describing this aspect of constitutive behavior. Instead, relaxation occurs during compaction and is inherently integrated in simulations in the same way that fibers reorganize continuously during compaction experiments.

Figure 5 shows curves of normalized pressure as a function of normalized number of runs, which is assimilated to time. The relaxation curves shown on the figure were generated at $v_f = 0.64$ for 10 000 and 5 000 runs respectively.

The trend seen with both simulated curves is representative of experimental relaxation data reported in the literature [7]. Compaction pressure decreases faster initially and progressively tends towards a stable value. Relaxation levels obtained for the simulated curves are 80% and 85% respectively. These numbers compare well with relaxation levels reported in the literature; however, no formal relation between the number of runs, movement step and time has been established yet. The general relaxation behavior predicted by the simulations is encouraging but it is not predictive at this stage. The curves show less relaxation in fiber assemblies compacted at a higher rate, which corresponds to trends observed experimentally [7]. Fiber positions at the onset and end of 10 000 runs are shown in Figure 6. Differences are barely visibly to the eye; the reduction in pressure results from small fiber displacements.

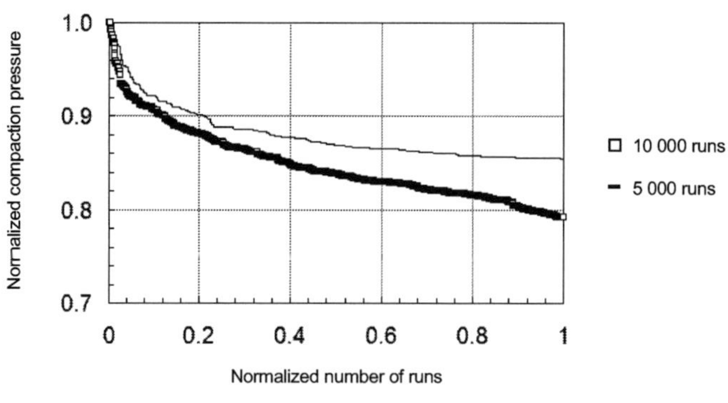

Figure 5. Normalized relaxation curves, 53 fibers, domain constrained along x and z.

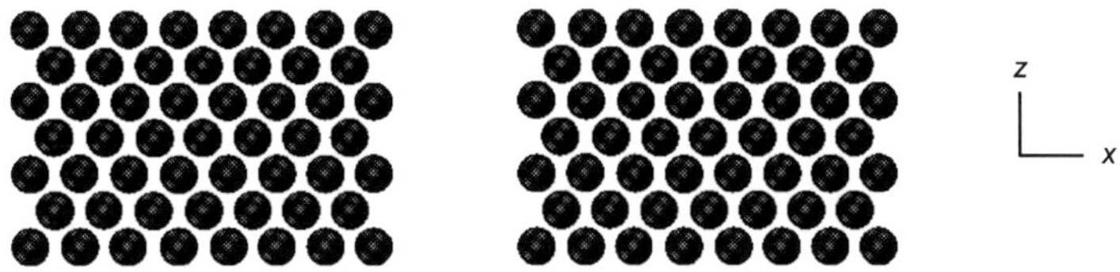

Figure 6. Steps in 2D relaxation of a 53 fiber assembly in a domain constrained along x and z.

COMPACTION IN A DOMAIN UNCONSTRAINED HORIZONTALLY

Yarns compacted between platens exhibit lateral spreading of fibers resulting from the occurrence of contact forces between fibers that are not parallel with the direction of compaction. However, in the process sizeable compaction forces are generated whilst limited spreading occurs. In actual yarns limited spreading is observed as relative fiber motion is limited by a number of phenomena including fiber intertwining and incorrect alignment.

The compaction and relaxation simulations reported above were performed in domains that are constrained along both x and z. In this case it is possible to generate compaction pressures using only inter-particle repulsive forces. However, in the case where fiber displacements are not constrained along x, using inter-particle repulsive forces only leads to spreading of the yarn until a domain height equivalent to one fiber diameter is reached, which is unsatisfactory.

Intertwining and imperfect alignment of fibers restrict relative fiber motion. In the following simulation this restriction was modeled by locking the distance of selected pairs of particles below a set maximum. The number of fibers in restricted pairs was chosen arbitrarily hence the results presented are primarily descriptive. However, the flexibility of particulate methods and their ability to model diverse types of interaction between fibers using simple inter-particle interaction rules are demonstrated. Figure 7 shows fiber positions during compaction generated from the modified particulate method featuring restricted displacement betweens pairs of particles; these pairs of particles are identified by dotted lines extending between them.

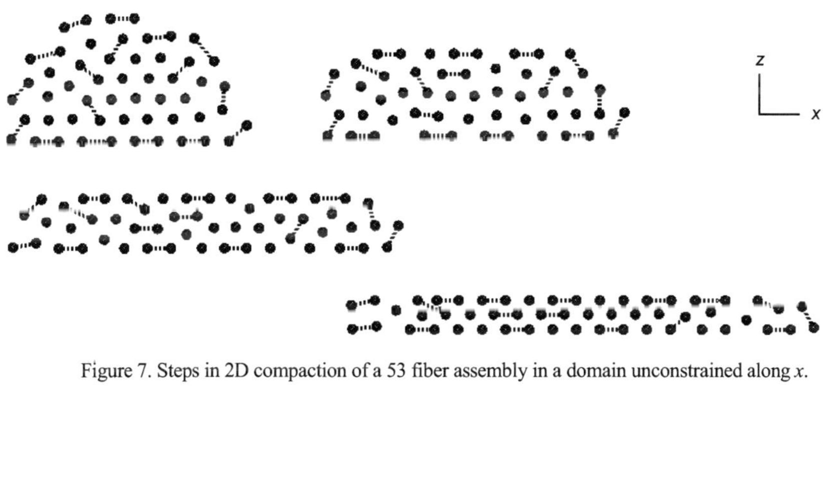

Figure 7. Steps in 2D compaction of a 53 fiber assembly in a domain unconstrained along x.

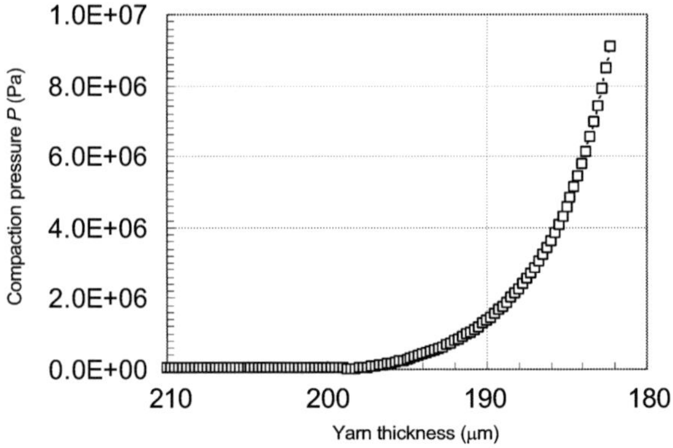

Figure 8. Compaction curves, 53 fibers, domain unconstrained horizontally.

Figure 8 shows a curve of compaction pressure P as a function of yarn thickness for a 53 fiber assembly similar to the one seen in Figure 7 over a selected range of thickness. Pressure is calculated based on the original width of the yarn. The fiber volume fraction v_f was not calculated. Typically this is derived from the initial fiber volume fraction, initial preform thickness and evolution of preform thickness during compaction. In the case of lateral fiber spreading, v_f in the yarn cannot be derived in this way unless assumptions are made with regard to tow width. The same is true in textile preforms and one must differentiate between tow, nominal and overall v_f.

Figure 8 shows that compaction pressure increases with a reduction of the yarn thickness. The particulate method used in generating the curve leads to a stiffening behavior which replicates actual behavior; however, at this stage the trend is qualitative. Compaction pressure arises as the lateral motion of fibers along x is impeded due to the interaction of pairs of fibers for which the inter-fiber distance was restricted by the additional fiber interaction defined in the modified particulate method. The number of additional interactions will affect the overall compaction behavior of the yarn. Prospectively, one may relate this number of interactions to the level of fiber twisting present in a yarn, for example. However, in practice parameters relating simulations and experimental behavior will be determined empirically by testing on yarns or reinforcements. Such relations remain to be developed. Furthermore, their definition is likely to change once 3D models including fiber tension are implemented. The preliminary results presented here are encouraging as they offer a unified framework to model different phenomena observed with fiber assemblies; however, formal interactions remain to be quantified.

DISCUSSION AND CONCLUSION

The applicability of particulate methods for modeling the mechanical behavior of assemblies of fibers was discussed for simple cases of 2D compaction, relaxation and compaction in an unconstrained domain. Stiffening compaction behavior was obtained from a well-known model of inter-fiber force, demonstrating that established and validated elements of constitutive behavior can be implemented using these methods. In this specific case particulate methods produced more realistic definitions of the geometry and inter-fiber force orientations whilst the overall behavior remained similar. Credible values of compaction pressure P and fiber volume fraction v_f were obtained. Effects of parameters such as the compaction speed could be replicated qualitatively; formal relations between the quantities involved in particulate methods and time remain to be identified; empirical parameters are likely to be involved.

The modeling of relaxation was demonstrated. Simulated and experimental trends correspond well; both show progressively slower reduction of compaction pressure. Credible amplitudes of simulated pressure reduction were obtained and experimental trends related to compaction speed were replicated. As mentioned above, relations between quantities involved will require further attention and empirical parameters are likely to be used. Relaxation and compaction are simulated jointly hence it is envisaged that parameters used in relating both aspects of simulations to physical behavior will be derived from the same tests.

Fiber spreading and the build-up of compaction pressure in tows that are not constrained laterally were demonstrated. A simple particle interaction relation was used, demonstrating that new models of constitutive behavior can be implemented in the methods. The simulation is not quantitative but it replicates actual behavior.

Immediate development work pertaining to the simulation of compaction and relaxation in 2D includes the establishment of formal relations between simulation parameters and experiments, and the identification of parameters related with basic fiber interaction phenomena. Reduction of the number of particles used in representing a given number of fibers is also investigated. Longer term objectives of this research include 3D modeling featuring more modes of particle interaction such as tension, friction and shear, and the integration of particulate methods in existing textile modeling software.

REFERENCES

1. Lomov, S.V., D.S. Ivanov, I. Verpoest, M. Zako, T. Kurashiki, H. Nakai and S. Hirosawa. 2007. “Meso FE Modelling of Textile Composites: Road Map, Data Flow and Algorithms”, *Compos. Sci. Tech.*, 67(9): 1870–1891.
2. Lin, H., M. Sherburn, J. Crookston, A.C. Long, M.J. Clifford and I.A. Jones. 2008. “Finite Element Modelling of Fabric Compression”, *Modelling Simul. Mater. Sci. Eng.*, 16: #035010.
3. Miao, Y., E. Zhou, Y. Wang and B.A. Cheeseman. 2008. “Mechanics of textile composites: Micro-geometry”, *Compos. Sci. Technol.* 68: 1671–1678.
4. Samadi, R. and F. Robitaille. 2008. “Compaction of Aligned Fibre Assemblies Using Particulate Methods,” presented at the 9th International Conference on Flow Processes in Composite Materials, Montreal, Canada, July 8–10, 2008.
5. Gutowski, T.G., T. Morigaki and Z. Cai. 1987. “Consolidation of Laminate Composites”, *J. Compos. Mater.*, 21(2): 172–188.
6. Robitaille, F. and Gauvin, R. 1998. “Compaction of textile reinforcements for composites manufacturing. I: Review of experimental results”, *Polym. Compos.*, 19(2): 198–216.
7. Robitaille, F. and Gauvin, R. 1998. “Compaction of textile reinforcements for composites manufacturing. II: Compaction and relaxation of dry and $H_2$0-Saturated woven reinforcements”, *Polym. Compos.*, 19(5): 543–557.
8. Robitaille, F. and Gauvin, R. 1999. “Compaction of textile reinforcements for composites manufacturing. III: Reorganization of the fiber network”, *Polym. Compos.*, 20(1): 48–61.

Identifying Sources of Variability in the Mechanical Performance of Resin Infused Textile Composites

J. TIMMS, H. MULCHANDANI, Q. GOVIGNON and S. BICKERTON

ABSTRACT

Variability in measured mechanical properties of composites contributes to increased design safety factors, potentially leading to overweight structures. Sources of variability can be introduced via the constituent materials, and the manufacturing processes employed. The long term aim of this program is to identify and quantify the links between materials and process variability, and mechanical performance. The current study addresses resin infusion of glass reinforced epoxy laminates, focusing on performance variability due to reinforcement type, nesting between layers, and preform compaction history. Multi panel infusions were completed with several fibre reinforcements; a plain weave, two stitched fabrics, and a chopped strand mat. In this manner, the influence of layer nesting has been studied, with six otherwise identical panels being manufactured from each fabric. In addition, the influence of multiple vacuum compaction cycles has been considered. A statistical analysis of tensile and compressive strength data is presented.

INTRODUCTION

Resin infusion (a.k.a. VARTM) is a closed moulding process for manufacture of fibre reinforced thermosets, differing from other Liquid Composite Mould (LCM) processes through application of a fully flexible upper mould. Although resin infusion is considered a simpler process than those using rigid and semi-rigid moulds, free deformation of the upper mould introduces an added dimension of variability. In addition to the complications of a variable cavity thickness, process control is further reduced by the use of cavity vacuum to provide both reinforce-

Jamie Timms, Hiten Mulchandani, Quentin Govignon, Simon Bickerton, Centre for Advanced Composite Materials, The University of Auckland, Private Bag 92019, Auckland 1020, New Zealand.

ment compaction, and the pressure gradient which draws the resin through the preform. As a consequence, there is considerable scope for variability in the quality of laminates produced under nominally identical resin infusion processes.

The aim of this study was to investigate the consequences of variation in reinforcement type, reinforcement layer interactions (nesting), and preform compaction history on the mechanical performance of glass-epoxy laminates manufactured via resin infusion. Four E-glass reinforcements—a plain weave, two stitched fabrics, and a chopped strand mat—were used. The aim of the nesting investigation was to quantify variation in mechanical properties of nominally identical laminates infused under the same conditions. The specific question posed was: do random variations in reinforcement layer interactions significantly affect mechanical performance, and does this variability occur predominantly between laminates or within them? The preform compaction history investigation sought to determine the effect of repeated application and release of vacuum before infusion on mechanical performance. This was an attempt to simulate a common industry practice, in which pre-filling compaction process is interrupted to fix leaks, re-seat preforms, and so on.

EXPERIMENTAL

Seven experiments were conducted in which six laminates 450 by 230 mm were infused under a single vacuum bag. The panels were laid three either side of a single filling line of porous distribution tape. Typical dimensions of the layout are presented in Figure 1, and photographs taken during layup and part way through infusion are provided in Figure 2. By infusing six panels concurrently, uniform pre-filling and post-filling conditions were maintained for all panels. The University of Auckland resin infusion facility features a temperature controlled table, pressure and temperature monitoring, and a stereophotogrammetry system capable of providing full-field thickness measurements of laminates to a resolution of 0.01mm and a frequency of 0.2 Hz [1]. Stereophotogrammetry data was collected, providing a complete record of changes in laminate thickness during the filling and post-filling stages. Resin pressure was also monitored at one point in each of the panels. Only three panels were monitored because of the large size of the infusion layout relative to the stereophotogrammetry cameras' fields of view.

The mechanical testing program included tensile (ASTM D3039[2]) and combined loading compression (ASTM D6641[3]) tests. Allowance was also made for fibre volume fraction and void content measurements, though only tensile and compressive strength are documented here. Each panel was processed by waterjet cutting, yielding six 250 by 25 mm tensile, six 143 by 14 mm compressive, and four 25 by 25 mm fibre volume fraction specimens. Five tensile and compressive specimens were tested for each panel, providing enough data points to perform a comparative quantitative analysis of mechanical performance. Layout of the test specimens within each panel is depicted in Figure 1.

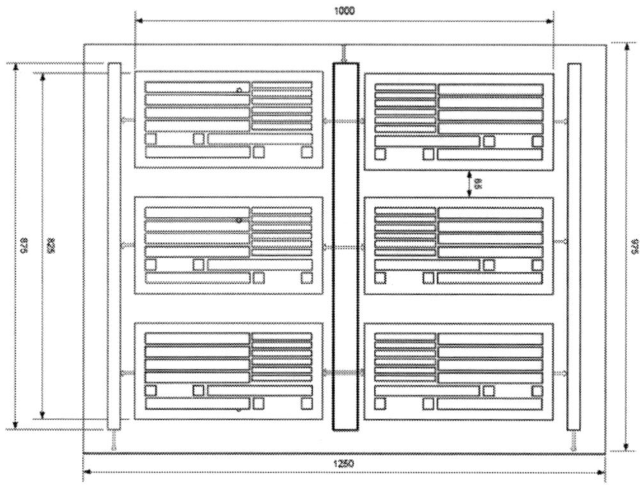

Figure 1. Schematic layout of the panels for infusion experiment. The layout of the mechanical test specimens is also presented.

Reinforcement Types and Stacking Sequence

Four E-glass reinforcements were used: 800 g/m^2 woven roving (EWR), 825 g/m^2 0°/90° biaxial stitched (EB), 820 g/m^2 ±45° double bias stitched (EDB), and 450 g/m^2 chopped strand mat (CSM). Each preform comprised of 8 layers of the heavier fabrics, or 14 layers of the CSM, giving final laminate thicknesses of approximately 5mm. The layers were stacked symmetrically in accordance to ASTM D3039 (i.e. $[0/90/0/90/0/90/0/90]_S$ for EB). The EDB preforms were cut at 45° to the weft direction, providing the same fibre alignment as the EB preforms.

Resin Infusion Procedures and Program

The infusion table was maintained at 30°C via a heat exchanger connected to a PC running LabView. An ADR-246/ADR-145 epoxy resin/hardener system from Adhesive Technologies (NZ) Ltd was used. A sheet of peel ply, the preforms, a top sheet of peel ply, and a flow-enhancing distribution medium were laid on the table in that order. The bottom sheet of peel ply ensured both laminate surfaces had a similar finish. A strip of tacky tape was then placed on each side of the preforms to prevent racetracking. The inlet, outlets, and vacuum tubing were connected, with the inlet line abutting the panels, and the outlets offset by 50 mm. The setup was completed by laying a border of tacky tape and placing a vacuum bag over the entire layout.

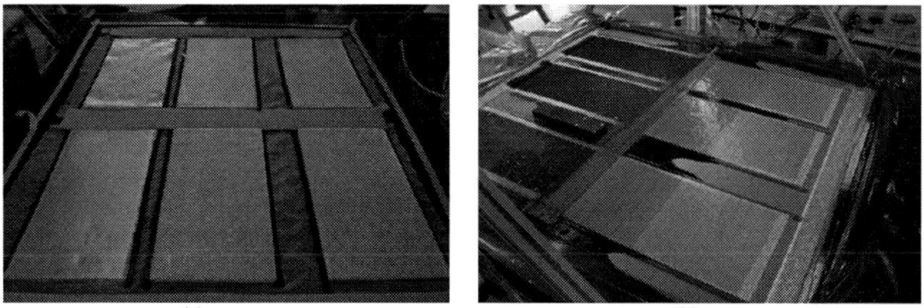

(a) (b)

Figure 2. a) Placement of six preforms during layup for an experiment, and b) part way through filling. Three panels are covered by the speckle pattern required for stereophotogrammetry.

The experiments began with the pre-filling compaction cycles. One cycle was defined as 10 minutes of approximately full vacuum (5 mBar or less) followed by 5 minutes with vacuum released, and this was repeated as required. The pre-filling stage was defined as the time between opening of the resin gate, and the resin flow fronts reaching the end of all six panels. Once filling was complete, vacuum pressure was raised to 200 mBar as recommended by the resin manufacturer. The curing panels were maintained at 30°C for approximately12 hours, and within an oven at 60°C for 24 hours to complete the curing process.

Table I summarises the experimental program. Seven 6 panel infusions were performed, of which four were for the nesting study, with one 6 panel infusion for each of the four reinforcement types. The remaining three infusions completed the preform compaction history study. All four reinforcement types were present for each of these infusions, which, when combined with the results of the nesting study, gave four levels of vacuum cycles: 0, 1, 3, and 10.

TABLE I. DESCRIPTION OF EXPERIMENTAL PROGRAM.

Infusion	Vacuum Cycles	Panel Reinforcement Type					
		1	2	3	4	5	6
1	1	EWR	EWR	EWR	EWR	EWR	EWR
2	1	EB	EB	EB	EB	EB	EB
3	1	CSM	CSM	CSM	CSM	CSM	CSM
4	1	EDB	EDB	EDB	EDB	EDB	EDB
5	0	EB	EWR	CSM	EDB	CSM	EWR
6	3	EB	EWR	CSM	EDB	CSM	EWR
7	10	EB	EWR	CSM	EDB	CSM	EWR

STATISTICAL ANALYSIS

Raw tensile and compressive strength data was analysed for statistically significant variations between and within groups, the results being quantified using the statistical data analysis package 'R' [4]. The underlying principle of this analysis is based on one-way analysis of variance (ANOVA), which assumes the data sets are normally distributed, and have equal variance from their respective sample means [5]. It is a single step comparison procedure applied simultaneously to the set of all pair wise comparisons (i.e. panel 1 to 2, 1 to 3, etc.) to calculate a "Tukey interval" for each sample. A direct overlap of these intervals between any two different samples indicates no difference in their respective sample means (i.e. variations within the samples are as large as variations between the samples), whereas the reverse is true if there is no overlap [6].

RESULTS AND DISCUSSION

The tensile and compressive strength data is presented in Figures 3 to 6, diamonds representing individual strength measurements, squares the sample means, and the vertical bars the Tukey intervals. Two panels of the EB material have been excluded from the tensile data analysis, as these specimens were tested without sandpaper in the grips. The use of sandpaper was required to prevent unacceptable failure at the grips for the remaining four panels. Furthermore, the data in Figure 5 did not satisfy the assumptions of the one-way ANOVA analysis, and hence meaningful Tukey intervals could not be obtained to quantify any differences in average strength between the different vacuum cycles. Tables II to IV include the results for which there was a statistically significant difference between sample means. The "P value" quantifies the significance of any difference between sample means (the closer to 0.0, the stronger the difference), and allows assignment of the labels provided in the evidence column [5]. For each panel pair, a difference in failure stress is provided, being the range in which the true difference can be stated with 95% confidence.

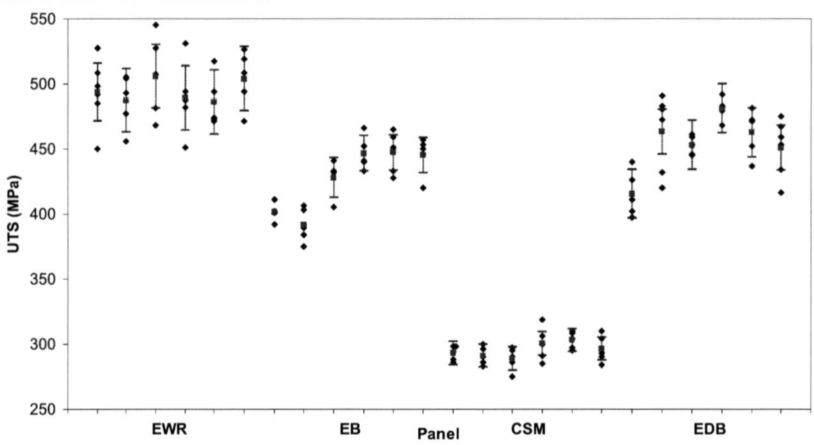

Figure 3. Tensile strength data for all panels at infused after 1 vacuum pressure cycle. Tukey intervals are represented by the vertical bars.

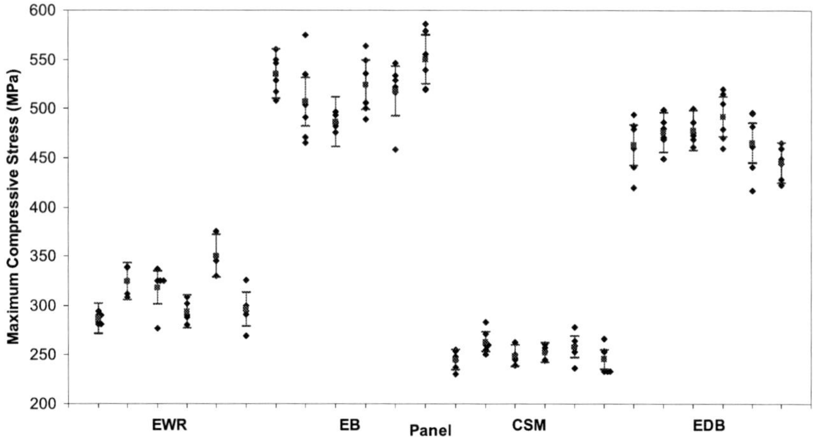

Figure 4. Compressive strength data for all panels at infused after 1 vacuum pressure cycle. Tukey intervals are represented by the vertical bars.

TABLE II. DIFFERENCES IN SAMPLE MEANS FOR TENSILE STRENGTH, FOR PANELS INFUSED AFTER 1 VACUUM PRESSURE CYCLE

Fabric	Combination	P value	Difference (MPa)	Evidence
EWR	--	--	--	--
CSM	--	--	--	--
EB	--	--	--	--
EDB	2-1	0.0048	12-84	Strong
EDB	4-1	0.0002	28-103	Very Strong
EDB	5-1	0.0079	10-85	Strong

TABLE III. DIFFERENCES IN SAMPLE MEANS FOR COMPRESSIVE STRENGTH, FOR PANELS INFUSED AFTER 1 VACUUM PRESSURE CYCLE.

Fabric	Combination	P value	Difference (MPa)	Evidence
EWR	2-1	0.0253	3-72	Average
EWR	5-1	0.0004	26-101	Very Strong
EWR	5-4	0.0020	18-95	Strong
EWR	5-6	0.0032	15-93	Strong
EB	6-3	0.0073	13-113	Strong
CSM	--	--	--	--
EDB	4-6	0.0166	6-87	Average

While there are clear and expected differences of strength between the reinforcement types, it is interesting to quantify these differences using ANOVA. From initial examination of Figures 3 and 4, EB and EDB are between 50 to 75 MPa stronger in compression, than in tension. It is also notable that the EWR is ~200 MPa weaker in compression, possibly due to the significant crimp of the fibre tows. From Figure 4 and ANOVA performed on EWR, EB, and EDB, we have very strong evidence that EB is between 34-66 MPa stronger, on average, in compression than EDB, when infused after 1 vacuum pressure cycle. Furthermore, EB and EDB are between 195-230 MPa and 145-180 MPa stronger respectively, on average, than

EWR. CSM was not included in the analysis because the data did not satisfy the assumptions of ANOVA, but it is clearly weaker than EB and EDB.

Considering tensile strength (Figure 3), EWR performs much better, and is actually stronger than EB and EDB. From ANOVA, we have very strong evidence that EDB is between 12-44 MPa stronger than EB, on average, when infused after 1 vacuum pressure cycle. Under the same processing conditions, EWR is between 52-83 MPa stronger, on average, than EB and between 24-55 MPa stronger, on average, than EDB. Again, CSM was not included for the same reason, but it is clearly the weakest of all four fabrics.

Variability Due to Nesting

From the results of the ANOVA analyses (see Figures 3 and 4), there are some statistically significant differences in the failure stress (both in tension and compression) between panels of the same fabric reinforcement. Variability in fabric nesting (the interaction between adjacent layers in a panel) is a possible explanation for these differences in mechanical properties. It can be inferred that EWR, EB, and CSM show no overall variability between panels for tensile tests, whereas EWR, EB, and EDB all show some statistically significant differences in sample means between panels in compression tests. For example, from Table III, EWR panel 5 is between 26 to 101 MPa stronger than panel 1.

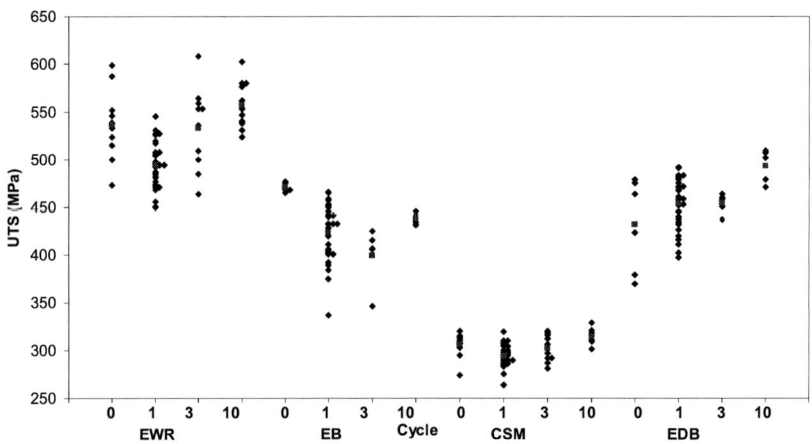

Figure 5. Distribution of all tensile strength data, organised by number of vacuum pressure cycles.

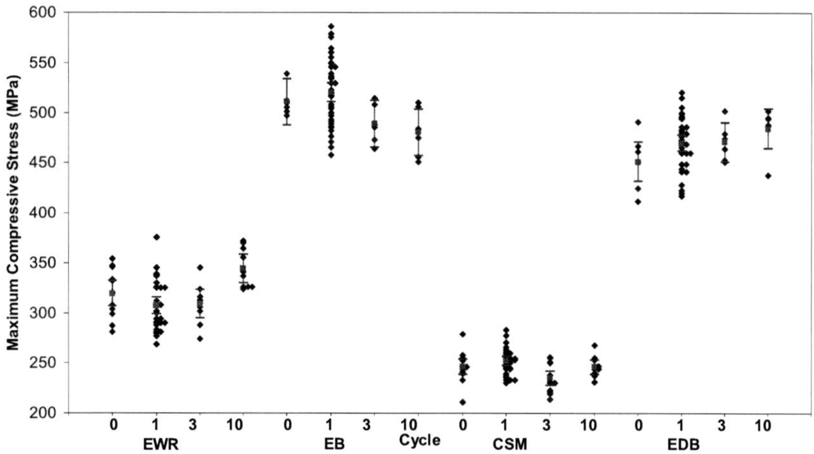

Figure 6. Distribution of all comprehensive strength data, organised by number of vacuum pressure cycles. Tukey intervals are represented by the vertical bars.

TABLE IV. DIFFERENCES IN SAMPLE MEANS FOR COMPRESSION TESTS UNDER DIFFERENT VACUUM PRESSURE CYCLES.

Fabric	Combination	P value	Difference (MPa)	Evidence
EWR	10-1	0.0004	14-60	Very Strong
EWR	10-3	0.0075	7-63	Strong
EB	1-10	0.0217	4-75	Some
CSM	1-3	0.0017	5-29	Strong
EDB	--	--	--	--

Between the fabric reinforcement types, the differences in failure stress can be attributed to a combination of nesting and the mechanism of failure, amongst other effects [7]. Between panels of the same fabric reinforcement, fabric nesting appears to be the predominant cause of variability. CSM is an exception to this, as the localized nesting effects are averaged out throughout each specimen, and therefore between panels, due to the nature of the shorter fibre tows and their random arrangement. This leads to similar results for stress at failure in tension and compression, with little overall variability between panels. Fibre tows are continuous, with a larger fibre count in the other three fabrics, and thus the tows can interact in a number of different permutations on a scale that is larger than the size of each testing specimen, producing the greater variability (within panel and between panels) in failure stress that is observed, particularly in the case of compression tests.

It is important to consider variability in mechanical properties within a panel, relative to variability observed between panels. For example, EWR exhibits relatively high variability in tensile strength within a panel, and this serves to mask any significant differences between panels. In contrast, the compressive strength within an EWR panel is typically more tightly distributed, leading to significant differences between panels as noted in the four cases presented in Table III. This research program will now focus on identifying the physical mechanisms responsible for strength variations, considering scales on the order of the testing specimen widths, to those approaching the full length of an infused panel. Differences between fabric layer interactions at these various scales remains a focus.

Variability Due to Compaction History

Considering Figures 5 and 6, there appears to be a moderate trend in maximum tensile and compressive stress between cycles, with panels infused after 10 vacuum pressure cycles performing better on average than the others (particularly for EWR and EDB). However, these small differences in failure stress appear to be practically insignificant in comparison to the within panel, and between panel variation of the respective fabrics. The results could also be affected by the fact that only one infusion was completed using 10 cycles, pointing to the need for a more in-depth experimental study of this issue.

CONCLUSIONS

An experimental study has been presented on the variability of tensile and compressive strength of glass fibre reinforcements manufactured by resin infusion. Considering four different reinforcements, a statistical analysis was presented on variability both within, and between nominally identical panels, manufactured under the same conditions. Fabric layer interactions (nesting) appear to be a predominant cause of variability in tensile and compressive strength within and between panels for all reinforcements except the CSM. Greater strength variation was identified between panels composed of EWR and EB fabrics in compression than in tension, for which within and between panel variations are comparable. The EDB panels showed greater variability between panels in tension, and there is no difference in sample means between CSM panels for either strength measure. Only small differences were noted due to increasing number of pre-filling vacuum cycles, increases in both tensile and compressive strength noted for EWR and EDB for 10 cycles. In future, void content and fibre volume fraction data will be incorporated to better define the role of fabric nesting on the observed variability in strength.

REFERENCES

1. Govignon, Q., S. Bickerton, J. Morris, and P. A. Kelly. 2008. "Full Field Monitoring of the Resin Flow and Laminate Properties During the Resin Infusion Process", *accepted to Composites Part A, available online May.*
2. ASTM Standard D 3039. "Standard Test Method for Tensile Properties of Polymer Matrix Composite Materials", ASTM International, West Conshohocken, PA, U.S.A., www.astm.org.
3. ASTM Standard D 6641. "Standard Test Method for Determining the Compressive Properties of Polymer Matrix Composite Laminates Using a Combined Loading Compression (CLC) Test Fixture", ASTM International, West Conshohocken, PA, www.astm.org.

4. Hornik, K. 2005. "The R FAQ", Retrieved online from http://cran.r-project.org/doc/FAQ/R-FAQ.html.
5. Crawley, M. J. 2005. *Statistics: An Introduction using R*. Hoboken, NJ: John Wiley & Sons, Inc., pp. 3–7.
6. Crawley, M. J. 2005. *Statistics: An Introduction using R*. Hoboken, NJ: John Wiley & Sons, Inc., pp. 155–170.
7. Guynn, E. G. 1992. "Parametric Study of Variables that Affect Fiber Microbuckling Initiation in Composite Laminates: Part 2—Experiments," *Journal of Composite Materials*, 26(11): 1617–1643.

TOPIC X

MECHANICAL PERFORMANCE

Long-term Durability of Plain Weave Polymer Composites

A. GUPTA and J. RAGHAVAN

ABSTRACT

Due to viscoelastic nature of polymer matrix time-dependent degradation in modulus, known as creep, has become a major concern in load-bearing structural applications utilizing polymer matrix composites. Despite the increasing use of woven polymer composites for such applications, published research on their long-term durability (creep) is very limited. Hence, long-term creep of polymer woven composites was investigated experimentally as well as analytically using Hexcel Corporation's F263/ T300 plain weave carbon fiber/ epoxy matrix composites. Constant load creep experiments were conducted over a wide range of temperatures, below the glass-transition temperature of polymer matrix, and stresses. While the composite exhibited linear viscoelasticity at lower stress levels at shorter creep times, its creep behavior became viscoelastically nonlinear at longer times. Since experimental determination of long-term creep for the entire service period is not

INTRODUCTION

Due to viscoelastic nature of polymer matrix time-dependent degradation in modulus, characterized as creep of polymer composites, has become a major concern in load bearing polymer composite structural parts. During the past three decades, considerable research has been carried out to understand the creep of polymer matrices, unidirectional composites and multi-directional composites made using unidirectional tapes [1, 2]. However, such studies on woven polymer composites are very limited [3-7] and requires further attention to generate sufficient knowledge in this area.

Abhishek Gupta, Composite Materials and Structures Research Group, University of Manitoba, Winnipeg, MB, Canada- R3T 5V6; Email: umgupta5@cc.umanitoba.ca
J. Raghavan, Associate Professor, Composite Materials and Structures Research Group, University of Manitoba, Winnipeg, MB, Canada- R3T 5V6; Email: rags@cc.umanitoba.ca

Bathgate et al. [3] have studied experimentally the creep of woven composites, while a micromechanical creep model presented by Shrotriya and Sottos [4-6] was not validated with experimental results. A lamination theory based model presented by Govindarajan et al. [7] validated the creep of woven composites well within an experimental time window of 5 hours. However, they did not verify its long-term predictive capability.

In summary, the published studies have focused on experimental characterization and modeling of the creep of polymer woven composites for a short experimental time window only. However, ability of these models to predict the creep for the entire service life of composites, well beyond the experiment time window, is unknown but crucial for their application in material selection and part design. Therefore, the present work is focused on both analytical and experimental study of the long-term creep of plain weave polymer composites.

CREEP MODEL

The crimp model, developed by Ishikawa & Chou [8] to predict time-independent properties of woven composites, was modified to predict the creep of plain weave woven composites using the experimental creep data for unidirectional composites. The creep of plain weave composites was predicted using a two-dimensional unit cell (see Figure 1), due to its repeat geometry. This two-dimensional unit cell was further considered to be an assembly of one-dimensional sub-unit cells (see Figure 2). Two different types of sub-unit cells, Parallel-Series Crimp (PSC) and Series-Parallel Crimp (SPC), were considered (see Figure 2b and 2c).

In the assembly using PSC sub-unit cells, the repeating cells, consisting of parallel assembly of blocks of warp and fill sections, are assembled together in series to form complete unit cell. The time-dependent stiffness of the unit cell is given by the time-dependent stiffness of the sub-unit cell and is calculated as:

$$A_{xxPSC}(t) = \frac{h_0}{2}(Q_{xx}^{F}(t) + Q_{xx}^{W}(t)) \tag{1}$$

where Q(t) is time-dependent stiffness of the warp and fill sections determined using the creep data for unidirectional tape lamina and h_0 is the total thickness of the cell.

In the assembly using SPC sub-unit cells, the repeating cells, consisting of series assembly of blocks of warp and fill sections, are assembled together in parallel to form complete unit cell. The time-dependent stiffness of the unit cell is given by the time-dependent stiffness of the sub-unit cell and is calculated as:

$$A_{xxSPC}(t) = \frac{h_0}{2}(2Q_{xx}^{F}(\theta,t) + Q_{xx}^{W}(t)) \tag{2}$$

Where θ is the local angle of orientation of the fill section with x-axis and

$$\tan(\theta) = \frac{\partial h_1(x)}{\partial x}$$

Using equations (1) and (2) and classical lamination theory [9], the longitudinal creep compliance of composite was determined as,

$$S_{xx}(t) = [A_{xx}(t)]^{-1} h_0 \tag{3}$$

SIMULATION DETAILS

Time-dependent properties of plain weave composites were computed using time-dependent properties of unidirectional tape lamina. Time-dependent properties (creep) of tape lamina were modeled by fitting the experimental creep data with modified Kohlraush William Watts Equation (KWW equation) [2]

$$S(t) = g_0 S_0 + g_1 g_2 X \left[1 - \exp\left[-\left[\frac{t}{a_\sigma B} \right]^C \right] \right] \tag{4}$$

where, X, B and C are the constants (see Table I) determined by fitting the linear creep data of tape lamina using equation (4). g_0, g_1, g_2 and a_σ are the nonlinear parameters defined by the Schapery's nonlinear viscoelastic model [10].

$$\{g_0, g_1, g_2, a_\sigma\} = \begin{cases} 1, & \text{Linear viscoelastic region} \\ f(\sigma, T), & \text{Nonlinear viscoelastic region} \end{cases} \tag{5}$$

Longitudinal, transverse and shear creep experiments were conducted for the tape lamina under tensile loading. The experiments were conducted at various temperatures and stresses in the nonlinear viscoelastic region for the epoxy matrix (stress > 7MPa [1]) of the composite. These experimental data were used to calculate the non-linear parameters following the data reduction method described by Zaoutsos et al. [11-13]. The parameters were calculated for S_{11}, S_{12} & S_{22} and modeled as a function of temperature and stress.

A lamination theory based MATLAB program was written to calculate the creep of plain weave composites. The undulation parameters used in the crimp model were obtained from a micrograph of polished edge of the plain weave fabric.

EXPERIEMNTAL DETAILS

F263-8 epoxy resin reinforced with T300 celion plain weave (2H) carbon fabric, from Hexcel Corporation, was used in this study. The composite panel was prepared using a 50-ton Carver hydraulic hot press and a mold simulating autoclave curing at a cure temperature of 177 ^{0}C and a pressure of 85 psi. The panel was post cured at 220 ^{0}C for 4 hours to ensure complete cross-linking of any un-reacted monomers during the curing process. The dimensions of the test specimens with end tabs were 50$\pm$1 mm in gage length, 17.5$\pm$1 mm in width, and 1$\pm$0.1 mm in thickness. Tensile creep tests were conducted at stress levels in the range of 100-450 MPa (up to 70% of ultimate tensile strength of composite) and at temperatures in the range of 80-160 ^{0}C. Instron's 5500R screw driven test frame, equipped with $\pm$ 25 KN load cell and an environment chamber, was used. A strain rate of 10^{-4} s^{-1} was used, during loading, for all tests. A single specimen was used per stress level for all creep tests in the entire temperature range. Each specimen was allowed to recover between two successive creep tests. Recovery time was twice the preceding creep time at each stress level.

RESULTS & DISCUSSION

Creep results for post cured plain weave composite samples measured at 200 MPa for various temperatures are provided in Figure 3. Creep results at a temperature of 80 ^{0}C for various stress levels are provided in Figure 4. It can be observed that the instantaneous and creep compliances increase with temperature. However, the composite showed a small increase in compliance within the experimental time window at any give stress. These composites had fibers along the loading direction, which resulted in lower creep with time. Moreover, the creep compliance did not change with stress up to stress of 400 MPa (see Figure 4). Since, stress shared by the matrix in the composite, for the tested stress levels, was in the linear viscoelastic range of the matrix (7 MPa), the composite's creep response remained in the linear viscoelastic region. However, the creep curve at 450 MPa did not superpose with other curves suggesting a nonlinearity in the creep response of the composite. Total increase in creep strain was 500 microstrains, within the experimental time window, at the highest stress and temperature levels tested. This is not sufficient to validate, with confidence, the creep model presented previously. Hence, Time-Temperature-Superposition Principle (TTSP) was used to generate a master creep curve for a time period beyond the experimental time window. Creep curves for the composite obtained at different temperatures for a single stress level were shifted horizontally along time axis to superpose on to a reference creep curve at 80 ^{0}C (see Figure 5). Master curves for various stress levels, in the range of 100-400 MPa superposed well up to 200 minutes. However, curves deviated beyond 200 minutes suggesting a transition from linear to nonlinear creep response of plain weave composites. Creep curve at 450 MPa did not superpose with other master curves due to nonlinearity.

Predicted creep results, in the linear visco-elastic region, obtained using PSC and SPC assembly are plotted with experiment results (see Figure 4). It is observed that the prediction accuracy of PSC based model is about 12% while it is 2% for

SPC based model. Since SPC idealization resembles plain weave geometry more closely (taking undulation into account) than PSC idealization, predictions are more accurate. Predicted results are also compared with master curves at each stress level (see Figure 6). Predictions compare reasonably well with experimental master curves within an error range of 1-2 %. However, model under-predicts nonlinearity at 450 MPa for longer time periods.

SUMMARY

Long-term creep behavior of carbon fiber/epoxy plain weave composites was investigated experimentally and analytically. Creep was found to be in the linear viscoelastic range for stress levels up to 400 MPa, though it transitioned in to non-linear creep at longer time. A modified crimp model using Series – Parallel Crimp (SPC) sub-unit cells resulted in predictions that compared well (within an error range of 2%) with experimental results at all stress levels except 450 MPa.

ACKNOWLEDGEMENT

This work was supported by NSERC (Natural Science and Engineering Research Council of Canada). The authors would like to acknowledge with gratitude donation of material by Hexcel Corporation, USA.

TABLE I. CONSTANTS OF KWW EQUATION TO CALCULATE TIME-DEPENDENT COMPLIANCE OF UNIDIRECTIONAL TAPE IN LINEAR VISCOELASTIC REGION.

Equation Constants	S_{11U} (t)	S_{22U} (t)	S_{66U} (t)
S_0 (MPa^{-1})	$7.11*10^{-6}$	$1.32*10^{-4}$	$1.96*10^{-4}$
X	0.005	0.004	0.88
B	$1.0*10^{22}$	$2.15*10^{20}$	$0.69*10^{13}$
C	0.37	0.16	0.39

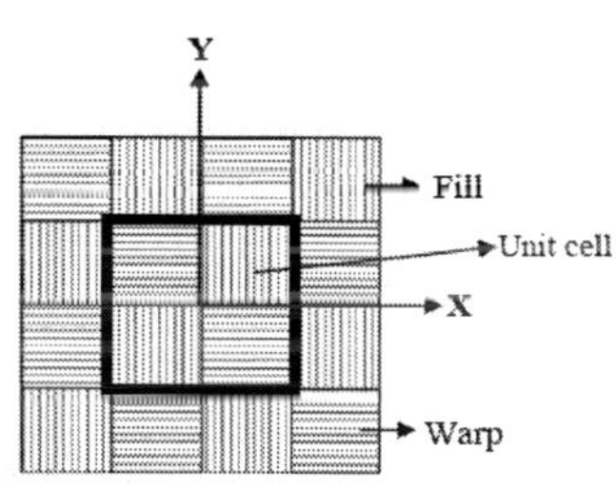

Figure 1. Plain weave fabric.

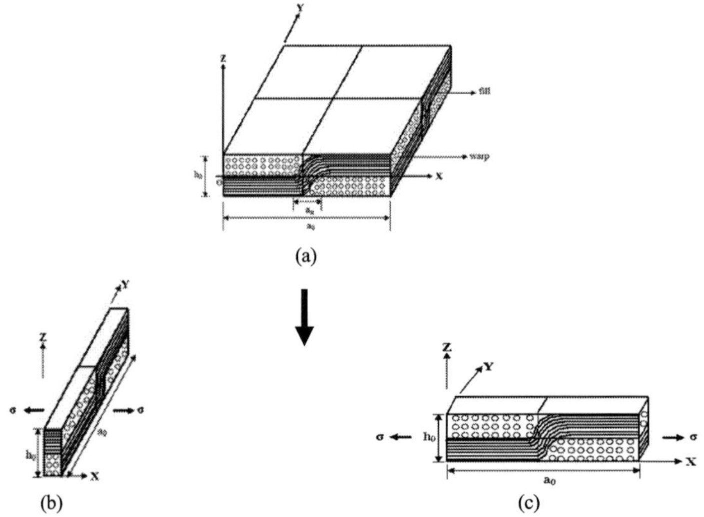

Figure 2. (a) 2-dimesional unit cell for Crimp model. (b) Parallel-Series Crimp (PSC) sub-unit cell. (c) Series-Parallel Crimp (SPC) sub-unit cell.

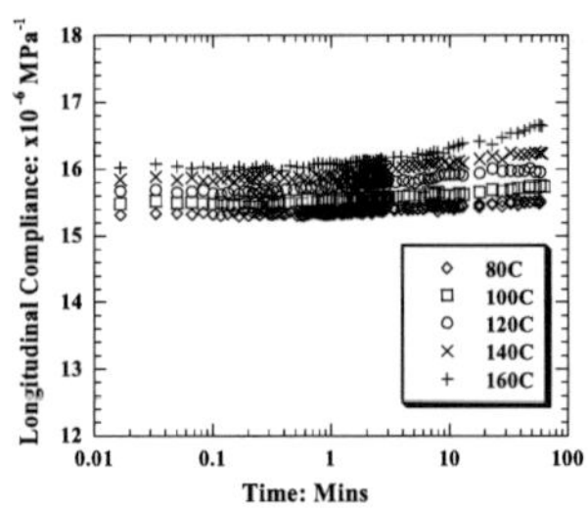

Figure 3. Experimental creep results at various temperatures for a stress level of 200 MPa.

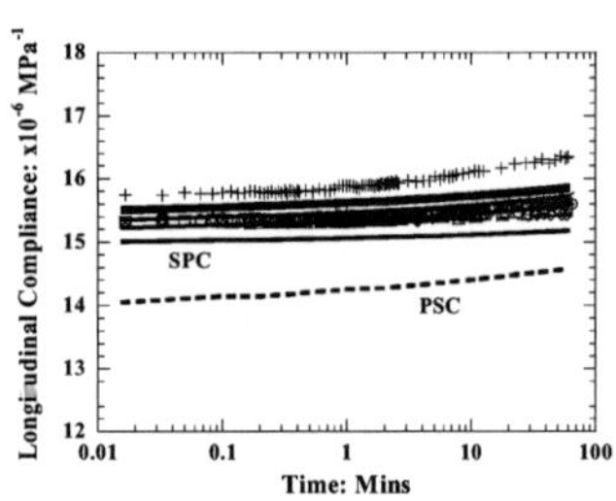

Figure 4. Experimental and predicted creep results at 80^0C various stress levels in the range of 100–450 MPa.

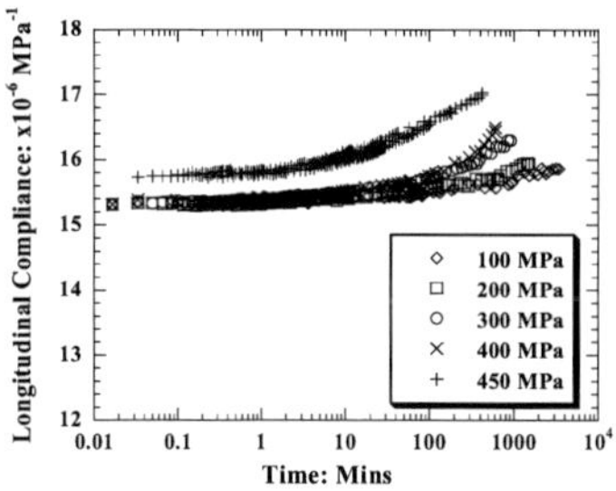

Figure 5. Master creep curves at 80^{0}C for various stress levels in the range of 100-450 MPa.

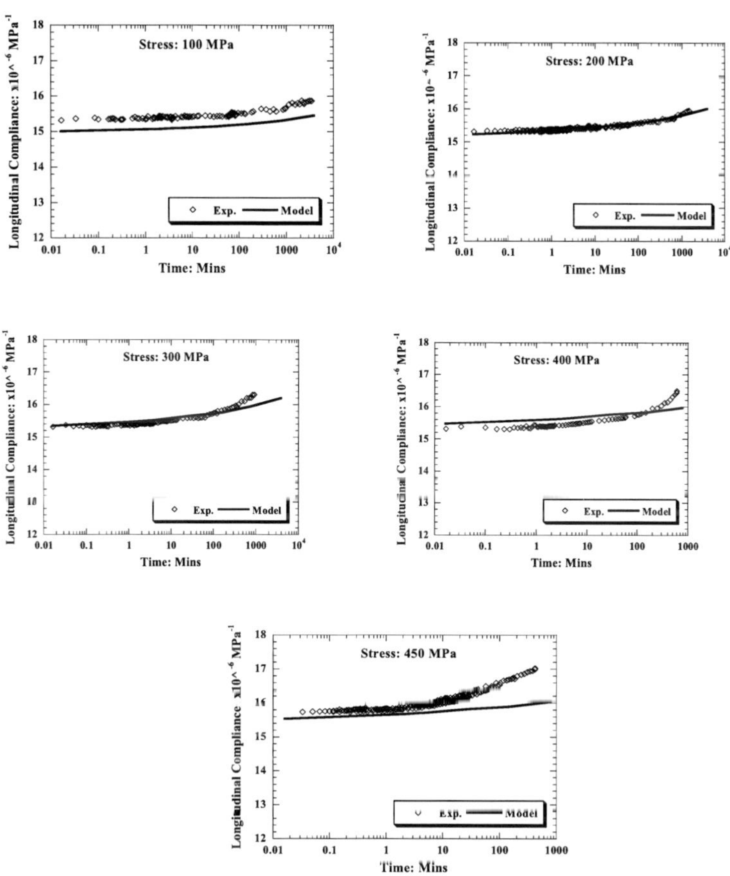

Figure 6. Comparison of experimental (master) creep curves with predictions at 80^{0}C for various stress levels.

REFERENCES

1. Viswanathan C. 2001. "Individual and interactive influence of temperature, stress, physical aging, moisture and fiber volume fraction on creep, creep-rupture and fracture of epoxy matrix and its composites," MSc. Thesis, University of Manitoba, Canada.
2. Balachander M.A. 2001. "On prediction of creep in multi-directional polymer composites," MSc. Thesis, University of Manitoba, Canada.
3. Bathgate R.G., F. Pang, and C.H. Wang. 1997. "Effects of temperature on the creep behavior of woven and stitched composites," *Comp. Str.,* 38(1): 435–445.
4. Shrotriya P., and N.R. Sottos. 1998. "Creep and relaxation behavior of woven glass/ epoxy substrates for multilayer circuit board applications," *Poly. Comp.*, 19(5): 567–578.
5. Shrotriya P., and N.R. Sottos. 2005. "Viscoelastic response of woven composite substrates," *Comp. Sci. Tech.*, 65: 621–634.
6. Zhu Q., P. Shrotriya, N.R. Sottos, and P.H. Geubelle. 2005. "Three-dimensional Viscoelastic simulation of woven composite substrates for multilayer circuit boards," *Comp. Sci. Tech.*, 63: 1971–1983.
7. Govindarajan S., N.A. Langrana, and G. J. Weng. 1996. "An Experimental and Theoretical Study of Creep of a Graphite/epoxy Woven Composite," *Poly. Comp.,* 17(3): 353–361.
8. Chou T.W. 1992. *Microstructural design of fiber composites.* Cambridge Solid State Science Series.
9. Daniel I.M., and O. Ishai. 1994. *Engineering Mechanics of Composite Materials.* Oxford University Press, New York.
10. Lou Y.C., and R.A. Schapery. 1971. "Viscoelastic characterization of a nonlinear fiber-reinforced plastic," *J. Comp. Mat.*, 5: 208–234.
11. Zaoutsos S.P., G.C. Papanicolaou, and A.H. Cardon. 1998. "On the non-linear viscoelastic behavior of polymer-matrix composites," *Comp. Sci. Tech.*, 58: 883–889.
12. Papanicolaou G. C., S. P. Zaoutsos, and A.H. Cardon. 1999. "Prediction of the non-linear viscoelastic response of unidirectional fiber composites," *Comp. Sci. Tech.*, 59: 1311–1319.
13. Papanicolaou G. C., S. P. Zaoutsos, and A.H. Cardon. 1999. "Further development of a data reduction method for the nonlinear viscoelastic characterization of FRPs," *Comps.Part A.*, 30: 839–848.

Introduction to the Morphological Characterization of Composites Reinforced by Stitched Non-woven UD Laminate

L. PIERRE-JACQUES, V. ALAIN, D. CHRISTINE, M. JÉRÔME and H. PATRICK

ABSTRACT

This study presents the morphological characterization of composites manufactured with a new reinforcement. The stitching of several UD plies for creating a multiaxial reinforcement presents several advantages but introduces a new meso-structure inside the laminates. To understand the mechanical effect of this new structure, we have to build a morphological model. In this purpose, we have used and adapted two methods : a destructive two dimensional one, optical observation, and a non-destructive three dimensional one, X-ray microtomography. The morphological model and parameters measured with these methods are presented and discussed.

INTRODUCTION

New high performance composites are presently being developed for advanced aerospace applications. These materials aim to minimize preconditioning to ease processing and decrease storage and raw material costs. In this study, we focus on composites reinforced by multiaxials, i.e. stitched non-woven unidirectional carbon fiber plies. These products are non-crimped, allowing to take full advantage of the high stiffness of carbon fibers, which represent the main part of the raw material cost. These types of reinforcements can be utilized in LCM processes – injection, infusion - aimed to lower manufacturing costs compared to prepregs (easier storage, higher throughput,…). However, the stitching introduces some new local parameters into the laminate morphology, which can induce some effects on the durability of these materials.

Liotier Pierre-Jacques[1], Vautrin Alain[1], Delisée Christine[2], Malvestio Jérôme[2], Henrat Patrick[3]
[1]LTDS/UMR 5513/ECL/ENISE/ENSM-SE SMS, ENSM-SE, 158, cours Fauriel 42023 Saint Etienne cedex 2 – France
[2]US2B-UMR5103, CNRS/INRA/Université Bordeaux 1, Bâtiment A11, 351 cours de la Libération, 33405 TALENCE CEDEX
[3]HEXCEL Reinforcements, ZI les Nappes 38630 Les Avenières

Our objective is to study the impact of the heterogeneities due to the stitching on the behavior of the composite during hygrothermal aging. This research is of particular interest to the aerospace industry since aircraft materials are often submitted to harsh environnements characterized by varying ambient humidity and temperatures. Those variations may induce some microcracking. Hence, efficient and reliable techniques must be developed to characterize the microcracking of a complex three-dimensional material mesostructure.

MATERIALS

The multiaxial material is composed of several plies with different orientations stitched together. The stitching has no mechanical purpose other than easing the handling and the processing of the reinforcement (figure 1). The stitching yarns introduce ellipsoidal perturbations in the fiber alignment [1]. These perturbations have different sizes and shapes according to the yarn size, stitching type and plies orientations. Nevertheless, these elliptical perturbations have positive consequences since they serve as channels to facilitate resin flow during resin infusion [2]. However, after processing, these channels become resin rich inclusions in the composite. In most cases, the final laminate is composed of several layers of stitched plies. For example, the stitching process allows assembling several plies into bound stacks of two to four plies per stack. So, a twelve-ply laminate will be composed of three to six stacks, each having a different resin inclusion distribution. Moreover, the stitched plies are not oriented in the same way. The ellipsoidal inclusion rotates with the fiber orientation [3] (figure 2). It becomes a periodic three-dimensional inclusion. The spatial period is known for each individual stack, but in an assembly, it becomes impossible to control the relative inter-stack offsets.

Figure 1. Picture of the two sides of a stitched stack : the I side (up) and the loop side (down).

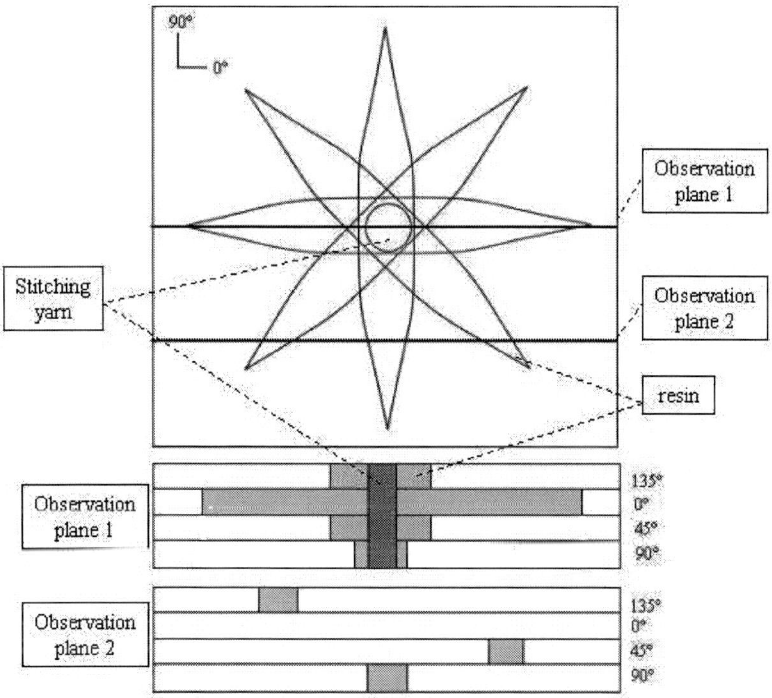

Figure 2. Resin inclusion representation.

METHODOLOGY

Optical Observations

To study and characterize the stitching area, we used optical micrography since it is the only direct method for looking inside the material [4]. This destructive 2D method has been used as a starting point to understand the material. The principle is to cut the sample and to polish the chosen observation plane to allow optical observation. For composite materials, we must use an abrasive cutting blade in order to prevent damage, such as cracking in the matrix of the observation plane.

The sample is then embedded in a cold coating epoxy resin to simplify handling and to allow automatic polishing.

The selected cutting plane is then polished with abrasive paper of digressive grain size. The automatic polisher is divided in two parts, the plate, where we place the abrasion paper or fabric and the head, where we can place up to six samples and apply an independent pressure on each of these samples. Each part has an independent rotation speed and direction. In the entire polishing process, the head rotation speed is fixed at 100 rotations per minute. The rest of the protocol is described in table 1. The principal advantage of this method is that we can control the observation plane location by adjusting the polishing time and thus the abrasion of the sample. This method is suitable to analyze specific zones like resin inclusions.

Table 1. Polishing protocol.

Stage N°	paper granulometry	pressure by sample	minimal time
1	P240	250 g/cm^2	40s
2	P600	250 g/cm^2	facultatif
3	P1000	250 g/cm^2	50s
3min in an ultrasound bath			
Stage N°	diamond suspension	pressure by sample	time
4	$3\mu m$	100 g/cm^2	7min
3min in an ultrasound bath			
5	$1\mu m$	100 g/cm^2	5min

This method gives a suitable surface state for optical observations (figure 3).

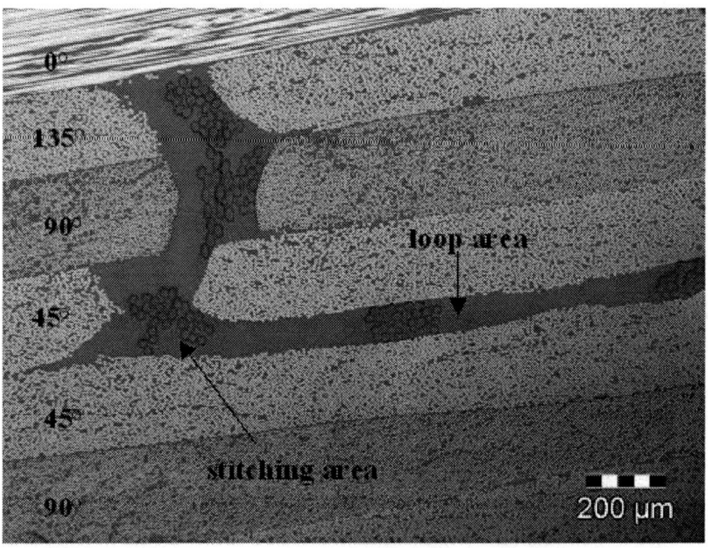

Figure 3. Optical observation example (angles are reported to the observation plane).

X-Ray Microtomography

The X-ray microtomography method relies on 3D reconstruction based on a series of X-ray projections. The sample analyzed is placed in an X-ray beam and the signal transmitted through the specimen is recorded by a CCD sensor panel. The sample rotates and the X-ray projection is recorded at several angles. The quality of the 3D reconstruction depends on the number of projections.
In this study we used at the laboratory USBB (Bordeaux, France) a Phoenix X-ray Nanotom for the volume acquisition. Its particularity is that the spot on the target that manages the X-ray emission has an approximate size of 0.5µm. It is one of the smallest ones for laboratory use.

The other specific feature of this device is the number of sensors in the CCD panel, namely 2304 by 2304 instead of the more common panels having 1200 by 1200 sensors. The combination of these two factors, minimum spot size and maximum number of sensors, allows for very good resolution in a simple laboratory microtomograph.

After the acquisition of all the projections, the reconstruction is made by the Phoenix X-ray commercial software. To maximize contrast between the different areas of the sample, the acquisition is made during a 360° rotation with 2000 projection planes (i.e. a projection for every 0.18°).

Nevertheless, the precaution taken could be insufficient because the response of the epoxy resin and the carbon fibers to an X-ray beam is rather similar. That is why the reconstructed volume has to be carefully treated to assess reliable measurements (figure 4).

We have used first a method based on an anisotropic diffusion filter [5-6] to make the volume treatable (figure 4). After filtering the volume is thresholded with an adequate gray spectrum enlargement, giving a binary 3D image. Finally, classical mathematical morphology operations [7, 8] (here an opening by reconstruction) are used to remove the noise.

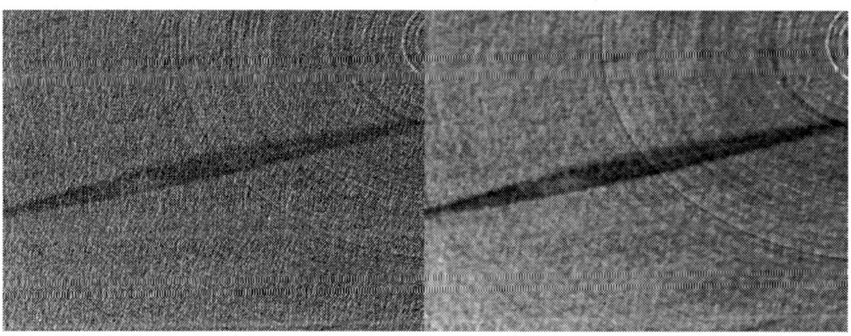

Figure 4. Illustration of the filtering effect (raw: left/filtered : right).

RESULTS

Measurement of the Volume Fraction of Resin Inclusions

To assess further calculations, we have to measure the resin inclusion extent, i.e. the volume ratio occupied by inclusions in the entire material. To obtain this value, we have to define an elementary reference volume. In this material, the reference volume is conditioned by the distances between stitching points and stitching yarns. After treatment and threshold of this volume, the total volume fraction is thus represented by the local ratio of voxels of the two colors. Measurements have been made on three quadriaxial samples with the same stitching patterns. For each sample, we characterize two stitching yarn. We also estimate the volume fraction without the opening by reconstruction. It gave us a basis to identify the main sources of experimental scatter that would allow us to specify the measurement accuracy. The results are summarized in table 2.

We can see that the average volume fraction of inclusions is 3% for this type of material i.e. 8.8% of the total resin volume for a material with a fiber volume fraction of 66%. The mean difference is around 0.5% which indicates that the measurement method is reliable. We have also studied the influence of different parameters on the values and mean differences. On table 2 we notice that the effect of the opening by reconstruction is to suppress the noise without any influence on the mean difference. It only tends to modify the measured volume fraction. We can conclude that this operation did not introduce any variation in the data.

The same tests have been conducted for the others operations of morphological mathematics, all of them seem to have very little effect on both the volume fraction and the mean difference. The principal source of variation is the threshold level. We minimize the possible errors of threshold by the spectrum enlargement during conversion from 32bits to 8bits but it is still very important. If we measure the same volume with three different threshold levels, we can obtain an average value of 3% and a mean difference of 3%. Threshold is thus the dominant factor in volume fraction variation. It requires a calibration which integrates all the treated volumes to avoid invalidation of the measurement by finding a mean difference near from the average value.

Table 2. Inclusion volume fraction measurements.

specimen.measure	entire treatment (%)	Variation (%)	without opening (%)	Variation (%)
1.1	3.6	0.56	3.,8	0.76
1.2	4.1	1.06	4.6	1.56
2.1	2.7	-0.34	3.4	0.36
2.2	3	-0.04	3.7	0.66
3.1	2.6	-0.44	3	-0.04
3.2	2.25	-0.79	2.95	-0.09
average	3.04	0.54	3.58	0.58

It is important to note that the volume fractions measured with stitching yarns from the same specimen tend to be closer to each other than stitching yarns from two different specimens. We attribute those effects depicted in figure 5 to the material spatial variability.

We must note the importance of the yarn loop areas in term of inclusion volume fraction. It represents approximately 1% of the entire material volume, or a third of the volume of resin inclusions. Those areas are difficult to characterize and are sources of variations between the measurements performed on the same specimen. This adds to the disparities of the inclusion sizes as depicted on figure 5.

To conclude on volume fraction measurement, we manage to develop a promising method of quantification of the inclusions volume fraction even with the large number of difficulties inherent to the material and method.

Morphological Model Corrections

There are two principal differences between the initial model (figure 2) and what is observed in reality. These differences could be explained in terms of materials stitching yarn morphology, i.e. in yarn shape modifications. These modifications have thus an effect on the inclusion morphology. To visualize these modifications, we can work either on several stacks of the volume or on a projection of all stacks in the transverse direction which is representative of the entire volume. To assess a good projection, after the treatment and threshold, we made an average calculation of the grey level in the transverse direction. On the projection, each pixel grey level thus represents the average of all the voxels of the volume in the transverse direction. This projection for a quadriaxial laminate produces images (figure 5).

We can see in the projection that the width of the lenticular zones is not constant. It tends to increase progressively to reach a maximum near the yarn loops. This value varies from 0.15 mm to 0.4 mm on the loop side for the considered stitching conditions in quadriaxial laminates. This could be translated into the morphological model as an enlargement of the stitching yarn.

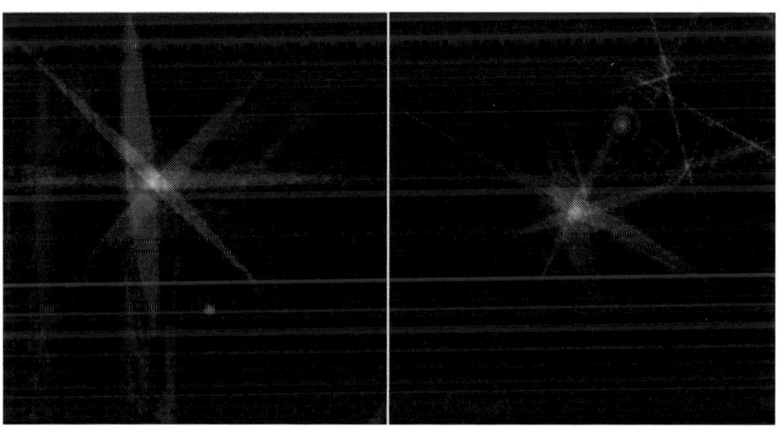

Figure 5. Average transverse projection of an inclusion.

We also notice on the projection that the stitching yarn is not normal to the plane of the plies. The angle from the normal can be measured on the 3D images. 3D measurements allow for determination of the real angle of the plane normal. This angle cannot be determined from 2D measurements. For example, the mean angle for this material is of 25° with a mean difference of 5.5°. The main interest of this measurement is to evaluate the inclusion shape, not the inclination of the yarn. The shape will permit us to improve the current morphological model for further analysis. For other stitching conditions, the lenticular zone loses its symmetrical with respects to the stitching yarn. This has an effect on the 3D distribution and is currently under study.

CONCLUSION

In this study, we have developed a new method to characterize the meso-structures morphology inside a composite laminate. We adapted the optical observation method to control the location of the observation plane and to analyze the resin rich regions which are specific to stitched multiaxial laminates. We have coupled this 2D observation method to a 3D method to obtain a good morphological model for further calculation of the mechanical state of laminates made with this type of reinforcement. We also manage to develop a treatment on 3D volumes that allows measurements of the resin volume fraction in the entire material.

We now focus on the durability of the laminate subjected to hygrothermal cycles. The resin rich regions could have an effect on kinetics water absorption, the residual stress profile and thus on the microcracking of this type of laminates. These topics are currently under study in our laboratory.

REFERENCES

1. S.V. Lomov et al. "Carbon composites based on multiaxial multiply stitched preforms. Part 1: Geometry of the perform", *Composite : Part A*, vol 33, 2002, p. 1171–1183.
2. S. Drapier, A. Pagot, A. Vautrin, P. Henrat, "Influence of the stitching density on transverse permeability of non-crimped new concept (NC2) multiaxial reinforcements : measurements and predictions", *Composite Science and Technology.*, vol 62, 2002, p. 1971–1991.
3. Heoung-Jae Chun, Hyung-Woo Kim, Joon-Hyung Byun, "Effects of through-the-thickness stitches on the elastic behavior of multi-axial warp knit fabric composites", *Composite Structures,* vol. 74, 2006, p. 484-494.
4. Hayes, B.S.and Gammon L.M. , "Composites", volume 21, *ASM Handbook*, 2001, p. 964–972.
5. Krissian, K., Malandain, G., et Ayache, N. Directional anisotropic diffusion applied to segmentation of vessels in 3D images. *Technical Report 3064, INRIA* ; 2001.
6. Perona, P. et Malik, J. Scale-space and edge detection using anisotropic diffusion. IEEE *Transaction on Pattern Analysis and Machine Intelligence*, 12(7): 629–639; 1990.
7. J. Serra : "Image analysis and mathematical morphology", volume1. *London academic Press*, Olando, FL, USA, 1982.
8. G. Matheron "Elements pour une théorie des milieux poreux". *Paris, Masson*, 1967.

Enhanced Mechanical Strength and Water Resistant of BFRP Composites Using Modified Bamboo Fibers

P. K. KUSHWAHA and R. KUMAR

ABSTRACT

In the present work epoxy resins have been reinforced with chemically treated bamboo strip matting to develop bamboo fiber reinforced plastic (BFRP) composites. The mechanical properties (tensile strength, elastic modulus, flexural strength and flexural modulus) and water absorption properties of the treated bamboo composites were determined. Improved mechanical properties were obtained for bamboo composite treated with 5% NaOH solutions. The morphology analysis reveals that the bamboo treated with 5% of NaOH exhibited better compatibility with the epoxy resins than the untreated bamboo. The improvements in tensile strength and flexural strength were higher by 55.15 % and 43.92 % respectively. At 5% of alkali treatment, the bamboo surface is further chemically modified with silane (γ-Aminopropyltriethoxysilane), benzoyl chloride and maleic anhydride. The mechanical and water absorption properties of the treated bamboo-epoxy composite was studied. And it was found that silane and benzoyl chloride improved the water resistance properties.

KEYWORDS: Bamboo mats; NaOH; Silane; Benzoyl Chloride; Maleic Anhydride.

INTRODUCTION

Bamboo is a natural material, and is an abundantly available natural resource in Asia and South America. Based on its anatomical properties, ultra structure and plant fracture mechanism bamboo establishes itself as a superior natural fiber among other known natural fibers (like jute, coir, sisal, straw, banana etc.) [1]. Jain et al. [2, 3] reported that bamboo has 60% cellulose with a considerably higher per-

Instrument Design development center, Indian Institute of Technology Delhi, New Delhi-110016, India centage of lignin (~32%), its microfibrillar angle being relatively small (2–10°). These facts about bamboo support its high tensile strength.

Bledzki[4] pointed out that the treatment of wood fiber with silane product A-175 improves wood dimensional stability. Various studies and review have also been done on the effects of alkali and silane treatment on natural fiber (flax, henequen, sisal and jute) [4-9], however, no such studies have been carried out on bamboo-epoxy composites.

Wang et.al [10] reported decrease in tensile strength of benzoylated flax fibers. Benzoylation improved sisal fiber/polystyrene interactions, which produces additional intermolecular bonding between fiber and matrix and hence tensile strength and modulus of the composite [11, 12].

Wood flour was chemically modified with maleic anhydride in order to improve dispersion\adhesion with unsaturated polyester resin. It was found that the hygroscopicity decreased so as the mechanical properties [13]. The fiber must be pretreated with aqueous solution of alkali before esterifying with maleic anhydride [14].Good interactions exist between maleic anhydride treated short palm tree fibers and epoxy matrix [15].

In the present work we try to overcome the problem of compatibility by alkalization, silane treatment, benzoylation and acetylazation of the fiber, which makes the fiber more hydrophobic and compatible with resin.

EXPERIMENTAL

Materials

Orthogonal bamboo strip mats were acquired from the local market. The cross-section of the strip used in the mats is 4.25mm × 0.5mm. Benzoyl chloride (Merck Limited), maleic anhydride (CDH, New Delhi), 3-aminopropyltrimethoxy silane (Acros Organics) and Epoxy resin (CY-230) and hardener (HY-951) (CIBA-GEIGY) were purchased.10% by weight hardener was added in the resin and stirred to mix thoroughly and subsequently applied on the bamboo mats.

Chemical Treatments

Bamboo mats were cleaned, dried and then dipped in 5% NaOH solution for 30 min at room temperature. Earlier investigations by the authors revealed that optimized percent concentration of NaOH treatment (mercerization) of bamboos for bamboo-epoxy composite was 5% [16]. These mats were thoroughly washed in distilled water and then neutralized in 2% HCL solution. Litmus paper test was carried out to check the neutrality. The mats were then dried in oven at 110 ^{0}C for 24 h.

1% silane (with respect to mat weight) was dissolved in a solution of acetone and water (50/50 by volume). The pH was adjusted to 3-4.5 using acetic acid and stirred continuously for 20 min. Then the mats were immersed in the solution for 2h. Finally, they were dried at 80°C for 24h [8].

For benzoylation, bamboo mats were soaked in 5% alkali solution for half an hour. Rinsed and cleaned with distilled water. 150 ml benzoyl chloride was dissolved in 2L of above alkali solution and the pretreated bamboo mats were again soaked in the solution for 15 minutes. The mats were filters out and soaked in ethanol for 30 min. they were cleaned with de-ionized water and then with distilled water. Finally, they were dried in oven at 80^0C for 24 h.

For esterification of bamboo with maleic anhydride, 5% alkali treated bamboo mats were dried for 24 h and then esterified using 2% maleic anhydride in xylene, keeping the fibers to solvent ratio at 1:10(wt/v). The soaking was allowed for 24 hrs at room temperature. The mats were filtered out and dried in oven at 80^0C for 24 h [15].

Fabrication of Composite

Perspex sheets of size 300×200 mm^2 were used as mould to prepare the composites. Hand lay-up technique was used to fabricate the composite. Epoxy resin mixed with the hardener was spread over the mats, piled one over another to seven layers, placed in between two Perspex sheets and pressed in the hydraulic press under a pressure of 170 KN for 24 h. The composites were cured at 80 °C for 4 h. The nomenclature used for different composites is given in Table 1.

Mechanical Test

The tensile test was performed according to the ASTM D638. Dog bone shaped specimens of the composites were cut. Cross-head speed was 2 mm/min.

Three-point bending test was performed according to ASTM D790 to measure the flexural properties. Specimens of dimension 96×13 mm^2 were prepared. Support separation was 70 mm and test speed was 10mm/min At least five replicate specimens were tested for each formulation on universal testing machine INSTRON 3369.

Water Absorption Test

Water absorption test was carried out according to ASTM D 570 -98. The two procedures adopted to ascertain water absorption are:

1. Long-Term Immersion
2. Two-Hour Boiling Water Immersion

Graph is plotted for percentage increase in weight of the specimen with time. Increase in weight is calculated as:

$$M\% = (M(t) - M_0) / M_0 \times 100 \quad (1)$$

where , M_0 is the mass of the dried specimen and M (t) is the mass of the specimen as a function of immersed time.

Scanning Electron Microscopy (SEM)

Tensile fracture surface of the composite samples were coated with silver and then analyzed using CARL ZEISS EVO 50 scanning electron microscope.

RESULTS AND DISCUSSION

Mechanical Properties

TABLE I. MATERIALS ANALYZED.

Material	Specification
UB	Bamboo-epoxy composite
ALK	5% NaOH treated bamboo-epoxy composite
SIL	5% NaOH pretreated and Aminopropyltrimethoxy silane treated bamboo-epoxy composite
BENZ	5% NaOH treated and benzoyl chloride treated bamboo-epoxy composite.
MAN	5% NaOH treated and maleic anhydride treated bamboo-epoxy composite.

TABLE II. MECHANICAL PROPERTIES OF BAMBOO-EPOXY COMPOSITES.

	Tensile Strength (MPa)	Elastic Modulus (MPa)	Flexural Strength (MPa)	Flexural Modulus (MPa)
UB	86	6736	119	11901
ALK	135	8204	149	9500
SIL	91	5674	161	8158
BENZ	105	4019	124	11620
MAN	108	2730	104	6463

Fabricated composites are tabulated in Table 1. Mechanical properties of the composites are given in Table 2 .During tensile test of untreated fiber reinforced composite, first the matrix failed and then the fiber. Failure is preceded by delamination indicating less adhesion between the untreated fiber and the matrix. In the case of composite reinforced with treated fibers, tensile failure shows a transverse fracture without any lateral propagation or separation of layers, delamination was nearly zero. The fracture surface is uniform and both fiber and matrix seem to have failed simultaneously. This indicates an improved interfacial adhesion between the fiber and the matrix

Tensile strength of BN is 135 MPa whereas tensile strengths of composite reinforced with mercerized fiber treated with silane is 91MPa. Tensile strength of alkali treated composite is 55% greater than that of untreated composite. The superior tensile properties may be attributed to the fact that this alkali treatment results in an improvement in the interfacial bonding by giving rise to additional sites of mechanical interlocking, hence promoting more resin-fiber interpenetration at the interface [5]. The alkali treatment improves the surface adhesive characteristics, by removing natural and artificial impurities, thereby producing rough surface topography.

Silane is a chemical compound with chemical formula SiH_4. Silane coupling agents may reduce the number of cellulose hydroxyl groups in the fiber-matrix interface. In the presence of moisture, hydrolysable alkoxy group leads to the formation of silanols. The silanol then reacts with the hydroxyl group of the fiber, forming stable covalent bonds to the cell wall that are chemisorbed onto the fiber surface [9]. The reaction schemes are given as follows [9].

$$CH_2CHSi\,(OC_2H_5)^3 + H_2O \rightarrow CH_2CHSi(OH)_3 + 3C_2H_5OH \quad (2)$$

$$CH_2CHSi\,(OH)_3 + \text{Fiber-OH} \rightarrow CH_2CHSi\,(OH)_2\text{O-Fiber} + H_2O \quad (3)$$

Alkaline treatment increments the amount of cellulose exposed on the fiber surface, thus increasing the number of possible reaction sites [5]. With more reaction sites on the surface, efficiency of silane deposition is enhanced. The silane is effectively deposited on the fiber surface [6]. Silane treatment may be blocking the impregnation of epoxy resin in the voids or reducing the surface roughness, resulting in the loss of mechanical component of adhesion. The fiber –surface silanization results in better interfacial load transfer efficiency but do not seem to improve the wetting of the fiber [5]. This may be the cause of decrease in the tensile strength and improved modulus.

Factors contributing to improve the properties are the reduction in hydrophilicity of the fiber and the rough surface of the benzoylated fiber. The reduction in hydrophilic nature makes the fiber more compatible with hydrophobic matrix. And the rough surface improved the adhesion between fiber and matrix [11]. The improvement in tensile strength is attributed to the better adhesion between the treated fiber and matrix [12]

Compared to the untreated composite, it was found that maleic anhydride treated composite, the tensile strength increased from 86 MPa to108 MPa .This behavior could be attributed to the mechanical interlocking, reduction in hemicellulose and lignin due to dissolution and increase in the density of cellulose fiber

Flexural strength is a combination of the tensile and compressive strengths, which directly varies with the interlaminar shear strength. In flexural testing various mechanism such as tension, compression, shearing etc. take place simultaneously. In three point flexural test, failure occurs due to bending and shear failure. During the flexural test, fracture in alkali treated composite samples starts by tensile failure on the tensile side of the loading plane, and then the crack runs towards the neutral plane. In untreated composite, failure starts with separation of layers, indicating lack of bonding. In the case of treated composites, flexural strengths were improved except with MAN composite. This increment in the flexural properties of composites after surface treatment of the bamboo mats can be attributed to the modification in the molecular level of the fibers due to chemical bonding between fiber and the silanes. Such linkage might have lead to better interfacial bondage and better adhesion.

Water Absorption

The moisture gain by the composite as a percentage of sample dry weight has been plotted against time in Figure 2. The water absorption for all the composites increased up to 5th week and then leveled-off.

5% alkali treated bamboo is better to resist the moisture uptake than the untreated one. Mercerization has greatly improved resin pick-up, or wettability, of the fibers because of the increase in density of the composite. It is believed that, by providing additional sites of mechanical interlocking, this treatment leads to improvement of interfacial bonding, hence promoting resin/fiber interpenetration at the interface. This hydrophobic resin pick-up could also account for the reduction in water absorption

Silane treatment after mercerization further reduced the water uptake by the composites. Composites having mercerized fibers coupled with silane treatment exhibit better behavior, suggesting that changes in surface property have reduced the hydrophilic nature of the fibers. Mercerization increments the amount of cellulose exposed on the fiber surface, thus increasing the number of possible reaction sites [9]. As observed by Valadez-Gonzalez et al. [7] the amount of silane adsorbed onto the alkali treated fibers is higher than that absorbed by the non-treated. Higher silane deposition reduces the amount of hydroxyl groups which were free to bind moisture. Fibers are modified in the molecular level due to chemical bonding between fiber and the silanes. Such linkage might have lead to better interfacial bondage, better adhesion and which has reduced the water absorption.

Benzoylation of bamboo provides the least water uptake by the composite.The absorption of water by different fiber-based composites is largely dependent on the availability of free—OH groups on the surface of the reinforcing fiber. The least water absorption may be attribute to the fact that benzoylation have reduced the free hydroxyl groups on the fiber surface to minimum resulting in increased hydrobhobicity of the fiber and increased the interfacial bonding.

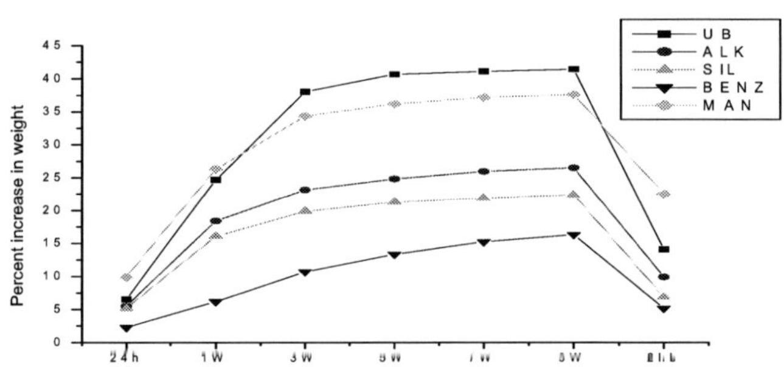

Figure 1. Water absorption behaviors of composites.

The initial absorption of water by the MAN composite is relatively high when viewed against the same for the untreated fiber-based composite up to one week. After the first week percent increase in water absorption reduces below that of untreated composite due to the attainment of a saturation effect.

Scanning Electron Microscopy

Fig 2(a) shows SEM micrograph of tensile fracture surface of untreated bamboo composite. The fibers get debonded from the resin. We can't see adherence of resin over the surface of the fibers It can also be observed from the SEM micrographs, Fig (b, c, d, e) that layer of resin is adhered over the surface of the bamboo fiber, thus indicating a closer contact between fiber and the matrix and a better wetting of the fibers. Failure of the laminate showed tearing and shearing, there is no indication of fiber pullout and this explains the presence of matrix on the fiber surface [6]. It shows the fiber and the matrix are working as a single entity during the tensile fracture and hence the higher tensile strength. Then, it can be said from these failure modes that the increase in tensile strength of the composite is an effect of the fiber-matrix interface increases. It can also be said that a low fiber-matrix adhesion, results in a failure mode dominated by fiber pull-out and matrix failure.

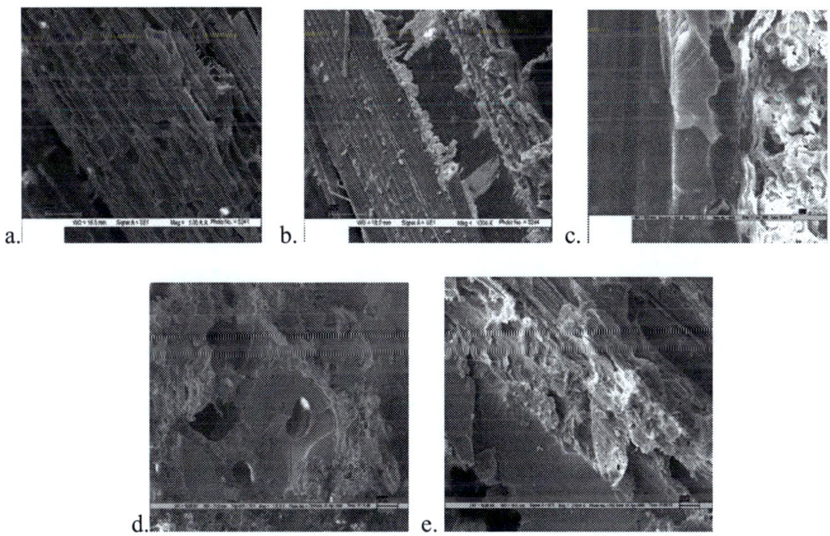

Figure 2. Photo- micrographs of tensile fracture surfaces of bamboo composites for (a) UB, (b) ALK, (c) SIL, (d) SIL and (e) MAN.

CONCLUSION

It is seen that 5% alkali treated fiber composite has improved overall properties in terms of mechanical and water resistant properties. Treatment of mercerized fibers with silanes reduces the tensile strength but considerably improves the moisture resistance and flexural strength of the composite. In the case of benzoylation, the water uptake is least of all the tested composites. Treatment with maleic anhydride after mercerization did not yield any improvement with respect to water resistance and mechanical properties as compared to mercerized composite.

REFERENCES

1. Jindal U.C.1986. "Development and testing of bamboo-fibers reinforced plastic composites," *J. Compos. Mater.* ,20:19–29.
2. Jain S., Kumar R. and Jindal U.C. 1993. "Mechanical behavior of bamboo and bamboo composite," *Journal of Material Science*, 27: 4598–4604.
3. Jain S. and Kumar R.1994. "Processing of bamboo fiber reinforced plastic composites," *Materials and Manufacturing processes*, 9(5): 813–828.
4. Bledzki A.K., Reihmane S. and Gassan J. 1996. "Properties and modification methods for vegetable fibers for natural fiber composites," *J Appl. Polym. Sci.*, 59: 1329–1336.
5. Valadez-Gonzalez A., Cervantes-Uc J.M., Olayo R. and Herrera-Franco P.J. 1993. "Effect of fiber surface treatment on the fiber-matrix bond strength of natural fiber reinforced composites," *Composites: Part B*, 30: 309–320.
6. Herrera-Franco P.J.andValadez-GonzalezA. 2005. "A study of the mechanical properties of short natural-fiber reinforced composites," *Composites: Part B*, 36: 597–608.
7. Valadez-Gonzalez A., Cervantes-Uc J.M., Olayo R. and Herrera-Franco P.J. 1999. "Chemical modification of henequen fibers with organosilane coupling agent," *Composites: Part B*, 30: 321–331.
8. Van de Weyenberg I., Ivens J., De Coster A., Kino B. E Baetens and I Verpoest. 2003. "Influence of processing and chemical treatment of flax fibers on their composites", *Compos. Sci. Technol.*, 63: 1241–1246.
9. Agrawal R., Saxena N.S., Sharma K.B., Thomas S. and Sreekala M.S. 2000. "Activation and crystallization kinetics of untreated and treated oil palm fiber reinforced phenol formaldehyde composites", *Mater Sci Eng: A*, 277: 77–82.
10. Wang B., Panigrahi S., Tabil L., Crerar W. and Sokansanj. 2003. "Modification of Flax fibers by chemical treatment", *CSAE/SCGR Meeting*, Montreal, Quebec.
11. Nair K.C.M., Thomas S. and Groeninckx G. 2001. "Thermal and dynamic mechanical analysis of polystyrene composites reinforced with short sisal fibers", *Compos Sci and Technol.*, 61: 2519–2259.
12. Nair K.C.M., Diwan S.M. and Thomas S. 1996. "Tensile properties of short sisal fiber-reinforced polystyrene composites", *J applied Polym Sci.*, 60:1483–1497.
13. Marcovich N.E., Roboredo H. and Aranguren M.I. 1998. "Mechanical properties of woodflour unsaturated polyester composites", *J Appl Polym Sci.*, 70:2121–2131.
14. Marcovich N.E., Aranguren M.I. and Roboredo H. 2001. "Modified woodflour as thermoset fillers Part I. Effect of the chemical modification and percentage of filler on the mechanical properties", *Polymer*, 42: 815–825.
15. Kaddami H., Dufresne A., Khelifi B. and Bendahou A., Taourirte M., Raihane M. 2006. "Short palm tree fibers – Thermoset matrices composites", *Composites: Part A* ,37(9): 1413–1422.
16. Kushwaha P.K. and Kumar R. "Enhanced mechanical strength of BFRP composites using modified bamboos", *J Reinf Plast Compos*, Accepted for publication, May, 2008.

Generation for Circumferential Distribution of Braiding Angle on Cylindrical Tubular Braided Fabrics

H. NISHIMOTO, A. OHTANI, A. NAKAI and H. HAMADA

ABSTRACT

Mechanical properties of composite materials reinforced by braided fabric are affected by fiber bundle orientation angle (called "braiding angle"). Controlling the braiding angle is a very important issue in order to create stiffness distribution according to the requirement.

The braiding angle on cylindrical tubular braided fabrics under unsteady-state condition is decided by four parameters; a longitudinal velocity of mandrel, an angular rate of spindle, a radius of round cross-sectional mandrel and a radius of creating circle. In order to create distribution of braiding angle on a braided fabric, generally, the longitudinal velocity of mandrel and the angular rate of spindle will be changed. By using this approach, longitudinal stiffness distribution can be made, but circumferential stiffness distribution cannot be created.

Thus, in order to obtain circumferential stiffness distribution, a new method for generating circumferential distribution of braiding angle on cylindrical tubular braded fabrics was needed. One of the solutions for creating circumferential distribution of braiding angle was by using a creating ellipse instead of a round-shaped creating circle. The dynamically change of a radius on the creating ellipse could make a circumferential distribution of braiding angle.

Hiroyuki NISHIMOTO, Akio OHTANI, Asami NAKAI, Hiroyuki HAMADA
Faculty of Textile Science, Kyoto Institute of Technology, Japan
Nishimoto_Hiroyuki@takeda.co.jp

Furthermore, the theoretical braiding angle was compared with the experimental results in order to validate this method. As a consequence, the proposed method was effective for creating circumferential stiffness distribution on the braided composite pipe.

KEY WORDS
STIFFNESS DISTRIBUTION, BRAIDING ANGLE, BRAIDED COMPOSITE

INTRODUCTION

One feature of braided fabric as reinforcements of composite materials is capability of changing braiding angle. The braiding angle affects the mechanical properties. Thus controlling the braiding angle is a very important issue.

The braiding angle on cylindrical tubular braided fabric is decided by four parameters; a longitudinal velocity of mandrel, an angular rate of spindle, a radius of cylindrical mandrel and a radius of creating circle as shown in Fig. 1. In case of creating longitudinal distribution of braiding angle on the braided fabric, the longitudinal velocity of mandrel and the angular rate of spindle will be changed. By using this approach, longitudinal stiffness distribution can be created, but circumferential stiffness distribution as illustrated in Fig. 2 cannot be made. Thus, in order to obtain the circumferential stiffness distribution, a new method was needed.

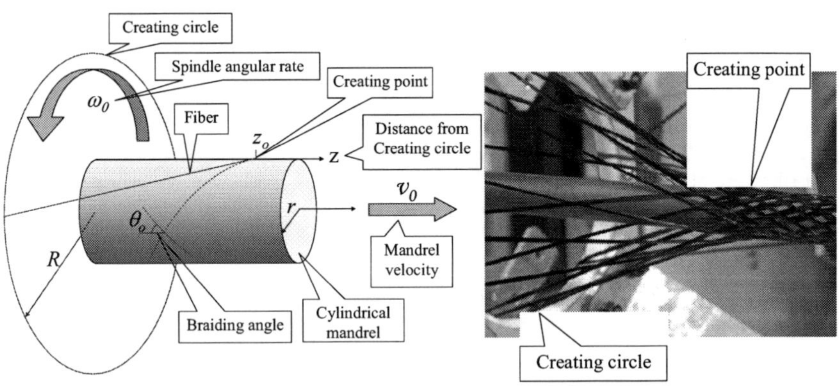

Figure 1. Braiding mechanism with a cylindrical mandrel.

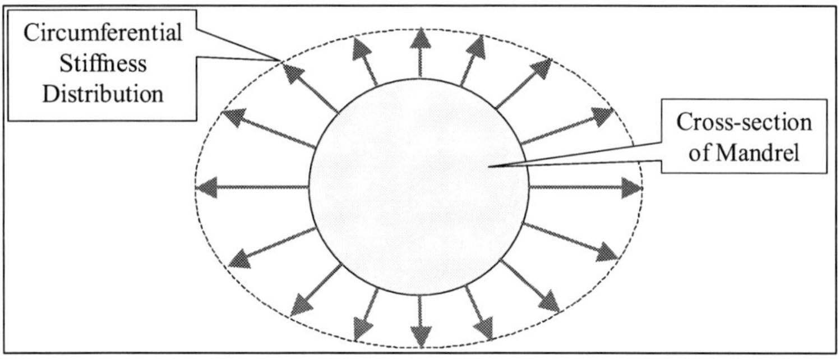

Figure 2. Circumferential stiffness distribution on a cylindrical braided fabric.

In this study, one of the solutions was proposed by using a creating ellipse instead of a round-shaped creating circle. The dynamically changing a radius of creating circle made circumferential distribution of braiding angle.

Firstly, mathematical principle of braiding angle was explained in the section 2.1. Secondly, additional useful parameters for describing a relation between a fiber providing point and the creating point were defined in the section 2.2 and 2.3. Thirdly, a new method for generating circumferential distribution of braiding angle on tubular braided fabrics was described in the section 2.4. Lastly, using this method, the theoretical braiding angles generated by creating ellipse with various axes were simulated in the section 3. Furthermore, the theoretical braiding angles were compared with the experimental results in the section 4.

MATHEMATICAL PRINCIPAL

Braiding Angle on Tubular Braided Fabric

Mathematical principle in case of a round cross-sectional mandrel is mentioned in this section. The braiding mechanism on a cylindrical mandrel is shown in Fig. 1. The mandrel is moving along the longitudinal direction. The spindles, moving bobbins of braiding yarns, are going on S-shaped intersectional rails to the circumferential direction. These movements make a tubular braided fabric on the mandrel.

As illustrated in Fig. 1, a creating point is defined as a fiber contact point on the mandrel first. A creating circle is depicted as an orbit of a fiber providing point

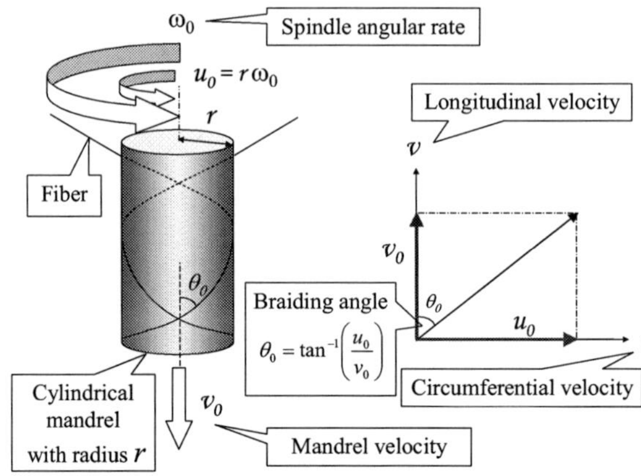

Figure 3. Geometrical derivation for a braiding angle of a cylindrical pipe.

between the creating point and the fiber supply point of spindle. It is the circular hole as shown in the picture of Fig.1.

The braiding angle is defined as an angle between the longitudinal direction and a braiding yarn. The geometrical derivation of braiding angle under steady-state condition is shown in Fig. 3. The braiding angle (θ_0) is derived from the ratio of the circumferential velocity ($u_0 = r\omega_0$) to the longitudinal velocity (v_0) as the following equation.

$$\theta_0 = \tan^{-1}\left(\frac{u_0}{v_0}\right) = \tan^{-1}\left(\frac{r\omega_0}{v_0}\right) \tag{1}$$

For changing the braiding angle, the ratio of the longitudinal velocity to the circumferential velocity should be changed.

Delay Angle ( )

As illustrated in Fig. 4, a delay angle ( ) is defined as a center angle from a fiber providing point on the creating circle to the creating point. The delay angle can be written as the following equation;

$$\cos\varphi = \frac{r}{R} \tag{2}$$

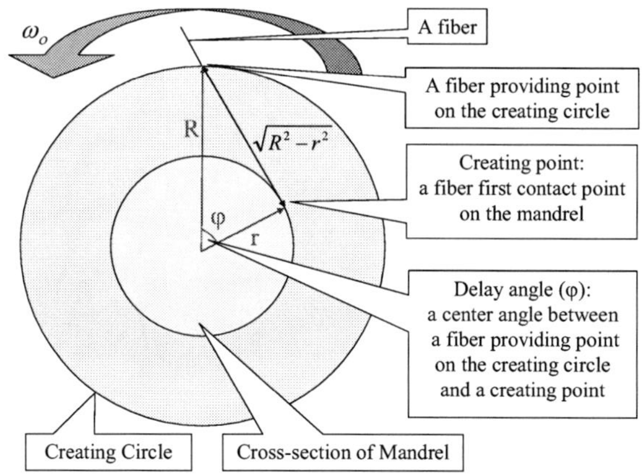

Figure 4. Delay angel ( ).

Alteration of the delay angle will affect the circumferential velocity of the creating point.

Distance from the Creating Circle to a Creating Point

A distance from the creating circle to a creating point is defined as variable z. As illustrated in Fig. 5, the distance z can be written as the following equation:

$$z = \frac{\sqrt{R^2 - r^2}}{\tan \theta_0} \quad (3)$$

Alteration of the distance z will affect the longitudinal velocity of the creating point.

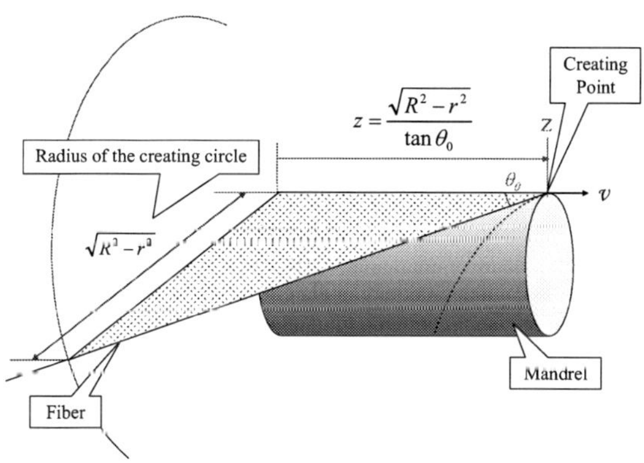

Figure 5. Distance (z) from the creating circle to a creating point.

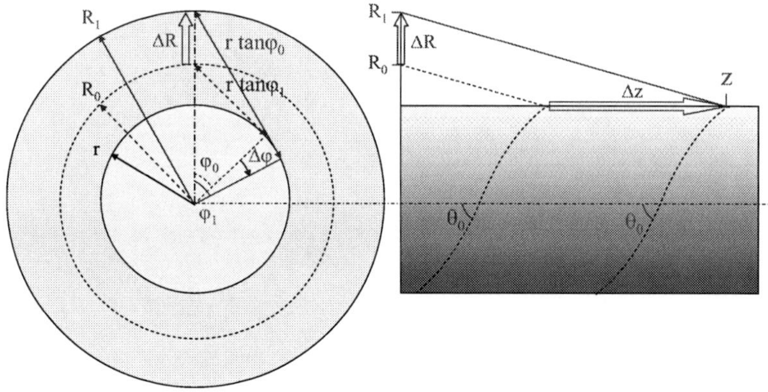

Figure 6. Changes of delay angle ( ) and distance of creating point (z) caused by R.

Generating Circumferential Distribution of Braiding Angle on Cylindrical Tubular Braided Fabric

Generating circumferential distribution of braiding angle on cylindrical tubular braded fabric, one of the solutions is by using a creating ellipse instead of a round-shaped creating circle. The dynamically changing a radius of the creating circle will make circumferential distribution for braiding angle.

As illustrated in Fig. 6, in order to predict the braiding angle, we need to figure out changes of delay angle ( ) and distance of creating point (z) caused by R.

As illustrated in Fig. 7, a braiding angle under unsteady-state condition can be described as the following equation:

$$\theta(t) = \tan^{-1}\left(\frac{u_0 - r\dfrac{d\varphi}{dt}\times\kappa_1}{v_0 - \dfrac{dz}{dt}\times\kappa_2}\right) = \tan^{-1}\left(\frac{u_0 - r\dfrac{d\varphi}{dR}\dfrac{dR}{dt}\times\kappa_1}{v_0 - \dfrac{dz}{dR}\dfrac{dR}{dt}\times\kappa_2}\right) \tag{4}$$

where factors κ_1 and κ_2 are decided by condition of braiding conjunction.

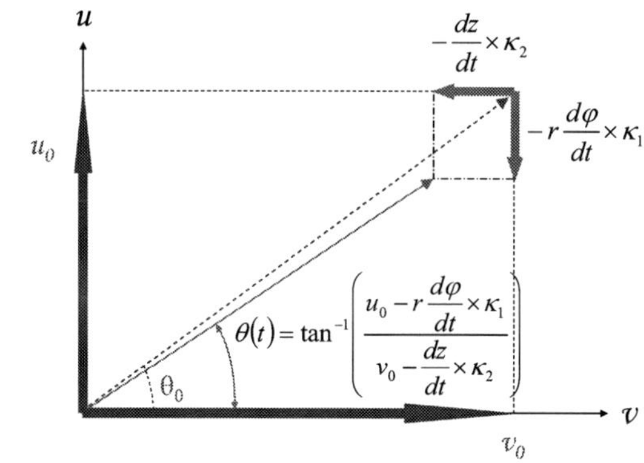

Figure 7. Braiding angle (t) under unsteady-state condition.

The derivative of delay angle ( ) with respect R can be written as the following equation:

$$\frac{d\varphi}{dR}=\frac{r}{R^2}\cdot\frac{1}{\sin\varphi}=\frac{r}{R\sqrt{R^2-r^2}} \tag{5}$$

The derivative of z with respect R can be written as the following equation:

$$\frac{dz}{dR}=\frac{d}{dR}\left(\frac{\sqrt{R^2-r^2}}{\tan\theta_0}\right)=\frac{1}{\tan\theta_0}\frac{R}{\sqrt{R^2-r^2}} \tag{6}$$

As illustrated in Fig. 8, assuming a function of creating ellipse (R(t)) with the major axis is $2R_a$ and the minor axis is $2R_b$, R(t) will be represented in polar coordinates as the following equation:

$$R(t)=\sqrt{R_a^{\ 2}\cos^2(\omega\cdot t)+R_b^{\ 2}\sin^2(\omega\cdot t)} \tag{7}$$

The derivative of R(t) with respect time can be written as the following equation:

$$\frac{dR}{dt}=\left(R_a^{\ 2}-R_b^{\ 2}\right)\frac{-\frac{1}{2}\omega\cdot\sin(2\omega\cdot t)}{\sqrt{R_a^{\ 2}\cos^2(\omega\cdot t)+R_b^{\ 2}\sin^2(\omega\cdot t)}} \tag{8}$$

Substituting Equation (5), (6) and (8) to Equation (4), the distribution of braiding angle can be calculated.

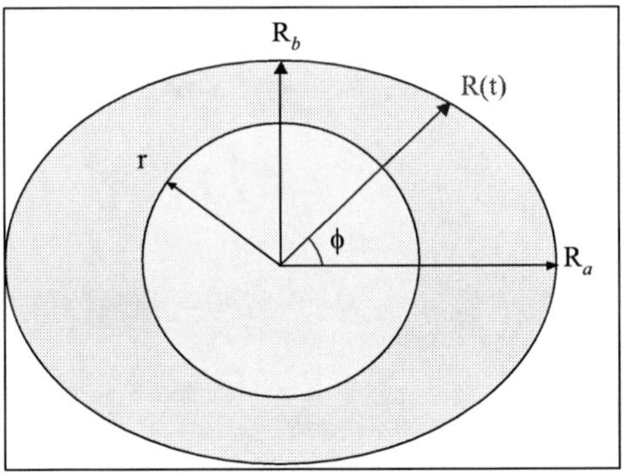

Figure 8. Function of creating ellipse: (R(t)).

SIMULATION FOR DISTRIBUION OF BRAIDING ANGLE BY USING CREATING ELLIPSE

Fig. 9 shows simulation for distribution of braiding angle by using creating ellipse with the following parameters:
The factors κ_1 and κ_2 were 0.06; The radius of mandrel cross-section (r) was 10.0 mm; The targeted braiding angle derived from equation (1) was 40 degrees. The ratio of major axis (R_a) to minor axis (R_b) are constant. The result shows larger R_a made larger amplitude of the distribution.

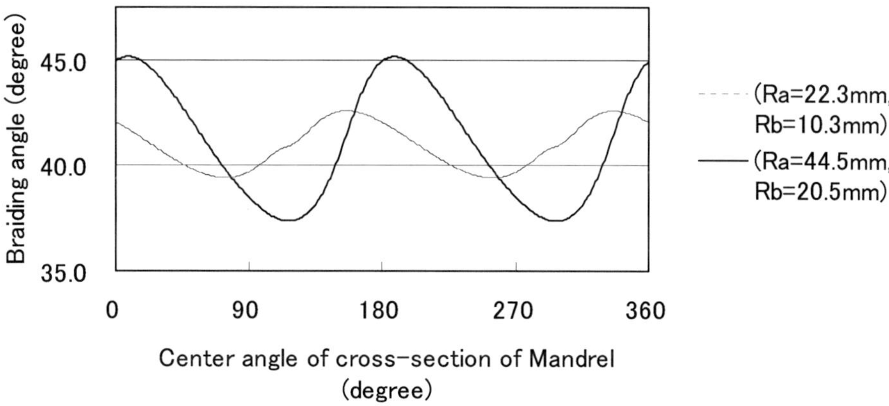

Figure 9. Simulation for distribution of braiding angle by using creating ellipse.

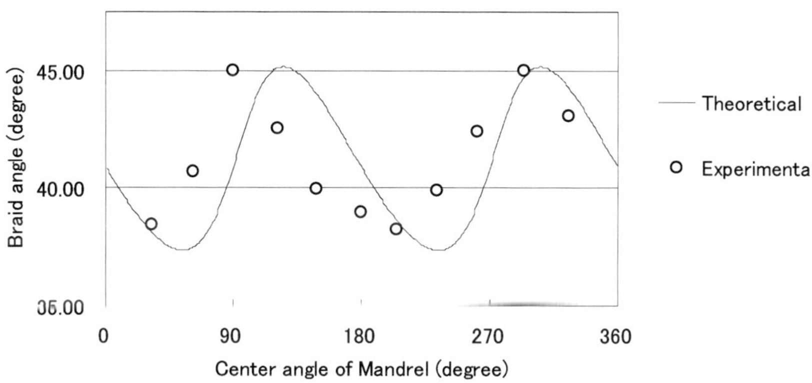

Figure 10. Experimental and theoretical distributions of braiding angle by using creating ellipse.

EXPERIMENTAL RESULTS

The distribution of braiding angle by using creating ellipse was examined in case of the major axis (R_a) equals 44.5 mm; the minor axis (R_b) equals 20.5 mm; radius of mandrel cross-section (r) equals 10.0 mm; The targeted braiding angle derived from equation (1) was 40 degrees. In order to verify the method, the result was compared with the theoretical values based on section 2. The experimental data and the theoretical curve derived by equation (4) are shown in Fig. 10. The experimental data had some deviations, although the cycle and amplitude were fitted with the theoretical curve.

CONCLUSION

In this study, by using a round-shaped creating circle instead of a creating ellipse, a circumferential distribution of braiding angle was generated on cylindrical tubular braided fabric. Furthermore, the proposed predicting method figured out a simple distribution of braiding angle. Consequently, a circumferential stiffness distribution will be created on cylindrical braided composite pipes caused by the distribution of braiding angle.

Internal Structure and Mechanical Properties of Silk Crape Composite

H. TAMOI, A. ODAWARA, T. SUGIE, A. NAKAI and H. HAMADA

ABSTRACT

Chirimen is the crape which is one of the textiles of Japanese traditional artifact. This textile is made from the silk and has used as a "kimono" in the old days. In chirimen, the hard twisted yarn is used as a weft fiber. The number of twist is from 2500 to 3500 per 1m. After weaving process, the textile is degummed to wash out the glue. After this process, the specific wrinkles appeared on the surface of this textile, which is called "Shibo".

The tensile strength and modulus of silk fiber are higher than those of other natural fibers, for example cotton and wool. Therefore it is expected that the silk fiber reinforced composite has high tensile properties. Moreover there is potential that the weaving structure with shibo affects on the mechanical properties. The purpose of this study is to clarify the internal structure of the textile composites used the silk crape and the effects of the internal structure on the tensile properties. In this study, the composites with silk crapes used as reinforcements were fabricated by hand-lay up method. The mechanical properties of composites were evaluated by using tensile test. Then the observations of internal structure and fracture mechanism of composites were evaluated.

1. INTRODUCTION

"*Chirimen*" which is the degummed silk fabrics are the Japanese traditional artifacts. Chirimen has the special texture, which is called "*Shibo*". Shibo is a kind of embossing and the characteristic texture in chirimen. Shibo creates the unique feel and the beautiful drape. By taking advantage of these features, since early times, chirimen has been used as the part of the high-grade kimono etc.

Hayato Tamoi*, Akiko Odawara*, Tomohiko Sugie*, Asami Nakai*, Hiroyuku Hamada*
*Kyoto Institute of Technology
Development of Advanced Fiber-Science, Kyoto Institute of Technology, Matsugasaki, Sakyo-ku, Kyoto 606-8585, Japan, E-mail: m7651007@edu.kit.ac.jp

The manufacturing method of chirimen is as follows; Ffirst, in the state of raw fiber, the fiber is twisted from 2500 to 3500 per 1m. The hard twisted fiber is called "hard twist yarn". In the case of chirimen, the hard twist yarn is used as the fiber in warp direction. By using these fibers, the fabrics are woven. Then the woven fabric was degummed. The degumming is to wash out the sericin from the fabric. After these processes, shibo appears on the surface of chirimen. In this study, to develop the application of chirimen in the industrial filed, chirimen was regarded as the reinforcement of the composite materials.

The purpose of this study is to clarify the internal structure of the textile composites used the chirimen and the effects of the internal structure on the tensile properties. The composites with chirimen used as reinforcements were fabricated by hand-lay up method. The mechanical properties of composites were evaluated by using tensile test. Then the observations of internal structure and fracture mechanism of composites were evaluated.

2. MATERIAL

Fig.1 shows two types of the Chirimen, before the degumming and after the degumming. In this study, the chirimen after the degumming was called "degummed silk" and before the degumming was called "raw silk". From the photographs, it was found that there were many "shibo" on the surface of the degummed silk.

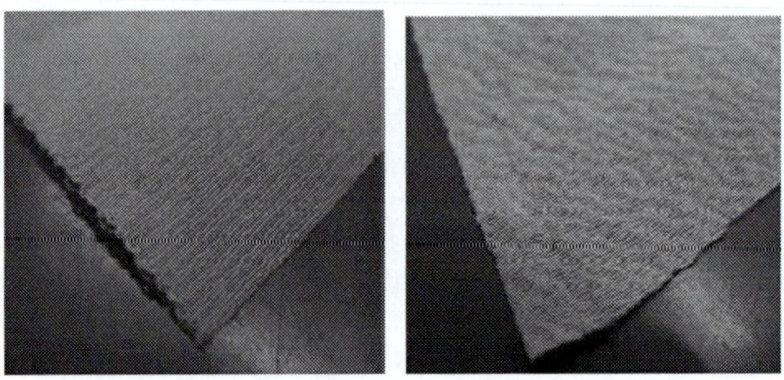

Before the degumming After the degumming

Figure 1. Photograph of before the degumming and after the degumming.

In order to examine the internal structure of chirimen as the reinforcements of the composite, the cross sectional observations were performed. Fig.2 and Fig.3 show the cross section of each specimen along the weft and warp direction. It was found that the textile structure was changed between before and after the degumming because the fabrics were shrunk by washing out the sericin. This phenomenon is unique property of Chirimen and results in the above-mentioned "shibo".

In the weft fibers of both specimens there were no crimps of fiber bundle as shown in Fig.2. From Fig.2 the one warp fiber bundle of the raw silk was separated into three fiber bundles. Moreover the dispersion of the fibers was insufficiency. On the other hand, in the case of degummed silk, the warp fiber bundle seems to be composed of only one bundle and the dispersion of the fiber was sufficiency. It is considered that the sericin played a bond role to connect the filaments.

Generally in chirimen to control the degree of the shibo, two types of fiber bundles were used as the warp fibers as shown in Fig.3. One of them was the hard twist yarn and the other was the weak twist yarn. The hard twist yarn was consisted of the two small bundles in the fabric used in this study. It is considered that fiber volume fraction (Vf) of hard twist was higher than that of the other fiber bundles. Therefore it can be said that chirimen was regarded as one of the fiber structural hybrid material. It is defined that the fiber structural hybrid has the different fiber structure on the one textile fabric.

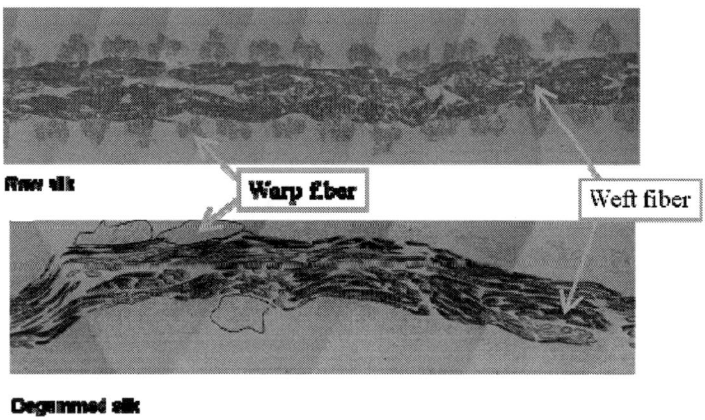

Figure 2. Cross section of each specimen along the weft fiber.

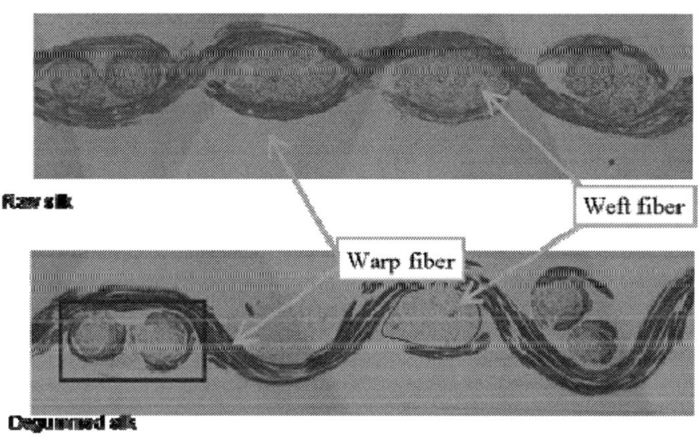

Figure 3. Cross section of each specimen along the warp fiber.

3. TEST METHOD

Tensile testing was performed by Instron universal testing machine (model 55R4206). Testing speed was 1mm/min. The strain was measured by the strain gage (type: KFG-5-120-C1-11: KYOWA ELECTRONIC INSTRUMENTS CO., LTD).

Generally the initial fracture of the composite is the micro-fracture such as interface fracture and the final fractures are caused by the accumulation of the micro-fractures. A characteristic property of composite materials is a knee point in the stress-strain curve as shown in Fig.4. In the case of woven fabrics, it is known that the knee point appears in consequence of the initial micro-fracture as transverse cracks. After the tests, the stress at knee point of each specimen was calculated and the observations of cross-section at knee point were carried out.

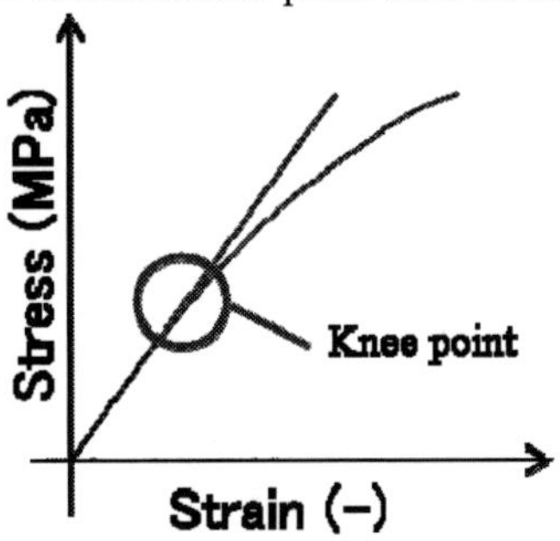

Figure 4. Calculation method of stress at knee point.

4. RESULT AND DISCUSSION

Fig.5 shows the stress-strain curves. The maximum tensile stress and the fracture strain of the raw silk were higher than those of degummed silk. Table I shows the tensile properties of each specimen. Both the modulus and the strength of degummed silk showed the higher than that of the raw silk. This study paid attention to the knee points on the stress-strain curves. Table II shows the stress at knee point of each specimen. As well as the modulus and strength, the stress at knee point of raw silk was higher than that of degummed silk.

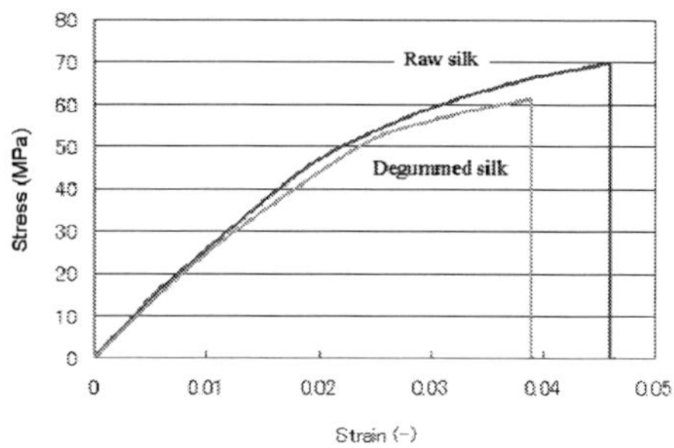

Figure 5. Stress-strain curves.

Table I . Summary of tensile properties.

	Modulus (GPa)	Strebgth (MPa)
Raw silk	4.74	66.8
Degummed silk	4.2	62.5

Table II . Stress at knee point.

	Raw silk	Degummed silk
Stress at knee point (MPa)	17.1	16.0

It was considered that the differences in the tensile properties were caused by the existence of the surface treatment, the sericin. Therefore next, the observations of Scanning Electronic Microscope (SEM) were carried out. The photographs of the raw silk and degummed silk were shown in Fig.6 and Fig.7. In the case of raw silk, there was the sericin around the silk. Therefore there are no space in the interface between the fiber and the matrix. On the other hand, in the degummed silk, the shape of each filament was visible since the sericin was washed out by the degumming. Moreover there were some spaces between the fiber and matrix. It was thought that the sericin kept the good interface between the fiber and the matrix. From these results, it is considered that that the sericin sffects the tensile modulus, stress at knee point and strength.

To investigate the initial fracture aspect, the cross sections of the composite after tensile testing were observed (Fig.8 and Fig.9). From the observation, it was found that the initial fracture occurred in the hard twist yarn as a transverse crack. Therefore effects of twist in fiber bundle also should be considered.

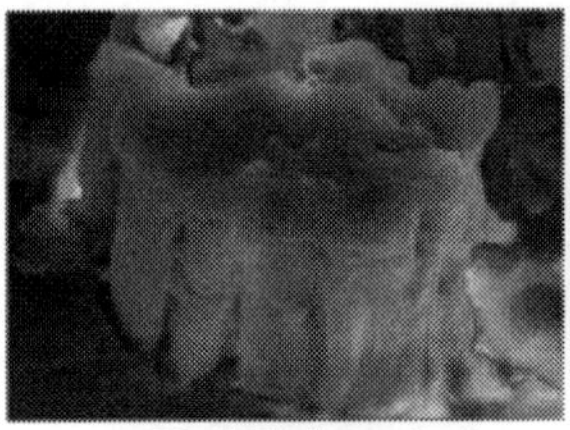

Fig 6 Photograph of the raw silk

Fig 7 Photograph of the degummed silk

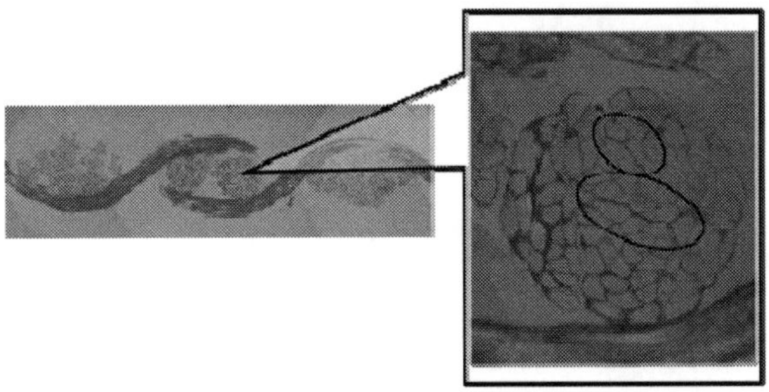

Figure 8. Cross section of raw silk.

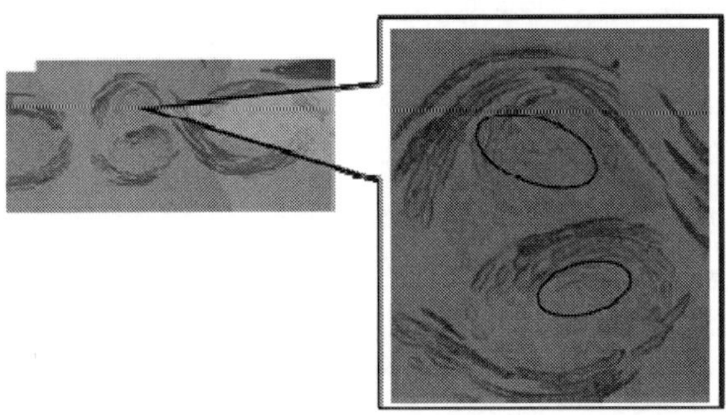

Figure 9. Cross section of degummed silk.

5. CONCLUSION

In this research, chirimen which is one of Japanese traditional artifact was used as a reinforcement of a composite. Chirimen consists of the weak twisted yarn and the hard twisted yarn. Therefore we can say that chirimen is the fiber structural hybrid composite. Two kinds of chirimen, before the degumming and after the degumming were used. The degumming is to wash out the sericin from the fabric. The tensile tests were carried out. From the result, it was found that the sericin affected the tensile properties. The sericin makes the interface between the fiber and the matrix good. By the observation of the initial fracture aspect, it was clarified the initial fracture occurred only at the hard twisted yarn, absolutely. Thus effects of twist in fiber bundle also should be considered.

Identification of a Damage Law for a 2.5D Woven C/C Composite using Displacement Fields

J.-N. PÉRIÉ, S. ROUX and F. HILD

ABSTRACT

The aim of this paper is to show how non conventional tests and full field measurements may help to identify anisotropic damage behaviours, in particular when sampling is an issue. The example of a biaxial shear test performed on a cruciform specimen made of a 2.5D woven C/C composite material is considered. The chosen damage law, based upon Continuum Damage Mechanics, accounts for shear softening. A Digital Image Correlation technique based on Finite Element shape functions is first used to extract the in-plane displacement fields from pictures taken during the experiment. The Equilibrium Gap Method is then used to retrieve a damage law. The identified damage maps are similar to those of post processed maps using classically identified parameters. The relative standard deviation of the displacement difference between measured and identified fields is then of the order of 5%.

INTRODUCTION

The complex anisotropic behaviour of composite materials, induced by matrix cracking and fibre breakage, may be modelled within the framework of Continuum Damage Mechanics [1]. The damage variables representing the relative loss of modulus are introduced at distinct scales. Models written at the constituent (micro) scale [2] or at the ply (meso) scale [3, 4] account for a loading induced anisotropy. The overall (macro) behaviour of a given stacking sequence then usually results from a generalized laminate theory. The identification of the numerous parameters related to the state and growth laws usually requires many tensile tests to be performed on various stacking sequences. Unfortunately, in the specific case of ceramic-matrix composites (CMC), such a sampling is not always possible given the lack of available material and of stacking sequences. Furthermore, the validity of the identification is usually difficult to check against experimental data.

Périé, J-N, Roux, S., and Hild, F., LMT-Cachan; ENS de Cachan/CNRS UMR-8535/Université Paris 6, 61 avenue du Président Wilson, F-94235 Cachan Cedex, France
Périé, J-N, LGMT (Laboratoire de Génie Mécanique de Toulouse); INSA, UPS, 135 avenue de Rangueil, F-31077 Toulouse, France

In the past, full-field measurements were essentially seen as a complementary means of comparison between experiments and FE simulations. The recent advances of these techniques [5] and of simulation tools are opening the way for convenient and robust identification methods. The richness of the kinematics resulting from the geometry of the tested specimen, the non proportional and multi-axial loadings, and latter from localizations, are exploited as an entry of inverse methods. One may then envision using these methods not only to identify parameters but also to evaluate the real benefit of a proposed refinement. A continuous pathway from images to mechanical properties now appears possible. The use of full-field data to feed an identification procedure for a damage law has already been proposed in the past. A first attempt was proposed by Claire *et al.* [6, 9]. In this case, the relationship between damage and an equivalent scalar strain was fitted to the collection of all element damage estimated from the Equilibrium Gap Method. Another approach was proposed by Chalal *et al.* [7] based on the virtual fields method. A unidirectional composite loaded in shear was described by a linear increase of damage with the strain. Based on a new Digital Image Correlation scheme (Q4-DIC [8]) and on the Equilibrium Gap Method (EGM [9]), the so-called Digital Image Mechanical Identification (DIMI [10]) approach allows one to retrieve an isotropic damage law directly from pictures acquired during a single biaxial test. In the following, it is proposed to extend DIMI to an anisotropic damage case.

The first part of this paper details the identification procedure. A two-step approach is proposed to go from digital images to an anisotropic damage law. First, the basis of the Q4-DIC technique is explained to retrieve the displacement fields from the experiments. Then, it is shown how the DIMI framework can be tuned when considering an anisotropic damage law. The second part of the paper presents some results of such approach applied to a biaxial test performed on a 2,5D C / C composite. A flat cruciform specimen made of [±45°] woven plies is subjected to a shear test. Non linear simulations are first used to check the ability of the method to identify a given damage law.

DIGITAL IMAGE MECHANICAL IDENTIFICATION

When performing an experiment, pictures are taken at different levels of loading. By matching pixels in different pictures, it is possible to measure displacement fields. The latter are subsequently used to identify the parameters of a damage law.

From Images to Displacement Fields

Let us consider two digital images taken at two stages. The reference and deformed pictures are denoted by $f(x)$ and $g(x)$. Classical Digital Image Correlation techniques consist in looking for the maximum correlation between small Zones of Interest sampled in both images [11]. An alternative scheme is used in the following [8]. The two images are related assuming the conservation of the optical flow. The problem consists in *identifying* the best displacement field $u(x)$ by minimizing the surface integral of the square of the correlation residual φ

$$\varphi(u) = |g(x) - f(x + u(x))| \tag{1}$$

After integration of φ^2 over the domain Ω, a global residual Φ is defined as

$$\Phi(u) = \iint [g(x) - f(x + [\psi(x)]\{u\})]^2 d\underline{x} \tag{2}$$

The minimization of this objective functional is intrinsically a non-linear and ill-posed problem. For these reasons, a discrete and weak format is preferred by adopting a general discretization scheme

$$u(x) = \sum_n u_n \psi_n(x) = [\psi(x)]\{u\} \tag{3}$$

where ψ_n are the vector shape functions, and a_n their associated degrees of freedom. In a matrix-vector format, $[\psi]$ is a row vector containing the values of the shape functions ψ_n and $\{u\}$ the column vector of the degrees of freedom.

If the amplitude of the displacements is small enough (typically less than 1 pixel), a linearised form of the functional Φ can be used

$$\Phi_{lin}(u) = \iint [f(x) + [\psi(x)]\{u\}.\nabla f(x) - g(x)]^2 d\underline{x} \tag{4}$$

A coarse to fine multiscale strategy then allows one to cope with the limitations induced by this last assumption. At this level of generality, many choices of trial basis can be made to measure displacement fields. In the following, classical bilinear shape functions associated with quadrilateral 4-node elements (or Q4) [8] are chosen. It is referred to as Q4 Digital Image Correlation (or Q4-DIC). The minimization of Φ_{lin} finally leads to a linear system $\boldsymbol{Ma} = \boldsymbol{b}$, where $\boldsymbol{a}$ represents the unknown nodal displacements, $\boldsymbol{M}$ and $\boldsymbol{b}$ are respectively a matrix and a vector computed from both the images and the shape functions.

From Displacement Fields to an Anisotropic Damage Law

MATERIAL, CONSTITUTIVE LAW AND STATE VARIABLES

The material studied herein is a 2.5D C / C material manufactured by SNECMA Engine Division. It is obtained by stacking plies of satin layers made of carbon yarns. A needling process transfers some fibres along the third direction, perpendicular to the layer (see Figure 1b), in order to forbid delamination propagation. A Chemical Vapour Infiltration deposits the matrix in the preform made of discontinuous fibres. This type of material is designed for thermostructural applications such as brakes or structural parts of rocket engines. For the studied $[0,90]_n$ laminate, tensile tests show an anisotropic behaviour with damage and inelastic strains at 45° but also at 0°. Different damage kinetics between tension and compression in both loading directions has been observed.

Because the laminate thickness is negligible in comparison with other dimensions (Figure 1: typically 10 mm versus more than 100 mm), a state of plane stress and Kirchhoff-Love kinematics are assumed. The material is assumed to be initially homogeneous. Indices (*1*, *2*) refer to the ply coordinate system (*i.e.* material frame), here coinciding with the fibre directions. With these notations, E_1 and E_2 denote the initial Young moduli (here in the fibre directions), G_{12} the initial shear modulus and ν_{12} the Poisson's ratio. The angle between this frame and that of the camera coordinate system (x, y) is θ. It is first assumed that damage is mainly dictated by the fibre orientation. The damage model considered herein [12] derives from an approach originally introduced for unidirectional plies [13]. Three damage variables are used. As a first step, the damages d_1 and d_2 accounts for a brittle fracture in the fibre direction. On the contrary, d_{12} reflects a gradual damage affecting the shear modulus. Many [0,90] carbon / epoxy woven composites, as a first order approximation, and at a certain scale, behave in such a way. A continuum thermodynamics framework is used [1]. Gibbs' free energy density ψ of the woven ply reads

$$\psi = \frac{1}{2}\left(\frac{\langle\sigma_{11}\rangle_+^2}{E_1(1-d_1)} + \frac{\langle\sigma_{11}\rangle_-^2}{E_1} + \frac{\langle\sigma_{22}\rangle_+^2}{E_2(1-d_2)} + \frac{\langle\sigma_{22}\rangle_-^2}{E_2} - 2\frac{\nu_{12}}{E_1}\sigma_{11}\sigma_{22} + \frac{\sigma_{12}^{\ 2}}{G_{12}(1-d_{12})} \right) \tag{5}$$

The use of positive parts $<\sigma_{ii}>_+$ of the stresses σ_{ii} allows one to model a deactivation of the damages related to the closure / opening of cracks depending on the sign of the applied stress ($<\sigma_{ii}>_- = <-\sigma_{ii}>_+$). From the potential Ψ, the state laws are derived, and in particular the driving forces Y_{d_i} (damage energy release rate) associated with the damage variables d_i

$$Y_{d_i} = \frac{\partial \Psi}{\partial d_i} = \frac{1}{2}\frac{\langle\sigma_{ii}\rangle_+^2}{E_i(1-d_i)} \text{ for } i = 1 \text{ or } 2, \text{ and } Y_{d_{12}} = \frac{\partial \Psi}{\partial d_{12}} = \frac{1}{2}\frac{\sigma_{12}^{\ 2}}{G_{12}(1-d_{12})} \tag{6}$$

Next, we consider that d_1 and d_2 remains null during the experiment. In the present case, we assume that the growth of d_{12} is simply related to its associated driving force $Y_{d_{12}}$. For the sake of simplicity, the following short-hand notations are used in the sequel $d = d_{12}$ and $Y = Y_{d_{12}}$. One can now define an equivalent strain ε_{eq}

$$\varepsilon_{eq}^2 = \frac{Y}{G_{12}} = 2\varepsilon_{12}^2 = \frac{1}{2}\left((\varepsilon_{yy} - \varepsilon_{xx})\sin 2\theta + \varepsilon_{xx}\cos 2\theta\right)^2 \tag{7}$$

This equivalent strain only depends on the shear strain in the material frame and is computed from the (measured) displacement fields. The goal is then to identify the damage law relating the shear damage variable d and the equivalent strain ε_{eq}.

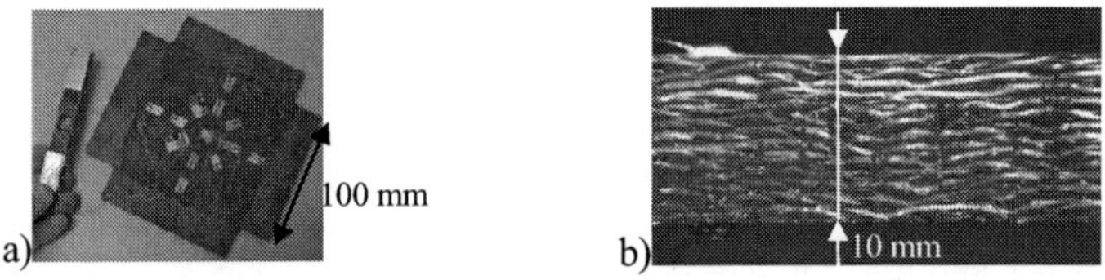

Figure 1. a) Tensile coupon and cruciform specimen. b) Cross section of the laminate.

IDENTIFICATION OF A DAMAGE LAW

The Equilibrium Gap Method (EGM) is followed herein. It consists in minimizing the force residuals associated with a mismatch of local elastic properties from elements to elements. It is written in a weak form by using a Finite Element discretization as the minimization of

$$[\mathbf{K}(\{d\})]\{u_{meas}\} - \{f_{body}\} = \{f_{res}\} \tag{8}$$

where $\{f_{body}\}$ are body forces at nodes, and $\{f_{res}\}$ is the residual vector associated with measured displacement fields $\{u_{meas}\}$.

Unlike classical FE problems, the aim is to determine the damage fields $\{d\}$ from known (*i.e.* measured) displacement fields $\{u_{meas}\}$ (*e.g.*, by Digital Image Correlation) and the nodal force vector $\{f_{body}\}$ assumed to vanish since only interior nodes are considered. To be consistent with measured displacements, Q4 elements are used again. The damage variable d is assumed to be element-wise uniform. Here, the elementary stiffness matrix K^{el} is expressed as an affine function of the damage variable d, namely $K_{ij}^{el} = M_{ij}^{0} + M_{ij}^{1}(1-d)$ where $[M^0]$ and $[M^1]$ are matrices dependent upon the initial elastic parameters (of the undamaged element). To simplify the expressions, an element-wise decomposition of the stiffness matrix is introduced $M_{ij}^{n} = \sum_e M_{ije}^{n}$ (for n = 0 or 1), where the sum runs over all elements e. The corresponding contribution of element e to the nodal force at the internal node j is $L_{je}^{n} = \sum_i M_{ije}^{n} u_i$ (n = 0 or 1). We next enforce that damage grows according to the same expression *everywhere* in the region of interest. The regularization of the problem consists in using, at the very beginning of the procedure, a specific decomposition of the damage law

$$d = \sum_k C_k \varphi_k(\hat{\varepsilon}_{eq}) \text{ with } C_k > 0,\ \varphi_k(\hat{\varepsilon}_{eq}) = 1 - \exp\left(\frac{\hat{\varepsilon}_{eq}}{\varepsilon_k}\right) \text{ and } \hat{\varepsilon}_{eq} = \max_{0<\tau<t}(\varepsilon_{eq}(\tau)) \tag{9}$$

In order to limit the sensitivity to noise, it is also proposed to introduce the operator S such that $[S]\{L\} = \{u_{meas}\}$, where S solves the homogeneous elastic problem for the undamaged medium, with Dirichlet boundary conditions as determined experimentally, and body forces $\{L\}$. A "reconditioned" equilibrium gap objective function $\widetilde{E}_g$, depending quadratically on the coefficients C_k (defining the damage law for the given set of characteristic strains ε_k), is proposed

$$\widetilde{E}_g = \sum_i \left(\sum_j S_{ij} \sum_e \left(L^0_{je} + L^1_{je} \left(1 - \sum_k C_k \varphi_k \left(\hat{\varepsilon}_{eq} \right) \right) \right) \right)^2 = \sum_i \left(u_i - \sum_k C_k \sum_j S_{ij} \sum_e L^1_{ie} \varphi_k \left(\hat{\varepsilon}_{eq} \right) \right)^2 \quad (10)$$

that can be read as the quadratic norm of a nodal vector homogeneous to a displacement field. In practice, the inverse operator S is not computed, but rather the "vector" $[S]\{L\varphi_k\}$, which is simply obtained as the solution of a homogeneous elastic problem for the undamaged solid. A remarkable feature of this procedure is that the identification of the entire (non-linear) damage evolution law is reduced to the solution of one single *linear* system (no iteration required). It has been first validated by using nodal displacement fields computed with a known damage law [14].

FIRST APPLICATION

The test has been carried out on the multiaxial machine ASTRÉE (Figure 2c). A flat cruciform specimen, considered as a [±45°] laminate, is subjected to a shear test (Figure 2a). This test was designed, in particular thanks to FE simulations [15], to induce a high value of shear damage in the central part of the specimen. The coupon is subjected to tension with respect to y and to compression with respect to x. In the following, we focus on identifying the damage part of a model in which inelastic effects are not described. Tabs glued on the arms allow for a transmission of the loads to the specimen.

The digital images of the surface (1016 × 1008 pixels, 8-bit digitization), taken at various steps of the loading (Figure 2b), are analyzed by using the Q4-DIC algorithm. The element size is set to 32 pixels (ca. 3 mm). When using the first image as the reference, the displacement fields result not only from damage but also from inelastic related effects. Similarly to classical identification procedures (based on tensile tests), one uses the 11 loading/unloading cycles. The hysteretic effects are neglected and the unloading is considered as elastic. In the following, only 11 pairs of pictures are used (Figure 2b). The first picture is taken at the maximum shear loading and the second one at the following unloaded state (in terms of resultant force in each arm). The displacement fields used as entries of the EGM correspond to the difference between the measured displacement fields associated with these two pictures. The elastic parameters are identified using a classical procedure based on tensile tests. Figure 3a presents the damage map corresponding to the identified damage law identified at the last load level.

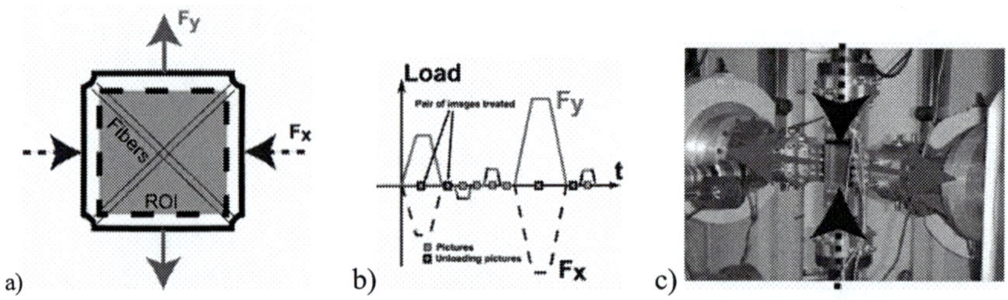

Figure 2. a) Principal of a shear test on a +/-45° cruciform specimen b) Loading path and corresponding pictures c) Detail of the experimental setup in ASTRÉE.

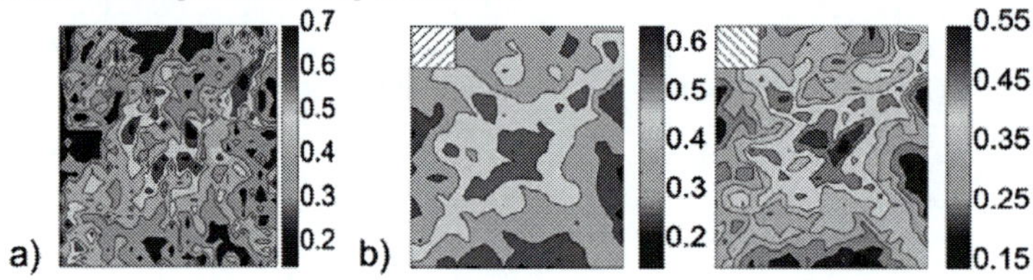

Figure 3. a) Identified damage map for the last loading step. b) Post-processed maps (in both plies).

The damage field within the ROI was also computed by using a damage post-processor [15]. The non linear parameters were identified using the same tensile tests. The woven ply is then modelled as a [0,90] laminate made of unidirectional plies. The shear damage of each ply is reported in Figure 3b. One notes a very good agreement between these post-processed damage maps and that determined by following the present procedure. In addition, Figure 4 shows a comparison between the measured and reconstructed displacement fields for the last loading step. One can, for example, define a residue R as the ratio of the root mean square of the difference between the measured and reconstructed displacement field, normalized by the root mean square of the measured displacement field. The corresponding residue is here estimated to be $R = 5\%$. It is to be underlined that unlike the classical tensile test, the present approach allows one to identify the damage law corresponding to post-peak data.

SUMMARY AND PERSPECTIVES

A way of identifying anisotropic damage laws using images shot during a mechanical test is presented. The proposed approach is based on recent developments of two (inverse) methods, namely, finite element digital image correlation and identification based on the equilibrium gap method. The first technique allows one to retrieve full-field (FE formatted) displacement fields from images along the loading history. The second one consists in solving an FE problem for which the data are measured displacements and the unknowns the parameters of the chosen trial damage law. The procedure is applied to analyze a biaxial test performed on a woven composite. It is possible to identify in a reliable way a damage pattern similar to the one obtained by post-processing the measurements with a classically identified model. The reconstructed displacement field is then very close to the measured one.

This work corresponds to the very first step towards the identification of more general constitutive laws when considering anisotropic materials. Future developments will include coupled multiaxial damage and inelasticity, which are important for a full account of the behaviour of, say, composite materials.

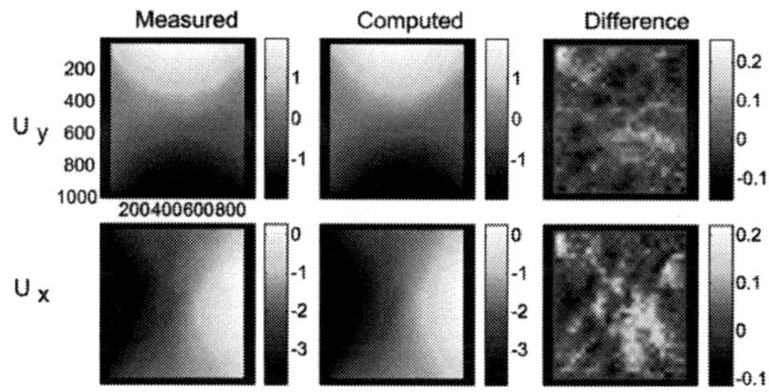

Figure 4. Comparison between measured and reconstructed displacements (expressed in pixels), and corresponding differences for the last loading level.

ACKNOWLEDGMENTS

This work was performed as part of the “Photofit” project supported by the Agence Nationale de la Recherche.

REFERENCES

1. Germain, P., Q. S. Nguyen and P. Suquet. 1983. “Continuum Thermodynamics,” *ASME J. Appl. Mech.*, 50:1010–1020.
2. Burr, A., F. Hild and F. Leckie. 1998. “On the mechanical behaviour under cyclic loading of ceramic-matrix composites,” *Mat. Sci. Eng.*, A250(2):256–263.
3. Chaboche, J.-L. and J.-F. Maire. 2002. “A new micromechanics based CDM model and its application to CMC's,” *Aerospace Sci. Tech.*, 6(2):131–145.
4. Ladevèze, P., 1995. “A damage computational approach for composites: Basic aspects and micromechanical relations,” *Comput. Mech.*, 17(1-2):142–150.
5. Rastogi, P.K. (ed.). 2000. “Photomechanics, *Springer*, Berlin.
6. Claire, D., F. Hild, and S. Roux. 2002. “Identification of damage fields using kinematic measurements,” *C. R. Mécanique*, 330:729–734.
7. Chalal, H., F. Meraghni, F. Pierron and M. Grédiac. 2004. “Direct identification of the damage behaviour of composite materials using the virtual fields method,” *Comp. Part A*, 35(7-8):841–848.
8. Besnard, G., F. Hild and S. Roux. 2006. “Finite-element displacement fields analysis from digital images: Application to Portevin-Le Châtelier bands,” *Exp. Mech.*, 46:789–803.
9. Claire, D., F. Hild and S. Roux. 2004. “A finite element formulation to identify damage fields: The equilibrium gap method,” *Int. J. Num. Meth. Engng.* , 61(2):189–208.
10. Roux, S. and F. Hild, 2008. “Digital Image Mechanical Identification (DIMI),” *Exp. Mech.*, [accepted].

11. Sutton, M.A., W.J. Wolters, W.H. Peters, W.F. Ranson, and S.R. McNeill. 1983. "Determination of displacements using an improved digital correlation method," *Im. Vis. Comput.*, 1(3):133–139.
12. Hochard, C., N. Lahellec and C. Bordreuil. 2007. "A ply scale non-local fibre rupture criterion for CFRP woven ply laminated structures," *Comp. Struct.,* 80(3):321–326.
13. Ladevèze, P. and E. LeDantec. 1992. "Damage Modelling of the Elementary Ply for Laminated Composites," *Comp. Sci. Tech.*, 43:257–267.
14. Leclerc, H. 2007. "Plateforme Metil : optimisations et facilités liées à la génération de code," *Proc. 8e Colloque National en Calcul des Structures*, Giens, France.
15. Périé, J.-N., S. Calloch, C. Cluzel and F. Hild. 2002. "Analysis of a Multiaxial Test on a C/C Composite by Using Digital Image Correlation and a Damage Model," *Exp. Mech.*, 42(3):318–328.

TOPIC XI

NUMERICAL METHODS AND SIMULATIONS

Material Damping Simulation of Textile Composites by Finite Element Analysis

Y. NAKANISHI, K. MATSUMOTO, M. ZAKO, T. KURASHIKI,
S. V. LOMOV and I. VERPOEST

ABSTRACT

In this paper, we have proposed an estimation method that the damping properties of textile composites are obtained by finite element analysis. The proposed method has been applied to the plain woven fabric composites. Experiments have been conducted to evaluate the validity of the proposed method. It is recognized that the computational results have agreed with the experimental ones. Since an effect of angle of fiber bundle on material damping of plain woven fabric composites can be simulated, the material damping properties for various woven fabric composites will be estimated the effect of woven architecture. From these points of view, the proposed simulation method is very useful for the estimation of material damping characteristics for woven fabric composites.

INTRODUCTION

Material damping represents the cumulative contributions of the viscoelastic response of the constituents, cyclic heat flow and the friction at the fiber/matrix interface. Recent work on the material damping of FRP has shown that the composite damping depends on an array of micromechanics and laminate parameters, including constituent material properties, fiber volume fraction, stacking sequence and so on [1-3]. These studies, however, are mostly limited to unidirectional composites.

On the other hand, textile composites are applied to many fields such as the space structures, sports items. Recently, many researchers have studied the identification of mechanical properties of textile composites by finite element Modeling to predict me-

---

Yasumasa Nakanishi & Kin'ya Matsumoto, Mie University, 1577 Kurima-Machiya, Tsu, Mie 5148507, Japan
Masaru Zako & Tetsusei Kurashiki, Osaka University, 2-1 Yamadaoka, Suita, Osaka 5650871, Japan
Stepan V. Lomov & Ignaas Verpoest, Katholieke Universiteit Leuven, Kasteelpark Arenberg 44, B-3001 Leuven, Belgium

chanical properties of textile composites. Though many researchers have reported the static characteristics of woven composites, the material damping of those materials have not been investigated [4-6]. And, the practical modeling technique and its systematization considering the complicated non-homogeneity of textile composites have not been developed completely. Furthermore, it is very difficult to evaluate the mechanical behavior of composites because of the complicated geometrical shape.

The purpose of this study is to establish an estimation program of damping properties for textile composites. The proposed procedure was applied to the simulation study of the material damping of plain woven fabric composites.

FINITE ELEMENT ANALYSIS FOR TEXTILE COMPOSITES

Finite Element Modeling

Figure 1 shows the scheme of the structural analysis of textile composites. The topology of a weave architecture is determined by the input data like a filament diameter, the architecture of strand, etc. The structural model of internal geometry of the fabrics is generated by WiseTex [4]. FE model with yarns and matrix by the hexahedron element is generated by the developed numerical modeling program MeshTex. The mechanical characteristics of textile composites can be calculated by the finite element analysis software SACOM.

The geometrical and mechanical model of textile can be provided by WiseTex, which is obtained the geometry of centerline of yarns [4]. The techniques of mesh generation by using WiseTex and MeshTex are shown in Figure 2. It is recognized that the precise FEM model is generated in process of technique and development of softwares.

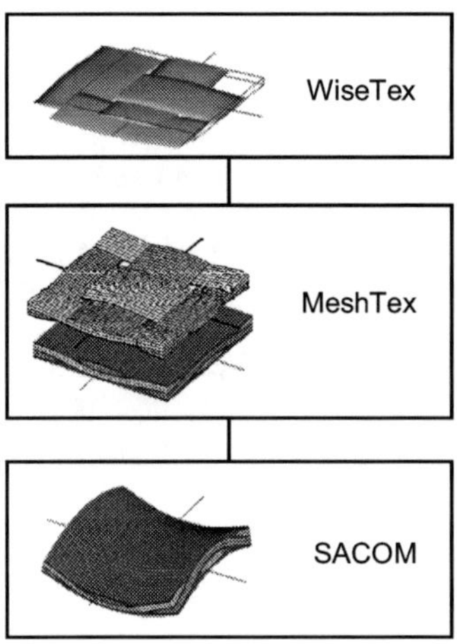

Figure 1. Numerical system for mechanical behaviors of textile composites.

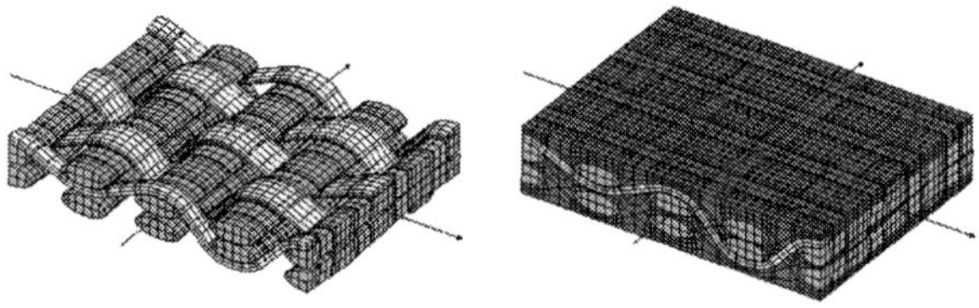

Figure 2. An example of finite element mesh model of textile composites.

FINITE ELEMENT ANALYSIS OF MATERIAL DAMPING

For unidirectional FRP, Adams and Bacon have proposed that the material damping energy is simply the sum of separable damping energies in the material due to each stress components [1]. In their theory, a modal damping ratio, the ratio of the dissipated energy during one complete cycle to the maximum stored energy from the beginning of the loading to the maximum, is expressed as

$$\zeta_n = \frac{1}{4\pi} \cdot \frac{\Delta U_n}{U_n} \tag{1}$$

where ΔU_n is the total damping energy per cycle of vibration, U_n is the maximum strain energy and n is the modal number.

Now, for woven composites, we employed a three-dimensional heterogeneous finite element model, which consists of fiber bundles and matrix resin. Fiber bundles and matrix resin are treated macroscopically as anisotropic and isotropic homogeneous bodies, respectively. The fibers were arranged unidirectionally within lamina of the composite laminates, but they are in the form of fiber bundles in textile composites. Damping analysis of woven composites involves the determination of contribution of two constituents i.e. fiber bundle and matrix resin.

The total damping energy of a cycle under the excitation ΔU_n is defined as

$$\Delta U_n = \frac{1}{2}\left\{ \int_V \psi^{\mathrm{fb}} \sigma_n^{\mathrm{fb}} \varepsilon_n^{\mathrm{fb}} \, dV + \int_V \psi^{\mathrm{r}} \sigma_n^{\mathrm{r}} \varepsilon_n^{\mathrm{r}} \, dV \right\} \tag{2}$$

where σ is the stress vector, ε is the strain vector, the suffix "fb" and "r" refer to the part of fiber bundles and matrix resin, respectively. ψ^{fb} and ψ^{r} are the specific damping capacities (SDC) which are expressed as follows:

$$\psi^{\mathrm{fb}} = \begin{bmatrix} \psi_L & 0 & 0 & 0 & 0 & 0 \\ 0 & \psi_T & 0 & 0 & 0 & 0 \\ 0 & 0 & \psi_Z & 0 & 0 & 0 \\ 0 & 0 & 0 & \psi_{TZ} & 0 & 0 \\ 0 & 0 & 0 & 0 & \psi_{ZL} & 0 \\ 0 & 0 & 0 & 0 & 0 & \psi_{LT} \end{bmatrix} \tag{3}$$

$$\psi^{\mathrm{r}} = \begin{bmatrix} \psi_x & 0 & 0 & 0 & 0 & 0 \\ 0 & \psi_y & 0 & 0 & 0 & 0 \\ 0 & 0 & \psi_z & 0 & 0 & 0 \\ 0 & 0 & 0 & \psi_{yz} & 0 & 0 \\ 0 & 0 & 0 & 0 & \psi_{zx} & 0 \\ 0 & 0 & 0 & 0 & 0 & \psi_{xy} \end{bmatrix} \tag{4}$$

On the other hand, U_n is the maximum strain energy which defined as

$$U_n = \frac{1}{2}\left\{ \int_V \sigma_n^{\mathrm{fb}} \varepsilon_n^{\mathrm{fb}}\, dV + \int_V \sigma_n^{\mathrm{r}} \varepsilon_n^{\mathrm{r}}\, dV \right\} \tag{5}$$

FINITE ELEMENT ANALYSIS PLAIN WOVEN FABRIC COMPOSITES

Material and Modeling

Test specimens have been fabricated by the hand-lay up method, using vinylester resin (supplied from Showa polymer Co. LTD.: R-806) reinforced by E-glass woven cloth fabric with 3 bundles (supplied from Asahi fiber glass Co. LTD.: WR570B). Figure 3 shows a part of cross section of the fiber bundles by a laser microscope. From this image, the volume fraction for a fiber bundle can be measured with a digital image processing technique. In case of the plain woven fabric composites, as shown in Figure 3, there is the distribution of volume fraction in a fiber bundle. The center parts have higher volume fraction 69.0% and edge parts have lower value of 58.7%.

Fiber bundles were treated as unidirectional FRP, and the mechanical properties were calculated by the rule of mixture based on the obtained fiber volume fractions. To estimate the woven architecture, finite element model has been generated and the mechanical properties have been calculated. The number of total nodes and elements of finite element mesh are 43,819 and 36,960, respectively.

Figure 3. Laser microscope image of test specimen.

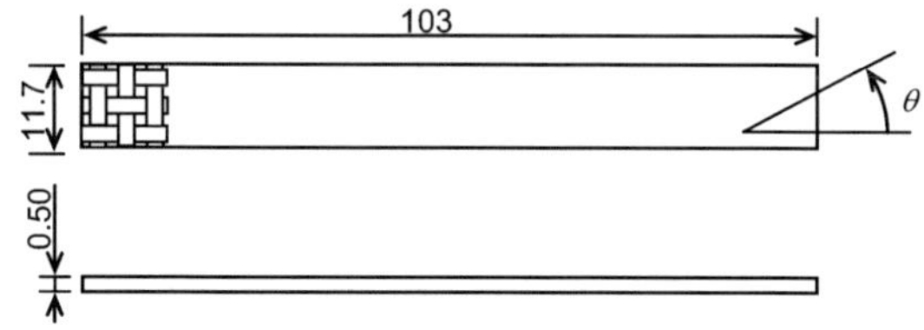

Figure 4. Dimensions of test specimen.

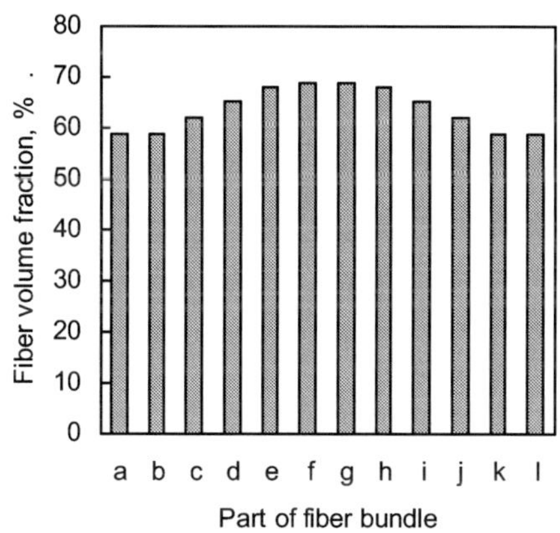

Figure 5. Fiber volume fraction of a fiber bundle.

RESULTS AND DISCUSSIONS

To estimate the material damping, finite element analysis has been carried out. Figure 6 shows simulated vibration modes. Computational results of plain woven composite ($\theta = 0°$) are shown in Figures 7 and 8. As shown in Figure 6, it is recognized that the damping ratio does not depend on frequency. The effects of the angle of the fiber bundle θ on natural frequency and damping ratio are shown in Figure 8. The natural frequency has maximum at $\theta = 0°$, $90°$. Because the stiffness becomes maximum at $\theta = 0°$, $90°$. On the other hand, the maximum damping ratio appears at $\theta = 45°$.

To confirm the validity of the numerical results, vibration tests have been carried out in the low-pressure condition (40Pa) in order to consider effect of aerodynamic force. The numerical results both natural frequency and damping ratio have a good agreement with the experimental ones as shown in Figures 7, 8.

From these results, it can be recognized that the material damping of plain woven composites can be estimated with the proposed numerical method.

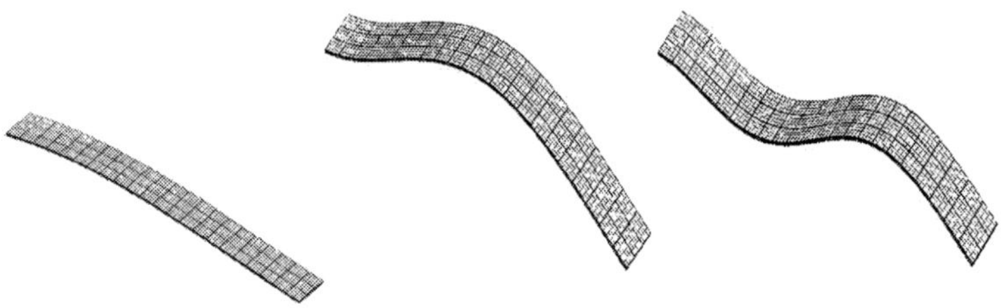

Figure 6. Vibration modes of plain woven fabric composites.

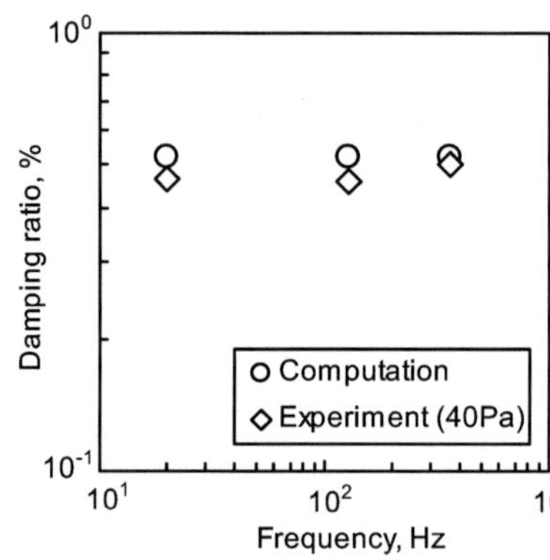

Figure 7. Comparison of experimental and computational results ($\theta = 0°$).

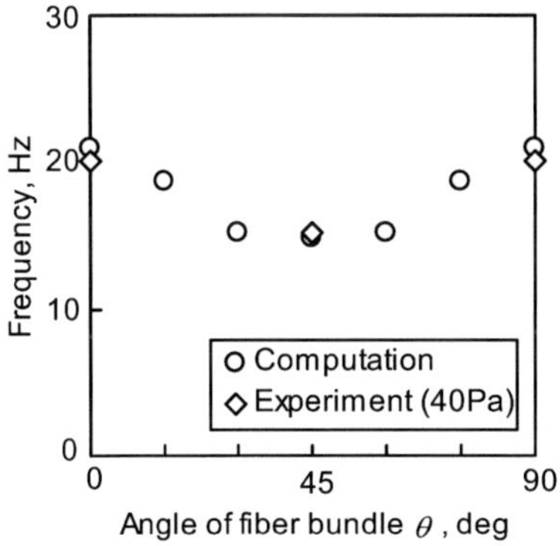

(a) Natural frequency

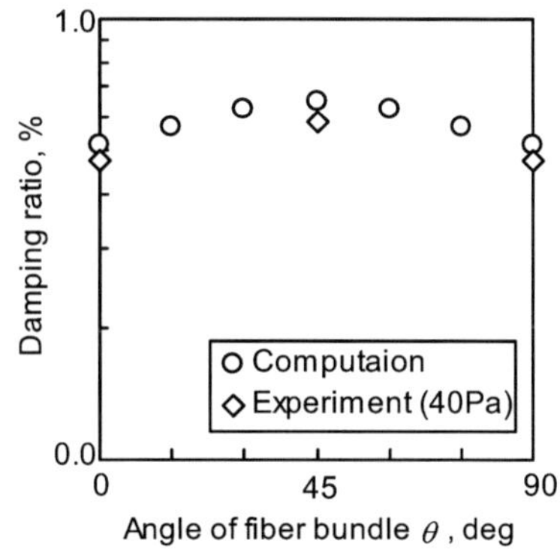

(b) Damping ratio

Figure 8. Relationship between vibration characteristics and angle of fiber bundle θ.

CONCLUSIONS

The estimation method of material damping for woven composites has proposed. Comparing the numerical results with the experimental ones, it is recognized that the proposed method is capable to estimate the material damping of woven composites. Furthermore, we can estimate the material damping of woven composites from damping characteristics of unidirectional FRP and the matrix resin.

REFERENCES

1. Saravanos, D.A. and Chamis, C.C. 1990. "Unified micromechanics of damping for unidirectional and off-axis fiber composites," *Journal of Composite Technology Research*, 12, pp. 31–40.
2. Adams, R.D. and Bacon, D.G.C. 1973. "Effect fiber-orientation and laminate geometry on properties of CFRP", *Journal of Composites Materials*, 7, pp. 402–428.
3. Chandra, R., Singh, S.P. and Gupta, K. 1999. "Damping studies in fiber-reinforced composites—A review", *Composite Structures*, 46, pp. 41–51.
4. Verpoest, I. and Lomov, S.V. 2005. "Virtual textile composites software WiseTex: Integration with micro-mechanical, permeability and structural analysis", *Composites Science and Technology*, 65, pp. 2563–2574.
5. Whitcomb, J. and Sriengan, K. 1996. "Effect of various approximations on predicted progressive failure in plain weave composites", Composite structures, 34, pp. 13–20.
6. Carvelli, V. and Poggi, C.A. 2001. "A homogenisation procedure for the numerical analysis of woven fabric composites", *Composites Part A*, 32, pp. 1425–1432.

Simulating the Impact of Multi-Layer Fabric Targets using a Multi-Scale Model and the Finite Element Method

G. NILAKANTAN, M. KEEFE, J. W. GILLESPIE, JR.
and T. A. BOGETTI

ABSTRACT

The size of a fabric model with a yarn level resolution that can be studied using a finite element analysis (FEA) is restricted by the computational requirements of the simulation. Such models are very useful in gaining insight into the mechanics of interactions, deformation, and energy absorption during the impact of woven composites. However such an analysis of systems with realistic dimensions is presently not possible. The focus of this paper is to present a multiscale model wherein a high level of resolution is used around the impact zone and this level decreases with distance away from the impact zone. Therefore yarns modeled using solid and shell elements are used at near-field regions while a homogenized resolution with shell elements that neglects the yarn architecture is used at far-field regions. Acoustic impedance matched interfaces are created between the various finite element formulations at the various scales of modeling. This approach is referred to as the '*Hybrid Element Analysis (HEA)*' and it has been shown to reduce the computational requirements in terms of both processor and memory requirements, while still being able to reproduce baseline simulation results. This will enable the simulation of larger fabric systems that were previously computationally prohibitively expensive.

INTRODUCTION

The computational requirements of a numerical simulation include two components, the processing power and memory requirements. With the advent of high performance computing (HPC), the use of numerical methods such as the finite element (FE) analysis have become very popular in simulating complex, high rate deformation events such as the impact of woven fabrics and composites.

Gaurav Nilakantan, Center for Composite Materials, University of Delaware, DE, USA
Michael Keefe, Dept. of Mechanical Engineering, University of Delaware, DE, USA
John W. Gillespie Jr., Center for Composite Materials, University of Delaware, DE, USA
Travis A. Bogetti, Army Research Laboratory, Aberdeen Proving Grounds, MD, USA

However due to intricate architecture of fabrics, fine FE meshes are needed to discretize the filament or yarn geometry, and complex contact algorithms are required to prevent interpenetrations between the yarns. This demands tremendous computing requirements in order to simulate the impact of a multi-layer fabric with realistic dimensions (e.g. a stack of 24 plies with in-plane dimensions of 1 ft x 1ft). Today's high performance clusters, such as a Beowulf cluster, consist of many nodes linked together, where each node is not very powerful by itself, but linked together they form a high performance computing system. As a side note, it is also important to mention that the cluster performance is also governed by the file handling system and communication technology used between the nodes, and not by just the processing power of each node. Dynamic finite element codes such as LS-DYNA® tend to perform the initialization phase of a simulation on a single processor before redistributing the job to multiple processors across multiple nodes. Depending on the architecture, the amount of memory available to each processor per node varies. In special systems such as the Dell Precision Workstations® and SGI Altix® with NUMA architecture, the memory is globally shared and therefore available to all processors. In most clusters however, the maximum memory available per node is around 2-8 GB. This means that if the simulation requires more memory than this during the initialization phase, even if it is run on a very powerful cluster where the combined memory of all nodes is more than sufficient, the simulation will not run since the memory available to one node during the initialization is insufficient. The next issue is the processing power needed by a simulation and what is actually available. While a simulation can be run for many weeks or months at a time, such an approach may not be efficient unless absolutely required or no other options are available. It is far more desirable to be able to get results quickly, because a faster turnaround means more productivity. In this paper, we present a multiscale model entitled the '*Hybrid Element Analysis*' method henceforth referred to as HEA, which reduces both the memory and processing power requirements of a simulation of the impact of woven composites or fabrics. Preliminary work dealing with the impact of a single layer of fabric has been presented in our previous work [1]. In this paper, we extend our work to the impact simulation of multiple layers of fabric. The basics of the HEA approach have been repeated here for the convenience of the reader.

THE HEA METHOD

We define the HEA as '*the finite element analysis of a structure by combining different finite element formulations at both a single and multiple scales of modeling*'. Figure (1) displays the FE mesh of a yarn modeled using solid and shell elements. The model with solid elements is considered to be the baseline model. Figure (2a) displays the old method of modeling a yarn with shell elements [2]. The thickness jumps seen between adjacent elements lead to rough contacting surfaces and lost yarn material which has an adverse effect on the results.

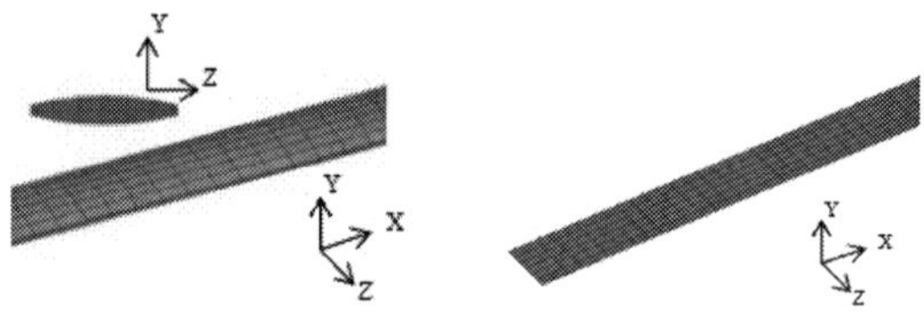

Figure 1. FE mesh of a yarn modeled using (on left) Solids (on right) Shells.

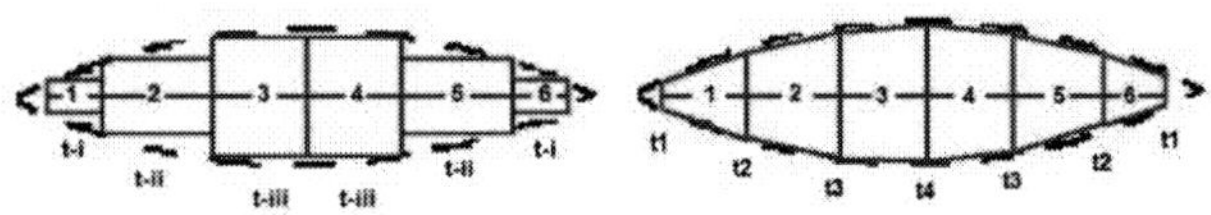

Figure 2. Yarn cross section using shell elements with (on left) Uniform nodal thicknesses—OLD (on right) Non-uniform nodal thicknesses—NEW.

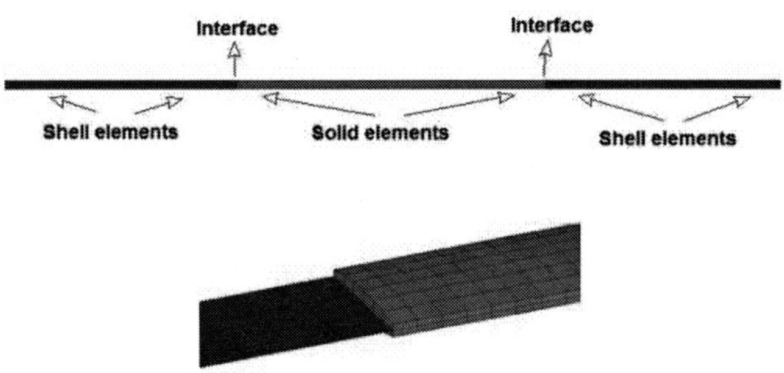

Figure 3. (top) HEA of a yarn using shells and solids (below) Close up at the interface.

Figure (2b) displays our new method wherein non-uniform nodal thicknesses have been assigned to each shell element so that the elliptical yarn cross section is exactly modeled as in the baseline and there is no lost yarn material between adjacent elements. Figure (3) displays the HEA applied to a yarn. Solid elements are used in the region where the yarn will be impacted by a projectile and shell elements are used elsewhere, wherein the predominant mode of deformation is tensile in nature. Clearly this model is computationally less intensive than the baseline model however it is able to accurately reproduce the baseline response [1]. The matched acoustic impedance between the solid-shell interface ensures no spurious longitudinal wave reflections at the interface which would otherwise adversely affect the results. Figure (4) displays the configuration of a typical HEA fabric model. Yarns with solid elements are used beneath the projectile impact zone, yarns with shell elements are used around this region, and a homogenized membrane with shell elements is used at far field regions. Thus the level of resolution decreases with distance away from the impact zone, in agreement with baseline observations where the physics and interactions to be captured become less pronounced away from the impact zone. Once again, impedances are matched across all interfaces and an approach to determining the material properties of the homogenized far field region is outlined in [1]. The region with yarn architecture is referred to as the *local* region while that of the homogenized architecture is referred to as the *global* region. Figure (5) displays some of the many possible HEA configurations.

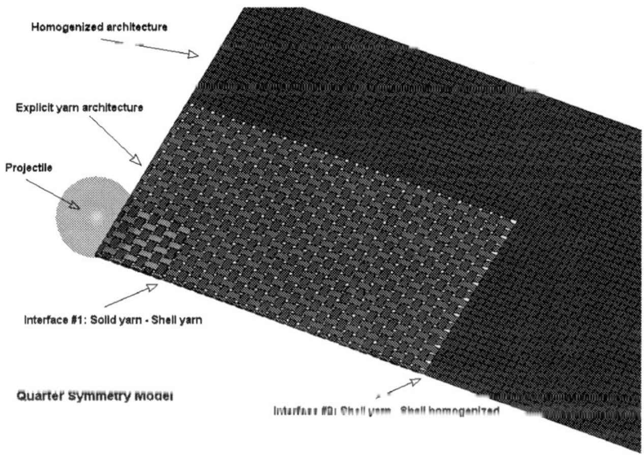

Figure 4. Setup of a fabric model using the HEA approach (yellow/brown region around the projectile—solid yarns, surrounding blue/green region - shell yarns, and far field grey region—homogenized shells).

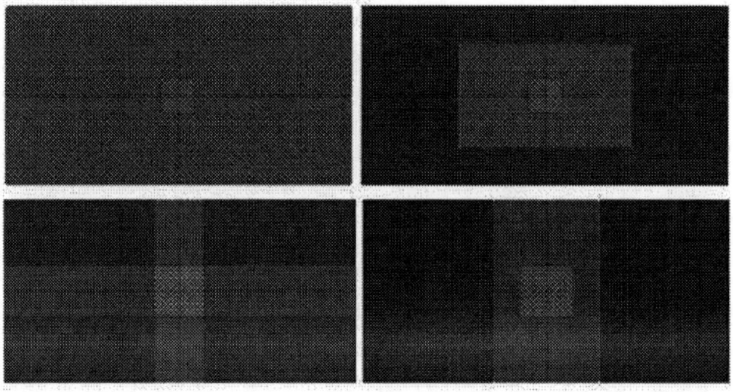

Figure 5. Possible HEA configurations: (top left) Single-scale Central-Patch; (top right) Multi-scale Central-Patch; (bottom left) Multi-scale Center-Cross; (bottom right) Multi-scale Center-Strip.

MULTILAYER FABRIC IMPACT SIMULATIONS USING THE HEA

When dealing with the impact of multiple plies of fabric, there are clearly many possible HEA configurations to choose from, wherein different plies may either use the same or different HEA configurations. The choice will depend on various factors such as the impact velocity and boundary conditions. For example, from experimental and baseline numerical observations of a two or four ply Kevlar KM2 fabric system held on two sides and impacted around the V_{50} velocity, it is seen that the top ply usually fails first and separates from the rest of the fabric plies when there is no coupling between the layers. In such a case, using a multiscale Central-Patch model with the least amount of local region for the top layer makes intuitive sense as initially only a small region around the impact zone plays a major role in the impact event and once this layer separates from the rest of the plies, we are no longer interested in capturing any further response of that layer, so a larger global region is appropriate. As we go down the layers, increasing amounts of local region may be necessary since they fail at later time instants. Further, since the global regions can currently only handle small degrees of yarn reorientation, a Center-Cross and Center-Strip model are more amenable for plies gripped only on two sides. However if a user defined material model is used in the global region that can handle large degrees of yarn reorientation, this preference may be relaxed and any of the HEA models may be used.

RESULTS AND DISCUSSION

The impact of two multi-layer plain-weave fabric targets is studied. The first target has 2 plies and is impacted at 190 m/s, while the second target has 4 plies and is impacted at 300 m/s. The plies are identically stacked with perfectly aligned unit cells. The impactor is a 0.63 gm rigid spherical projectile of diameter 5.55 mm moving along the *y*-direction. Each fabric ply has dimensions 101.6 mm x 50.8 mm and is gripped on the two longer sides.

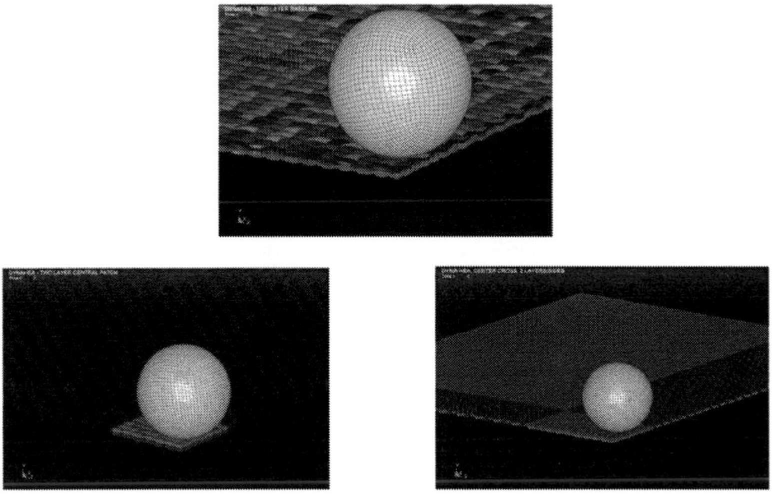

Figure 6. Two-ply fabric targets: (top) Baseline; (bottom left) Single-Scale Central-Patch; (bottom right) Multi-Scale Center-Cross.

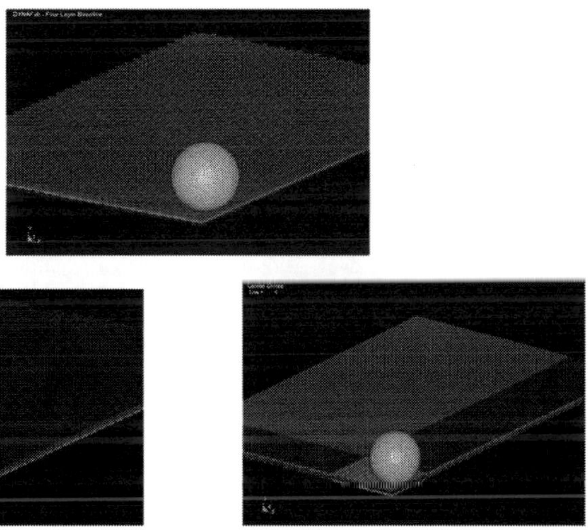

Figure 7. Four-ply fabric targets: (top) Baseline; (bottom left) Single-Scale Central-Patch; (bottom right) Multi-Scale Center-Cross.

It is comprised of Kevlar KM2 yarns with an areal density of 180 g/m^2. The effective material properties of the yarn are assigned as follows: longitudinal elastic modulus (E_x) of 62 GPa, transverse elastic moduli (E_y) and (E_z) of 6200 MPa, shear modulus (G) of 3280 MPa, zero Poisson ratios, density (ρ) of 1310 kg/m^3, and a maximum principal stress to failure of 3400 MPa. Due to symmetry, only one quarter of the fabric has been modeled. A static frictional coefficient of 0.18 is specified for both the yarn to yarn and projectile to yarn contact algorithms. The two HEA models chosen are the Single-Scale Central-Patch and Multi-Scale Center-Cross (see Figure 5). The terminology and model setup of the multi-layer Baseline and HEA models remain similar to the single layer models outlined in [1], with the exception that the shell element formulation has been switched from LS-DYNA *Type 16* to *Type 2* for computational reasons. The material properties of the homogenized or global region for the Multi-Scale Center-Cross model are as follows: longitudinal in-plane elastic modulus (E_x) and (E_z) of 9008MPa, transverse elastic moduli (E_y) of 901 MPa, shear modulus (G) of 50 MPa, zero Poisson ratios, and density (ρ) of 745.04 kg/m^3. Further details about how these properties were obtained are provided in Ref. [1].

In Ref. [1], we had noted the effect the contact algorithms had on the system response and computational requirements for a single layer fabric target. The setting up of the contact pairs in a multi-layer fabric target is further complicated since interpenetrations need to be prevented both between yarns in the same layer, as well across multiple layers. To investigate the effect that different contact algorithms may have on the layer-layer interactions of the first target, two baseline cases have been set up. Here the '*baseline*' case refers to the fabric model with explicit yarn architecture using only single integration point solid elements. Baseline #1 uses a computationally more intensive and robust eroding type contact (*type 14*) while Baseline #2 uses a computationally less intensive non-eroding type contact (*type a3*) to handle layer-layer contact interactions. Baseline #1 will provide more realistic results, but at a much increased computational expense. However this increase in computational expense can be limited by using the eroding type interlayer contact only for a small area around the impact region where element erosion is likely to occur, and a non-eroding type interlayer contact for the rest of the ply. Details of the working of these contact algorithms can be obtained in [3]. The four-ply targets all use the robust eroding type contact, since with an increasing number of layers and impact velocities, there is a greater extent of transverse yarn compression and crushing of the top layers, which increases the chances of element erosion occurring between layers. Without an eroding type contact, the top layers may penetrate the layers below them. Figure (6) displays the setup of the 2 ply targets, while Figure (7) displays that of the 4 ply targets. The same frictional coefficient used for yarn to yarn interactions is used between the global regions of all plies in the Multi-Scale Center-Cross model.

Figure (8) compares the results of the first target which consists of 2 plies. There is a very good agreement in the projectile velocity predictions. The effect of including an eroding type contact is clearly observable in the difference between the responses of Baseline #1 and Baseline #2 for the interlayer contact force and interlayer frictional sliding energy. The Multi-Scale Center-Cross model that also uses an eroding type layer-layer contact agrees well with Baseline #1, while the Single-Scale Central-Patch model that doesn't use the eroding type layer-layer contact agrees well with Baseline #2. The frictional sliding energies show a sudden increase as the projectile begins to penetrate through the fabric and the layers start to elastically spring back to their original shape. It is interesting that even though the Multi-Scale Center-Cross uses a homogenized global region, the interlayer frictional sliding energies and contact forces follow those of Baseline #1. This implies that the extent of layer to layer interactions at the far field regions is much smaller than those occurring around the impact region for this impact scenario. By suitably determining the effective frictional coefficient for the global-global region contact that is equivalent to the contact between undulating yarn surfaces of two layers, there will be a better agreement in the predicted interlayer frictional energy. Both HEA models overestimate the fabric internal energy during the post-yarn failure response. However more importantly, the pre-yarn failure response agrees well with the baselines. Figure (9a) compares the memory requirements and run times for the first target impact case. Figure (9b) displays the relative areas of each region in the models. The run time and memory requirements of Baseline #1 are slightly higher than Baseline #2 because of the addition of the eroding interlayer contact for a small area around the impact region. For the 2 ply target, the Single-Scale HEA model is 2.5 times faster than both Baselines and the Multi-Scale HEA model is 5 times faster than both Baselines. The details of the computing system used for the LS-DYNA simulations are outlined in [1].

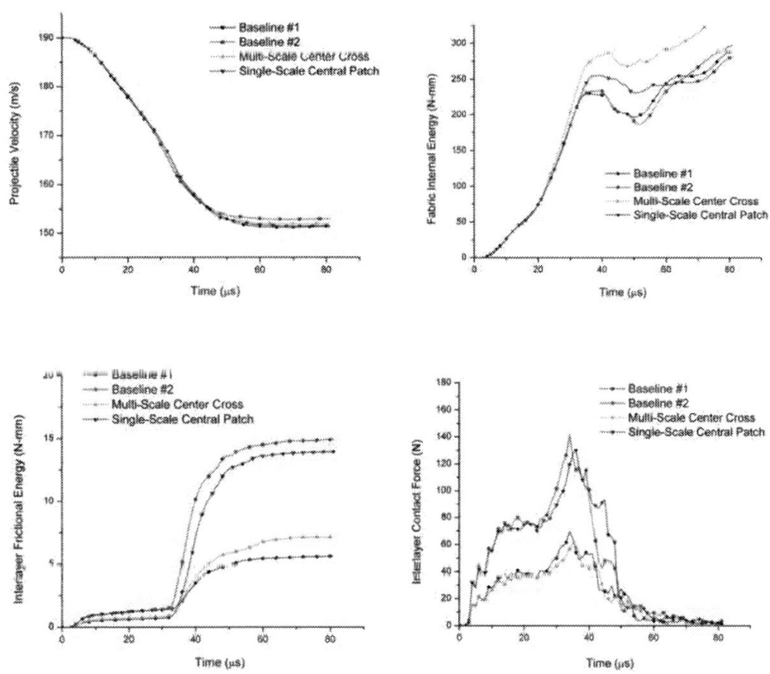

Figure 8. Baselines (#1 and #2) Vs. HEA for the 2 ply target: (top left) Projectile velocity; (top right) Fabric internal energy; (bottom left) Interlayer frictional sliding energy (bottom right) Interlayer contact force.

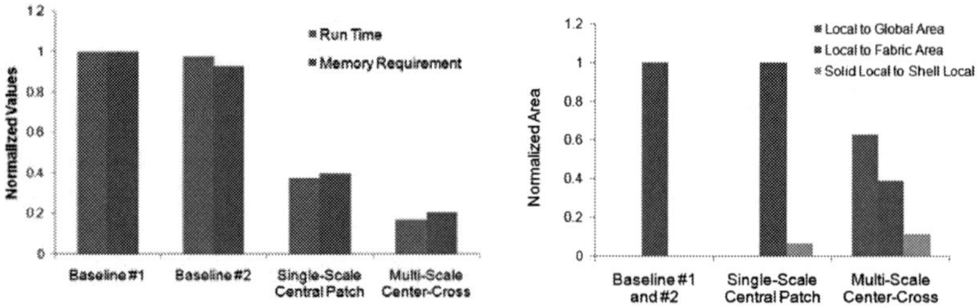

Figure 9. (on left) Simulation memory requirements and run times (on right) Relative areas of each region.

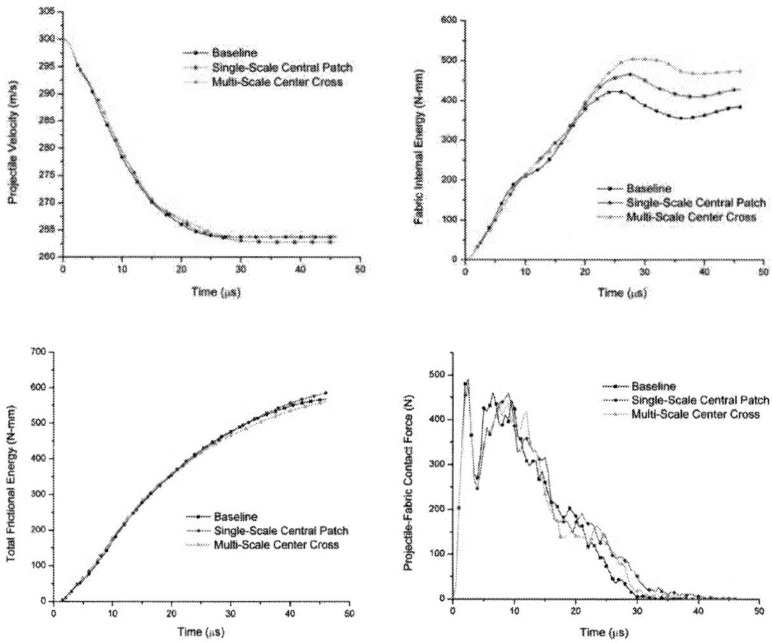

Figure 10. Baseline vs. HEA for the 4 ply target: (top left) Projectile velocity; (top right) Fabric internal energy (bottom left) Total frictional sliding energy (bottom right) Projectile to fabric contact force).

Figure (10) compares the results for the second target which consists of 4 plies. There is only one baseline case here, which uses the eroding type contact for layer to layer interactions. Both HEA models also use this eroding type contact for layer to layer interactions. There is a very good agreement seen in the projectile velocity predictions and a similar trend to the 2 ply targets in the fabric internal energy predictions between the baseline and HEA models. There is also a very good agreement between the total frictional energy dissipated from all contact interactions including projectile to fabric target, yarn to yarn within a layer, and layer to layer. The predictions of the projectile to fabric target contact force for all models also agree very well with each other.

The results reported are still preliminary. One goal of the research is to capture the residual projectile velocity history. In this case, the HEA models have served their purpose extremely well even at this preliminary stage. However if other parameters need to be accurately predicted as well, there is still much work that needs to be done. This includes deducing the optimal sizes for each region in the HEA model based on the deformations and interactions that need to be captured, the coarsest possible mesh density for the global region that still preserves the propagation of the transverse displacement wave through the local region, the optimal configuration for multi-layer fabric targets where the HEA model may vary from layer to layer depending on the boundary conditions and impact velocity, and the optimal set of contact algorithms with more robust algorithms at regions of high deformations and interactions and computationally less intensive algorithms at far field regions, to name a few. Our work will be continually improved in stages.

CONCLUSIONS

A large savings in both run time and memory requirements has been realized through the HEA approach presented herein to simulate the impact of fabric targets. Much work still needs to be done in order to extend the predictive capabilities of the HEA approach to larger fabric targets as well as to further improve on the savings in computational expense. The results obtained so far are encouraging. It is envisioned that the HEA approach will provide an invaluable tool to simulating the impact of large fabric target systems within the framework of available computing resources.

ACKNOWLEDGEMENTS

The financial support of the US Army Research Laboratory at the Aberdeen Proving Grounds, MD, USA, and the Center for Composite Materials at the University of Delaware, DE, USA is gratefully acknowledged.

REFERENCES

1. Nilakantan, G., Keefe, M., Gillespie Jr., John W., "Novel Multi-scale Modeling of Woven Fabrics for use in Impact Studies," *10th International LS-DYNA Users Conference, Dearborn, MI, USA,* June 8-10, 2008.
2. Barauskas, R., "Multi-scale modeling of textile structures in terminal ballistics," *6th European LS-DYNA Users' Conference, Gothenburg, Sweden,* May 28-30, 2007
3. Hallquist, J.O., "LS-DYNA Theory Manual," *Livermore Software Technology Corporation,* March 2006.

Simulation and Verification of Local and Global Forces Exerted on Rigid LCM Tools

W. A. WALBRAN, S. BICKERTON and P. A. KELLY

ABSTRACT

The SimLCM code is being developed at the University of Auckland as a generic Liquid Composite Moulding (LCM) simulation. A comprehensive tooling force analysis has been implemented, providing local and global predictions. Complex geometries require the consideration of both the normal and tangential stress components acting on mould surfaces, due to the compaction of fibre reinforcement and internally generated resin pressures. This approach is required as SimLCM will be extended from the rigid tool (i.e. RTM, I/CM) to flexible (i.e. Resin Infusion, VARTM) and semi-rigid tool (i.e. RTMLight) LCM techniques. Several elastic and viscoelastic reinforcement compaction models have been developed to address both dry and wet response, and are implemented within SimLCM. A simple friction based model is included to model the tangential stress component, allowing consideration of complex non-planar part geometries. This paper will demonstrate current capability of the package, and present experimental verification studies.

INTRODUCTION

The term Liquid Composite Moulding (LCM) encompasses processes such as Resin Transfer Moulding (RTM), Injection/Compression Moulding (I/CM), and Resin Infusion (a.k.a. VARTM). As various factors drive industry from open to closed mould manufacturing techniques, an increasing range of LCM processes are finding application. To date, various simulations have been developed to model filling during rigid tools processes (i.e. RTM, I/CM) [1,2], and recently these have been extended to cover the total forces exerted on these moulds [3]. The SimLCM code is being developed at the University of Auckland as a generic LCM simulation. A thorough tooling force analysis is required to accurately simulate processes within flexible (i.e. VARTM) and semi-rigid tools (i.e. RTMLight), considering the complex interactions between the injected resin, the fibre reinforcement, and the mould components.

William A. Walbran, Simon Bickerton, Piaras A. Kelly, Centre for Advanced Composite Materials, The University of Auckland, Private Bag 92019, Auckland 1020, New Zealand.

This paper demonstrates the current capabilities of SimLCM, addressing a selection of RTM and I/CM experiments. Trials have been completed using both a simple planar, and a non-planar mould cavity. By considering radial injection into a flat plate, measurements were made of the normal stress distribution exerted on the mould platens through entire moulding cycles. The TekScan distributed pressure sensing system was used to achieve this, highlighting spatial variability in reinforcement compaction response, and confirming fluid pressure fields generated. A non-planar part geometry has also been addressed experimentally, providing further verification for SimLCM. Total clamping forces have been monitored during processing, being affected by stress components exerted normal and tangential (due to friction) to each face of the component.

PROCESS MODELLING

Application of a complete filling and force analysis requires models for resin flow and complex reinforcement compaction response. A brief description of the modelling approach employed within SimLCM is provided below.

Flow and Modelling

Resin flow is considered to follow Darcy's Law, assuming all resin velocities will remain in-plane, and the wetted portion of the preform is fully saturated with resin;

$$\mathbf{q} = -\frac{\mathbf{K}}{\mu}\nabla p . \tag{1}$$

$\mathbf{q}$ is the Darcy velocity, p the local fluid pressure, $\mathbf{K}$ the permeability tensor, and μ is the fluid viscosity. The continuity equation appropriate for a thickness-varying (planar) LCM process is;

$$\nabla \cdot (h\mathbf{q}) = -\frac{\partial h}{\partial t}, \tag{2}$$

where h is the local mould cavity thickness. Combining Equations 1 and 2 leads to the governing equation for fluid pressure and cavity thickness;

$$\nabla \cdot \left(h\frac{\mathbf{K}}{\mu}\nabla p \right) = \frac{\partial h}{\partial t}. \tag{3}$$

The glass fibre chopped strand mat (CSM) used in this study is in-plane isotropic. An exponential relationship represents the in-plane permeability well,

$$K(V_f) = Ae^{BV_f}, \tag{4}$$

with $A = 2.20x10^{-8}$ m^2 and $B = -12.40$.

Force Analysis

Normal stresses exerted from the laminate to the mould surface, σ_{ext}, are due to fluid pressure, and the compressive stress taken up by the preform σ_f. These quantities are related by Terzaghi's equation,

$$\sigma_f + p = \sigma_{ext} . \tag{5}$$

σ_f is often taken to be a non-linear elastic law but can also express the viscoelastic properties of fibrous materials [4,5] and permanent deformations which are known to occur [6]. In moulds of complex shape, significant shear stresses are expected to develop at the laminate/mould interfaces. The shear stress is assumed to be related to the normal compaction stress through the following simple model;

$$\tau = \begin{cases} 0, & \alpha = 0 \\ \mu_f \sigma_f & \alpha > 0 \end{cases}, \tag{6}$$

where μ_f is a coefficient of friction and α is the angle between the local outward normal and the closing direction for the mould. The total clamping force that must be supplied to the mould, F_{clamp}, is computed as a summation of the force due to fluid pressure, F_{fluid}, and the force required to compress the preform to the current volume fraction, F_{fibre}.

Viscoelastic Compaction Model

Previous studies have shown that compaction response of reinforcements varies with V_f and compaction speed [5]. Compaction curves can be normalised, time with respect to the time taken to reach the final V_f, and stress with respect to stress at the final V_f. When applied to the CSM used in this study, these normalised curves collapse onto a single master curve. A compaction model for materials which respond in this way was developed previously [7]. The principal features of this model are briefly described here. The "collapsing curves" response mentioned above implies that the stress, a function of V_f and compaction speed (v), decomposes multiplicatively;

$$\sigma(V_f, v) = \frac{1}{\lambda}\left[\sigma_\alpha(v)\sigma_\beta(V_f)\right]. \tag{7}$$

One method of deriving expressions for $\sigma_\alpha(v)$, $\sigma_\beta(V_f)$, is to examine compaction curves at a reference V_f and a reference compaction speed. The choice of references which are typical for the process under study is preferable, but not critical. Let $\sigma_\beta(V_f)$ be the stress response (compaction curve) corresponding to the reference compaction speed v^{ref}; it is a smooth monotonic curve and is well represented using a polynomial function;

$$\sigma_\beta(V_f) = \sum_{i=0}^{N_\beta} \beta_i \left(V_f - V_f^{dat}\right)^i , \tag{8}$$

where V_f^{dat} is some chosen (small) datum volume fraction corresponding to a measured nominal stress. Determination of σ_β requires only a single experiment. The second function, $\sigma_\alpha(v)$, is the stress at the reference volume fraction V_f^{ref}. Due to the rapid initial increase in stress with compaction speed, a polynomial fit is inadequate. One possibility is to model this behaviour using a function of the form

$$\sigma_\alpha(v) = \sigma_\alpha(\infty) - (\sigma_\alpha(\infty) - \sigma_\alpha(0))\frac{1}{N}\sum_{i=1}^{N_\alpha} e^{-\alpha_i\left(\frac{v}{v^{ref}}\right)}, \tag{9}$$

where $\sigma_\alpha(\infty)$ and $\sigma_\alpha(0)$ are the stresses at 'very fast' and 'very slow' compaction speeds, at V_f^{ref}, respectively. Let $\lambda = \sigma_\alpha(v^{ref}) = \sigma_\beta(V_f^{ref})$. The stress at any given V_f and $\dot{h}$ can then be determined from Equation 7.

Stress Relaxation Response

A few models have been proposed which incorporate the viscoelastic response of fibrous materials, e.g. [5,8]. A viscoelastic model which incorporates the "collapsing curve" response has also been presented in [7]. The stress can be expressed as a summation of $\sigma_{eq}(V_f)$, the equilibrium stress, i.e. the stress at infinitely slow compaction velocity, and q, the viscous stress. In general, q can be determined by solving a differential equation [5]. However, at constant V_f, the differential equation can be solved exactly, and incorporating the collapsing behaviour gives;

$$\sigma(\bar{t}) = \frac{\sigma^R}{\sigma_\beta(V_f)}\left[\sigma_{eq}(V_f) + \left(A\left(\bar{t}\frac{v}{v^{ref}}\right)^n + \left(\sigma_\beta(V_f) - \sigma_{eq}(V_f)\right)^{-m}\right)^{-\frac{1}{m}}\right], \tag{10}$$

where $\bar{t}$ is the time since the onset of relaxation, σ^R is the stress at the onset of relaxation, v the compaction speed prior to relaxation, $m = (n-1)/n$, $A = nE^{1/n}/\eta$, n, E and η are material parameters [5,7]. $\sigma_{eq}(V_f)$ can be determined using Equation 7 and the three relaxation parameters are A, m and n. This approach models the collapsing of relaxation curves at a certain V_f. As the relaxation behaviour also depends on V_f one of the parameters must vary with V_f. In this case it is assumed to be A, which can be expressed as a polynomial. The parameters for the CSM used in this study were determined previously in [9].

PLANAR PART GEOMETRY

Shown in Figure 1 is a schematic of the experimental facility used for this study. The mould is mounted in an Instron 1186 universal testing machine which provides accurate displacement and velocity control. A spherical crosshead mount allows accurate parallel alignment of the platens, providing consistent V_f across the part. Test fluid was injected via a pressure pot, pressure being maintained constant via an electronic regulator. The mass of fluid injected was monitored continuously in all cases using electronic weigh scales. A laser gauge was used to measure the cavity thickness when under load. Injection pressure, mass of fluid, cavity thickness and applied load were all recorded using the Instron's Bluehill software. The Tekscan system provides dynamic measurement of pressure distribution at rates up to 8.0 Hz. This study utilised 238mm square sensors having a grid of 44 by 44 sensels. Two sensors were used, with pressure ranges of 0-200 kPa and 0-1303 kPa.

One RTM and I/CM experiment are presented, being a subset of a larger study. The final cavity thickness in each case was 3.5mm, a V_f of 0.50 being achieved using 10 layers of the 450 g/m^2 CSM reinforcement. The preforms were cut to a diameter of 200mm, with a 15mm diameter hole punched at the centre. The test fluid in all cases was Mobil DTE AA mineral oil (viscosity 1.16 Pa.s at 20°C), injected at a nominal pressure of 400 kPa. For the compression stage of I/CM, the platens were brought together at a constant speed of 10 mm/min, from an initial cavity thickness of 4.4mm.

Figure 2a presents clamping force traces for RTM, comparing experiment to predictions made using the complete compaction model, and a simplified model ignoring stress relaxation. Maximum clamping force is experienced at the end of initial compaction (t=0.0s). The experimental force then reduces due to stress relaxation within the reinforcement, raising again as the force component due to fluid pressure increases. During filling the full compaction model provides a closer fit to the experimental data. This is due to the nature of the purely elastic model, which assumes all relaxation occurs instantaneously at t=0.0s. Figure 2b presents the I/CM force comparisons, which are in good agreement with predictions during injection. During secondary compaction, agreement is generally good, but both compaction models are in significant error during the last few seconds of filling. Comparison of the predicted and recorded gate pressures during this period highlighted similar levels of error. After secondary compaction, the viscoelastic model provides much better agreement with experiment. While this is less significant for speed-controlled I/CM, the improved accuracy of the viscoelastic model during relaxation may have significant benefits when modeling force-controlled processes.

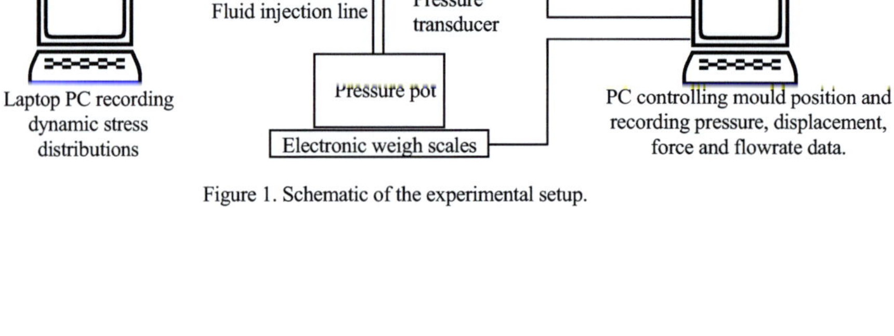

Figure 1. Schematic of the experimental setup.

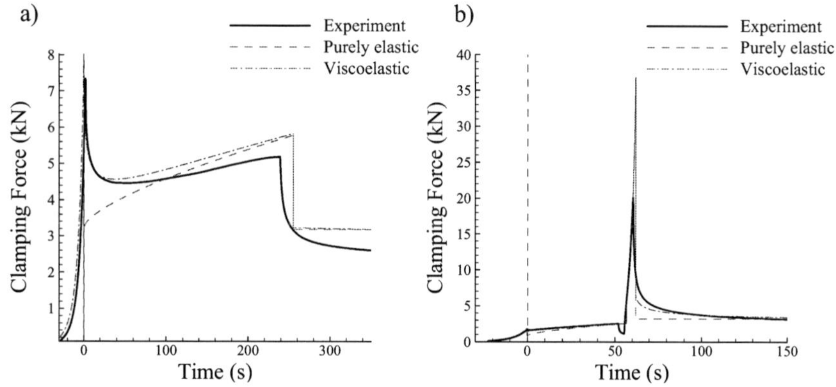

Figure 2. Clamping force traces for the a) RTM, and b) I/CM cases.

Shown in Figure 3 are circumferentially averaged normal stresses measured using Tekscan, at two instances in each process. Predicted normal stress data are also presented. For RTM, stress near the inlet is over predicted at both time instances. When the flow front has reached a radius of 0.05m, the viscoelastic model provides a very good fit to the experimental data from approximately $r = 0.02$ m. At the end of filling, the majority of stress relaxation has taken place, both models predicting the same stress distribution. In all cases, there is a drop in observed stress at the outer edge of the preform (approaching $r = 0.1$ m). This could be attributed to irregularities in the unconstrained preform edge, reducing compaction stress in this region.

Figure 3b presents similar distributions for I/CM, half way through, and at the completion of the secondary compaction phase. Both models exhibit good qualitative agreement with experiment. However, the predicted fluid pressure levels are significantly greater than for the experiment, confirming observations on the total clamping force (Figure 2b). Stresses predicted by both models are identical, as during this phase of filling the reinforcement is being compacted.

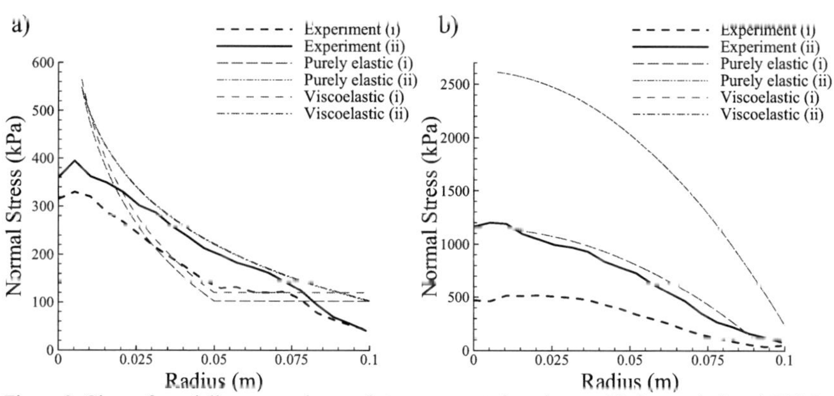

Figure 3. Circumferentially averaged normal stresses exerted on the mould platens during a) RTM at i) $r_f = 0.05$m, and ii) end of filling. b) I/CM at i) half way through secondary compaction phase, and ii) end of filling.

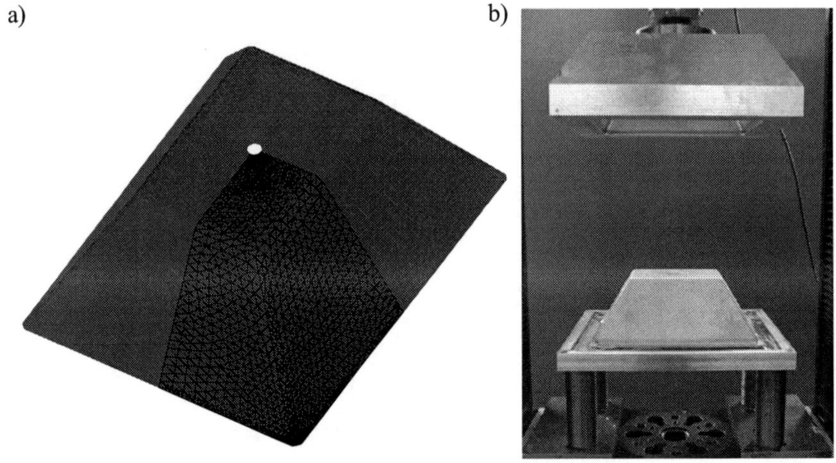

Figure 4. a) Non-planar part geometry and finite element mesh. b) Two-piece aluminium mould.

NON-PLANAR PART GEOMETRY

To experimentally verify SimLCM for a non-planar geometry, the mould cavity depicted in Figure 4a has been addressed. The part is a four-sided truncated pyramid, with uneven width sides, of constant 4mm thickness throughout. The two piece mould shown in Figure 4b was constructed, and employed within the experimental setup described in Figure 1. One RTM and I/CM experiment are presented, a V_f of 0.46 being achieved using 11 layers of CSM. The test fluid was injected at a nominal pressure of 200 kPa, and the compression phase of I/CM was completed at 2 mm/min from an initial cavity thickness at the peak of 5 mm. Only total clamping force traces are compared, as the Tekscan sensors were difficult to install within this mould.

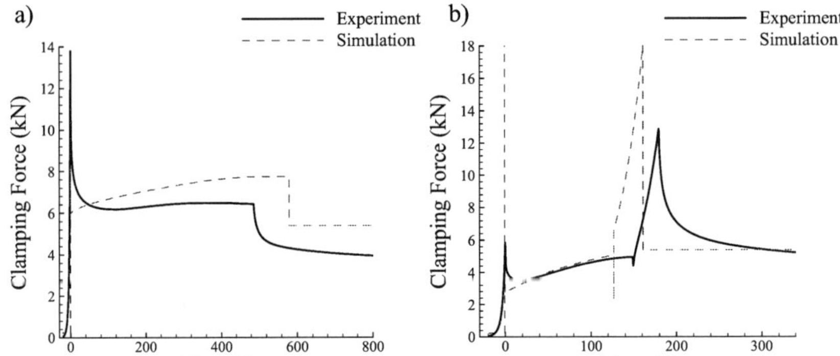

Figure 5. Clamping force traces for the non-planar geometry, for a) RTM, and b) I/CM.

Figure 5a presents clamping force trace data for RTM undertaken using the non-planar mould, comparing experimental data to predictions made using the purely elastic compaction model (the stress relaxation model is yet to be implemented in the full 3D SimLCM code). As with the flat plate experiments, maximum clamping force is experienced at the end of reinforcement compaction, the model providing good agreement with experiment during this phase. The magnitude of the fluid force at the end of injection is also predicted well, however fill time is over predicted by 20%. Figure 5b presents the I/CM force comparisons, which again are in good agreement during initial compaction and fluid injection, although in this case the fill time is under-predicted. However, the clamping force is significantly overpredicted during secondary compaction, as was observed with the flat plate experiment. The source of errors during secondary compaction is the focus of ongoing research.

CONCLUSION

Mould tools used for LCM processes such as RTM and I/CM must withstand the forces generated by the fluid and reinforcement within. SimLCM is under development as a generic LCM simulation, and the tooling force prediction capabilities have been compared to experiment in this paper. A series of LCM trials have been completed, a subset of which is presented here. Observations have been made of total clamping force, and normal stress distributions acting on a flat mould tool. A non-planar part geometry has also been addressed, considering fill times and clamping force histories. Local and global force predictions are in good agreement during phases of injection, while fluid pressure fields (and force components arising from them) are overpredicted during compression driven phases of mould filling.

REFERENCES

1. Simacek, P. and S. G. Advani. 2004. "Desirable Features in Mold Filling Simulations for Liquid Composite Molding Processes," *Polymer Composites*, 25(4):355–367.
2. Trochu, F., E. Ruiz, V. Achim, and S. Soukane. 2006. "Advanced Numerical Simulation of Liquid Composite Molding for Process Analysis and Optimization," *Composites Part A*, 37(6):890–902.
3. Bickerton, S. and P. A. Kelly. 2007. "Application of a Complete Tooling Force Analysis for Simulation of Liquid Composite Moulding Processes," *Key Engineering Materials*, 334-335:17–20.
4. Cai Z. 1995. "A Nonlinear Viscoelastic Model for Describing the Deformation Behaviour of Braided Fiber Seals," *J. of Textile Research*, 65(8):461–470.
5. Kelly P. A., R. Umer, and S. Bickerton. 2006. "Viscoelastic Response of Dry and Wet Fibrous Materials during Infusion Processes," *Composites Part A*, 37(6): 868–873.
6. Somashekar A. A., S. Bickerton, and D. Bhattacharyya. 2007. "Exploring the Non-elastic Compression Deformation of Dry Glass Fibre Reinforcements," *Composites Science and Technology*, 67(2): 183–200.
7. Kelly, P. A. 2007. "A Fibre Compaction Model for Liquid Composite Moulding", *presented at the 6th Canadian International Composites Conference (CANCOM)*, Winnipeg, Canada.
8. Breard, J., Y. Henzel, F. Trochu, and R., Gauvin. 2003. "Analysis of Dynamic Flows Through Porous Media Part II: Deformation of a Double-Scale Fibrous Reinforcement," *Polymer Composites*, 24(3):409–421.
9. Walbran, W. A., S. Bickerton, and P. A. Kelly. 2008. "Predicting Stress Distributions Exerted on LCM Tools during Filling Phases", *presented at the 9th International Conference on Flow Processes in Composite Materials (FPCM)*, Montreal, Canada.

Mechanical and Manufacturing Optimization of a Large Vessel through Simulation

M. GASCONS, N. BLANCO and K. MATTHYS

ABSTRACT

Computational modelling to perform mechanical stress analysis has matured into an essential step of the design process for high performance composite components. Numerical simulations allow, by modifying the geometry, part optimization under given mechanical strength requirements, minimizing the material content and reducing the final cost. In more recent years, manufacturing simulation techniques are also finding their way into the field of virtual design and product development.

Simulation of the injection moulding process not only allows for optimized injection and gate locations, control and minimized filling times, so as to improve the production process itself, but also generates feedback to improve mould and part geometry and material content.

However, the optimization strategy for mechanical strength cannot be expected to coincide with the optimization strategy for the injection moulding process and a compromise between the two is needed.

In this study, we investigate the optimization of the production of a large composite pressure vessel under flexible tooling. The vessel geometry is fixed, but the material content can be controlled as the elaboration of the reinforcement fabric is made using a robotized spray-up system that can define the thickness and volume fibre content of any section of the vessel in a quick and precise way.

The infusion process takes place in a mould set-up that consists of a fixed external vessel mould and a flexible rubber tooling bag as the internal vessel mould. The bag can be pressurized under controlled conditions in order to provide a compression force to the reinforcement fabrics against the external tooling before infusion. This paper investigates how localized fibre volume fraction, permeability and tooling compression affect the vessel infusion process as well as its final mechanical strength, in order to find an optimized production strategy that suits both the mechanical and manufacturing design criteria.

Marc Gascons, Universitat de Girona. Av. Lluis Santaló s/n, Lab. AMADE. 17071 Girona
Norbert Blanco Universitat de Girona. Av. Lluis Santaló s/n, Lab. AMADE. 17071 Girona
Koen Matthys, Brunel University. School of Engineering and Design. Uxbridge, Middlesex UB8 3PH

INTRODUCTION

The creation of a new composite part is one of the most important stages in terms of investment and effort where a wide range of points need to be solved in a short amount of time. Traditionally, some of those points are solved by trial-and-error method, which notably increases the cost and effort in the realization of experimental tests and prototypes. The introduction of virtual tools and numerical simulations in the industry is every time more accepted for the amount of valuable information that is capable to offer with a relatively low investment. Though nowadays those techniques are not accurate enough to completely replace the prototype and test stage, their introduction into a development process can save money and time.

In our specific application, the process and product development of a full composite vessel is proposed. The selected production system for the vessel is fibreglass preform infusion with a polyester resin. The reinforcement preform is obtained from fibreglass roving, which is chopped and projected jointly with a binder emulsion over a perforated mould, which is under partial vacuum. After compaction, a dry fibre preform is generated ready for the infusion stage. Infusion takes place in a singular mould setup with an internal inflatable bag, which allows the production of full vessels in one shoot, without a middle joint as are currently made. Vacuum pressure is used to conduct the resin all along the part, and the pressure of the inflatable bag is set to the point that its deformation during the infusion process can be negligible, hence, the setup can be assimilate to an RTM setup. In this context, a full optimization process that involves material definition, part design and implementation of a production process can be improved. All the optimization process has been conducted over a single parameter, the fibre volume of the material, which will act as a guide line for the stages improved along the process.

In the first step, the elastic properties of the material for a selected range of fibre volumes will be estimated. These values will be subsequently used to conduct the simulations of the mechanical behaviour of the vessel, using a finite element method code. Thickness of the part will be optimized for each value of fibre volume, assuming it constant all along the part. Finally, infusion process simulations will be conducted to study the resin flow through the part. A strategy based on materials cost and production rates will be used to finally determine the best combination of parameters and feedback each step to improve the rest of variables and finally obtain a combination which fits all requirements.

NUMERICAL PREDICTION OF THE ELASTIC PROPERTIES

At present no experimental data is available for the complete characterisation of the elastic properties of the material considered for the production of the pressure vessel. Thus, a study among a wide variety of analytical and numerical models has been conducted to predict the required elastic properties. In the preform production process, chopped fibres are randomly projected over the mould, so his configuration can be taken in as fibreglass mat. Not so much has been published relating to this type of material, mainly for being associated to a low technology and low mechanical requirements applications. Numerical efforts have been conducted to determine properties of unidirectional or knitted fabrics, which are typically used for applications with strong mechanical requirements and needs for improvements. The material has been considered transversely-isotropic, so only 6 different elastic properties need to be determined: in-plane and out-of-plane longitudinal and shear modulus and Poisson ratios. A wide variety of theories are present in the literature [1, 2], which have been compared. Most of these theories have been initially formulated for unidirectional or bidirectional laminates, such as the ones based on Halpin-Tsai, Rule of Mixtures, Matrix Dominated or Periodic Microstructure approaches. Theories developed for continuous strand mats have also been considered and contrasted with previous ones.

Not all the mentioned theories can be used to predict all the needed properties and not all the models correlate with the experimental data obtained for some of the properties. After a comparison between all the available models, and for coherency, one analytical model has been chosen to predict all the required properties. The work from Luciano and Barbero [3] has been used to implement and analyse the elastic behaviour of the material all along the range of fibre volume. According to equations (1)-(6), the terms C_{ij} correspond to the positions *i-j* of the stiffness matrix determined using an approach based on the periodic microstructure supposition.

$$E_{in-plane} = C^*_{11} - \frac{2C^*_{12}}{C^*_{22} + C^*_{23}} \tag{1}$$

$$E_{Transverse} = \frac{(2C^*_{11}C^*_{22} + 2C^*_{11}C^*_{23} - 4C^{*2}_{12})(C^*_{22} - 2C^*_{23} + 2C^*_{44})}{3C^*_{11}C^*_{22} + C^*_{11}C^*_{23} + 2C^*_{11}C^*_{44} - 4C^{*2}_{12}} \tag{2}$$

$$G_{in-plane} = C^*_{66} \tag{3}$$

$$G_{transverse} = \frac{C^*_{22}}{4} - \frac{C^*_{23}}{4} + \frac{C^*_{44}}{2} = \frac{E_T}{2(1+\upsilon_T)} \tag{4}$$

$$\upsilon_{in-plane} = \frac{C^*_{12}}{C^*_{22} + C^*_{23}} \tag{5}$$

$$\upsilon_{transverse} = \frac{C^*_{11}C^*_{22} + 3C^*_{11}C^*_{23} - 2C^*_{11}C^*_{44} - 4C^{*2}_{12}}{3C^*_{11}C^*_{22} + C^*_{11}C^*_{23} + 2C^*_{11}C^*_{44} - 4C^{*2}_{12}} \tag{6}$$

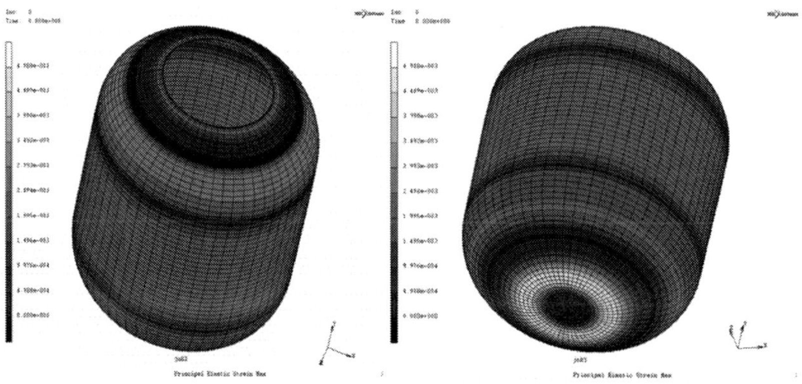

Figure 1. Plot of the strain distribution all along the vessel.

STRUCTURAL OPTIMIZATION OF THE VESSEL

The properties generated for each material fibre volume are implemented on the FEM code MSC.MARC to perform a structural analysis of the vessel.

Displacement of the vessel, as well as strains and stresses are analyzed to accomplish the design limit taken from the Germanischer Lloyd Regulations [4]. Several iterations where conducted under constant thickness all along the vessel to put the material at its limit under the pressure used on service and on leak test cycle. Figure 1 shows a typical strain distribution obtained for the vessel. A critical zone is observed on the bottom of the vessel, which limits the thickness improvement. Strain in this zone doubles the value of the strain on the rest of the vessel.

PRODUCTION PROCESS OPTIMIZATION

The mechanical properties previously obtained also have an impact on the manufacturing properties. The most important of these properties is permeability. Permeability is traditionally determined experimentally, because of the influence of properties inherent to the fabric.

There are several experimental techniques for the determination of the permeability which combine the use of reusable material with similar viscosity to thermoset and different infusion configurations. The radial airflow proposed by Pomeroy et al. [5] or the liner flow proposed by Lee et al. [6], among others, are examples of this variability. There is no unanimity on the definition of a test which clearly determines the permeability values for all materials, due to the lack of precision appreciated on all of them. This is why the evaluation of the change in permeability all along the variation of the fibre content has been trusted to numerical methods.

Most authors agree in using the Kozeny-Carmann equation (7) as the best model to predict the permeability, although there is no a clear way to determinate the value of the Kozeny constant (*A*).

$$K = \frac{\varepsilon^3 d_g^2}{A(1-\varepsilon)^2} \tag{7}$$

In (7), ε is the porosity and d_g is the equivalent fibre diameter. Although the variation prompted by the use of different values for the Kozeny constant is clearly significant, it is not so significant in terms of relative increase. Hence, for a full range of fibre volumes, the relative increase from the first to the last value using the same Kozeny value does not notice important variation.

This expression has been contrasted with experimentally fitted curves, obtained from [7] which are based in results obtained on tests of similar material. The difference on the results caused by the difference between the experimental material and the one that will be used are reflected in [8].

After comparison, the experimental fitting proposed by Cai Z and Molnar et al. has been rejected and equation (7) has been selected as the one that best fits our proposals. Results obtained are shown in Figure 4. Due to the special characteristic of our reinforcement, it has been considered as a 5mm thick mat with constant flow through the thickness and 2D analysis through an isotropic media is considered.

PART PROCESS AND COST IMPROVEMENTS

As expected, the elastic properties of the material increase with the volume fibre fraction, as it can be seen in Figure 2. As a result of the mechanical optimization of the process, the percentage in the reduction of the thickness of the vessel that can be achieved increasing the fibre volume is shown in figure 3. This variation in the thickness has been obtained by means of a finite elements analysis with the premise that no material point in the vessel can exceed a strain of 5%.

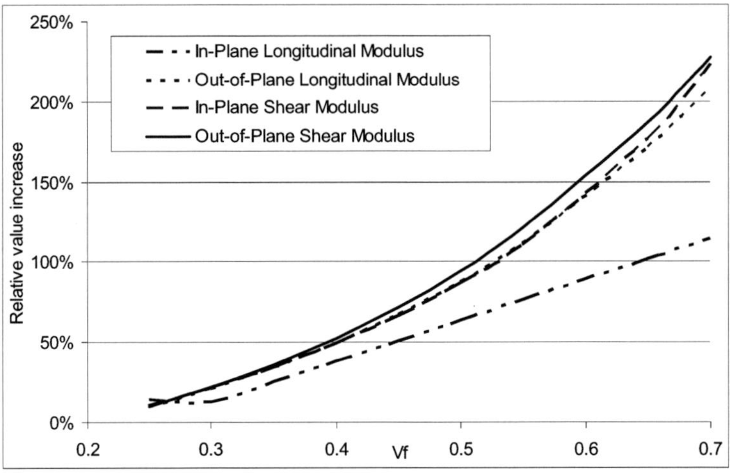

Figure 2. Variation of the material elastic properties vs. fibre volume.

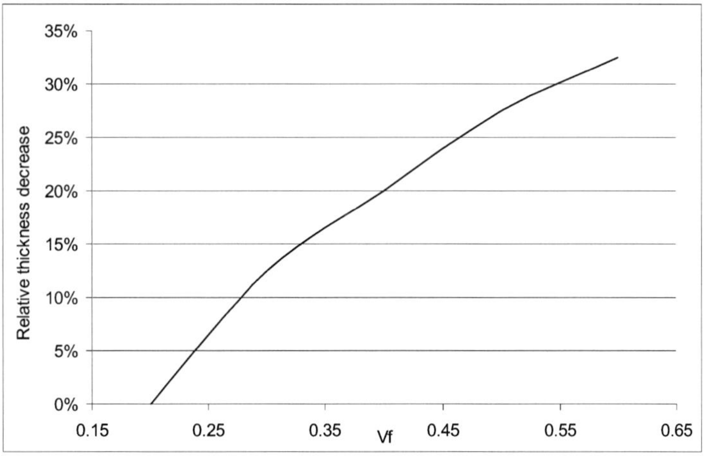

Figure 3. Relative variation of the thickness of the vessel vs fibre volume.

The simulations conducted over the production stage clearly reflect the influence of the permeability variation on the final infusion time. On Figure 4, the relative decrease of the permeability and relative increase of the infusion time are shown. A special point is detected around 40% of fibre volume, where there is an important change in the pace of infusion time.

In this point of the analysis, the material costs criteria have been introduced to search an improvement point in preform manufacturing. The achievable reduction is plot on Figure 5, where it is showed that a save of 17% can be obtained inside the production available range.

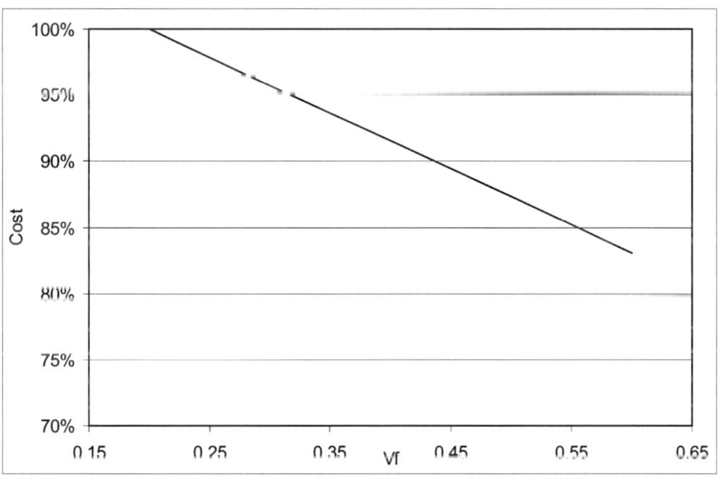

Figure 4. Production time and permeability vs. fibre volume.

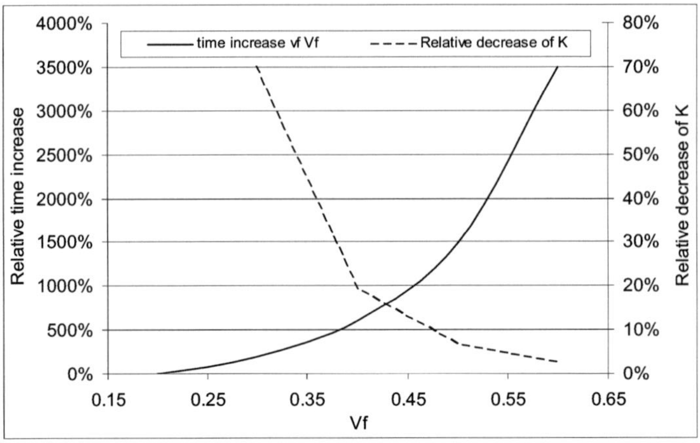

Figure 5. Cost variation vs. fibre volume.

Three different productions scenarios that reflect the product life cycle have been also considered to fit infusion times on desired range, with series of 500, 2000 and 3500 vessels per year that reflect the introduction in the market introduction stage, the growth stage and the mature stage.

The cost decrease of the preform as the level of fibre increases had to be evaluated against the notable increase in the infusion time to search for an optimal point. Under the observations coming from the structural analysis, where a critical zone is clearly defined in the bottom of the vessel, and the possible thickness reduction with the increase of fibre volume delimitated by production rates, an improve alternative consisting on differentiating two areas in the vessel has been taken into account.

The traditional method of increasing the thickness has been compared to a peculiarity of the preform production method, where the Vf can be tailored in different points. Then, for the fixed strain limit, and keeping the fibre volume constant, the thickness of the material has to be increased in a 25% on the critical area. If the same thickness is employed everywhere, Vf in the critical area has to be increased in a 40% to achieve the same strain limit.

Resin flow simulation conducted with both possible solutions show that time infusion values for the extra thickness solution present slightly variations with the original ones, while the solution of the increment in fibre volume makes infusion time reach to nearly three times the initial value.

CONCLUSIONS

Numerical methods have been used to obtain material elastic properties, structural behaviour and resin flow simulations for a new production process. Feedback between the structural and manufacturing analysis has been used to search for an optimal working point where the different design requirements for strength and manufacturing are best met.

The typical reinforcement method of adding thickness to a part has been valorised against the modification of the properties of the material. Its effects on production have been valued to ensure the viability of the constant thickness method, which can reduce costs and simplify the mould production and design process.

ACKNOWLEDGEMENTS

The authors want to acknowledge the assistance of the Center for Composite Materials of the University of Delaware and Poltank SAU

BIBLIOGRAPHY

1. E.J. Barbero, "Introduction to Composite Materials Design" Taylor & Francis, Philadelpha, PA, 1999, pp. 67–80.
2. *Liquid Moulding Technologies*, by C.D. Rudd, A.C. Long, K.N. Kendall, and C.G.E. Mangin, Woodhead Publishing and SAE International, (1997) pp 150–155.
3. R.Luciano, E.J.Barbero, Formulas for the stiffness of composites with periodic microstructure. Int J. Solids Structures. Vol 31 No 21 pp 2933–2944 (1994).
4. Germanischer Lloyd, Regulations for the certification of wind energy conversion system, Germanischer Lloyd, Germany (1999).
5. R.Pomeroy et al. Measurement of permeability of continuous filament mat glass-fibre reinforcements by saturated radial airflow. Composites: Part A 38(2007) 1439–1443.
6. Y.J Lee et al. A prediction method for in-plane permeability of Mat/roving fibres laminates in Vacuum Assisted Resin Transfer Moulding. Polymer Composites 2006 pp. 665–670.
7. *Liquid Moulding Technologies*, by C.D. Rudd, A.C. Long, K.N. Kendall, and C.G.E. Mangin, Woodhead Publishing and SAE International, (1997) pp. 203–254.
8. K A Olivero et al. Effect of Preform Thickness and Volume Fraction on Injection Pressure and Mechanical Properties of Resin Transfer Moulded Composites. J. of Composite Materials, Vol 38 No 11 (2004).

Author Index